Smart Manufacturing Factory

Artificial Intelligence (AI) technologies enable manufacturing systems to sense the environment, adapt to external needs, and extract process knowledge, including business models such as intelligent production, networked collaboration, and extended service models. This book therefore focuses on the implementation of AI in customized manufacturing (CM).

The main topics include edge intelligence in manufacturing, heterogeneous networks, intelligent fault diagnosis and maintenance, dynamic resource scheduling in manufacturing, and the construction mode of the smart factory. Based on the insights of CM and AI, the authors demonstrate the implementation of AI in the smart factory for CM, including architecture, information fusion, data analysis, dynamic scheduling, flexible production line construction, and smart manufacturing services.

This book will provide important research content for scholars in artificial intelligence, smart manufacturing, machine learning, multi-agent systems, and industrial Internet of Things.

Smart Manufacturing Factory
Artificial-Intelligence-Driven Customized Manufacturing

Jiafu Wan, Baotong Chen and Shiyong Wang

CRC Press
Taylor & Francis Group
Boca Raton London New York

CRC Press is an imprint of the
Taylor & Francis Group, an **informa** business

This book is published with financial support from the Joint Fund of the National Natural Science Foundation of China and Guangdong Province (No. U1801264), the National Natural Science Foundation of China under Grant 62203340 and Grant 52275503, and the Natural Science Foundation of Hubei Province under Grant 2022CFC006.

Designed cover image: ©Blue Planet Studio

First edition published 2024
by CRC Press
2385 NW Executive Center Drive, Suite 320, Boca Raton FL 33431

and by CRC Press
4 Park Square, Milton Park, Abingdon, Oxon, OX14 4RN

CRC Press is an imprint of Taylor & Francis Group, LLC

ISBN: 978-1-032-60896-9 (hbk)
ISBN: 978-1-032-60901-0 (pbk)
ISBN: 978-1-003-46099-2 (ebk)

DOI: 10.1201/9781003460992

Typeset in Minion
by codeMantra

Contents

Preface

With the rapid development of electric and electronic technology, information technology and advanced manufacturing technology, the production mode of manufacturing enterprises is being transferred from digital to intelligent. The traditional production paradigm of large batch production does not offer flexibility toward satisfying the requirements of individual customers. A new generation of smart factories is expected to support new multi-variety and small-batch customized production modes. For this, artificial intelligence (AI) is enabling higher value-added manufacturing by accelerating the integration of manufacturing and information communication technologies, including computing, communication, and control. The characteristics of a customized dynamic reconfiguration, and intelligent decision-making. The AI technologies will allow manufacturing systems to perceive the environment, adapt to the external needs, and extract the process knowledge, including business models, such as intelligent production, networked collaboration, and extended service models. This book focuses on the implementation of AI in customized manufacturing (CM). Details of intelligent manufacturing devices, intelligent information interaction, and construction of a flexible manufacturing line are showcased. The state-of-the-art AI technologies of potential use in CM, that is, machine learning, multiagent systems, Internet of Things, big data, and cloud-edge computing, are surveyed. This book covers the breakthroughs of AI technologies and their practical beneficial results in smart manufacturing factory. For senior students and graduate students, they will be able to learn state-of-the-art AI technologies and AI algorithms in theoretical aspects. For research scholar and industrial engineer, this book will broaden their horizon and develop their abilities to deal with industrial problems.

The text is organized into six chapters. Chapter 1 mainly provides introduction to smart manufacturing factory. This chapter gives the background of smart manufacturing factory, customized manufacturing mode, and the overview and the frontier technologies of the AI-driven customized manufacturing factory. Chapter 2 mainly introduces edge intelligence and its architecture in customized manufacturing. The key technologies of edge computing in customized manufacturing are discussed in detail. This chapter enlightens its features of low latency, real-time interaction, location awareness, mobility support, geo-distribution, etc. over cloud. Chapter 3 analyzes the cause, classification, and current solution framework of smart factory heterogeneous network. SDN and EC-based heterogeneous networks framework and AI-Enabled QoS optimization of heterogeneous networks in smart manufacturing factory are proposed. Chapter 4 mainly proposes the important topic both in practice and research, namely intelligent fault diagnosis and maintenance in smart manufacturing factory. It introduces how to effectively use advanced machine learning algorithms for supervised, semi-supervised and unsupervised intelligent fault diagnosis to deal with the existing dilemma. Chapter 5 fully explores the resource dynamic scheduling based on edge cloud collaboration and resource reconfiguration in smart manufacturing. This chapter fully explores the application potential of the Industrial Internet of Things from the aspects of edge computing and cloud computing. Chapter 6 introduces the components of the platform in detail and reflect the key links and core technologies of custom manufacturing through mobile services and conveyor belt control. The implement of customized manufacturing factory is provided to inspire the manufacturing enterprises to realize the transformation of smart manufacturing factories.

The authors would like to acknowledge the invaluable cooperation and suggestions from many collaborators, both in China and other countries, on their research works on AI-Driven customized manufacturing in smart manufacturing. In particular, the authors express their gratitude for the invaluable contributions to this book from the members in: Precision Manufacturing Laboratory at South China University of Technology Prof. Di Li and Assoc. Prof. Shiyong Wang. Prof. Haidong Shao of Hunan University, Prof. Min Xia of Lancaster University, Prof. Guangjie Han of Hohai University, Prof. Hong-Ning Dai of Hong Kong Baptist University, and Assoc. Prof. Xiaomin Li of Zhongkai University of Agriculture and Engineering.

Some contents of this book were financially supported by the following research projects from China: The Key Program of National Natural Science Foundation of China (No. U1801264). The authors wish to acknowledge with great appreciation the colleagues and graduate students for their collaboration or suggestions in the presented research. Specific acknowledgment goes to the contributions from Dr. Chun Jiang for Chap. 2, Dr. Dan Xia for Chap. 3, Dr. Zijie Ren for Chap. 4, Dr. Xiangdong Wang for Chap. 5, Dr. Jinbiao Tan for Chap. 6. The authors would also like to thank Zixu Fan from Taylor & Francis and all the committee members for their efforts to organize this book series which makes publication of this book possible.

Jiafu Wan
School of Mechanical and Automotive Engineering,
South China University of Technology, China

Baotong Chen
School of Mechanical Automation,
Wuhan University of Science and Technology, China

Shiyong Wang
School of Mechanical and Automotive Engineering,
South China University of Technology, China

About the Authors

Jiafu Wan is a Professor in School of Mechanical & Automotive Engineering at South China University of Technology, China. He has directed 20 research projects, including the National Key Research and Development Program of China and the Joint Fund of the National Natural Science Foundation of China and Guangdong Province. Thus far, he has published more than 170 scientific papers, including 130+ SCI-indexed papers and 50+ IEEE Trans./Journal papers. According to Google Scholar Citations, his published work has been cited more than 23,000 times. He was listed as a Clarivate Analytics Highly Cited Researcher (2019–2022).

Baotong Chen received a PhD degree in mechanical engineering from the Department of Mechanical & Electrical Engineering, South China University of Technology (SCUT). He is an Associate Professor in the Department of Industrial Engineering and Manufacturing, Wuhan University of Science and Technology (WUST). He is also associated with the Key Laboratory of Metallurgical Equipment and Control Technology, Ministry of Education, Wuhan University of Science and Technology, and the Hubei Key Laboratory of Mechanical Transmission and Manufacturing Engineering, Wuhan University of Science and Technology. Thus far, he has published more than 10 scientific papers, including 10+ SCI-indexed papers. According to Google Scholar Citations, his published work has been cited more than 2,000 times. His research interests include equipment preventive maintenance, multi-robot collaboration, and mixed model assembly.

Shiyong Wang is an Associate Professor in the School of Mechanical & Automotive Engineering at South China University of Technology, China. He has directed 20 research projects, including the National Key Research and Development Program of China and the National Natural Science Foundation of China. Thus far, he has published more than 50 scientific papers, including 30+ SCI-indexed papers. According to Google Scholar Citations, his published work has been cited more than 5,000 times. He was listed as a Highly Cited Chinese Researcher by Elsevier (2020 and 2022).

CHAPTER 1

Introduction to Smart Manufacturing Factory

THE EMERGING TECHNOLOGIES [E.G., artificial intelligence (AI), Internet of Things (IoT), wireless sensor networks, big data, cloud computing, embedded system, and mobile Internet] are being introduced into the manufacturing environment, which ushers in a fourth industrial revolution. Consequently, "Industrie 4.0" was proposed and was adopted as part of the "High-Tech Strategy 2020 Action Plan" of the German government [1]. Countries around the world are actively engaging in the new industrial revolution. The United States has launched the Advanced Manufacturing Partnership and "Industrial Internet" [2], Germany has developed the strategic initiative "Industrie 4.0," and the United Kingdom has put forward the UK Industry 2050 strategy. In addition, France has unveiled the New Industrial France program, Japan has a Society 5.0 strategy, and Korea has started the Manufacturing Innovation 3.0 program. The Made in China 2025 plan, formerly known as China Manufacturing 2025, has specifically set the promotion of intelligent manufacturing as its main direction, and "Internet +" was also proposed from China [3]. This paradigm describes a production-oriented Cyber-Physical Systems (CPSs) that integrate production facilities, warehousing systems, logistics, and even social requirements to establish the manufacturing value creation.

With the rapid development of electric and electronic technology, information technology, and advanced manufacturing technology, the production mode of manufacturing enterprises is being transferred from digital to intelligent. The new era that combines virtual reality technology based on the CPS is coming [4–7]. Due to the new challenges, the advantages of traditional manufacturing industries have been gradually diminished. Consequently, the intelligent manufacturing technology is one of the high-technology areas where industrialized countries highly pay more attention to. Europe 2020 strategy [8], Industry 4.0 strategy [9], and China Manufacturing 2025 [10] have been proposed. The United States has gradually accelerated the speed of reindustrialization and manufacturing reflow [11]. The transformation of intelligent manufacturing intrigued the profound and lasting effect on the future manufacturing

DOI: 10.1201/9781003460992-1

worldwide. In the context of Industry 4.0 and AI, the intelligent manufacturing attracts enormous interest from government, enterprises, and academic researchers [12]. Therefore, the construction patterns of smart manufacturing factory are widely discussed.

The standards for smart factory implementation have not been established yet. Benkamoun et al. [13] proposed a class diagram that can be used to represent the manufacturing system from different perspectives of entities and functions. Radziwon et al. [14] expounded former research from the concept of smart factory, and they pointed out that smart factory is actually an exploration of adaptive and flexible manufacturing. Lin et al. [15] proposed an architecture for cloud manufacturing systems oriented to aerospace conglomerate, which facilitates optimal configuration of manufacturing resources. The above-mentioned authors provided a guidance architecture for smart factory. In summary, the smart factory, which is based on digital and automated factory, uses information technology [e.g., cloud platform and Industrial Internet of Things (IIoT)] to improve the management of manufacturing resources and Quality-of-Service (QoS) [16,17]. In order to build the smart factory, manufacturing enterprises should improve production and marketing, enhance controllability of production process, and reduce manual intervention in workshop. Through the analysis of manufacturing data, the smart factory can realize flexible manufacturing, dynamic reconfiguration, and production optimization, which are aimed to adapt the system to the changes of business model and consumer shopping behavior [18].

In the implementation of smart factory, the IIoT is employed to integrate the underlying equipment resources. Accordingly, the manufacturing system has abilities of perception, interconnection, and data integration. The data analysis and scientific decision are used to achieve production scheduling, equipment service, and quality control of products in smart factory. Further, the Internet of services is introduced to virtualize the manufacturing resources from a local database to the cloud server. Through the human–machine interaction, the global collaborative process of intelligent manufacturing oriented to the order-driven market is built. Therefore, the smart factory represents an engineering system that mainly consists of three aspects: interconnection, collaboration, and execution.

In the context of intelligent manufacturing, it is important to establish the smart manufacturing factory to achieve advanced manufacturing based on network technologies and manufacturing data. In addition, the implementation of smart manufacturing factory should take into account the status quo and manufacturing requirements. Due to the different characteristics of manufacturing field and information field, there are still many technical problems to be solved in order to accelerate the path of smart factory.

In the physical resources layer, physical equipment needs to have support for real-time information acquisition, and communication devices should provide a high-speed transmission of heterogeneous information. The workshop should ensure fast reconfiguration and adaptability. In addition, the intelligence of underlying equipment should be enhanced in order to meet the requirements of IoT.

In the network layer, IIoT should support new protocols and new data format with high flexibility and scalability, whereas the Industrial Wireless Sensor Networks (IWSNs) bring new opportunities for development of industrial network. Additionally, the other relevant technologies (e.g., OLE for Process Control Unified Architecture (OPC UA), Software-Defined Networks (SDNs), and Device-to-Device (D2D) communication) should be introduced to guarantee QoS of the network, reliable communication, and cooperation among equipment.

In the data application layer, the cloud platform should be able to analyze the semantics of various data. Therefore, ontology is being employed in modeling of the smart factory, which can provide the abilities of self-organization, self-learning, and self-adaption. Moreover, data analysis could provide the scientific basis for decision-making, while data mining could be used to ensure design optimization and active maintenance.

1.1 BACKGROUND OF SMART MANUFACTURING FACTORY

The human society desires a progressive improvement of life quality. The industry has been advancing to keep pace with this kind of requirements. By now, it has experienced three revolutionary stages, that is, three industrial revolutions. The industry can continue to improve people's living standard by providing customized and high-quality products to consumers and setting up a better work environment for employees. The Industry 4.0 initiative is advocating smart manufacturing as the industrial revolution leading to global economic growth [9,19–21]. Many countries, corporations, and research institutions have embraced the concept of Industry 4.0, in particular, the United States, the European Union, and East Asia [22]. Some industries have begun a transformation from the digital era to the intelligent era. Manufacturing represents a large segment of the global economy, while the interest in smart manufacturing is expanding [23]. The progress in information and communication technologies, for example, the IoT [24,25], AI [26,27], and big data [28,29] for manufacturing applications, has impacted smart manufacturing [30]. In the broad context of manufacturing, customized manufacturing (CM) offers a value-added paradigm for smart manufacturing [31], as it refers to personalized products and services. The benefits of CM have been highlighted by multinational companies.

Today, information and communication technologies are the base of smart manufacturing [32,33], and intelligent systems driven by AI are the core of CM [34]. With the development of AI technologies, new theories, models, algorithms, and applications—toward simulating, extending, and enhancing human intelligence—are continuously developed. The progress of big data analysis and deep learning (DL) has accelerated AI to enter the 2.0 era [35–37]. AI 2.0 manifests itself as a data-driven deep reinforcement learning intelligence [38], network-based swarm intelligence [39], technology-oriented hybrid intelligence of human–machine and brain–machine interaction [40–42], cross-media reasoning intelligence [43,44], and so on. Therefore, AI 2.0 offers significant potential to smart manufacturing, especially CM in smart factories [45].

Typically, AI solutions can be applied to several aspects of smart manufacturing. AI algorithms can run the manufacturing of personalized products in a smart factory [46,47]. The AI-assisted CM (AIaCM) is to construct smart manufacturing systems supported by cognitive computing, machine status sensing, real-time data analysis, and autonomous decision-making [48,49]. AI permeates through every link of CM value chains, such as design, production, management, and service [50,51]. Based on these insights of CM and

AI, the focus of this chapter is on the implementation of AI in the smart factory for CM involving architecture, manufacturing equipment, information exchange, flexible production line, and smart manufacturing services. Most studies have focused on information communications [52] or big data processing [53–55]. So far, research proposing generic AI-based CM frameworks is limited. System performance metrics, for example, flexibility, efficiency, scalability, and sustainability, can be improved by adopting AI technologies, such as machine learning (ML), knowledge graphs, and human–computer interaction (HCI). This is especially true in sensing, interaction, resource optimization, operations, and maintenance in a smart CM factory [56,57]. Since cloud computing, edge computing, and local computing paradigms have their unique strengths and limitations, they should be integrated to maximize their effectiveness. At the same time, the corresponding AI algorithms should be redesigned to match the corresponding computing paradigm. Cloud intelligence is responsible for making comprehensive, time-insensitive analysis and decisions, while the edge and local node intelligence are applicable to the context or time-aware environments. Intelligent manufacturing systems include smart manufacturing devices, realize intelligent information interaction, and provide intelligent manufacturing services by merging AI technologies. An AIaCM framework includes smart devices, smart interaction, AI layer, and smart services. We then explain this framework in detail as follows:

- **Smart devices**: It includes robots, conveyors, and other basic controlled platforms. Smart devices serve as "the physical layer" for the entire AIaCM. Especially, different devices and equipment, such as robots and processing tools, are controlled by their corresponding automatic control systems. Therefore, it is crucial to meet the real-time requirement for the device layer in an AIaCM system. To achieve this goal, ML algorithms can be implemented at the device layer in low-power devices such as FPGAs. The interconnection of the physical devices, for example, machines and conveyors, is implemented at the device layer [58,59] using edge computing servers.

- **Smart interaction**: It links the device layer, AI layer, and services layer [60,61]. It represents a bridge between different layers of the proposed architecture. The smart interaction layer is composed of two vital modules. The first module includes basic network devices, such as access points, switches, routers, and network controllers, which are generally supported by different network operating systems or equipped with different network functions. The basic network devices constitute the core of the network layer [62,63]. Different from the first module that is fixed or static, the second module consists of the dynamic elements, including network/communications protocols, information interaction, and data persistent or transient storage. These dynamic elements are essentially information carriers to connect different manufacturing processes. The dynamic module is running on top of the static one. AI is utilized in the prediction of wireless channels, optimization of mobile network handoffs, and control network congestion. Recurrent neural networks (RNNs) or reservoir computing (RC) are candidate solutions due to the advantages of them in analyzing temporal network data.

- **AI layer**: It includes algorithms running at different computing platforms, such as edge or cloud servers [54,64]. The computing environment consists of cloud and edge computing servers running MapReduce, Hadoop, and Spark. AI algorithms are adopted at different levels of computing paradigms in AIaCM architecture. For instance, training a DL model for image processing can be conducted in the cloud. Then, edge computing servers are responsible for running the trained DL model and executing relatively simple algorithms for specific manufacturing tasks.

- **Smart manufacturing services**: It includes data visualization, system maintenance, predictions, and market analysis. For example, a recommender system can provide customers with details of CM products and the information including the performance of a production line, market trends, and efficiency of the supply chain.

1.2 INTEGRATED INTELLIGENCE OF SMART MANUFACTURING FACTORY

In order to preferably implement Industry 4.0, the following three key features should be considered: (1) horizontal integration through value networks, (2) vertical integration and networked manufacturing systems, and (3) end-to-end digital integration of engineering across the entire value chain. The setting for vertical integration is the factory, so the vertical integration means implementing the smart factory that is highly flexible and reconfigurable. Therefore, the smart factory is believed to be able to produce customized and small-lot products efficiently and profitably.

Prior to the smart manufacturing of Industry 4.0, many other advanced manufacturing schemes have already been proposed to overcome the drawbacks of the traditional production lines, for example, the flexible manufacturing and the agile manufacturing. Among these schemes, the multi-agent system (MAS) is the most representative. The manufacturing resources are defined as intelligent agents that negotiate with each other to implement dynamical reconfiguration to achieve flexibility. However, it is difficult for the MAS to handle the complexity of manufacturing system, so it still lacks a generally accepted MAS implementation. In our view, the cloud-assisted industrial wireless network (IWN) can suitably support the smart factory by implementing IoT and services. By this means, the smart artifacts can communicate and negotiate with each other through the IWN to implement self-organization, and the massive data can be uploaded to and processed by the cloud that has scalable storage space and powerful computing ability to implement system-wide coordination.

This section focuses on building the overall architecture of the smart factory, exploring the operating mechanisms of the technology components involved in the organization. First, a smart factory framework is proposed to integrate IWN, cloud, and terminal with smart workshop equipment such as machines, products, products, and conveyors, and the key technologies involved in it and their applications are explained, and the main technical challenges are discussed and possible solutions proposed. Second, a macro-closed-loop model is proposed to describe the operating mechanism of the architecture, that is, smart workshop artifacts form a self-organizing system, and big data analysis provides global feedback and coordination. Finally, the technical characteristics and beneficial outcomes of smart factories are outlined.

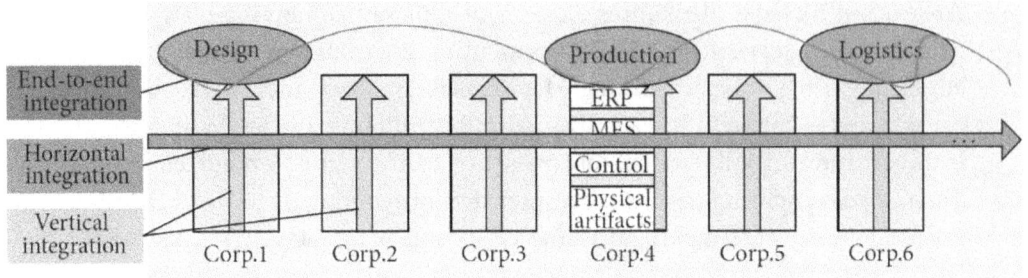

FIGURE 1.1 Illustration of three kinds of integration and their relationship.

As we know, the main features of Industry 4.0 include (1) horizontal integration through value networks to facilitate inter-corporation collaboration, (2) vertical integration of hierarchical subsystems inside a factory to create flexible and reconfigurable manufacturing system, and (3) end-to-end engineering integration across the entire value chain to support product customization. Figure 1.1 depicts the relationship of the three kinds of integration. The horizontal integration of corporations and the vertical integration of factory inside are two bases for the end-to-end integration of engineering process. This is because the product life cycle comprises several stages that should be performed by different corporations.

Horizontal integration: One corporation should both compete and cooperate with many other related corporations. By the inter-corporation horizontal integration, related corporations can form an efficient ecosystem. Information, finance, and material can flow fluently among these corporations. Therefore, new value networks as well as business models may emerge.

Vertical integration: A factory owns several physical and informational subsystems, such as actuator and sensor, control, production management, manufacturing, and corporate planning. It is essential to vertical integration of actuator and sensor signals across different levels right up to the enterprise resource planning (ERP) level to enable a flexible and reconfigurable manufacturing system. By this integration, the smart machines form a self-organized system that can be dynamically reconfigured to adapt to different product types; and the massive information is collected and processed to make the production process transparent.

End-to-end engineering integration: In a product-centric value creation process, a chain of activities is involved, such as customer requirement expression, product design and development, production planning, production engineering, production, services, maintenance, and recycle. By integration, a continuous and consistent product model can be reused by every stage. The effect of product design on production and service can be foreseen using the powerful software tool chain so that the customized products are enabled.

1.3 CUSTOMIZED MANUFACTURING MODE

Due to the production paradigm of single variety and large batches, traditional manufacturing factories cannot meet the requirement of individualized consumption. The new generation of CM system of smart factory should support the multi-variety and small batch customized production. With the rapid development of electric and electronic technology, information technology, and advanced manufacturing technology, the production mode of manufacturing enterprises is being transferred from digital to intelligent. The new era that combines virtual reality technology based on the CPS is coming [5].

1.3.1 Concept of Customized Manufacturing

Due to the rapid development of information technology, computer science, and advanced manufacturing technology, the manufacturing production has been changing from automated production to digitalized and intelligent production [65]. Nowadays, the traditional single and mass production cannot meet market demands for the multiple varieties, small batches, and personalized customization [12]. Therefore, the change from the traditional manufacturing model to the intelligent production model is an urgent issue. In the context of Industry 4.0, the main way to realize the intelligent manufacturing is to establish a smart factory based on the CPS [6]. The CPS needs the technical support in various aspects, such as IoT, big data, cloud computing, and AI technologies [66,67].

Smart factories based on the cloud computing have a large number of low-cost resources of storage and computing, which can enable the dynamic reconstruction and optimized distribution, as well as provide reliable support for the application of industrial big data [68]. Shu et al. [69] proposed a cloud-integrated CPS that provides solutions for complex industrial applications from three aspects: virtual resource management, cloud resource scheduling, and life cycle management. Wang et al. [21] demonstrated a cloud-based personalized smart factory application for candy packaging. By using the private cloud and IWN, the smart production devices can be directly connected to the client terminals to achieve product customization and production. A large number of studies have shown that cloud computing provides an effective solution for resource sharing and information exchange in intelligent manufacturing systems, but only a few studies have specifically integrated the AI technologies into the systems.

Recently, the AI has attracted a lot of attention in various fields including smart manufacturing. Namely, significant progress has been achieved in many fields, such as image processing, natural language processing, and speech recognition [70,71]. The development of a new generation of AI technologies has also brought new opportunities and challenges to the smart factories. Considering a smart factory as a large information system, it is possible to apply the AI technologies at different levels of the CaSF. Thus, the AI can be applied to the smart factories to a large extent. Deploying the AI technologies in smart factories has produced many significant changes including the following: (1) smart devices that integrate the AI technologies, such as machine vision, are more accurate and reliable; (2) collaborative mechanisms with autonomous decision-making and reasoning capabilities exhibit more reasonable dynamic behaviors; and (3) data processing methods based on the advanced AI algorithms, such as DL, are more accurate and efficient. Therefore, the application of AI technologies has provided a new construction direction of smart factories.

1.3.2 Characteristics of Customized Manufacturing

Despite the progress made, the manufacturing industry faces a number of challenges, some of which are as follows: traditional mass production is not able to adapt to the rapid production of personalized products; resource limitations, environmental pollution, global warming, and an aging global population have become more prominent. Therefore, a new manufacturing paradigm to address these challenges is needed. The customer-to-manufacture concept reflects the characteristics of customized production where a manufacturing system directly interacts with a customer to meet his/her personalized needs. The goal is to realize the rapid customization of personalized products. The new generation of intelligent manufacturing technology offers improved flexibility, transparency, resource utilization, and efficiency of manufacturing processes. It has led to new programs, for example, the Factory of the Future in Europe [8], Industry 4.0 in Germany [9], and Made in China 2025 [10]. Moreover, the United States has also accelerated research and development programs [11].

Compared with mass production, the production organization of CM is more complex, quality control is more difficult, and the energy consumption needs attention. In classical automation, the production boundaries were rigid to ensure quality, cost, and efficiency. Compared with traditional production, CM has the following characteristics:

- **Smart interconnectivity**: Smart manufacturing embraces a cyber-physical environment, for example, processing/detection/assembly equipment and storage, all operating in a heterogeneous industrial network. The IIoT has progressed from the original industrial sensor networks to the narrow-band IoT (NB-IoT), LoRa WAN, and LTE Cat M1 with increased coverage at reduced power consumption [24]. Edge computing units are deployed to improve system intelligence. Cognitive technology ensures the context awareness and semantic understanding of the IIoT [72]. Intelligent IIoT as the key technologies is widely used for intelligent manufacturing.

- **Dynamic reconfiguration**: The concept of a smart factory aims at the rapid manufacturing of a variety of products in small batches. Since the product types may change dynamically, system resources need to be dynamically reorganized. A MAS [73] is introduced to negotiate a new system configuration.

- **Massive volumes of data**: An intelligent manufacturing system includes interconnected devices generating data, such as device status and process parameters. Cloud computing and big data science make data analysis feasible in failure prediction, active preventive maintenance, and decision-making.

- **Deep integration**: The underlying intelligent manufacturing entities, cloud platforms, edge servers, and upper monitoring terminals are closely connected. Data processing, control, and operations can be performed simultaneously in the CPSs, where the information barriers are broken down, thereby realizing the deep integration of physical and information environments.

1.4 AI-DRIVEN CUSTOMIZED MANUFACTURING FACTORY

1.4.1 Overview of the AI Technologies Framework

AI embraces theories, methods, technologies, and applications to augment human intelligence. It includes not only AI techniques, such as perception, ML, DL, reinforcement learning, and decision-making, but also AI-enabled applications, such as computer vision, natural language processing, intelligent robots, and recommendation systems, as shown in Figure 1.2a. ML has outperformed traditional statistical methods in tasks such as classification, regression, clustering, and rule extraction [74]. Typical ML algorithms include decision tree, support vector machines, regression analysis, Bayesian networks, and deep neural networks. As a subset of ML algorithms, DL algorithms have superior performance than other ML algorithms. The recent success of DL algorithms mainly owes to three factors: (1) the availability of massive data; (2) the advent of computer capability achieved by computer architectures and hardware, such as graphic processing units (GPUs); and (3) the advances in diverse DL algorithms, such as a convolutional neural network (CNN), long short-term memory (LSTM), and their variants. Different from ML methods, which require substantial efforts in feature engineering in processing raw industrial data, DL methods combine feature engineering and learning process, thereby achieving outstanding performance.

However, DL algorithms also have their own disadvantages. First, DL algorithms often require a huge amount of data to train DL models to achieve better performance than other ML algorithms. Moreover, the training of DL models requires substantial computing resources (e.g., expensive GPUs and other computer hardware devices). Third, DL algorithms also suffer from poor interpretability, that is, a DL model is like an uncontrollable "black box," which may not obtain the result as predicted. The poor interpretability of DL models may prevent their wide adoption in industrial systems, especially in critical tasks such as fault diagnosis [75], despite recent advances in improving the interpretability of DL models [76].

1.4.2 AI-driven Frontier Technologies

As AI technologies have demonstrated their potential in areas such as customized product design, customized product manufacturing, manufacturing management, manufacturing maintenance, customer management, logistics, after-sales service, and market analysis, as shown in Figure 1.2b, industrial practitioners and researchers have begun their implementation. For example, the work in [77] presents a Bayesian network-based approach to analyze the consumers' purchase behavior via analyzing RFID data, which

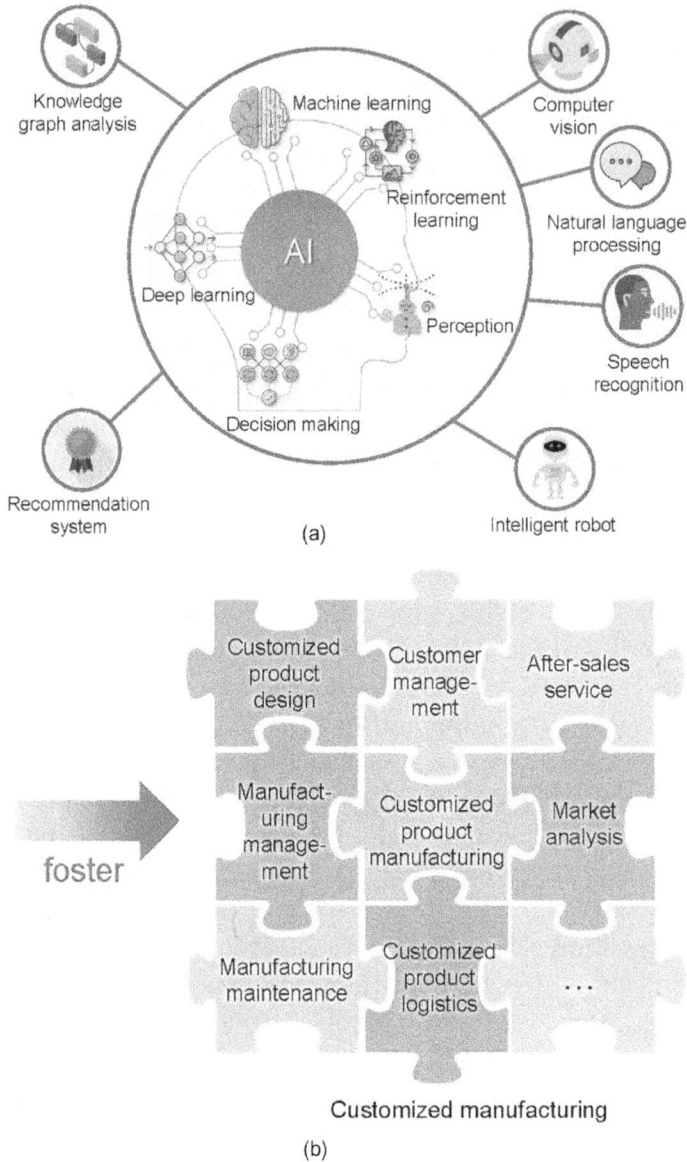

FIGURE 1.2   AI and CM. (a) AI technologies include perception, ML, DL, reinforcement learning, and decision-making. (b) AI can foster CM in the following aspects.

is collected from RFID tags attached to the in-store shopping carts. Moreover, a DL method is adopted to identify possible machine faults through analyzing mechanic data collected from the real industrial environments, such as induction motors, gearboxes, and bearings [78].

Therefore, the introduction of AI technologies can potentially realize CM. We name such AI-driven CM as AI-driven CM. In summary, AI-driven CM has the following advantages [79,80]:

- **Improved production efficiency and product quality**: In CM factories, automated devices can potentially make decisions with reduced human interventions. Technologies such as ML and computer vision are enablers of cognitive capabilities, learning, and reasoning (e.g., analysis of order quantities, lead time, faults, errors, and downtime). Product defects and process anomalies can be identified using computer vision and foreign object detection. Human operators can be alerted to process deviations.

- AI-enabled edge intelligence aims to combine edge computing with AI and other applications. It syncs the data processing capability based on cloud computing to the edge nodes to provide advanced data analysis, scene perception, real-time decision-making, and other service functions in the edge network.

- AI-driven CM aims at the rapid manufacturing of a variety of products in small batches. Since the product types may change dynamically, system resources need to be dynamically reorganized. AI-driven technologies of resource dynamic scheduling have become one of the most prominent characteristics. AI-enabled MAS can be deployed to negotiate resource scheduling and reconfiguration.

- **Facilitating predictive maintenance**: Scheduled maintenance ensures that the equipment is in the best state. Sensors installed on a production line collect data for analysis with ML algorithms, including CNNs. For example, the wear and tear of a machine can be detected in real time, and a notification can be issued.

- **Development of smart supply chains**: The variability and uncertainty of supply chains for CM can be predicted with ML algorithms. Moreover, the insights obtained can be used to predict sudden changes in customer demands.

In short, the incorporation of AI and IIoT brings benefits to smart manufacturing. AI-assisted tools improve manufacturing efficiency. Meanwhile, higher value-added products can be introduced to the market.

However, we cannot deny that AI technologies still have their limitations when they are formally adopted to real-world manufacturing scenarios. On the one hand, AI and ML algorithms often have stringent requirements on computing facilities. For example, high-performance computing servers equipped with GPUs are often required to fasten the training process on massive data [81], while exiting manufacturing facilities may not fulfill the stringent requirement on computing capability. Therefore, the common practice is to outsource (or upload) the manufacturing data to cloud computing service providers who can conduct the computing-intensive tasks. Nevertheless, outsourcing the manufacturing data to the third party may lead to the risk of leaking confidential data (e.g., customized product design) or exposing private customer data to others. On the other hand, transferring the manufacturing data to remote clouds inevitably leads to high latency, thereby failing to fulfill the real-time requirement of time-sensitive tasks.

1.4.2.1 AI-driven Intelligent Manufacturing Device

1.4.2.1.1 Edge-Computing-Assisted Intelligent Agent Construction   In the customized production paradigm, manufacturing devices should be capable of rapid restructuring and reuse for small batches of personalized products [82,83]. However, it is challenging to achieve elastic and rapid control over massive manufacturing devices. The agent-based system was considered a solution to this challenge [84,85]. Agents can autonomously and continuously function in a collaborative system [86]. A MAS can be constructed to take autonomous actions. Different types of agents have been constructed in [87–89].

Although a single agent may have sensing, computing, and reasoning capabilities, it alone can only accomplish relatively simple tasks. Smart manufacturing may involve complex tasks, for instance, the image-based personalized product recognition, expected from the emerging MASs [90,91]. However, the multiple agents are deficient in processing massive data. Recent advances in edge computing can meet this emerging need [92–94]. As shown in Figure 1.3, a variety of decentralized manufacturing agents are connected to edge computing servers via high-speed industrial networks. The edge-computing-assisted manufacturing agents embrace the device layer, agent layer, edge computing layer, and AI layer.

An agent is equipped with a reasoning module and a knowledge base, offering basic AI functionalities such as inferencing and computing. Moreover, with the support of new communication technologies (e.g., 5G mobile networks and high-speed industrial wired networks), all agents and edge computing servers can be interconnected. Agents

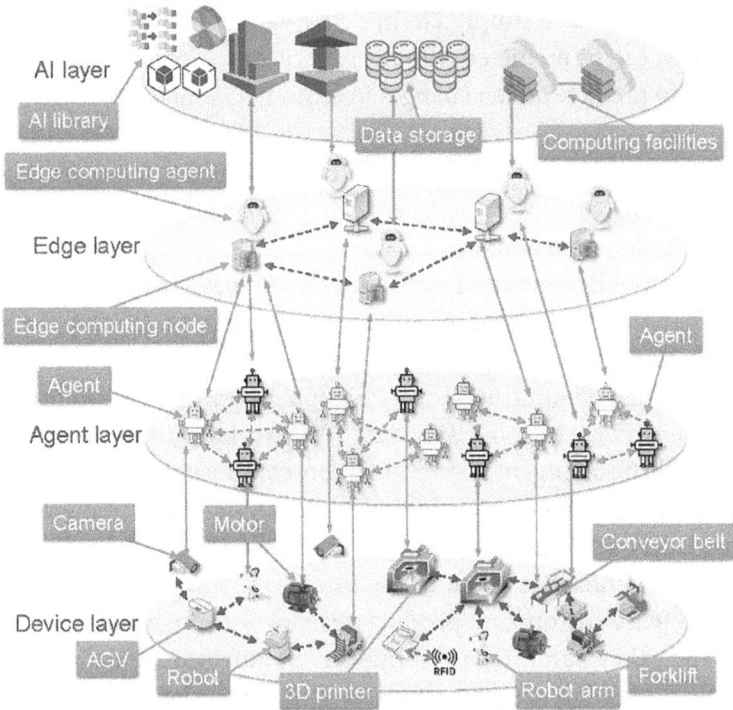

FIGURE 1.3   Edge-computing-assisted manufacturing devices.

run on edge computing servers to guarantee low-latency services for data analytics. The agent edge servers are connected by high-speed IIoT to achieve low latency. Generally, edge computing servers support a variety of AI applications. An example of such a system is a personalized product identification based on DL image recognition. First, a multiple agent subsystem is constructed for producing personalized products. Then, a single agent records image or video data at different stages of the CM process. Next, the edge computing server runs the image recognition algorithms, such as a CNN, R-CNN, Fast R-CNN, Faster R-CNN, YOLO, or single-shot detection (SSD), all of which have demonstrated their advantages in computer vision tasks. The identification results are rapidly transferred to the devices. When the single edge computing server cannot meet the real-time requirements, the multiple agent edge servers may work collaboratively to complete the specific tasks, such as product identification. Indeed, during the process, the master–slave or auction mode can be adopted for coordination according to the status analysis of each edge server.

In addition, with the help of edge computing, it is possible to establish a quantitative energy-aware model with a MAS for load balancing, collaborative processing of complex tasks, and scheduling optimization in a smart factory [95]. The above procedure can also optimize the production line with better logistics while ensuring flexibility and manufacturing efficiency.

1.4.2.1.2 Manufacturing Resource Description Based on Ontology Intelligent manufacturing will be greatly beneficial to the integration of distributed competitive resources (e.g., manpower and diverse automated technologies) so that resource sharing between enterprises and flexibility to respond to market changes is possible (i.e., CM). Therefore, in smart manufacturing, it is imperative to realize dynamic configurations of manufacturing resources [96,97]. CM can optimize lead time and manufacturing quality under various real-world constraints of dynamic nature (resource and manpower limitations, market demand, and so on).

There are several strategies in describing manufacturing resources, such as databases, object-oriented method [98], and the unified manufacturing resource model [99]. In contrast to the conventional resource description methods, the ontology-based description is one of the most prominent methods. An ontology represents an explicit specification of a conceptual model [100] by way of a classical symbolic AI reasoning method (i.e., an expert system). Modeling an application domain knowledge through an expert system provides a conceptual hierarchy that supports system integration and interoperability via an interpretable way [101,102].

In our previous work [103], the device resources of smart manufacturing were integrated by the ontology-based integration framework to describe the intelligent manufacturing resources. The architecture consisted of four layers: the data layer, the rule layer, the knowledge layer, and the resource layer. The resource layer represented the entity of intelligent manufacturing equipment [e.g., manipulators, conveyor belt, and programmable logic controller (PLC)], which was essentially the field device. The knowledge layer was essentially the information model composed of intelligent devices, which was integrated into the domain knowledge base through the OWL language [104]. The rule layer was used to gather the intelligent characteristics of intelligent equipment, such as decision-making and reasoning. The data layer included a distributed database for real-time data storing, and the relational database was used to associate the real-time data.

Due to the massive amount of data generated from the manufacturing devices, it is nearly impossible to consider all the manufacturing device resources. Thus, it is important to construct a new manufacturing description model to realize the reconfiguration of various manufacturing resources. In this model, the resources can be easily adjusted by running the model. Therefore, ontology modeling is conducted on a device and related attributes of an intelligent production line in CM. The manufacturing resources are mapped to different functions with different attributes. For instance, the time constraint of a product manufacturing is divided into a number of time slots with consideration of features of processes and devices. Then, the CM resources of a product can be mapped into computing, cutting, conveying, and other functions with the limited time slot, as shown in Figure 1.4. Next, a customized product can be produced by different devices with different time constraints. Accordingly, a product can be represented by ontology functions.

Meanwhile, after making a reasonable arrangement of different manufacturing functions at different time slots, a DL algorithm can forecast time slots of working states. The time slots of working states are important for the reconfiguration of manufacturing resources. Therefore, in actual applications, a different attribution of a device and customized products can be employed as a constraint condition.

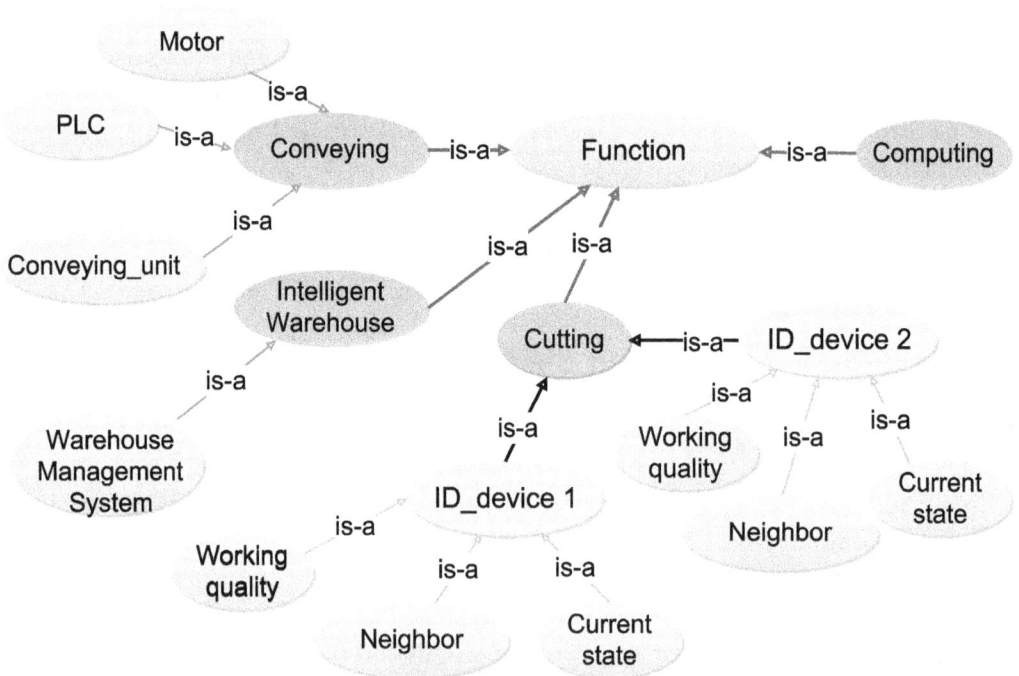

FIGURE 1.4   Manufacturing resource from the device function perspective.

1.4.2.1.3 Edge Computing in Intelligent Sensing   The concept of ubiquitous intelligent sensing is a cornerstone of smart manufacturing in the Industry 4.0 framework. Numerous research studies have been conducted in monitoring manufacturing environments [105–107]. Most published results adopt a precondition-sensing system that only accepts a static sensing parameter. It is obvious that this results in inflexibility, and the sensing parameters are difficult to be adjusted to fulfill different requirements. Second, although some studies claim dynamic parameter tuning, the absence of a prediction function is still an issue. Existing environment sensing (monitoring) cannot adjust the sensing parameters in advance to achieve a more intelligent manufacturing response.

As shown in Figure 1.5, the manufacturing environment intelligent sensing based on the edge AI computing framework includes two components: sensors nodes and edge computing nodes [108]. Generally, smart sensor nodes are equipped with different sensors, processors, and storage and communication modules. The sensors are responsible for converting the physical status of the manufacturing environment into digital signals, and the communication module delivers the sensing data to the edge server or remote data centers. The edge computing servers (nodes) include stronger processing units, larger memories, and storage space. These servers are connected to different sensor nodes and deployed in approximation to the devices, with the provision of the data storage and smart computing services by running different AI algorithms. Meanwhile, the edge computing servers are interconnected with each other to exchange information and knowledge.

Especially, the sensing parameters can be adjusted in a flexible monitoring subsystem in the manufacturing environment according to different application requirements and the

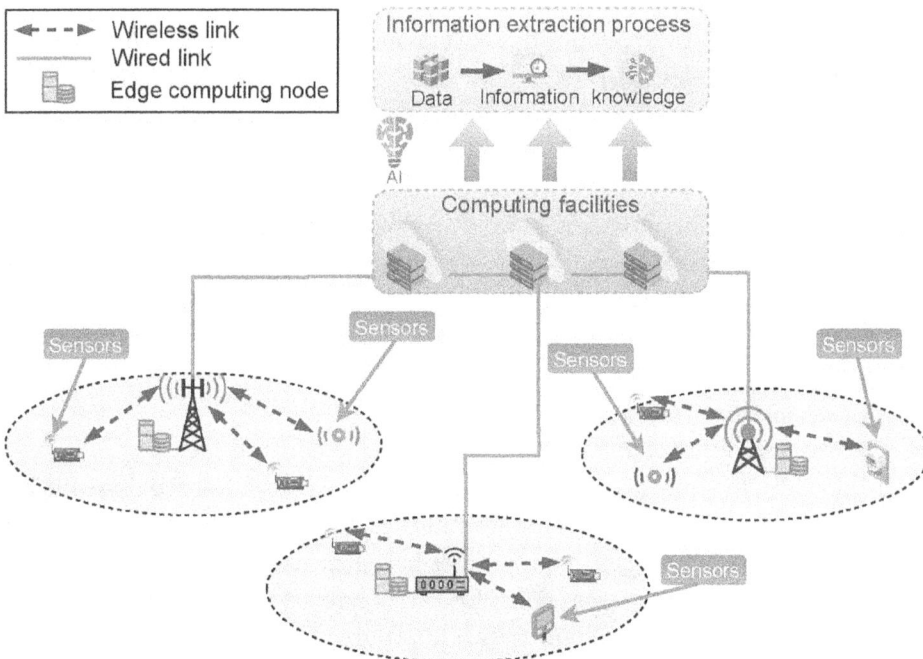

FIGURE 1.5   Intelligent sensing based on edge AI computing.

task priority. To achieve a rapid-response high-priority system, the edge AI servers should have access to the sensing data and capability to categorize the status of the CM environment. This can be done by processing the data features through ML classification algorithms such as logistic regression, SVM, and classification trees. When the data are out of the safety range, a certain risk may exist in the manufacturing environment. For instance, if an anomalous temperature event would happen in the CM area, the edge server could drive the affected nodes to increase their temperature sensing frequency in order to obtain more environmental details and make proactive forecasts and decisions.

The environmental sensing data delivery is another important component in CM. With the development of smart manufacturing, a sensing node not only performs sensing but also transmits the data. With the proliferation of massive sensing data, sensor nodes have been facing more challenges from the perspectives of data volume and data heterogeneity. In order to collect environment data effectively, it is needed to introduce new AI technologies. The sensor nodes can realize intelligent routing and communications by adjusting the network parameters and assigning different network loads and priorities to different types of data packets. With this optimized sensing transfer strategy, the AI methods can make adequate forecasts with reduced bandwidth usage.

> **Discussion**: We present intelligent manufacturing devices from edge computing-assisted intelligent agent construction, manufacturing resource description based on ontology, and edge computing in intelligent sensing. It is a challenge to upgrade the existing manufacturing devices to improve interoperability and interconnectivity. Retrofitting, instead of replacing all the legacy machines, may be an alternative strategy in this regard. The legacy manufacturing equipment can be connected to the Internet by additively mounting sensors or IoT nodes in approximation to existing manufacturing devices [109,110]. Moreover, monitors can be attached to existing machinery to visualize the monitoring process. It is worth mentioning that retrofitting strategies may apply for the sensing or monitoring scenarios, while they are not suitable or less suitable for the cases requiring to make active actions (such as control or movement). Furthermore, a comprehensive plan should be made in advance rather than arbitrarily adding sensors to the existing production line [111]. Retrofitting strategies also have their own limitations, such as a limited number of internal physical quantities that can be monitored in a retrofitted asset with respect to a newly designed smart machine.

1.4.2.2 AI-Driven Information Interaction in Smart Manufacturing Factory

In the CM domain, the information exchange system needs to fulfill the dynamic adjustment of network resources, so as to produce multiple customized products in parallel. In order to obtain optimal strategies, many studies have focused on this topic and proposed insightful algorithms and strategies [112]. However, there are still two open issues: a network framework to dynamically adjust network resources and the end-to-end (E2E) data delivery. In this section, we present software-defined industrial networks (SDINs) and AI-assisted E2E communication to tackle these two challenges.

1.4.2.2.1 Software-Defined Industrial Networks   Industrial networks are a crucial component in CM, and customized product manufacturing groups can be understood as subnets. Via an industrial network (consisting of base stations, access points, network gateways, network switches, network routers, and terminals), the CM equipment and devices are closely interconnected with each other and can be supported by edge or cloud computing paradigms [113]. Taking full advantage of AI-driven SDINs, relevant networking technologies are an important method to achieve intelligent information sharing in CM [114,115].

In conventional industrial networks, network control functions have been fixed at network nodes (e.g., gate ways, routers, and switches). Consequently, industrial networks cannot be adapted to dynamic and elastic network environments, especially in CM. The SDN technology can separate the conventional network into the data plane and the control plane. In this manner, SDN can achieve flexible and efficient network control for industrial networks. It has been reported that an SDIN can increase the flexibility of a dynamical network system while decreasing the cost of constructing a new network infrastructure [116].

The introduction of AI technologies to SDN can further bestow network nodes with intelligence. As demonstrated in Figure 1.6, AI technologies are introduced into traditional SDN, so as to form a novel SDIN. The proposed SDIN contains a number of mapping network nodes, SDIN-related devices, data centers, and cloud computing servers to support intensive computing tasks of AI algorithms. Manufacturing devices are connected

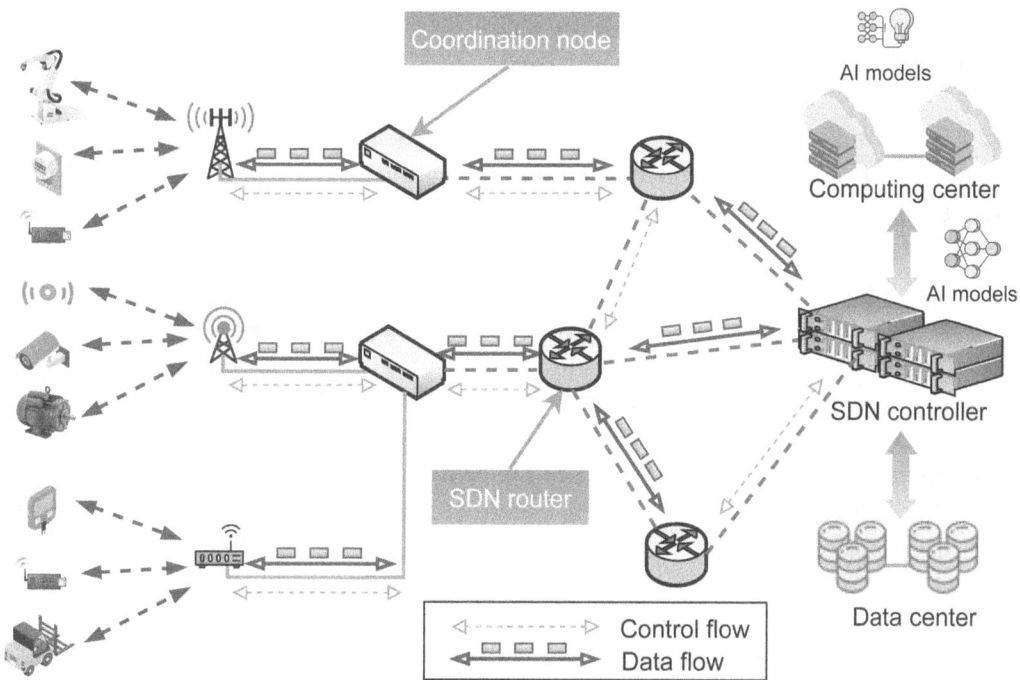

FIGURE 1.6 SDINs consist of network coordinated nodes, SDN routers, SDN controllers, data centers, and cloud computing servers that can support intensive computing tasks of AI algorithms.

by their communication modules and mapped to different network terminal nodes. On the SDIN level, key devices, such as coordinated nodes and SDIN controllers, construct the SDIN layer. First, coordinated nodes are linked with the ordinary nodes and deliver network control messages from other SDN devices. Second, the SDN routers are the key devices that realize the separation of data flow and control flow of the entire manufacturing network. In addition, the SDIN controller is directly connected to the AI server, and the AI server provides network decisions directly to the SDN controller.

In the network information process, AI algorithms, such as deep neural networks, reinforcement learning, SVM, and other ML algorithms, can be executed in a server according to the state of the network devices, such as load information, communication rate, received signal strength indicator, and other data. Then, the AI server returns the optimized results to the SDN controller, and the results are divided into different instructions for different network devices in the light of a specific CM task. Following the above steps, the SDN controllers send a set of instructions to the routers and the coordination nodes. Finally, network terminals readjust the related parameters (e.g., communication bandwidth and transmitted powers) to complete the data communication process.

Intelligent optimization algorithms (e.g., ant colony or particle swarm optimization) can find optimal data transfer strategies based on the network parameters provided by the SDIN or given by the constraints of data interaction. These algorithms can adjust the latency and energy consumption requirements. Thus, SDIN can improve the information management processes within a CM industry framework, reducing the cost of dynamically adjusting or reconfiguring network resources. Moreover, it can improve and propel the whole manufacturing intelligence. In addition, by adopting an AI-assisted SDIN, the production efficiency can be further improved.

1.4.2.2.2 End-to-End Communication   E2E or device-to-device communication between manufacturing entities is a convenient communication strategy in industrial networks [117,118]. E2E communication provides communication services with lower latency and higher reliability compared with a centralized approach [119]. With effective information interaction via E2E communication, the entire system can achieve full connectivity. In the context of CM, data transmission with different real-time constraints has become a critical requirement [120]. The E2E industrial communication approach optimizes the usage of network resources (e.g., network access and bandwidth allocation) through data communication of varying latency [121,122]. Meanwhile, in order to realize the E2E communication in the industrial domain, a hybrid E2E communication network—based on the AI technology and SDIN—is here constructed by exploiting different media, communication protocols, and strategies. The hybrid E2E-based communication mechanism with AI assistance can be divided into three layers: the physical layer, the media access controlling (MAC) layer, and the routing layer.

In the physical layer, according to the advantages and disadvantages of the involved communication technologies, different communication media include optical fiber [123], network cable [124], and wireless radio [125]. Generally, industrial communications can be divided into wired or wireless communications. On the one hand, wired communications typically exhibit high stability and low latency. A representative case is an industrial Ethernet that is based on a common Ethernet and runs improved Ethernet protocols, such as EtherCAT [126], EtherNet/IP [127], and Powerlink [128]. On the other hand, wireless networks have been adopted in applications with relatively high flexibility [129,130]. Nowadays, an increasing number of mobile elements have been incorporated in manufacturing systems; therefore, wireless media has been widely exploited in mobile communications [131]. Conventional strategies on fixed and static industrial networks may not fulfill the emerging requirements of flexible network configurations. The AI and related technologies, such as deep reinforcement learning, optimization theory, and game theory, can play significant roles in improving the communication efficiency in the physical layer, for example, determining the optimal communication between wired and wireless networks while achieving a good balance between network operational cost and network performance.

In the MAC layer, different devices have different requirements for E2E communications according to their specific functions. Although many different MAC protocols have been proposed [e.g., carrier sense multiple access (CSMA) [132], code-division multiple access (CDMA) [133], time-division multiple access (TDMA) [134], and their improved versions], these methods still lack flexibility and do not fulfill the emerging requirements of industrial applications. Generally, industrial E2E communications can be divided into two categories: periodic communications and aperiodic communications. Similarly, AI plays an important role in the MAC layer. An example is a hybrid approach that combines the CSMA and TDMA with an intelligent optimization method to improve the efficiency of the E2E communication. In particular, the two categories of communication requirements (high and low real-time or periodic and aperiodic communications) are classified by the AI-based method (e.g., naïve Bayes). Next, an improved hybrid MAC is constructed on top of the CSMA and TDMA. TDMA and CSMA schemes deal with the periodic and aperiodic data flows of the E2E communications. The size of this proposed mechanism can be adjusted in accordance with the AI-optimized results of a real application.

The network routing is also another key component of E2E communications. The key node of the routing path plays an important role in the E2E communications as well. However, the performances of routing key nodes are impacted by the workload, for instance, the amount of forwarded data. Similarly, AI plays a significant role in the routing layer. The predicted state parameters, such as communication rate and network loads of key nodes, can be obtained by using historical data from the network node status by algorithms, such as deep neural networks or deep reinforcement learning (e.g., deep Q-learning).

1.4.2.3 AI-Driven Flexible Manufacturing Line

A flexible manufacturing production line realizes customization. AI-driven production line strategies and technologies, such as collective intelligence, autonomous intelligence, and cross-media reasoning intelligence, have accelerated the global manufacturing process. Therefore, the subjects of cooperative operation between multiple agents, dynamic reconfiguration of manufacturing, and self-organizing scheduling based on production tasks are presented in this section.

1.4.2.3.1 Cooperative Multiple Agents   Cooperation among multiple agents is necessary to dynamically construct collaborative groups for the completion of customized production tasks [135]. As discussed in Section 1.4, multiple agents with edge computing provide a better option than a single device to build a collaborative operation to realize CM [95,136]. Therefore, by combining the edge computing-assisted intelligent agents and different AI algorithms, a novel cooperative operation can be constructed, as shown in Figure 1.7. The strategy of cooperative operation by multiples agents can be divided into the order of submission, task decomposition, cooperative group, and subgroup assignment.

The working process of a flexible manufacturing production line can be described as follows. First, according to the customers' requirements, the CM product orders are issued to the manufacturing system through the recommender system. After receiving the product orders, the AI-assisted task decomposition algorithms take the product orders as the input, the device working procedure as the output, and the product manufacturing time as a constraint; these algorithms are mainly executed at the remote cloud server. A product order can be divided into multiple subtasks that are sent to all the agents via the industrial network. After the negotiation, agents return the answers to the edge server, which handles the working subtasks according to corresponding conditions and constraints. Next, the AI-assisted cost-evaluation algorithm calculates the cost of a producing group (i.e., cooperative manufacturing group) from the historical data. Then, the edge agents intelligently select suitable device agents to finish the product order after considering the whole cooperative group performances, such as production time and product quality. Moreover, the edge agents send the selection result to the device agents, which are chosen to take part in the producing order. The main cooperative group is constructed based on the working steps.

The main cooperative group may not be well suited for real applications, especially for complicated CM tasks. Therefore, an AI-based method for constructing a suitable-size cooperative subgroup is an important step for dealing with the mentioned problem. A possible strategy is to use cognitive approaches, such as the Adaptive Control of Thought—Rational

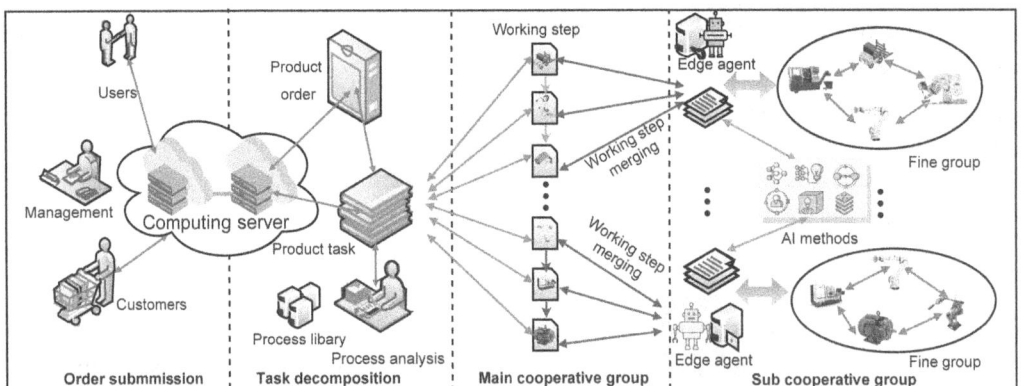

FIGURE 1.7 Cooperative multiple agents. The strategy of cooperative multiple agents can be divided into (1) the order of submission; (2) task decomposition; (3) cooperative group; and (4) subgroup assignment.

(ACT-R) model [137]. These subtask cooperative groups can be mapped to the digital space (i.e., edge agent) and form even lower-level subgroups, all interconnected by the conveyor, logistics systems, and industrial communication systems. Each subgroup can delegate the same edge agent to provide the management and customers with manufacturing services. The characteristics of the subgroups are partly derived from the process constraints and the physical constraints of the plant. In principle, the higher the constraints, the deeper the task tree will expand, from more abstract tasks to particular atomic targets achievable by the present devices. This structure can be replicated with a probabilistic graphical model or with a fuzzy tree.

After all the agents have been assigned with subtasks, they form two level-cooperative groups. The formation of these cooperative groups is beneficial to resource management. Then, according to the manufacturing task attributes, multiple agents complete the producing task. During this period, the corresponding device agents send their status data to edge servers timely, and the manufacturing process can be monitored by analyzing these data in the entire system. In contrast to the AI-driven cooperative operation between multiple agents, conventional methods often rely on human operators who participate in the whole process or computer-assisted operators also requiring human interventions. These methods inevitably result in huge operational expenditure.

1.4.2.3.2 Dynamic Reconfiguration of Manufacturing Systems   With the scientific development of the industrial market and manufacturing equipment, different industrial devices present different performance requirements representing multiple function trends [138]. For instance, the latest computer numerical control (CNC) machine tool can complete a wide range of tasks, from lathing to milling functions. On the other hand, a dedicated manufacturing line does not meet new industrial requirements, especially for customized production [139]. The trend today is toward reconfiguration and reprogrammability of manufacturing processes [140]. Although several studies have investigated the problem and presented meaningful results [141,142], most of them lack intelligent design to fulfill the emerging requirements of dynamic reconfiguration of manufacturing systems, especially for CM. In particular, the work in [141] focuses on the communications between agents, while the work in [142] investigates the relationship between manufacturing flexibility and demands. Thus, AI technologies have seldom been adopted in these studies. At present, ontology (as shown in Section 1.4) offers insights into dynamic reconfiguration of manufacturing resources [119,143].

A schematic of the dynamic reconfiguration process based on the ontology inference is shown in Figure 1.8. Each customized product invokes several processing procedures. First, a personalized product manufacturing-related device (such as a cutting and materials' handling device) is selected by ontology reasoning based on the device function. Then, the second selection of the devices involved in the manufacturing is completed according to ontology results with respect to the related manufacturing process, the manufacturing time, manufacturing quality, and other parameters of a device. Finally, a CM production line is constructed. Especially, when the production line receives a production task, the raw material for a specific type of product is delivered from an autonomous warehouse.

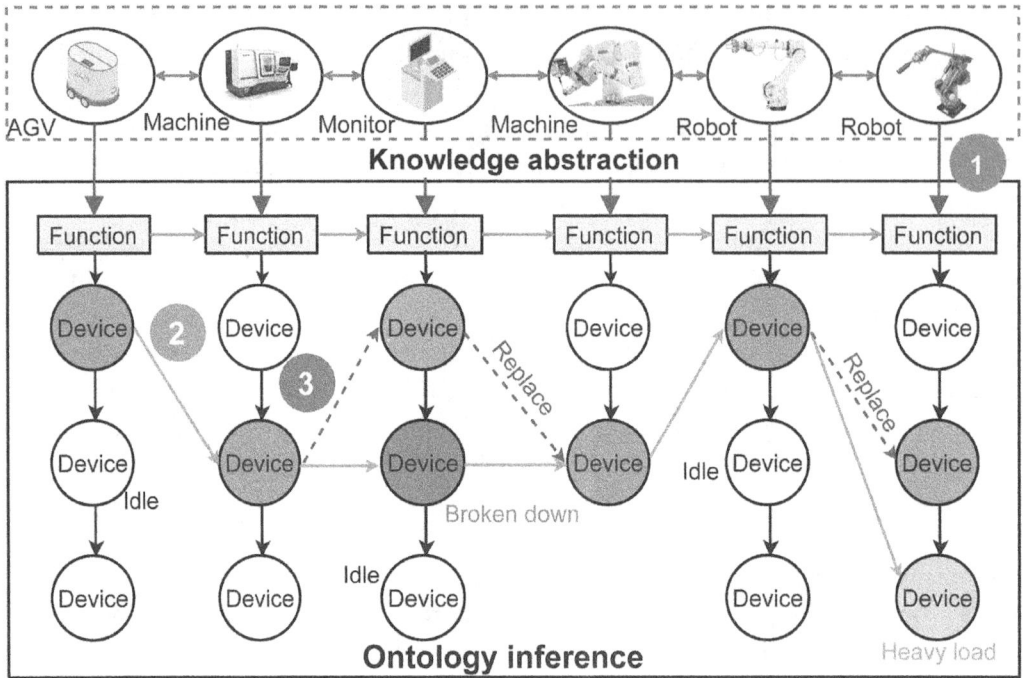

FIGURE 1.8 Dynamic reconfiguration of manufacturing resources.

Then, the production line completes the manufacturing tasks in the process sequence. Furthermore, when one of the manufacturing devices breaks down during the process, automatic switching of the related machining equipment by ontology inference is conducted. Meanwhile, the reasoning mechanism reflects the reconstruction function of a flexible production of the production line.

The presented approach leads to optimal process planning and functional reconstruction. Besides, it shows the strengths of ontology modeling and reasoning. In practice, only ontology and constraints need to be established according to the above description. According to Jena syntax, the corresponding API interface can be invoked to meet the task requirements of this scenario. In the future, other AI algorithms are expected to be integrated with ontology inference.

1.4.2.3.3 Self-Organizing Schedules of Multiple Production Tasks   Product orders generally have stochastic and intermittent characteristics as the arrival time of orders is usually uncertain [144]. This may result in having to share production resources among multiple tasks. Therefore, creating self-organizing schedules with a time slot based on multiple agents for multiple production tasks is paramount [88]. The mechanism of self-organizing schedules for multiple production tasks can be divided into three steps: task analysis, task decomposition, and task execution.

As shown in Figure 1.9, in terms of initialization, when a new production task is processed by the multitasking production line, the new production tasks are divided into multiple steps by an AI-based method executed at the cloud. In addition, according

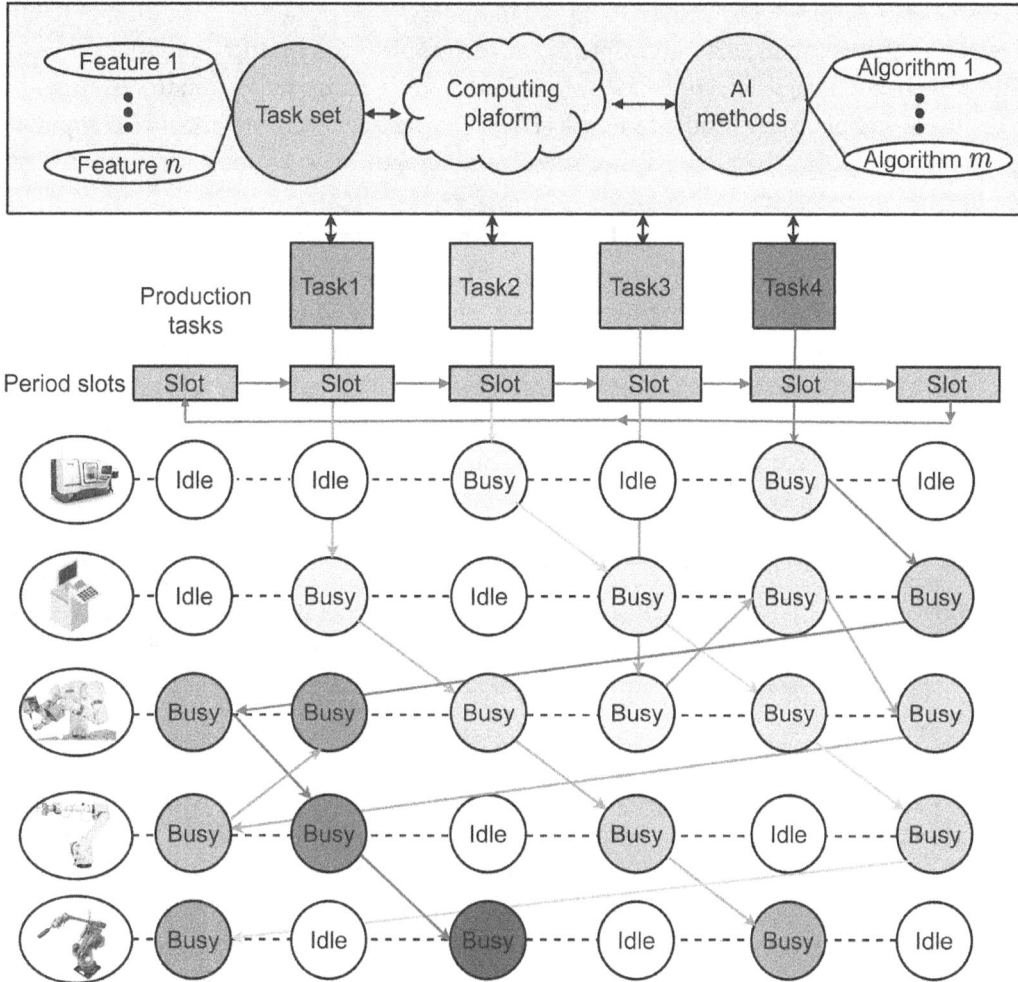

FIGURE 1.9 Self-organization of schedules of multiple production tasks consists of three steps: task analysis, task decomposition, and task execution.

to the process lead time, the production period can be decomposed into time slots of different lengths. Moreover, for one working step in a time slot, edge agents select all idle device agents by comparing the mapping relationship between working steps and device agent functions. This processed time slot information is then broadcasted to all the agents simultaneously. Then, idle device agents choose the working step by price bidding or negotiating with others according to the manufacturing requirements and self-conditions (e.g., manufacturing time and quality). These results are broadcasted to other agents, including different servers. Next, the edge agents update the working state of the idle device agent in the corresponding time slot. These procedures are repeated until the new task steps are allocated within a certain or fixed time. Finally, multiple agents finish the scheduling of the new production task in a self-organization manner.

Self-organization of schedules with multiple agents and time slots can effectively complete simultaneous production tasks using a flexible production. Furthermore, production line efficiency is improved. Consequently, all manufacturing resources, including different devices and subsystems, are more intelligent to complete the multiple production tasks autonomously. In contrast, conventional methods often require huge human resources in scheduling and planning production tasks [145]. Despite the recent advances in computer-aided methods [146], they still require substantial human interventions and cannot meet the flexible requirements.

However, we have to admit that AI-driven self-organization of schedules does not get rid of humans in the loop of the entire production process. The main goal of AI-driven methods is to save unnecessary human resource consumption and mitigate other operational expenditures. In this manner, human workers can concentrate on planning and optimizing the overall production procedure instead of conducting tedious and repetitive tasks. Meanwhile, an appropriate human intervention is still necessary when full automation is not achievable or is partially implemented. In this sense, AI-driven methods can also assist human workers to give intelligent determinations.

1.5 OPPORTUNITIES AND CHALLENGES

The convergence of AI technologies, IoT, SDN, and other new ICTs in smart factories not only significantly increases the flexibility, intelligence, and efficiency of CM systems but also poses challenges. In this section, we discuss these challenges and recent advances.

1.5.1 Smarter Devices in Customized Manufacturing

In a CM environment, the equipment does not only perform the basic functions of automation but also needs to be intelligent and flexible. From the point of view of standalone equipment, the devices should have the following functions: parameter sensing, data storage, logical inference, information interaction, self-diagnostics, hybrid computing support, and preventive maintenance. From the perspective of the physical layer, the realization of CM is increasingly complex, as standard devices or equipment cannot accomplish this complicated task. Therefore, devices need to have both collaborative ability and swarm intelligence to collaborate with each other to complete complex tasks. Devices with the above functions can meet the requirements of personalized production and adapt to the trend of achieving the automation and intelligence of manufacturing. Therefore, CM devices must be based on the integration and deep fusion of advanced manufacturing technologies, automation technologies, information technology, image technology, communication technologies, and AI technologies to realize the local and swarm intelligence. In this aspect, to be realizable in a generic context, further progress needs to be made in AI and AI/robotics integration. However, as shown in the above section, CM is already realizable in restricted environments.

Extensive studies have been conducted in the respective technologies to support CM [147–149]: intelligent connectivity, data-driven intelligence, cognitive IoT, and industrial big data. Unfortunately, most of these studies have mainly focused on single-device intelligence. Although these studies provide useful suggestions toward smarter devices, there is still a huge gap between realistic smart devices and the current solutions. We consider that an ideal method for realizing smart devices in manufacturing should integrate a series of methods to ensure the CM with edge intelligence that offers adaptability and response to a wide range of scenarios. These different methods or algorithms can be orchestrated by a central AI unit, which decides which algorithm is more useful for a particular task (combining a symbolic AI approach with ML). Meanwhile, a good balance between intelligence and cost is still a challenge that needs to be addressed in the future.

1.5.2 Information Interaction in CM

Most manufacturing frameworks represent the distributed systems, where highly efficient communications between different components are needed [150,151]. On the one hand, different components of CM need not only effective connections but also highly efficient information interactions. As shown earlier in this chapter, the CM systems must incorporate different information interaction technologies and algorithms. On the other hand, due to the device heterogeneity and the increased demands of communications, the CM information interactive systems have to transmit massive data with different latency requirements. Consequently, the networks of CM need to be optimized to fulfill the diverse requirements of different applications, such as media access, moving handover, and congestion control. In other words, CM needs as a basis of a highly efficient information interaction system, which integrates multiple information technologies, including the absorbing DL and other AI algorithms. This system particularly depends on efficient and reasonable information interaction. With the development of manufacturing techniques, an increased number of studies have focused on the massive volume of communications and a large number of connections [152]. Based on network performance optimization methods, researchers have developed several industrial networks [153,154]. As different information transfer protocols exist with different requirements, such as time-sensitive and -insensitive data transfer, the construction of a more efficient interaction industrial network represents another important challenge of future research. Existing studies mostly focus on high-speed communications rather than efficient interaction with intelligent networks. These methods usually only meet the basic requirements of industrial networks, such as bandwidth and delay.

Recently, wireless mobile communications (e.g., 5G systems) and optical fiber communications have achieved great development [155,156]. The 5G technology for smart factories is still in its infancy, and there are still a number of issues, such as deployment, accessing, and spectrum management, to be addressed. A hybrid industrial network with the latest communication technologies and AI (including DL, integrated learning, transfer learning, and so on) can be a promising solution for information interaction, which should consider different data flows according to different applications of intelligent CM.

1.5.3 Dynamic Reconfiguration of Manufacturing Resources

Intelligent manufacturing, especially CM, involves a dynamic reorganization of the available resources and extreme flexibility. The essence of CM is to provide customized products that have the characteristics of small batch, short processing cycle, and flexible production. Therefore, manufacturing resources need constant readjustment and reorganization [157]. In addition, customized products are thought of as a variety of processing crafts.

Moreover, the increasing complexity of the industrial environments makes the management of the resources even more difficult. A factory nowadays is a system consisting of many subsystems, which can produce emergent behaviors through the interaction of the subsystems. Therefore, the dynamic reconstruction of manufacturing resources is one of the main challenges toward achieving a generic CM facility. In this aspect, a number of strategies have been proposed [158,159] in this domain.

Recently, knowledge reasoning, knowledge graph, transfer learning, and other AI algorithms have attained great progress [160–162]. In our opinion, hybrid AI methods combining the latest knowledge reasoning technologies with swarm intelligence can be a promising solution for dynamic reconstruction, which may consider different application scenarios of CM. Manufacturing-based process optimization with ML technologies may be one of the effective methods to reorganize resources. Moreover, digital twin technologies may be the driving technologies to improve resource reconfiguration [163–165].

1.5.4 Practical Deployment and Knowledge Transfer

Although the AIaCM framework is promising to foster smart manufacturing, several challenges in industrial practice arise before the formal adoption of this framework. The first nontechnical challenge mainly lies in the cost of upgrading existing manufacturing machinery, digitizing manufacturing equipment, and purchasing computing facilities, as well as AI services. This huge cost may be affordable for small- and medium-sized enterprises (SMEs). With respect to upgrading manufacturing production lines for SMEs, retrofitting legacy machines can be an economic solution, as discussed in Section 1.4.2.3. In particular, diverse sensors and IoT nodes can be attached to existing manufacturing equipment to collect diverse manufacturing data. Those sensors and IoT nodes can be collected with the Internet, so as to improve the interconnectivity of legacy machines. For example, Raspberry Pi models mounted with sensors can be deployed in workrooms to collect ambient data [109,166]. Besides hardware upgrading, software tools and AI services should be also purchased and adopted by manufacturing enterprises. Similarly, the economic solution for SMEs is to outsource manufacturing data to cloud services providers or ML as a service (MLaaS) provider who can offer on-demand computing services. Nevertheless, outsourcing confidential data to untrusted third parties may increase the risks of security and privacy leakage. Thus, it is a prerequisite to enforce privacy and security protection schemes on manufacturing data before outsourcing. Moreover, the expenditure of system operating and training personnel should not be ignored in practical deployment.

Another challenge is the effective technology transfer from research institutions to enterprises. Technology transfer involves many nontechnical factors and multiple parties. The nontechnical issues of technology transfer include marketing analysis, intellectual property management, technical invention protection, commercialization, and financial returns. Many frontier technological innovations often end up with unsuccessful technology transfer due to ignorance of the nontechnical factors [167]. One of the main obstacles in technology transfer lies in the technology readiness level (TRL) gap between research and industrial practice. In particular, research institutions often focus on research results at TRL 1–3 implying basic feasibility and effectiveness, while industrial enterprises often require transferred technologies at TRL 7–8 or even higher levels meaning prototype demonstration and real deployment [168]. We admit that there is still a long way for the AIaCM framework before reaching TRL 7–8. The investigation of the technology transfer of AIaCM will be a future direction. Both researchers and industrial practitioners are expected to work together to realize AIaCM.

1.6 BOOK OUTLINE

Chapter 1 mainly provides introduction to smart manufacturing factory. This chapter gives the background of smart manufacturing factory, CM mode, and the overview and the frontier technologies of the AI-driven CM factory. The rationale of this book is explained, and a short overview of the chapters is provided. It also discusses the opportunities and challenges of the AI-driven CM in smart manufacturing factories.

Chapter 2 mainly introduces edge intelligence and its architecture in CM. The key technologies of edge computing in CM are discussed in detail. This chapter enlightens its features of low latency, real-time interaction, location awareness, mobility support, geo-distribution, etc. over cloud. It also gives the application of edge intelligence in CM environment.

Chapter 3 analyzes the cause, classification, and current solution framework of smart factory heterogeneous network. SDN- and EC-based heterogeneous networks framework and AI-enabled QoS optimization of heterogeneous networks in smart manufacturing factory are proposed. This work aims at network QoS optimization objectives, including low latency and high reliability, network load balancing, and high security and privacy protection.

Chapter 4 mainly proposes the important topic both in practice and research, namely intelligent fault diagnosis and maintenance in smart manufacturing factory. It introduces how to effectively use advanced ML algorithms for supervised, semi-supervised, and unsupervised intelligent fault diagnosis to deal with the existing dilemma, especially intelligent fault diagnosis of rotating machinery, fault prediction, and intelligent maintenance under Industry 4.0.

Chapter 5 fully explores the resource dynamic scheduling based on edge cloud collaboration and resource reconfiguration in smart manufacturing. This chapter fully explores the application potential of the IIoT from the aspects of edge computing and cloud computing. Meanwhile, compared with mass production, the new generation of intelligent manufacturing technology offers improved flexibility, transparency, resource utilization, and efficiency of manufacturing processes.

Chapter 6 introduces the components of the platform in detail and reflects the key links and core technologies of custom manufacturing through mobile services and conveyor belt control. The implementation of CM factory is provided to inspire the manufacturing enterprises to realize the transformation of smart manufacturing factories.

REFERENCES

[1] R. Neugebauer, S. Hippmann, M. Leis et al., "Industrie 4.0- From the perspective of applied research," *Procedia CIRP*, vol. 57, pp. 2–7, 2016.

[2] R. G. G. Caiado, L. F. Scavarda, D. L. D. M. Nascimento et al., "A maturity model for manufacturing 4.0 in emerging countries," in *Operations Management for Social Good*, eds: A. Leiras, C. A. González-Calderón, I. de Brito Junior, S. Villa, and H. T. Y. Yoshizaki. Springer International Publishing, Cham, 2020, pp. 393–402.

[3] J. Zhou, P. Li, Y. Zhou et al., "Toward new-generation intelligent manufacturing," *Engineering*, vol. 4, no. 1, pp. 11–20, 2018.

[4] F. Li, J. Wan, P. Zhang et al., "Usage-specific semantic integration for cyber-physical robot systems," *ACM Transactions on Embedded Computing Systems (TECS)*, vol. 15, no. 3, pp. 1–20, 2016.

[5] M. Brettel, N. Friederichsen, M. Keller et al., "How virtualization, decentralization and network building change the manufacturing landscape: An industry 4.0 perspective," *International Journal of Information and Communication Engineering*, vol. 8, no. 1, pp. 37–44, 2014.

[6] J. Wan, D. Zhang, Y. Sun et al., "VCMIA: A novel architecture for integrating vehicular cyber-physical systems and mobile cloud computing," *Mobile Networks and Applications*, vol. 19, no. 2, pp. 153–160, 2014.

[7] J. Wan, H. Yan, D. Li et al., "Cyber-physical systems for optimal energy management scheme of autonomous electric vehicle," *The Computer Journal*, vol. 56, no. 8, pp. 947–956, 2013.

[8] European Commission, "EUROPE 2020: A strategy for smart, sustainable and inclusive growth," *Working paper {COM (2010) 2020}*, 2010.

[9] H. Lasi, P. Fettke, H.-G. Kemper et al., "Industry 4.0," *Business & Information Systems Engineering*, vol. 6, no. 4, pp. 239–242, 2014.

[10] J. Zhou, "Intelligent manufacturing—Main direction of 'Made in China 2025'," *China Mechanical Engineering*, vol. 26, no. 17, p. 2273, 2015.

[11] J. Holdren, T. Power, G. Tassey et al., *A National Strategic Plan for Advanced Manufacturing*. US National Science and Technology Council, Washington, DC, 2012.

[12] J. Wan, M. Yi, D. Li et al., "Mobile services for customization manufacturing systems: An example of industry 4.0," *IEEE Access*, vol. 4, pp. 8977–8986, 2016.

[13] N. Benkamoun, W. ElMaraghy, A.-L. Huyet et al., "Architecture framework for manufacturing system design," *Procedia CIRP*, vol. 17, pp. 88–93, 2014.

[14] A. Radziwon, A. Bilberg, M. Bogers et al., "The smart factory: Exploring adaptive and flexible manufacturing solutions," *Procedia Engineering*, vol. 69, pp. 1184–1190, 2014.

[15] T. Lin, C. Yang, M. Gu et al., "Application technology of cloud manufacturing for aerospace complex products," *Computer Integrated Manufacturing Systems*, vol. 22, no. 4, pp. 884–894, 2016.

[16] J. Wan, S. Tang, Q. Hua et al., "Context-aware cloud robotics for material handling in cognitive industrial Internet of Things," *IEEE Internet of Things Journal*, vol. 5, no. 4, pp. 2272–2281, 2017.

[17] J. Wan, S. Tang, H. Yan et al., "Cloud robotics: Current status and open issues," *IEEE Access*, vol. 4, pp. 2797–2807, 2016.

[18] Y. Lyu and J. Zhang, "Big-data-based technical framework of smart factory," *Computer Integrated Manufacturing Systems*, vol. 22, no. 11, p. 2691–2697, 2016.

[19] V. Roblek, M. Meško, and A. Krapež, "A complex view of industry 4.0," *Sage Open*, vol. 6, no. 2, p. 2158244016653987, 2016.

[20] A. Kusiak, "Smart manufacturing," *International Journal of Production Research*, vol. 56, no. 1–2, pp. 508–517, 2018.

[21] S. Wang, J. Wan, M. Imran et al., "Cloud-based smart manufacturing for personalized candy packing application," *The Journal of Supercomputing*, vol. 74, no. 9, pp. 4339–4357, 2018.

[22] L. Li, "China's manufacturing locus in 2025: With a comparison of "Made-in-China 2025" and "Industry 4.0"," *Technological Forecasting and Social Change*, vol. 135, pp. 66–74, 2018.

[23] Y. Lu, "Industry 4.0: A survey on technologies, applications and open research issues," *Journal of Industrial Information Integration*, vol. 6, pp. 1–10, 2017.

[24] S. Li, L. D. Xu, and S. Zhao, "The internet of things: A survey," *Information Systems Frontiers*, vol. 17, no. 2, pp. 243–259, 2015.

[25] C. El Kaed, I. Khan, A. Van Den Berg et al., "SRE: Semantic rules engine for the industrial Internet-of-Things gateways," *IEEE Transactions on Industrial Informatics*, vol. 14, no. 2, pp. 715–724, 2017.

[26] N. J. Nilsson, *Principles of Artificial Intelligence*. Springer Science & Business Media, 1982.

[27] S. J. Russell, *Artificial Intelligence A Modern Approach*. Pearson Education, Inc., 2010.

[28] M. Chen, S. Mao, and Y. Liu, "Big data: A survey," *Mobile Networks and Applications*, vol. 19, no. 2, pp. 171–209, 2014.

[29] A. McAfee, E. Brynjolfsson, T. H. Davenport et al., "Big data: The management revolution," *Harvard Business Review*, vol. 90, no. 10, pp. 60–68, 2012.

[30] B. Chen, J. Wan, L. Shu et al., "Smart factory of industry 4.0: Key technologies, application case, and challenges," *IEEE Access*, vol. 6, pp. 6505–6519, 2017.

[31] V. Modrak, Z. Soltysova, P. Semanco et al., "Production scheduling and capacity utilization in terms of mass customized manufacturing," *International Scientific-Technical Conference Manufacturing*. Springer, 2019, pp. 295–306.

[32] J. R. Baldwin and D. Sabourin, "Impact of the adoption of advanced information and communication technologies on firm performance in the Canadian manufacturing sector," *OECD Science, Technology and Industry Working Papers, No. 2002/01*. OECD Publishing, Paris, 2002.

[33] H. Lightfoot, T. Baines, and P. Smart, "Examining the information and communication technologies enabling servitized manufacture," *Proceedings of the Institution of Mechanical Engineers, Part B: Journal of Engineering Manufacture*, vol. 225, no. 10, pp. 1964–1968, 2011.

[34] B.-H. Li, B.-C. Hou, W.-T. Yu et al., "Applications of artificial intelligence in intelligent manufacturing: A review," *Frontiers of Information Technology & Electronic Engineering*, vol. 18, no. 1, pp. 86–96, 2017.

[35] W. Li, W.-J. Wu, H.-M. Wang et al., "Crowd intelligence in AI 2.0 era," *Frontiers of Information Technology & Electronic Engineering*, vol. 18, no. 1, pp. 15–43, 2017.

[36] Y. Pan, "Heading toward artificial intelligence 2.0," *Engineering*, vol. 2, no. 4, pp. 409–413, 2016.

[37] Y.-T. Zhuang, F. Wu, C. Chen et al., "Challenges and opportunities: From big data to knowledge in AI 2.0," *Frontiers of Information Technology & Electronic Engineering*, vol. 18, no. 1, pp. 3–14, 2017.

[38] M. Mohammadi and A. Al-Fuqaha, "Enabling cognitive smart cities using big data and machine learning: Approaches and challenges," *IEEE Communications Magazine*, vol. 56, no. 2, pp. 94–101, 2018.

[39] Z. Zhang, K. Long, J. Wang et al., "On swarm intelligence inspired self-organized networking: Its bionic mechanisms, designing principles and optimization approaches," *IEEE Communications Surveys & Tutorials*, vol. 16, no. 1, pp. 513–537, 2013.

[40] M. Martinez-Garcia and T. Gordon, "A new model of human steering using far-point error perception and multiplicative control," *2018 IEEE International Conference on Systems, Man, and Cybernetics (SMC)*. IEEE, 2018, pp. 1245–1250.

[41] F.-Y. Wang, X. Wang, L. Li et al., "Steps toward parallel intelligence," *IEEE/CAA Journal of Automatica Sinica*, vol. 3, no. 4, pp. 345–348, 2016.

[42] M. Martínez-García, Y. Zhang, and T. Gordon, "Memory pattern identification for feedback tracking control in human-machine systems," *Human Factors*, vol. 63, no. 2, pp. 210–226, 2021.

[43] Y. Peng, X. Zhai, Y. Zhao et al., "Semi-supervised cross-media feature learning with unified patch graph regularization," *IEEE Transactions on Circuits and Systems for Video Technology*, vol. 26, no. 3, pp. 583–596, 2015.

[44] Y.-X. Peng, W.-W. Zhu, Y. Zhao et al., "Cross-media analysis and reasoning: Advances and directions," *Frontiers of Information Technology & Electronic Engineering*, vol. 18, no. 1, pp. 44–57, 2017.

[45] X. Yao, J. Zhou, J. Zhang et al., "From intelligent manufacturing to smart manufacturing for industry 4.0 driven by next generation artificial intelligence and further on," *2017 5th International Conference on Enterprise Systems (ES)*. IEEE, 2017, pp. 311–318.

[46] J. Wan, J. Yang, Z. Wang et al., "Artificial intelligence for cloud-assisted smart factory," *IEEE Access*, vol. 6, pp. 55419–55430, 2018.

[47] M. A. K. Bahrin, M. F. Othman, N. H. N. Azli et al., "Industry 4.0: A review on industrial automation and robotic," *Jurnal Teknologi*, vol. 78, no. 6–13, pp.137–143, 2016.

[48] M. Dopico, A. Gómez, D. De la Fuente et al., "A vision of industry 4.0 from an artificial intelligence point of view," *Proceedings on the International Conference on Artificial Intelligence (ICAI)*. The Steering Committee of the World Congress in Computer Science, 2016, p. 407.

[49] N. Mircică, "Cyber-physical systems for cognitive industrial Internet of Things: Sensory big data, smart mobile devices, and automated manufacturing processes," *Analysis and Metaphysics*, no. 18, pp. 37–43, 2019.

[50] V. Vasiliou and D. Milner, "Computer-integrated manufacture for cold roll forming," in *Advances in Manufacturing Technology*, ed: P.F. McGoldrick. Springer, Boston, MA, 1986, pp. 79–85.

[51] I. E. Hassani, C. E. Mazgualdi, and T. Masrour, "Artificial intelligence and machine learning to predict and improve efficiency in manufacturing industry," *arXiv preprint arXiv:1901.02256*, 2019.

[52] M. Wollschlaeger, T. Sauter, and J. Jasperneite, "The future of industrial communication: Automation networks in the era of the internet of things and industry 4.0," *IEEE Industrial Electronics Magazine*, vol. 11, no. 1, pp. 17–27, 2017.

[53] S. Ren, Y. Zhang, Y. Liu et al., "A comprehensive review of big data analytics throughout product lifecycle to support sustainable smart manufacturing: A framework, challenges and future research directions," *Journal of Cleaner Production*, vol. 210, pp. 1343–1365, 2019.

[54] Y. Lu and X. Xu, "Cloud-based manufacturing equipment and big data analytics to enable on-demand manufacturing services," *Robotics and Computer-Integrated Manufacturing*, vol. 57, pp. 92–102, 2019.

[55] J. Lee, B. Bagheri, and H.-A. Kao, "A cyber-physical systems architecture for industry 4.0-based manufacturing systems," *Manufacturing Letters*, vol. 3, pp. 18–23, 2015.

[56] K. S. Aggour, V. K. Gupta, D. Ruscitto et al., "Artificial intelligence/machine learning in manufacturing and inspection: A GE perspective," *MRS Bulletin*, vol. 44, no. 7, pp. 545–558, 2019.

[57] A. Kusiak, "Intelligent manufacturing: Bridging two centuries," *Journal of Intelligent Manufacturing*, vol. 30, no. 1, pp. 1–2, 2019.

[58] X. Li, D. Li, J. Wan et al., "A review of industrial wireless networks in the context of industry 4.0," *Wireless Networks*, vol. 23, no. 1, pp. 23–41, 2017.

[59] L. Liao, C.-F. Lai, J. Wan et al., "Scalable distributed control plane for On-line social networks support cognitive neural computing in software defined networks," *Future Generation Computer Systems*, vol. 93, pp. 993–1001, 2019.

[60] J. Wan, S. Tang, Z. Shu et al., "Software-defined industrial internet of things in the context of industry 4.0," *IEEE Sensors Journal*, vol. 16, no. 20, pp. 7373–7380, 2016.

[61] S. Zoppi, A. Van Bemten, H. M. Gürsu et al., "Achieving hybrid wired/wireless industrial networks with WDetServ: Reliability-based scheduling for delay guarantees," *IEEE Transactions on Industrial Informatics*, vol. 14, no. 5, pp. 2307–2319, 2018.

[62] J. Wan and M. Xia, "Cloud-assisted cyber-physical systems for the implementation of Industry 4.0," *Mobile Networks and Applications*, vol. 22, no. 6, pp. 1157–1158, 2017.

[63] A. H. Sodhro, S. Pirbhulal, and V. H. C. De Albuquerque, "Artificial intelligence-driven mechanism for edge computing-based industrial applications," *IEEE Transactions on Industrial Informatics*, vol. 15, no. 7, pp. 4235–4243, 2019.

[64] Q. Zhao, L. Wei, Z. Sheng et al., "Research on service-oriented application support software platform of manufacturing informatization," *2016 2nd IEEE International Conference on Computer and Communications (ICCC)*, 2016. IEEE, pp. 1402–1406.

[65] F. Tao, Y. Cheng, L. Zhang et al., "Advanced manufacturing systems: Socialization characteristics and trends," *Journal of Intelligent Manufacturing*, vol. 28, no. 5, pp. 1079–1094, 2017.

[66] C. Zhu, Z. Sheng, V. C. Leung et al., "Toward offering more useful data reliably to mobile cloud from wireless sensor network," *IEEE Transactions on Emerging Topics in Computing*, vol. 3, no. 1, pp. 84–94, 2014.

[67] X. He, K. Wang, H. Huang et al., "QoE-driven big data architecture for smart city," *IEEE Communications Magazine*, vol. 56, no. 2, pp. 88–93, 2018.

[68] J. Wan, S. Tang, D. Li et al., "Reconfigurable smart factory for drug packing in healthcare industry 4.0," *IEEE Transactions on Industrial Informatics*, vol. 15, no. 1, pp. 507–516, 2018.

[69] Z. Shu, J. Wan, D. Zhang et al., "Cloud-integrated cyber-physical systems for complex industrial applications," *Mobile Networks and Applications*, vol. 21, no. 5, pp. 865–878, 2016.

[70] S. S. Rautaray and A. Agrawal, "Vision based hand gesture recognition for human computer interaction: A survey," *Artificial Intelligence Review*, vol. 43, no. 1, pp. 1–54, 2015.

[71] E. Cambria and B. White, "Jumping NLP curves: A review of natural language processing research," *IEEE Computational Intelligence Magazine*, vol. 9, no. 2, pp. 48–57, 2014.

[72] B. Chen, J. Wan, Y. Lan et al., "Improving cognitive ability of edge intelligent IIoT through machine learning," *IEEE Network*, vol. 33, no. 5, pp. 61–67, 2019.

[73] S. Wang, J. Wan, D. Li et al., "Implementing smart factory of industrie 4.0: An outlook," *International Journal of Distributed Sensor Networks*, vol. 12, no. 1, p. 3159805, 2016.

[74] H.-N. Dai, H. Wang, G. Xu et al., "Big data analytics for manufacturing internet of things: Opportunities, challenges and enabling technologies," *Enterprise Information Systems*, vol. 14, no. 9–10, pp. 1279–1303, 2020.

[75] H. Wang, Z. Liu, D. Peng et al., "Understanding and learning discriminant features based on multiattention 1DCNN for wheelset bearing fault diagnosis," *IEEE Transactions on Industrial Informatics*, vol. 16, no. 9, pp. 5735–5745, 2019.

[76] X. Zhang, F.-L. Chung, and S. Wang, "An interpretable fuzzy DBN-based classifier for indoor user movement prediction in ambient assisted living applications," *IEEE Transactions on Industrial Informatics*, vol. 16, no. 1, pp. 42–53, 2019.

[77] Y. Zuo, "Prediction of consumer purchase behaviour using Bayesian network: An operational improvement and new results based on RFID data," *International Journal of Knowledge Engineering and Soft Data Paradigms*, vol. 5, no. 2, pp. 85–105, 2016.

[78] S. Shao, S. McAleer, R. Yan et al., "Highly accurate machine fault diagnosis using deep transfer learning," *IEEE Transactions on Industrial Informatics*, vol. 15, no. 4, pp. 2446–2455, 2018.

[79] J. Lee, H. Davari, J. Singh et al., "Industrial Artificial Intelligence for industry 4.0-based manufacturing systems," *Manufacturing Letters*, vol. 18, pp. 20–23, 2018.

[80] T. Hayhoe, I. Podhorska, A. Siekelova et al., "Sustainable manufacturing in Industry 4.0: Cross-sector networks of multiple supply chains, cyber-physical production systems, and AI-driven decision-making," *Journal of Self-Governance and Management Economics*, vol. 7, no. 2, pp. 31–36, 2019.

[81] V. Sze, Y.-H. Chen, T.-J. Yang et al., "Efficient processing of deep neural networks: A tutorial and survey," *Proceedings of the IEEE*, vol. 105, no. 12, pp. 2295–2329, 2017.

[82] E. Puik, D. Telgen, L. van Moergestel et al., "Assessment of reconfiguration schemes for reconfigurable manufacturing systems based on resources and lead time," *Robotics and Computer-Integrated Manufacturing*, vol. 43, pp. 30–38, 2017.

[83] D. Calvaresi, M. Marinoni, A. Sturm et al., "The challenge of real-time multi-agent systems for enabling IoT and CPS," *Proceedings of the International Conference on Web Intelligence*. 2017, pp. 356–364.

[84] W. Shen, L. Wang, and Q. Hao, "Agent-based distributed manufacturing process planning and scheduling: A state-of-the-art survey," *IEEE Transactions on Systems, Man, and Cybernetics, Part C (Applications and Reviews)*, vol. 36, no. 4, pp. 563–577, 2006.

[85] Z. A. Khan, M. T. Khan, I. Ul Haq et al., "Agent-based fault tolerant framework for manufacturing process automation," *International Journal of Computer Integrated Manufacturing*, vol. 32, no. 3, pp. 268–277, 2019.

[86] A. Giret, D. Trentesaux, M. A. Salido et al., "A holonic multi-agent methodology to design sustainable intelligent manufacturing control systems," *Journal of Cleaner Production*, vol. 167, pp. 1370–1386, 2017.

[87] S. Kumari, A. Singh, N. Mishra et al., "A multi-agent architecture for outsourcing SMEs manufacturing supply chain," *Robotics and Computer-Integrated Manufacturing*, vol. 36, pp. 36–44, 2015.

[88] S. Wang, J. Wan, D. Zhang et al., "Towards smart factory for industry 4.0: A self-organized multi-agent system with big data based feedback and coordination," *Computer Networks*, vol. 101, pp. 158–168, 2016.

[89] I. Kovalenko, D. Tilbury, and K. Barton, "The model-based product agent: A control oriented architecture for intelligent products in multi-agent manufacturing systems," *Control Engineering Practice*, vol. 86, pp. 105–117, 2019.

[90] B. Park and J. Jeong, "Knowledge-based multi-agent system for smart factory of small-sized manufacturing enterprises in Korea," *Computational Science and Its Applications-ICCSA 2019: 19th International Conference, Saint Petersburg, Russia, July 1-4, 2019, Proceedings, Part VI 19*. Springer, 2019, pp. 81–93.

[91] B. Chen, J. Wan, A. Celesti et al., "Edge computing in IoT-based manufacturing," *IEEE Communications Magazine*, vol. 56, no. 9, pp. 103–109, 2018.

[92] W. Shi, J. Cao, Q. Zhang et al., "Edge computing: Vision and challenges," *IEEE Internet of Things Journal*, vol. 3, no. 5, pp. 637–646, 2016.

[93] V. S. Shah, "Multi-agent cognitive architecture-enabled IoT applications of mobile edge computing," *Annals of Telecommunications*, vol. 73, no. 7–8, pp. 487–497, 2018.

[94] C. Yang, S. Lan, W. Shen et al., "Towards product customization and personalization in IoT-enabled cloud manufacturing," *Cluster Computing*, vol. 20, pp. 1717–1730, 2017.

[95] J. Wan, B. Chen, S. Wang et al., "Fog computing for energy-aware load balancing and scheduling in smart factory," *IEEE Transactions on Industrial Informatics*, vol. 14, no. 10, pp. 4548–4556, 2018.

[96] G.-J. Cheng, L.-T. Liu, X.-J. Qiang et al., "Industry 4.0 development and application of intelligent manufacturing," *2016 International Conference on Information System and Artificial Intelligence (ISAI)*. IEEE, 2016, pp. 407–410.

[97] S. Lemaignan, A. Siadat, J.-Y. Dantan et al., "MASON: A proposal for an ontology of manufacturing domain," *IEEE Workshop on Distributed Intelligent Systems: Collective Intelligence and Its Applications (DIS'06)*. IEEE, 2006, pp. 195–200.

[98] Y. Zhang, S. C. Feng, X. Wang et al., "Object oriented manufacturing resource modelling for adaptive process planning," *International Journal of Production Research*, vol. 37, no. 18, pp. 4179–4195, 1999.

[99] P. Vichare, A. Nassehi, S. Kumar et al., "A unified manufacturing resource model for representing CNC machining systems," *Robotics and Computer-Integrated Manufacturing*, vol. 25, no. 6, pp. 999–1007, 2009.

[100] Y. Lu, H. Wang, and X. Xu, "ManuService ontology: A product data model for service-oriented business interactions in a cloud manufacturing environment," *Journal of Intelligent Manufacturing*, vol. 30, pp. 317–334, 2019.

[101] H.-K. Lin and J. A. Harding, "A manufacturing system engineering ontology model on the semantic web for inter-enterprise collaboration," *Computers in Industry*, vol. 58, no. 5, pp. 428–437, 2007.

[102] Y. Alsafi and V. Vyatkin, "Ontology-based reconfiguration agent for intelligent mechatronic systems in flexible manufacturing," *Robotics and Computer-Integrated Manufacturing*, vol. 26, no. 4, pp. 381–391, 2010.

[103] J. Wan, B. Yin, D. Li et al., "An ontology-based resource reconfiguration method for manufacturing cyber-physical systems," *IEEE/ASME Transactions on Mechatronics*, vol. 23, no. 6, pp. 2537–2546, 2018.

[104] C.-H. Liao, Y.-F. Wu, and G.-H. King, "Research on learning owl ontology from relational database," *Journal of Physics: Conference Series*, vol. 1176, no. 2, p. 022031, 2019.

[105] W. Li and S. Kara, "Methodology for monitoring manufacturing environment by using wireless sensor networks (WSN) and the internet of things (IoT)," *Procedia CIRP*, vol. 61, pp. 323–328, 2017.

[106] Z. Sheng, C. Mahapatra, C. Zhu et al., "Recent advances in industrial wireless sensor networks toward efficient management in IoT," *IEEE Access*, vol. 3, pp. 622–637, 2015.

[107] Y. Chen, G. M. Lee, L. Shu et al., "Industrial internet of things-based collaborative sensing intelligence: Framework and research challenges," *Sensors*, vol. 16, no. 2, p. 215, 2016.

[108] X. Sun and N. Ansari, "EdgeIoT: Mobile edge computing for the Internet of Things," *IEEE Communications Magazine*, vol. 54, no. 12, pp. 22–29, 2016.

[109] K. Zhang, J. Long, X. Wang et al., "Lightweight searchable encryption protocol for industrial internet of things," *IEEE Transactions on Industrial Informatics*, vol. 17, no. 6, pp. 4248–4259, 2020.

[110] J. Huang, L. Kong, H.-N. Dai et al., "Blockchain-based mobile crowd sensing in industrial systems," *IEEE Transactions on Industrial Informatics*, vol. 16, no. 10, pp. 6553–6563, 2020.

[111] M. Pueo, J. Santolaria, R. Acero et al., "Design methodology for production systems retrofit in SMEs," *International Journal of Production Research*, vol. 58, no. 14, pp. 4306–4324, 2020.

[112] F. Tao, J. Cheng, and Q. Qi, "IIHub: An industrial Internet-of-Things hub toward smart manufacturing based on cyber-physical system," *IEEE Transactions on Industrial Informatics*, vol. 14, no. 5, pp. 2271–2280, 2017.

[113] D. Bruckner, M.-P. Stănică, R. Blair et al., "An introduction to OPC UA TSN for industrial communication systems," *Proceedings of the IEEE*, vol. 107, no. 6, pp. 1121–1131, 2019.

[114] A. Tsuchiya, F. Fraile, I. Koshijima et al., "Software defined networking firewall for industry 4.0 manufacturing systems," *Journal of Industrial Engineering and Management (JIEM)*, vol. 11, no. 2, pp. 318–333, 2018.

[115] M. Ojo, D. Adami, and S. Giordano, "A SDN-IoT architecture with NFV implementation," *2016 IEEE Globecom Workshops (GC Wkshps)*. IEEE, 2016, pp. 1–6.

[116] K. Kaur, S. Garg, G. S. Aujla et al., "Edge computing in the industrial internet of things environment: Software-defined-networks-based edge-cloud interplay," *IEEE Communications Magazine*, vol. 56, no. 2, pp. 44–51, 2018.

[117] V. Galetić, I. Bojić, M. Kušek et al., "Basic principles of Machine-to-Machine communication and its impact on telecommunications industry," *2011 Proceedings of the 34th International Convention MIPRO*. IEEE, 2011, pp. 380–385.

[118] S. Mascia, P. L. Heider, H. Zhang et al., "End-to-end continuous manufacturing of pharmaceuticals: Integrated synthesis, purification, and final dosage formation," *Angewandte Chemie*, vol. 125, no. 47, pp. 12585–12589, 2013.

[119] J. Wan, B. Chen, M. Imran et al., "Toward dynamic resources management for IoT-based manufacturing," *IEEE Communications Magazine*, vol. 56, no. 2, pp. 52–59, 2018.

[120] T. Stock and G. Seliger, "Opportunities of sustainable manufacturing in industry 4.0," *procedia CIRP*, vol. 40, pp. 536–541, 2016.

[121] Z. Lin and S. Pearson, "An inside look at industrial Ethernet communication protocols," *Texas Instruments, White Paper*, 2013.

[122] E. Sisinni, A. Saifullah, S. Han et al., "Industrial internet of things: Challenges, opportunities, and directions," *IEEE Transactions on Industrial Informatics*, vol. 14, no. 11, pp. 4724–4734, 2018.

[123] X. Li, M. Luo, Y. Qiu et al., "Independent component analysis based digital signal processing in coherent optical fiber communication systems," *Optics Communications*, vol. 409, pp. 13–22, 2018.

[124] H. Boyes, B. Hallaq, J. Cunningham et al., "The industrial internet of things (IIoT): An analysis framework," *Computers in Industry*, vol. 101, pp. 1–12, 2018.

[125] D. Li, X. Li, and J. Wan, "A cloud-assisted handover optimization strategy for mobile nodes in industrial wireless networks," *Computer Networks*, vol. 128, pp. 133–141, 2017.

[126] D. Jansen and H. Buttner, "Real-time Ethernet: The EtherCAT solution," *Computing and Control Engineering*, vol. 15, no. 1, pp. 16–21, 2004.

[127] A. Moldovansky, "Utilization of modern switching technology in ethernet/IP networks," *Proceedings of 1st Workshop on Real-Time LANs in the Internet Age*. Citeseer, 2002, pp. 25–27.

[128] G. Cena, L. Seno, A. Valenzano et al., "Performance analysis of Ethernet Powerlink networks for distributed control and automation systems," *Computer Standards & Interfaces*, vol. 31, no. 3, pp. 566–572, 2009.

[129] W. Liang, X. Zhang, Y. Xiao et al., "Survey and experiments of WIA-PA specification of industrial wireless network," *Wireless Communications and Mobile Computing*, vol. 11, no. 8, pp. 1197–1212, 2011.

[130] V. C. Gungor and G. P. Hancke, "Industrial wireless sensor networks: Challenges, design principles, and technical approaches," *IEEE Transactions on Industrial Electronics*, vol. 56, no. 10, pp. 4258–4265, 2009.

[131] X. Li and J. Wan, "Proactive caching for edge computing-enabled industrial mobile wireless networks," *Future Generation Computer Systems*, vol. 89, pp. 89–97, 2018.

[132] M. Bertocco, G. Gamba, A. Sona et al., "Performance measurements of CSMA/CA-based wireless sensor networks for industrial applications," *2007 IEEE Instrumentation & Measurement Technology Conference IMTC 2007*. IEEE, 2007, pp. 1–6.

[133] E. E. Petrosky, A. J. Michaels, and D. B. Ridge, "Network scalability comparison of IEEE 802.15. 4 and receiver-assigned CDMA," *IEEE Internet of Things Journal*, vol. 6, no. 4, pp. 6060–6069, 2018.

[134] D. Yang, J. Ma, Y. Xu et al., "Safe-WirelessHART: A novel framework enabling safety-critical applications over industrial WSNs," *IEEE Transactions on Industrial Informatics*, vol. 14, no. 8, pp. 3513–3523, 2018.

[135] C. L. Tan and M. Tracey, "Collaborative new product development environments: Implications for supply chain management," *Journal of Supply Chain Management*, vol. 43, no. 3, pp. 2–15, 2007.

[136] J. Queiroz, P. Leitão, J. Barbosa et al., "An agent-based industrial cyber-physical system deployed in an automobile multi-stage production system," *Service Oriented, Holonic and Multi-agent Manufacturing Systems for Industry of the Future: Proceedings of SOHOMA 2019 9*. Springer, 2020, pp. 379–391.

[137] N. A. Taatgen, C. Lebiere, and J. R. Anderson, "Modeling paradigms in ACT-R," in *Cognition and Multi-Agent Interaction: From Cognitive Modeling to Social Simulation*, ed: R. Sun. Cambridge University Press, Cambridge, 2006, pp. 29–52.

[138] H. A. ElMaraghy, "Flexible and reconfigurable manufacturing systems paradigms," *International Journal of Flexible Manufacturing Systems*, vol. 17, pp. 261–276, 2005.

[139] Y. Koren and M. Shpitalni, "Design of reconfigurable manufacturing systems," *Journal of Manufacturing Systems*, vol. 29, no. 4, pp. 130–141, 2010.

[140] S. M. Saad, "The reconfiguration issues in manufacturing systems," *Journal of Materials Processing Technology*, vol. 138, no. 1-3, pp. 277–283, 2003.

[141] J. W. Park, M. Shin, and D. Y. Kim, "An extended agent communication framework for rapid reconfiguration of distributed manufacturing systems," *IEEE Transactions on Industrial Informatics*, vol. 15, no. 7, pp. 3845–3855, 2018.

[142] Z. Wei, X. Song, and D. Wang, "Manufacturing flexibility, business model design, and firm performance," *International Journal of Production Economics*, vol. 193, pp. 87–97, 2017.

[143] E. M. Sanfilippo, S. Benavent, S. Borgo et al., "Modeling manufacturing resources: An ontological approach," *Product Lifecycle Management to Support Industry 4.0: 15th IFIP WG 5.1 International Conference, PLM 2018, Turin, Italy, July 2-4, 2018, Proceedings 15*. Springer, 2018, pp. 304–313.

[144] D. Mourtzis, E. Vlachou, C. Giannoulis et al., "Applications for frugal product customization and design of manufacturing networks," *Procedia CIRP*, vol. 52, pp. 228–233, 2016.

[145] K. N. McKay and V. C. Wiers, "Integrated decision support for planning, scheduling, and dispatching tasks in a focused factory," *Computers in Industry*, vol. 50, no. 1, pp. 5–14, 2003.

[146] K. Biel and C. H. Glock, "Systematic literature review of decision support models for energy-efficient production planning," *Computers & Industrial Engineering*, vol. 101, pp. 243–259, 2016.

[147] A. K. Tyagi, S. Sharma, N. Anuradh et al., "How a user will look the connections of internet of things devices?: A smarter look of smarter environment," *Proceedings of 2nd International Conference on Advanced Computing and Software Engineering (ICACSE)*, 2019.

[148] J. Zhang, J. Du, and X. Shi, "Data-driven intelligent system for equipment management in automotive parts manufacturing," *2019 58th Annual Conference of the Society of Instrument and Control Engineers of Japan (SICE)*. IEEE, 2019, pp. 1349–1354.

[149] J. Wan, J. Li, Q. Hua et al., "Intelligent equipment design assisted by Cognitive Internet of Things and industrial big data," *Neural Computing and Applications*, vol. 32, no. 9, pp. 4463–4472, 2020.

[150] G. B. Grant, T. P. Seager, G. Massard et al., "Information and communication technology for industrial symbiosis," *Journal of Industrial Ecology*, vol. 14, no. 5, pp. 740–753, 2010.

[151] S. Jeschke, C. Brecher, T. Meisen et al., "Industrial internet of things and cyber manufacturing systems," in *Industrial Internet of Things*, eds: S. Jeschke, C. Brecher, H. Song, and D. Rawat. Springer, Cham, 2017, pp. 3–19.

[152] D. Mourtzis, E. Vlachou, and N. Milas, "Industrial big data as a result of IoT adoption in manufacturing," *Procedia CIRP*, vol. 55, pp. 290–295, 2016.

[153] W. Liang, M. Zheng, J. Zhang et al., "WIA-FA and its applications to digital factory: A wireless network solution for factory automation," *Proceedings of the IEEE*, vol. 107, no. 6, pp. 1053–1073, 2019.

[154] A. E. Kalør, R. Guillaume, J. J. Nielsen et al., "Network slicing in industry 4.0 applications: Abstraction methods and end-to-end analysis," *IEEE Transactions on Industrial Informatics*, vol. 14, no. 12, pp. 5419–5427, 2018.

[155] J. Cheng, W. Chen, F. Tao et al., "Industrial IoT in 5G environment towards smart manufacturing," *Journal of Industrial Information Integration*, vol. 10, pp. 10–19, 2018.

[156] E. Udd, "Overview of fiber optic sensors," In *Fiber Optic Sensors*. CRC Press, 2017, pp. 1–34.

[157] J. Zhang, X. Yao, J. Zhou et al., "Self-organizing manufacturing: Current status and prospect for Industry 4.0," *2017 5th International Conference on Enterprise Systems (ES)*, 2017. IEEE, pp. 319–326.

[158] G. Jia, "Customers-focused strategy of manufacture and the reconstruction of manufacture systems," *Chinese Journal of Management*, vol. 1, pp. 70–75, 2006.

[159] V. R. S. Kumar, A. Khamis, S. Fiorini et al., "Ontologies for industry 4.0," *The Knowledge Engineering Review*, vol. 34, p. e17, 2019.

[160] E. Rajangam and C. Annamalai, "Graph models for knowledge representation and reasoning for contemporary and emerging needs-a survey," *International Journal of Information Technology and Computer Science (IJITCS)*, vol. 8, no. 2, pp. 14–22, 2016.

[161] A. R. Zamir, A. Sax, W. Shen et al., "Taskonomy: Disentangling task transfer learning," *Proceedings of the IEEE Conference on Computer Vision and Pattern Recognition*. 2018. pp. 3712–3722.

[162] F. K. Došilović, M. Brčić, and N. Hlupić, "Explainable artificial intelligence: A survey," *2018 41st International Convention on Information and Communication Technology, Electronics and Microelectronics (MIPRO)*. IEEE, 2018, pp. 0210–0215.

[163] F. Tao, J. Cheng, Q. Qi et al., "Digital twin-driven product design, manufacturing and service with big data," *The International Journal of Advanced Manufacturing Technology*, vol. 94, no. 9, pp. 3563–3576, 2018.

[164] Q. Qi and F. Tao, "Digital twin and big data towards smart manufacturing and industry 4.0: 360 degree comparison," *IEEE Access*, vol. 6, pp. 3585–3593, 2018.

[165] W. Kritzinger, M. Karner, G. Traar et al., "Digital Twin in manufacturing: A categorical literature review and classification," *IFAC-PapersOnLine*, vol. 51, no. 11, pp. 1016–1022, 2018.

[166] Y. Wang, T. L. Nguyen, Y. Xu et al., "Cyber-physical design and implementation of distributed event-triggered secondary control in islanded microgrids," *IEEE Transactions on Industry Applications*, vol. 55, no. 6, pp. 5631–5642, 2019.

[167] J. Pan, S. Ding, D. Wu et al., "Exploring behavioural intentions toward smart healthcare services among medical practitioners: A technology transfer perspective," *International Journal of Production Research*, vol. 57, no. 18, pp. 5801–5820, 2019.

[168] J. Rybicka, A. Tiwari, and G. A. Leeke, "Technology readiness level assessment of composites recycling technologies," *Journal of Cleaner Production*, vol. 112, pp. 1001–1012, 2016.

CHAPTER 2

Edge Intelligence in Customized Manufacturing

EDGE COMPUTING, OFTEN PROJECTED as an extension to cloud, renders its design to deal with challenges of traditional cloud-based Internet of Things (IoT). Edge enlightens its features of low latency, real-time interaction, location awareness, mobility support, geo-distribution, etc. over cloud. Edge by nature does not work on cloud, but instead on a network edge for facilitating higher speeds. Edge improves the ability to process tasks in real time. The industrial sector is revolutionized by ever-changing technical advancements and IoT, which is a young discipline embraced by manufacturing thereby bringing in Manufacturing Internet of Things (MIoT). Edge computing is viable to manufacturing processes. MIoT is well supported by the middleware edge computing as manufacturing process requires most of the task performed locally and securely at end points with minimum delay. Edge, deployed for manufacturing processes and entities which are part of internet, is gaining importance in recent times being titled as the edge for MIoT. Additionally, as manufacturing big data is often ill structured, it can be polished before sending it to cloud, resulting in an enhanced computing. This chapter presents the fundamentals for the heterogeneous network modeling and dynamic scheduling of resources in subsequent chapters. First, edge intelligence and its architecture in customized manufacturing are briefly introduced. Second, the key technologies of edge computing in customized manufacturing are discussed in detail. Finally, the validation of relevant key methods of edge intelligence in customized manufacturing environment is given.

2.1 OVERVIEW OF EDGE INTELLIGENCE IN CUSTOMIZED MANUFACTURING

In recent years, due to the rapid increase in the number of network terminal devices and the quantity of data they generate in customized manufacturing, artificial intelligence (AI) based solely on the cloud has had some shortcomings. Due to these shortcomings, edge intelligence came into existence. It deploys AI on the edge network and emphasizes being close to the source of the data to reduce the delay in the delivery of intelligent cloud

DOI: 10.1201/9781003460992-2

computing services [1]. In addition, it realizes the true sense of wireless situation intelligent perception, intelligent rapid decision-making, and real-time response [2].

Edge intelligence aims to combine edge computing with AI and other applications. Then, it syncs the data processing capability based on cloud computing to the edge nodes to provide advanced data analysis, scene perception, real-time decision-making, and other service functions in the customized manufacturing edge network. According to the development trend of edge intelligence technology, edge intelligence consists of the following six functions in the customized manufacturing.

2.1.1 Collection

Sensors are the front-end components of edge intelligent application systems in the customized manufacturing [3]. Combined with sensor technology, edge devices can obtain data [4]. efficiently. Especially in smart buildings and smart home systems, many edge devices are integrated with sensors to collect data in real time. For example, smart passive sensors (SPSs) [5] are wireless batteryless sensors that can monitor various parameters, such as temperature, pressure, humidity, or distance, in the edge network. iDAR (intelligent detection and ranging) [6] is a micro-optical-electromechanical-system (MOEMS) lidar integrated with a low-light camera, and it also has embedded AI algorithms for real-time sensing of the surrounding dynamic environment. The widespread use of these edge intelligent devices provides a rich data source for edge intelligence. However, edge sensor devices may be constrained by energy, bandwidth, or raw computing power. The bandwidth limits the maximum rate at which data can be collected from sensors and are transmitted downstream. This may bring challenges to programming and may sacrifice some benefits. Ideally, sensors should only send absolutely necessary information, and critical data can immediately make critical decisions. In addition, edge sensors that are small in size and unobtrusive can be easily deployed in space constrained environments. This kind of edge sensors can be placed anywhere that is possible to perceive valuable information, not just where existing communications and power infrastructure is located.

2.1.2 Communication

The large quantity of data collected in the customized manufacturing edge network poses a great challenge to the data transmission ability. As a result, many edge-oriented communication technologies have emerged. The device-to-device (D2D) approach is used to provide services such as real-time data transmission and sharing, which plays a key role in the edge intelligence framework. D2D communication technology, unlike the traditional network-centric data transmission mode, opens up a device-centric communication direction, which can effectively increase the communication capacity of the edge network [7,8] and provides the necessary data transmission and information sharing capability for edge intelligence. Communication theory and techniques can substantially bridge the capacity of the cloud and the requirement of devices by the network edges, thus accelerating content delivery and improving the quality of edge system services. To bring more intelligence to edge systems, deep Reinforcement Learning techniques and Federated learning frameworks have been integrated for optimizing edge computing, caching, and communication [9].

2.1.3 Computing

With the improvement of edge device performance, the edge network has sufficient computing resources. With the help of edge computing technology, data can be processed in the edge network in a timely and effective manner [10,11]. Edge computing has played a considerable role in many fields [12] and is widely used in medical care [13], mobile data analysis, the internet of vehicles [14,15], wireless sensor networks, and in other scenarios [16]. In particular, the use of edge computing can considerably reduce the waiting time of applications and services, thereby improving user experience in the customized manufacturing. Therefore, edge computing is regarded as an important component of edge intelligence, which aims to reduce response time, save bandwidth, reduce network traffic and energy consumption [17], and ensure data security and privacy.

2.1.4 Caching

In recent years, many studies have focused on the deployment of caches in edge networks [18] to relieve core network congestion and reduce end-to-end waiting time to improve network performance. Unlike traditional data storage, edge caching only temporarily stores real-time information on edge devices to improve the information distribution capabilities of edge devices. In particular, a study [19] advocated the use of predictive information requirements and active caching on the base station and user equipment, thereby greatly reducing the peak traffic demand. Therefore, edge caching provides powerful information sharing capabilities for edge intelligence and improves information utilization efficiency in the customized manufacturing.

2.1.5 Control

With the popularization of the industrial internet, autonomous driving, smart homes, smart transportation, smart cities, and other technologies, the requirements for real-time control and feedback capabilities of devices in the edge network environment are increasing. For example, in terms of public safety, the authors in [20] proposed a shared car unsafe event detection system based on edge computing, called the SafeShareRide system. Specifically, the mobile phones of drivers and passengers are used as the edge end. Through voice recognition and driving behavior detection, video recording and processing mechanisms (mobile terminal compression, cloud analysis) are triggered after abnormal behaviors are detected. Also, on-board video records are analyzed to monitor the safety situation of passengers. In edge networks, edge intelligence can autonomously control edge devices according to specific tasks and provide users with corresponding services, effectively avoiding the high-latency defects of traditional cloud-based control systems [21]. For example, the smart cruise control system is a set of environmental perceptions for planning and decision-making, multi-level driving assistance, and other functions in one of the comprehensive systems. The information collected by sensors, such as cameras and radars, is sent to the control computer to extract features. Additionally, driving data information is obtained through model calculation to master comprehensive and complete driving habits, helping semiautonomous driving vehicles make decisions and control [22]. In addition, for some mission-critical functions, edge computing cannot tolerate any delay. This requires high-precision real-time analysis and control to perform calibration to minimize defects. According to different applications and data sizes, solutions suitable for different control situations are required. Real-time data acquisition and millisecond-level control were committed to achieve real-time control of applications in the customized manufacturing.

2.1.6 Collaboration

With the support of edge intelligence technology, edge devices can make full use of the advantages of geographic proximity and segmentation (for example, changes in application scenarios and user requirements) and cooperate to accomplish a series of tasks, improving users' daily experiences and quality of life [23,24] in the customized manufacturing. The authors in [25] proposed social sensing-based edge computing (SSEC) using edge devices close to users (mobile phones, tablets, smart wearables, and the IoT) as pervasive sensors. A federation of computing nodes was formed to perform sensing, storage, networking, and computing tasks near the data source. The authors in [26] combined AI with edge computing and proposed the concept of collaborative intelligence to reduce the computing burden of edge devices in applications running on deep neural networks (DNNs). Collaborative Intelligence (CI) is an AI deployment strategy that leverages both edge-based and cloud-based resources to make DNN computing faster and more efficient [27].

In summary, edge intelligence can meet the key needs of customized manufacturing digitization in terms of agile connectivity, real-time services, data optimization, application intelligence, and security and privacy protection. Meanwhile, data convergence can effectively facilitate large searches in cyberspace and contribute to the construction of knowledge graphs in the edge intelligence domain [28], creating considerable social and economic benefits [29] in the customized manufacturing.

2.2 EDGE COMPUTING-ENABLED ARCHITECTURE OF INTELLIGENT MANUFACTURING FACTORY

Edge computing extends the capabilities of computation, network connection, and storage from the cloud to the edge of the network. It enables the application of business logic between the downstream data of the cloud service and the upstream data of the IoT. In the field of MIoT, edge computing provides added benefits of agility, real-time processing, and autonomy to create value for intelligent manufacturing. With a focus on the concept of edge computing, this subsection proposes an architecture of edge computing for IoT-based intelligent manufacturing [30]. In the edge intelligence architecture, intelligence is pushed from the cloud core to the edge devices to better support timely and reliable data analysis and service response.

With the goal of proposing a novel application scenario for edge computing, we present the system architecture of edge computing in an IoT-based manufacturing scenario. As shown in Figure 2.1, the architecture is divided into four fields, namely, (1) the device domain, (2) the network domain, (3) the data domain, and (4) the application domain.

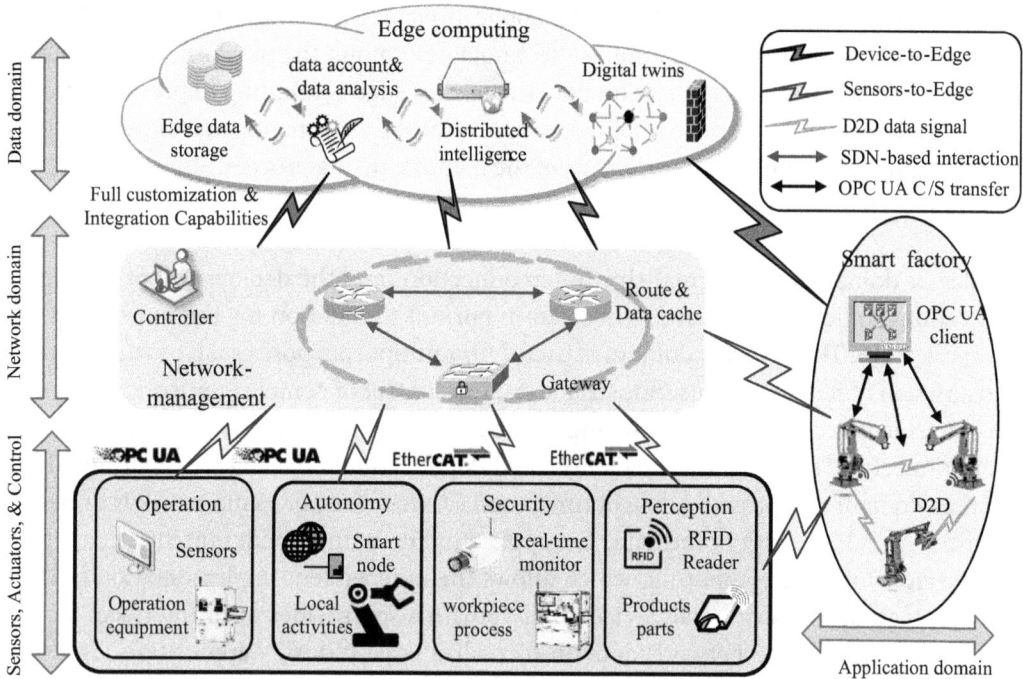

FIGURE 2.1   Architecture of an edge computing platform in IoT-based manufacturing.

- The device domain is either embedded in or located close to the field devices such as sensors, meters, robots, and machine tools. The device domain should support a flexible communication infrastructure and be able to establish standardized communication models to support various types of communication protocols. Nodes in an edge computing network must have the abilities to compute and store data to dynamically adjust the execution strategy of the industrial equipment based on sensor inputs. The information model is built on the edge computing nodes and includes mainstream protocols such as OLE for Process Control Unified Architecture (OPC UA) and Data Distributed Service (DDS). It is deployed to realize the unified semantics of information interaction and ensure data security and privacy.

- The network domain connects the field equipment to the data platform in a flat manner. In IoT-based manufacturing, the network domain uses Software Defined Networks (SDNs) to achieve separation between the network transmission and the control. Due to the necessary time sensitivity of the task data, Time Sensitive Network (TSN) protocol is applied to process the sequence of network information and to provide the general standards for maintaining and managing the sensitive time nodes.

- The data domain in the data origin provides services such as data cleaning and feature extraction, which improves the availability of the heterogeneous industrial data. This also allows the implementation of predefined responses which can be based on the real-time IoT data. The abstract data at the device terminal end is provided to the remote service center for the virtualization of the manufacturing resources.

- The application domain inherits the applications of the manufacturing cloud at the network edge. It integrates the key technologies among the network, data, computing, and control. The application domain enables edge computing to provide general, flexible, and interoperable intelligent applications. Components such as service composition, based on the requirements of the manufacturing process, allow the dynamic management [30] and optimal scheduling of field equipment.

The device domain supports real-time interconnections and the deployment of intelligent applications to field devices. It provides an important foundation for the upper layers of the application. The network domain is useful for multiple purposes such as the real-time transmission of heterogeneous industrial data, the control of complex network states, and the convenient access to manufacturing resources. It provides a platform for performing inter-connection, information interaction, and data fusion on the edge computing nodes. The data domain also provides data optimization services. One of its main task is to ensure consistency and integrity of the data. The application domain provides intelligent application services for edge computing, which allows the independent implementation of local business logic at the peripheral devices. The domain also provides open interfaces for the device domain and the network domain to realize edge industry application. The proposed edge computing architecture presented in Figure 2.1 makes full use of the embedded computing capabilities and ensures the autonomy of the system and the manufacturing equipment while following a distributed computing paradigm. Furthermore, the edge computing framework cooperates with the remote data center to realize intelligent terminals for the manufacturing system.

2.3 EDGE COMPUTING IN MANUFACTURING AND ITS KEY TECHNOLOGIES

The key techniques for manufacturing edge computing in this section include (1) edge computing node deployment method for smart manufacturing; (2) proactive caching for edge computing in manufacturing; (3) thing–edge–cloud collaborative computing decision-making method for smart manufacturing; (4) resource scheduling strategy for edge computing in smart manufacturing; and (5) cognitive ability of edge computing in customized manufacturing.

2.3.1 Edge Computing Node Deployment Method for Smart Manufacturing

With the rapid development of the mobile Internet, MIoT, cyber-physical systems, and the emergence of edge computing have provided an opportunity to realize the high computing performance and low latency of intelligent devices in the smart manufacturing environment. In this subsection [31], we propose and verify an edge computing node deployment method for smart manufacturing. First, the architecture of a smart manufacturing system used for implementing the edge computing node deployment methods is presented. Then, comprehensively balancing the network delay and computing resources deployment cost, and considering the influence of device spatial distribution, device function, and computing capacity of edge nodes on the above optimization objectives, the optimal deployment number of edge computing nodes is obtained by using an improved k-means clustering algorithm.

2.3.1.1 System Architecture of Deployment Method

An SDN-based edge computing architecture and its working process in smart manufacturing are presented. The architecture of an SDN-based edge computing system in a manufacturing environment is shown in Figure 2.2. The SDN system framework is divided into a centralized control plane (usually a controller) and a distributed data plane. The controller is located between the upper-layer application and the physical device. The controller first is responsible for abstracting and establishing various functions in the network. The specific operation model provides a programming interface to the upper layer. The upper-layer application focuses on interacting with the physical device through the controller according to business requirements. Devices on the network will need to pass messages to the application plane through the controller.

As shown in Figure 2.2, an SDN controller represents the core of the entire intelligent manufacturing system. It consists of SDN sub-controllers and performs real-time centralized control and resource scheduling of the entire edge network using the SDN control protocols, such as OpenDaylight (ODL) and Open Network Operating System (ONOS). There are a large number of network access points (APs) supporting the SDN protocol on the production line. These edge network infrastructures act as switches in the SDN data layer, connect with the SDN controller through the application programming interface (API), such as OpenFlow, and receive management information from the SDN control layer. In addition, a smart manufacturing workshop includes machines, conveyors, product, and other equipment necessary for the workshop production, which is all connected by a wired or wireless network to realize the data interaction of the workshop equipment.

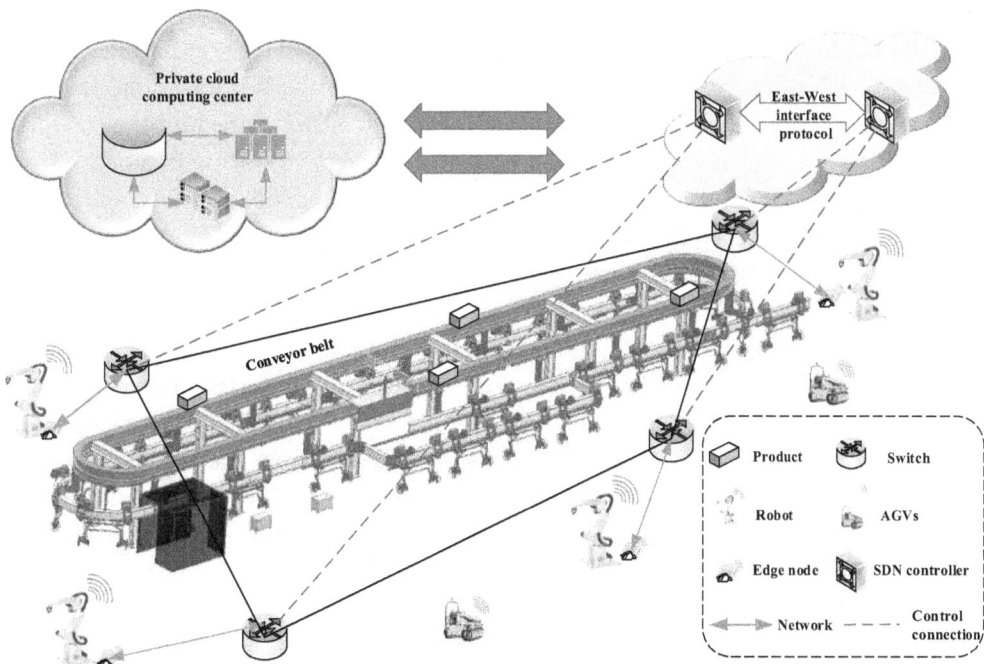

FIGURE 2.2   The SDN-based edge computing architecture in a manufacturing environment.

The workshop equipment transmits the required data processing request to the nearby AP through a one-hop wireless connection, and all APs transmit the current data processing request situation, as well as the resource utilization situation of the data link and the edge node in the edge network, to the SDN controller through the SDN protocol. Based on this real-time information, the SDN controller can intelligently evaluate and predict the congestion situation in the edge network and the load status of each edge node, and then send the control information to each AP using the SDN protocol; thus, each AP can optimize the routing of the current incoming data, thereby maximizing the utilization of the link resources in the entire edge network, and uniformly all the data processing requests from the plant equipment to different edge nodes, thereby satisfying the data processing real-time requirement of each device while maximizing the utilization of service resources in the edge network.

The private cloud computing center of the workshop serves as a source of application services in the entire edge network service architecture and provides services to users through Ethernet as a supplement to the edge service resources. As the number of computing requests of workshop equipment increases, the data transmission load in edge network will increase accordingly; however, computing resources of the edge nodes are limited, so the computing feedback quality of workshop equipment will decrease. Faced with such a situation, the SDN controller will reasonably allocate device data processing tasks to the edge node resources and the cloud computing center to optimize the overall computing service.

The biggest difference of the proposed SDN-based edge computing system from the previous system is that it can flexibly define the forwarding function of network equipment by writing software. In the previous system, the control plane functions were distributed among various network nodes (e.g., hubs, switches, and routers). Therefore, the deployment of new network functions requires the upgrade of all corresponding network equipment, which often makes it difficult to implement network innovation. The proposed SDN-based edge computing system separates the control plane and the forwarding plane of the network equipment, and implements the control plane in a centralized manner. In this way, the deployment of new network functions requires only centralized software upgrades at the control node, thereby enabling rapid and flexible customization of network functions. Using the presented architecture of the SDN-based edge computing system, this subsection focuses on the deployment of edge computing nodes in a workshop. In order to ensure that a large part of the workshop equipment can access the edge network to obtain the response at any time during the production process, a large number of APs need to be deployed. In the smart manufacturing scenario, an AP refers to the switch connecting machines. The transmission delay when the requested data is transferred from the APs to the edge nodes will greatly affect the QoS of edge nodes. The data transmission delay between AP and the edge node is mainly determined by the data load of each switch through which the transmitted data are routed. If the data load of a switch is too large, a severe delay will be produced in the edge network; thus, high real-time requirements of the smart manufacturing system for detection, control, and execution will not be met.

2.3.1.2 Edge Computing Node Deployment Specific Method

Assume an unknown edge network with the edge network topology $G=(V, E)$, where V denotes a set of network nodes in the edge network, E denotes a set of links in the edge network, L represents a set of edge nodes in the edge network, and U denotes a set of data source nodes. The edge network has n network nodes, and each node is denoted as v_j, where $i=1, 2, ..., n$. A link between every two nodes in the edge network is represented as (i, j), and the accessible bandwidth is B_{ij}, where (i, j) is within the range of E. When there is traffic in the network, the data flow from node v_i to node v_j is denoted as f_{ij}. Further, x_i represents the computing ability of a network node, where $x_i=0$ means that the node does not need to deploy computing resources and that it is a non-edge computing node; $x_i>0$ means that the node needs to deploy computing resources and that it is an edge computing node. In the edge network, there are m data source nodes $(m<n)$, so it can be assumed that the network node $v_i\{i=1,2,...,m\}$ is a data source node; the data amount collected by each data source node per unit time is $d_i\{i=1,2,...,m\}$; the response time of each data source node is t_i.

In addition, in a smart manufacturing environment, not only the influence of the edge network link topology but also manufacturing equipment space distribution and equipment function on the edge node deployment in the actual workshop production is considered. There are N equipment parts in the production workshop. The production equipment is represented by n_i and its spatial coordinate is p_i (X_i, Y_i, Z_i). Further, N equipment parts in the workshop collect data through sensors and transmit the collected data to the data source node, i.e., $N=m$, in which the data normalization principle is adopted; the influence of height is ignored, and only the average value of X_i and Y_i is considered, i.e., the larger the average value, the greater the deployment cost. The equipment in the workshop has M different functions, and $M<N$ (r_i, M, and N are integer values), r_i represents the function of the production equipment n_i, which can be quantified into $(0, 1/M]$, $(1/M, 2/M]$,..., $[(m-1)/M, 1]$. For instance, the value belonging to the range of $(0, 1/M)$ represents Function 1 and so on to Function M. The larger M is, the more complex the function of a device will be, so more computing resources of edge nodes will be required and the deployment cost will increase accordingly. In addition, for the computing ability of edge nodes in the edge network $x_i>0, (x_i=1,2,...,n)$, that is, the computing ability of network nodes should be non-negative. The description of notations used in the edge computing node deployment is given in Table 2.1. The deployment optimization of edge computing nodes is mainly aimed at edge network delay and the cost of deploying computing resources. Thus, the response time of a data source node should be as short as possible, as given by equation (2.1), while the cost of deploying computing resources should be minimalized, as given by equation (2.2):

$$\min \sum_{i=1,2,...,m} t_i \tag{2.1}$$

$$\min \left[\sum_{i=1}^{n} g(x_i) + \sum_{i=1}^{N} h(p_i) + \sum_{i=1}^{M} q(r_i) \right] \tag{2.2}$$

TABLE 2.1 Notations

Notations	Definition
G	Topology of edge network
V	Assembly of network node in edge network
E	Assembly of communication link between nodes in edge network
L	Assembly of edge computing nodes in edge network
U	Assembly of data source nodes in edge network
(i, j)	Connection between nodes v_i and v_j
b_{ij}	Maximum bandwidth allowed between nodes v_i and v_j
f_{ij}	Amount of data transmitted per unit time from node v_i to node v_j
d_i	Amount of data to be transmitted by the data source node. When node v_i is a non-data source node, $d_i=0$
x_i	Computing capability of node v_i
p_i	Spatial coordinate of production equipment n_i
r_i	Function of the production equipment n_i
g	Function between the cost of deploying edge computing nodes and the computing capability
h	Function between the cost of deploying edge computing nodes and the device spatial distribution coordinates
q	Function between the cost of deploying edge computing nodes and the device functionality
α	Weight parameter of delay in objective function
β	Weight parameter of edge node cost in objective function

The function g used in equation (2.2) is defined as follows:

$$g(x) = \begin{cases} 0, x = 0 \\ ax + A, x > 0 \end{cases} \tag{2.3}$$

where A and a are two constants. When no computing resources need to be deployed, the cost is equal to 0, but when computing resources need to be deployed, it is assumed that there is a certain cost A, and the cost increases linearly with the increase in computing resources, with the growth rate of a. Similarly, functions $h(p)$ and $q(r)$ are similar to $g(x)$. When the spatial distribution coordinate of a device is 0, the cost is 0. When the spatial distribution coordinates of device increase, the device volume also increases, so more space and more computing resources are needed due to the increase in the transmission distance. Function $q(r)$ indicates that when a device does not have any function, the cost is equal to 0. As the function of a device increases, it indicates that a node with a stronger edge computing capability needs to be deployed, which also increases the cost.

Because the two mentioned optimization objectives are contradictory, the weight parameters α and β are introduced into the two optimization objectives to constitute a new joint optimization objective. Therefore, a new objective function can be expressed as:

$$\min\left(\alpha * \sum_{i=1,2,\dots,m} t_i + \beta * \left[\sum_{i=1}^{n} g(x_i) + \sum_{i=1}^{N} h(p_i) + \sum_{i=1}^{M} q(r_i) \right] \right) \tag{2.4}$$

where α and β, respectively, represent the weight parameters corresponding to the edge network delay and computing resource deployment cost in the optimization process.

When we focus on the network time delay, the weight parameter α has a greater impact on the objective function. This is suitable for manufacturing scenarios with tight data transmission time and many tasks, so as the more edge computing nodes deployed, the better. In contrary, when the focus is on the cost of computing resource deployment, the weight parameter β has a great influence on the objective function, which is suitable for manufacturing scenarios with abundant time, fewer task amount, and low-cost budget; in that case, the number of deployed edge computing nodes should be as small as possible. Lastly, when weighing the impact of the edge network delay and computing resources deployment cost, the impact of α and β is equivalent, which is suitable for most manufacturing scenarios. The focus of this paper is to get a balance between the edge network delay and the computing resources deployment cost.

Based on the previous analysis, the response time of each task request is composed of two times, the data-sending time and the data processing time of edge computing node, which is given by:

$$t_i = \frac{d_i - x_i}{\sum\limits_{(i,j)\in E} \delta(f_{ij}) * b_{ij}} + t_p \tag{2.5}$$

$$\delta(x) = \begin{cases} 0, x = 0 \\ 1, x > 0 \end{cases} \tag{2.6}$$

where the numerator of equation (2.5) represents the amount of data that a data source node v_i needs to transmit through the network. When the denominator is not summed, the data source node selects the corresponding link to send data, and the whole denominator represents the sum of the output bandwidth of the data source node's connection. Therefore, the first term in equation (2.5) represents the time required by the data source node to send the data, while the second term represents the data processing time of the edge computing node. Equation (2.6) represents a unit step function. When the data amount f_{ij} of the unit time from node v_i to node v_j is 0, its value is 0; when f_{ij} is greater than 0, its value is 1.

Moreover, this paper uses the amount of data processed per unit time to characterize the processing capacity of network nodes. It is assumed that each edge node can be an edge computing node or a data source node. The computing capacity x_i of edge computing node may be equal to 0, so the node degenerates into a non-edge computing node. The data d_i generated by a data source node per unit time can also be equal to 0, so the node degenerates into a non-data source node. For each node, the total amount of data coming from a node should be equal to the total amount of data consumed by the node. The total amount of data coming from a node includes two parts: the total amount of data flowing into the node and the total amount of data that the node needs to send per unit time as a data source node. Also, the total amount of data consumed by a node also includes two parts:

the total amount of data through the node and the total amount of data processed by the node per unit of time as an edge computing node. According to the above, the relationship presented in equation (2.7) can be obtained.

Furthermore, according to the principle of flow conservation, this chapter defaults to processing all task-processing requests per unit time, that is, the processing time becomes constant, so when constructing the objective function, t_p item in equation (2.5) can be ignored, and the final objective function can be expressed by equation (2.8).

$$\forall k \in v, \sum_{(i,k) \in E} f_{ik} + d_k = \sum_{(k,j) \in E} f_{kj} + x_k \tag{2.7}$$

$$\min \left(\alpha * \sum_{i=1,2,\dots,m} \frac{d_i - x_i}{\sum_{(i,j) \in E} \delta(f_{ij}) * b_{ij}} + \beta * \left[\sum_{i=1}^{n} g(x_i) + \sum_{i=1}^{N} h(p_i) + \sum_{i=1}^{M} q(r_i) \right] \right) \tag{2.8}$$

The planning model constructed above denotes a convex optimization model. To find a solution to this model, an improved k-means clustering method is used to solve the problem. According to the basic k-means clustering algorithm, the clustering samples $p_i \{i = 1, 2, \dots, n\}$ are divided into different categories $C = \{C_1, C_2, \dots, C_k\}$, where p_i is a d-dimensional vector and k is the number of clusters. In this work, k denotes the number of edge computing nodes deployed in smart manufacturing. The clustering center and the criterion function of the k-means clustering are given by equations (2.9) and (2.10), respectively:

$$C_j = \frac{1}{C} \sum_{i=1}^{n} p_i; j = 1, 2, \dots, k \tag{2.9}$$

$$J = \sum_{j=1}^{k} \sum_{p_i \in C_j}^{n} d(p_i, C_j) \tag{2.10}$$

In equation (2.10), $d(p_i, C_j)$ represents the distance between the data p_i and the class center C_j; J represents the sum of the class distances. The basic k-means clustering algorithm is a clustering algorithm based on the partitioning idea. It has the advantages of simple ideas, fast clustering, and strong local search ability. When the position of the devices is relatively fixed, it is suitable to choose k-means as the basic methodology of our work for deployment problems. However, the k-means clustering has the following disadvantages: the location of the initial selection center is random, different random initial centers have different iterations, sometimes the algorithm can be calculated many times, and the obtained results can differ; thus, the initial clustering center should be determined artificially. Namely, different initial clustering centers may lead to completely different clustering results, which is not suitable for a direct deployment, such as the edge computing node deployment method in smart manufacturing. So, the standard k-means clustering has to be improved. Following the principle that the distance between centers should be as far as possible, the initial k-means clustering center is set and then the k-means clustering center is used to run the standard k-means clustering algorithm. An improved algorithm is proposed to optimize the number of edge computing nodes deployed in smart manufacturing and its pseudocode is presented in Algorithm 2.1.

In Algorithm 2.1, the steps can be mainly described as follows: First, the system searches for the initial clustering center according to each input variable. Then, according to equations (2.7) and (2.8), we can get the weighted sum of network delay and computing resource cost under the current number of edge computing nodes deployment. Third, we evaluate the current total cost to determine if it is less than the total cost from the previous cycle. Finally, according to the principle of minimum total node deployment cost, we update K to obtain the optimal number of edge computing nodes.

2.3.2 Proactive Caching for Edge Computing in Manufacturing

As manufacturing systems shift from automated patterns to smart frameworks such as smart factories in Industry 4.0, industrial wireless networks (IWNs) are serving as promising communication systems that can be applied to the manufacturing field. When the mobile elements and static nodes are introduced into the system, large amounts of data downloaded from mobile networks or tele-servers can be one of the greatest challenges for industrial mobile wireless networks (IMWNs). Mobility and industrial properties have rarely been considered by the previous research on download strategies and caching methods. In this subsection [32], we present a three-layer cache architecture based on edge computing and other heritage traditional networks. Then, useful spatial and temporal mobility properties are mapped using different groups and edge computing servers that contain mobile nodes. Then, according to the sojourn time, the capacity of edge computing servers, and other neighboring nodes, we propose a proactive caching strategy for large amounts of data downloaded by mobile networks that considers location and mobile trajectories.

2.3.2.1 Architecture of Manufacturing Mobile Edge Computing

In accordance with manufacturing system properties and previous work on the structure of IWNs, we first construct a single-hop clustered industrial wireless mobile network. Then, we present a three-layer storage system based on cloud and edge computing in the industrial domain for the delivery of large amounts of data. Moreover, we briefly review our strategies for different stages of big data traffic for working MNs. Figure 2.3 demonstrates the scene of a smart factory, where MIoT nodes, sensor nodes, automated guided vehicles (AGVs), workers, and machines are deployed in a factory. In contrast to randomly moving traditional networks, IMWN MNs must follow a certain or a fixed path to complete a specific task, such as workpiece conveyance, as shown in Figure 2.3. When higher real-time data delivery efficiency is required, single-hop and clustered network frameworks are widely adopted in the industrial domain. Therefore, our architecture is based on clustered networks. We deploy an edge computing server (CES) at every cluster head with a storage system.

ALGORITHM 2.1   Pseudocode of Edge Computing Node Deployment Based on Improved k-Means Clustering for Smart Manufacturing

---

Initialization: Input device spatial distribution p, device function category r, edge node computing ability x, network topology G, data source node set U, number of edge computing nodes to be deployed k

Output: number of edge computing node deployment optimizations K

Begin

> Randomly select k data points from the data set as the starting centroid
>
> **Do**
>
>> For each point, the distance $D(x)$ to the nearest starting center of mass is calculated and stored in an array
>> Add up these distances to get $Sum(D(x))$
>> Take a random value $Random$ that can fall in $Sum(D(x))$
>>
>>> **Do**   // set initial clustering center
>>>
>>>> $Random\ \text{-}= D(x)$
>>>
>>> **while** $Random <= 0$
>>
>> End while
>
> **while** k cluster centers are selected
>
> **End while**
>
> The k data points are used to form the initial cluster
>
> $Cost_{min} = \infty$
>
> **for** $v \leftarrow 1$ to N // v is the data source node;
>
>> **for** $i \leftarrow 1$ to k // k is the edge computing nodes number
>>
>>> Calculate the distance between the centroid and the data point
>>> Assign data points to the cluster closest to them according to equations (2.9) and (2.10)
>>> For each cluster: find the mean and update it to the centroid
>>> Find the values of f_{ij} and x_i by equation (2.7)
>>> Get $Cost$ according to equation (2.8)
>>> **if** $Cost < Cost_{min}$ **then**
>>>
>>>> $Cost_{min} = Cost$ // determine if the $Cost$ at this k nodes is the smallest
>>>> $K = k$
>>>> // k with lower $Cost$ is saved to the optimized quantity K;
>>>
>>> **End if**
>>
>> **End for**
>> **Return** K
>
> **End for**

End

The framework can be divided into three layers: cloud storage, cluster head edge storage, and local node storage. Using wireless links to connect nodes and cluster heads and using backbone network links to cloud and edge servers, each edge server can provide communication service within the coverage ranges for MNs and other nodes. Consequently, when

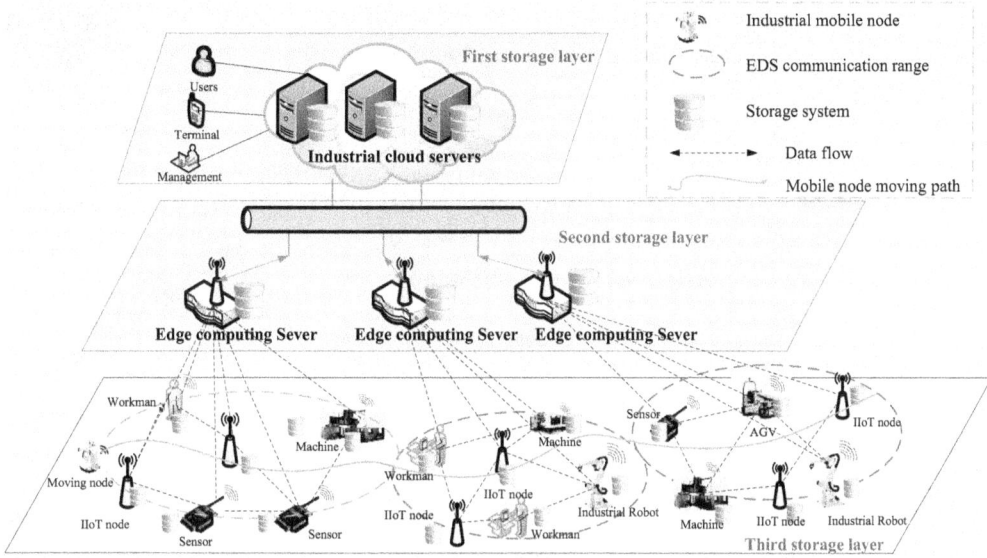

FIGURE 2.3   Three-layer architecture for big data traffic.

an MN must transmit a large amount of data from tele-cloud servers (TCSs), proactive caching methods can be used. The three-layer data storage layer working processes are as follows for data distribution for moving nodes. Wireless moving node needs big data storing in an industrial cloud server (ICS), namely the first storage layer. Edge computing server (ECS) caches parts of big data. That is the second storage layer. Because ECS cannot provide continuous wireless link service, sensor nodes are employed to store fragment data of big data to make up for ECS insufficiencies. Namely, moving node can fetch data from sensor nodes for guaranteeing real time.

2.3.2.2 Proactive Cache Optimization Problem

We present the system model of IMWNs for large amounts of data and any related assumptions considered in this study. We can compute the transmission time of the big data in an MN and in the corresponding constraints via the proposed model.

Network coverage model: We consider an IMWN with multiple equipment. The physical network devices include multiple cluster heads with CESs, static nodes (SNs) (e.g., sensors), fixed machines, MNs, and a TCS. Let L denote the path of the MNs. Let E_{all} and C_{all} be the set of CESs and clusters. Let $Cov(e_{all}^i, L)$ denote the coverage function for e_{all}^i that partly covers L; thus,

$$Cov(e_{all}^i, L) = \begin{cases} 1 & e_{all}^i \, \Lambda \, L \\ 0 & \text{otherwise} \end{cases}, \qquad (2.11)$$

where $e_{all}^i \wedge L$ is the partially covered L. Let $E \subseteq E_{all}$ be the set of edge clusters covering L. We assume that L is divided into k sections by clustering C and E. We use l_i to denote that the edge server e_i covers the ith section path, that is, $L = \{l_1, \ldots, l_k\}$, $l_i \cap l_j \neq \emptyset, \forall\, 1 \leq i < j \leq k$. Let S_i be the set of SNs in each $e_i \in E$, that is, $c_i = e_i \cup S_i$. For each CES $e_i \in E$, the cache storage capability is S_e, and the SN cache capability is S_s. Moreover, we denote the neighboring SNs in the cluster $c_i \in C$ by $N(e_i)$. We divide the big data F into m portions, that is, $F = \{f_1, \ldots, f_m\}, f_i \cap f_j \neq \emptyset, \forall\, 1 \leq i < j \leq m$. Therefore, the big data that include all the portions of data can be denoted as

$$F = \bigcup_{i \in m} f_i. \tag{2.12}$$

For the sake of simplicity and without loss of generality, we assume that each portion of big data f_i has the same size D_f, namely, $D_{fi} = D_{fj}, \forall\, 1 \leq i < j \leq m$. Thus, the data size of F can also be described as

$$D_F = \sum_{i \in m} Df = mD_f. \tag{2.13}$$

MN's data communication: One MN links with one edge server or one neighboring SN in the edge server. Let $a_m^{e_i}(t)$ denote the association between MN and edge servers e_i at time t; thus,

$$a_m^{e_i}(t) = \begin{cases} 1 & M \leftrightarrow e_i \ or\, S_i \\ 0 & \text{otherwise} \end{cases}, \tag{2.14}$$

where $a_m^{e_i}(t) = 1$ means that the MNs connect to the edge servers e_i to request the big data F; otherwise, $a_m^{e_i}(t) = 0$. Furthermore, each MN can only access one edge server or SN at a time; thus,

$$\sum_{e_i \in E} a_m^{e_i}(t) = 1. \tag{2.15}$$

We use $b_m^{e_i(si)}(t)$ to denote the association between the MN and the SN of s_i of the edge cluster e_i; thus,

$$b_m^{e_i(s_i)}(t) = \begin{cases} 1 & M \leftrightarrow S_i \\ 0 & \text{otherwise} \end{cases}. \tag{2.16}$$

In equations (2.14) and (2.16), A↔B denotes A linking to B. Thus, according to the discussion above, we arrive at the following proposition: if $b_m^{e_i(si)}(t) = 1$, then $a_m^{e_i}(t) = 1$. We let $V = \{v_1, \ldots, v_k\}$ be the set of moving speeds in an edge server E. Then, the sojourn time of MNs for edge server e_i is

$$t_i = \frac{l_i}{v_i} \ (1 \leq i \leq k), \tag{2.17}$$

where l_i indicates that the edge server e_i covers the section path of L. The term γ_i denotes the communication efficiency of an MN in an edge server e_i. Thus, the efficient sojourn time t_i' for the edge server can be expressed as

$$t_i' = \gamma_i t_i. \tag{2.18}$$

We use $T = \{t_1, t_2, ..., t_k\}$ to denote the sojourn time of different edge servers. In an IWN, the traffic loads are different for different clustered edge servers. Consider that the wireless channels between MNs and their connected edge servers or clustered members are space-varying channels in different edge servers. Therefore, $CR = \{cr_1, ..., cr_k\}$ is used to express the communication rate of edge servers. The MN communication rate space $mr(t)$ can be divided and quantized into k discrete levels: $mr_1 = cr_1$, if $0 \le t < t_1$; $mr_i = cr_i$, if $t_{i-1} \le t \le t_i$; and $mr_k = cr_k$, if $t_{k-1} \le t \le t_k$. According to the description, the caching data size D_{ei} for edge server e_i can be formulated as

$$D_{ei} = t_i' \cdot mr_i. \tag{2.19}$$

The sum caching data size of an edge server e_i cannot exceed its storage capacity; thus, the following requirement must be met:

$$D_{ei} \le Se + \sum_{i \in ni} Ss_i = Se + ni \cdot Ss, \tag{2.20}$$

where ni is the number of edge servers e_i. Moreover, the number of big data cache portions u_i of an edge server e_i can be given by

$$u_i = \left\lceil \frac{D_{ei}}{D_f} \right\rceil, \tag{2.21}$$

where $0 \le u_i \le m$. In other words, a cluster edge server can cache u_i portions of data from the cloud.

Data caching: Then, the u_i portions of data are stored in an edge server or SNs. Now, in the edge cluster e_i, the location where the SN can store these portions of data is the next problem to be solved. Let $s_j^i \in S_i$ be any member SN of e_i, and let $Cov(s_j^i, li)$ denote whether s_j^i covers li; thus,

$$Cov(s_j^i, li) = \begin{cases} 1 & s_j^i \wedge li \\ 0 & \text{otherwise} \end{cases}, \tag{2.22}$$

where $s_j^i \wedge li$ means that s_j^i covers li. Let Γ_j^i be the intersection of a covering circle s_j^i and path li $(f(x, y))$. The length of Γ_j^i is determined by

$$Len(e_i, s_j^i) = Cov(s_j^i, li) \int_{\Gamma_j^i} f(x, y) ds, \tag{2.23}$$

where $f_i(x, y)$ is a function of li. Thus, an MN sojourn time within s_j^i covering time can be described as

$$t(s_j^i) = b_m^{e_i(s_j)} \frac{Len(ei, s_j^i)}{V_i}.$$

(2.24)

Moreover, the condition that $s_j^i \in S_i$ can store part of the u_i portions of data must meet the time and cache storage capability. Thus, the constraints can be expressed via

$$t(s_i^j) \geq \frac{D_f}{mr_i}.$$

$$Se \geq D_f$$

(2.25)

Thus, in one edge cluster e_i, we find that the storage of the SN can be formulated by

$$\arg_{s_j \in S_i} t(s_j^i) \geq \frac{D_f}{mr_i},$$

(2.26)

$$\text{S.T. } Se \geq D_f$$

(2.26a)

$$b_m^{e_i(s_j)}(t) = 1,$$

(2.26b)

$$a_m^{e_i}(t) = 1,$$

(2.26c)

where constraints (2.26a) are constrained by the storage capability and (2.26b, c) assure the effective links. In equation (2.26), i is the ith cluster edge server and j is the jth SN that belongs to the ith cluster edge server.

Problem formulation: In one edge server cluster (ESC), e_i, which includes the Si SNs, and u_i portions of data are distributed and stored in the server and nodes. MNs fetching data f_u, $1 \leq u \leq u_i$, are associated with a cost fc_u^i that can be expressed by

$$fc_u^i = T_{delay} + pi \cdot T_{Handoer} + \frac{D_f}{mr_i},$$

(2.27)

where pi is the number of stored data portions of an edge server and its SN is an edge cluster ei. Thus, the sum of the fetching data cost of u_i portions of data in the edge cluster e_i can be described as

$$Fc_i = \sum_{u \in u_i} fc_u^i.$$

(2.28)

Furthermore, the complete fetching data file of F in the work is as follows:

$$Fc = \sum_{i \in k} \sum_{u \in u_i} fc_u^i. \tag{2.29}$$

Mathematically, the proactive caching of industrial mobile networks can be formulated via

$$\min Fc$$

$$\text{S.T.} \quad D_{ei} \leq Se + niSs$$

$$\sum_{e_i \in E} a_m^{e_i}(t) = 1 \qquad . \tag{2.30}$$

$$mr_k = cr_k$$

Problem (2.29) is to determine the minimum fetch data process cost for the entire mobile network. In other words, we need to find the minimum of equation (2.27) for every edge cluster. Thus, we now focus on one edge cluster e_i.

We assume that there are n ($1 \leq n \leq m$) data bits and that there are ns storage locations, including the edge server and MNs. This problem is a classic assignment problem. We assume that $n = ns$. If $n \neq ns$, then we adopt standardization enabling $n = ns$. According to equation (2.26), we can obtain the fetching value of the MN from the ith storage location fetching the jth portion of data. Thus, in one edge cluster, the problem can be described as follows:

$$\min Z = \sum_{i=1}^{n} \sum_{j=1}^{n} fc_{ij} x_{ij} \tag{2.31}$$

$$\sum_{i=1}^{n} x_{ij} = 1 \tag{2.31a}$$

$$b_m^{e(s_j)}(t) = 1 \tag{2.31b}$$

$$\sum_{j=1}^{n} x_{ij} = 1 \tag{2.31c}$$

$$x_{ij} = 0 \ or \ 1, i, j = 1, 2, ..., n \tag{2.31d}$$

where x_{ij} is the binary variable that determines whether the MN fetches the jth portion of data from the ith storage place. The objective of the optimization problem is to minimize the average fetching cost. The constraint of the optimization is specified in equations (2.31a–2.31d). Equality (2.31a) indicates that each portion of data can be fetched from only one storage place. Equation (2.31b) indicates that every neighboring node must cover the MN path for meeting the minimum communication time.

2.3.2.3 Wireless Mobile Edge Network Data Caching Strategy

Considering the previous architecture of the models and formulation, proactive cache approach for MNs in IWN edge computing includes two stages: a proactive cache is selected to store large amounts of data, and the steps to fetch the data for the MN are provided. Specifically, the sub-problem in equation (2.26) must be efficiently solved. We develop an efficient coverage segmentation algorithm to solve the second problem in equation (2.31) and present a distributed improved Hungarian algorithm.

> **Proactive data cache**: The caching location that can hold the big data F in edge clusters and their SNs is the critical part of proactive caching. Inspired by the previous work, we formulate a double layer caching or storage algorithm. Using the cover time and storage capacity as the evaluation condition, we select the optimal edge clusters and SN. A complete description of the overall proposed scheme for stage I is stated in Algorithm 2.2.
>
> The proposed algorithm is explained as follows. Line 1 is initialized. The caching strategy can be divided into two layers: the edge server cache and the SN cache. In the first caching layer, we repeat the first FOR loop of lines 2–12 until each edge cluster e_{all}^i in E_{all} is considered. Line 3 computes the coverage function. Then, lines 4–8 consider whether the edge cluster e_{all}^i can cover the MN path L and whether the storage capacity can meet the constraint. When the evaluation is complete, the algorithm stores the edge cluster in E by meeting the constraint in sojourn time and memory. Moreover, each edge cluster e_i caches u_i portions of data from the TCS. In addition, in the second caching layer, we repeat the second FOR loop in lines 13–23 to find the correct SN to cache part of the large amount of data from e_j in E. First, line 16 evaluates whether the edge cluster member covers the path L. Then, lines 18–19 determine whether the current static case meets the condition of communication time and storage capacity. Then, the correct SNs cache the corresponding amount of data from their edge servers. The time complexity of Algorithm 2.2 is analyzed as follows. The worst upper bound of the time complexity of Algorithm 2.2 is $O(\text{Max}|E_{all}| + \text{Max}|E| \cdot \text{Max}|N(e)|)$.

> **MN cache strategy in each edge cluster**: The proposed distributed Hungarian algorithm works with different edge clusters $e_i \in E$. First, each fetching data cost is computed using the current communication state of the edge cluster and its appropriate SN, which is determined using Algorithm 2.2. When the MN enters a range of the edge cluster, the DH algorithm then finds an optimal place (edge server e_i or an SN). Moreover, each portion of data is fetched from the optimal place. This step is repeated until the MN downloads the entire dataset F.
>
> Algorithm 2.3 states the proposed DH strategy. The key steps of the algorithm are explained as follows. Lines 1–5 are the initialization of the data downloading

ALGORITHM 2.2   Efficient Coverage Segmentation for the Storage Algorithm

Input: wireless network status ($\{E_{all}\}$, $\{C_{all}\}$, $\{S\}$, $\{MR\}$), mobile status ($\{V\}$, $\{L\}$) storage capacity $\{S_e\}$, $\{S_s\}$, big data status ($\{F\}$, $\{D_f\}$).

Output: $\{E\}$, $\{Sc\}$, $\{Ns\}$.

1      $E \leftarrow \emptyset$, $S_c \leftarrow \emptyset$, $Ns \leftarrow \emptyset$

2   **for** $i=1$ to $|E_{all}|$ **do**

3   calculate $Cov\,(e_{all}^i, L)$ by equation (2.11)

4   **if** $Cov\,(e_{all}^i, L) == 1$

5   find the covered section path li by e_{all}^i

6                                  $t_i \leftarrow li/vi$

7                                  $D_{e_{all}^i} \leftarrow t_i \gamma_i \cdot mr_i$,

8   **if** $D_{e_{all}^i} \leq S_e + niS_s$

9                                  $E = E \cup \{e_{all}^i\}$

10   cache $u_i = \left\lceil \dfrac{D_{e_{all}^i}}{D_f} \right\rceil$ portions of data from cloud server

11   **end if**

12   **end for**

13   **for** $j=1$ to $|E|$ **do**

14   **for** $k=1$ to $|S^j|$ **do**

15   calculate $Cov(s_k^j, l_j)$ by equation (2.22)

16   **if** $Cov(s_k^j, l_j) == 1$

17                          $t(s_k^j) \leftarrow b_m^{e_j(sl)} \cdot {Len(ej.s_k^j)}\big/{v_j}$

18   **if** $t(s_k^j) \geq {D_f}\big/{mr_j}$ && $S_e \geq D_f$

19                                  $S_C^j \leftarrow S_C^j \cup s_k^j$

20   cache $\left\lceil \dfrac{S_s^k}{D_f} \right\rceil$ portions of data from edge server e_j

21   **end if**

22   **end if**

23   **end for**

24                                  $S_C \leftarrow S_C \cup S_c^j$

25   **end for**

26   **Return** E, S_C

requirement, providing the value of the data finish flag *isDone*. Lines 7–20 are the main loop for fetching the data, which is repeated until the MNs fetch the complete data of F or move out of the range of edge cluster E. In the main loop, this approach consists of two sub-loops that calculate the fetching data cost of each portion of data in every edge cluster. Using a Hungarian algorithm, we obtain the optimal fetching data assignment. In lines 16–19, the MN data is fetched according to the optimal assignment. In line 21, when the fetching of F is completed, *isDone*=1.

ALGORITHM 2.3 Distribution Hungarian Algorithm

Input: edge cluster status ({E}, {Sc}, {Ns}, {MR}, {T_{delay}}), mobile status ({V}, {L}), big data status ({F}, {D_j}, {u_i}).

Output: Fetching data finish flag ***isdone***

1	**if** requirement of F is sending to servers
2	$isDone \leftarrow 0$
3	**Else**
4	$isDone \leftarrow 1$
5	**end if**
6	MN entering the range of E
7	**for** $k=1$ to $|E|$ do
8	MN entering edge cluster e_i
9	**for** $i=1$ to ui do
10	**for** $j=1$ to ui
11	Calculate fetching data cost fc_{ij} by equation (2.11)
12	**end for**
13	**end for**
14	Create the fetching cost matrix
15	Apply the Hungarian algorithm to identify the optimal fetching data assignment
16	**for** each data of ui portion of data
17	create wireless link to corresponding edge server or SN
18	fetching data f_{ui}
19	**end for**
20	**end for**
21	$isdone \leftarrow 1$
22	**Return** $isDone$

2.3.3 Thing–Edge–Cloud Collaborative Computing Decision-Making Method for Smart Manufacturing

With the development of Industry 4.0 and cloud computing technology, personalized customization as a new production mode is showing a trend of rapid development. Personalized customization has the characteristics of order-driven production, strict processing times, high dynamic external conditions, and large flexibility in the production process, all of which bring more uncertainty to the production system and great challenges to the edge computing processing of related tasks in personalized customization production. Aiming at the above problems, a thing–edge–cloud collaborative computing decision-making (TCCD) method [33] in customized production is proposed. First, the architecture of a personalized customized production system used for implementing the TCCD method is presented. Then, according to the number and type of products in the customer order received from the private cloud platform, the customer's personalized customized order is dynamically divided. Subsequently, a task priority sorting algorithm is proposed to optimize the waiting time of all tasks involved in the order. Furthermore, a discrete particle swarm algorithm is proposed to optimize the average execution time of all tasks and equipment utilization decision-making options [thing–edge collaborative (TEC) computing, edge-edge collaborative (EEC) computing, or edge-cloud collaborative (ECC) computing].

2.3.3.1 System Architecture of Thing–Edge–Cloud Collaborative Computing

To realize the collaborative computing environment of production equipment, edge computing node, and private cloud in the customized production system, a KubeEdge-based edge collaborative computing system architecture is constructed as shown in Figure 2.4. KubeEdge builds a homogeneous execution environment for cloud computing and edge computing and connects each edge node, cloud virtual machine, and network container as a VPN. The core of the KubeEdge architecture includes EdgeMetadata service and KubeBus. Edge collaborative computing based on KubeEdge can realize network communication and collaborative computing between production equipment, edge nodes, and the private cloud platform. The EdgeMetadata service is responsible for data storage and synchronization computing when the connection

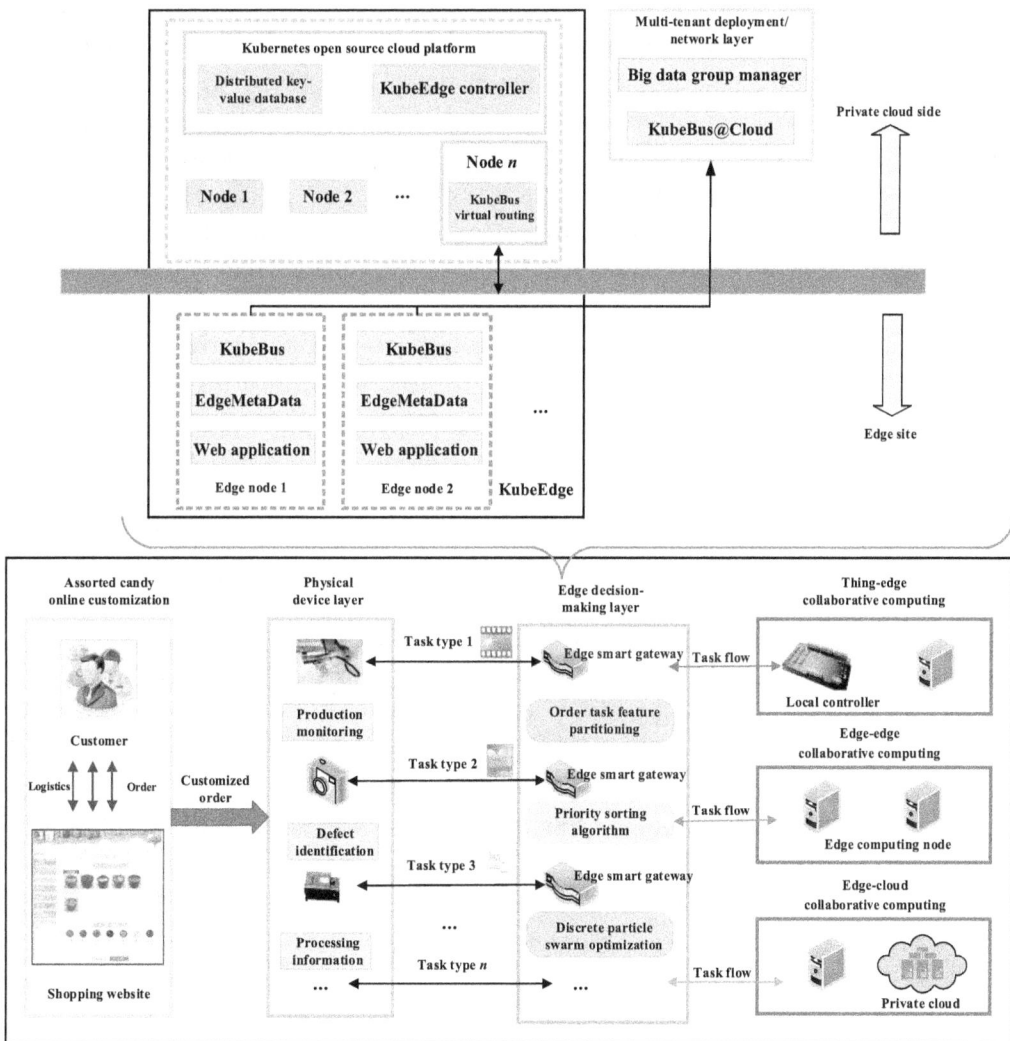

FIGURE 2.4 The KubeEdge-based edge collaborative computing architecture in a customized production environment.

between the edge node and the private cloud platform is unstable. KubeBus provides a software interface for the data communication link between the edge node and the private cloud platform.

The KubeEdge provides a multi-tenant edge infrastructure. On the private cloud data center side, it includes a multi-tenant management/data plane and has a cluster of multiple tenants. The multi-tenant management/data plane includes KubeBus in the cloud center and tenant management functions. A tenant cluster includes one or more edge nodes running in the edge area and a Kubernetes cluster running in the cloud. KubeEdge is deployed in the edge intelligent gateway and realizes the edge collaborative computing function through proxy services. In the private cloud, for each tenant's Kubernetes cluster, a KubeBus virtual router runs on a VM node to route traffic between the edge node subnet and the VM subnet/container network subnet.

Besides, in a private cloud, the server is mainly responsible for controlling edge nodes, receiving information uploaded by edge nodes, assigning tasks, providing IoT applications such as device management, intelligent production applications such as virtual factories, and intelligent service applications based on big data analysis. Operators are provided with a visual interface, and the supervision and scheduling of on-site resources are realized through collaborative computing with edge nodes. In the edge decision-making layer, industrial field resources are connected to the edge intelligent gateway device through the field network. The algorithm model, event management, message routing, and other services deployed in the edge intelligent gateway device perform real-time processing and analysis of the accessed data, on-site reasoning decision-making, conversion, transmission, and so on. Furthermore, the physical device layer, which includes the autonomous guided vehicle, sensor, and robot, is not only the executing part of the system but also the information acquisition part, and it is mainly responsible for completing the sensing and control work.

2.3.3.2 Thing–Edge–Cloud Collaborative Computing Decision-Making Method

Based on the proposed edge collaborative computing architecture, this section analyzes the TCCD method from the perspective of order task division and different edge collaborative computing methods. Moreover, it focuses on optimizing the task queue waiting time, average task execution time, and equipment utilization.

2.3.3.2.1 Representative Order Task Division for Personalized Customized Production   In the customized production environment, the edge intelligent gateway dynamically divides the customer's customized orders according to the number and types of products in the customer's order received from the private cloud platform. The customized order can be expressed as [$n1$ $n2$ $n3$ ...], where $n1$ represents the quantity of product Category 1, $n2$ represents the quantity of product Category 2, and so on. The number of products required by the customer is set to M, the type of products required by the customer is set to F, the number and upper limit of all categories are set to N, and the product category that can be produced by personalized customization is set to G. The steps of dynamic division can be described as follows:

1. When the order is submitted, it is divided into the following three categories according to the quantity of the products required by the customer. When $1<=M<N/3$, it is set to Class A; when $N/3<=M<2N/3$, it is set to Class B; and when $2N/3<M<=N$, it is set to Class C.

2. When the order is submitted, it is divided into the following three categories according to the types of products required by the customer. When $0<F<=G/3$, it is set to Category a; when $G/3<F<=2G/3$, it is set to Category b; and when $2G/3<F<=G$, it is set to Category c.

3. Based on the above two classification characteristics, (A, a), (A, b), and (B, a) are classified as Class I, which can be selected for TEC computing, EEC computing, and ECC computing; (A, c), (B, b), and (C, a) are classified as Class II, which can be selected for EEC computing or ECC computing; and (B, c), (C, b), and (C, c) are classified as Class III, which can be selected for ECC computing.

The diagram of the representative order task division for customized production is shown in Figure 2.5. The "thing" in TEC computing refers to end-of-things devices, which can also be described as IoT devices, mainly including sensor devices, cameras, and factory machinery equipment. The "edge" in EEC computing and ECC computing refers to edge computing nodes, which mainly include Raspberry Pi, gateways, routers, and servers. "Cloud" refers to a private cloud platform, mainly a computer cluster with powerful computing, storage, and analysis capabilities.

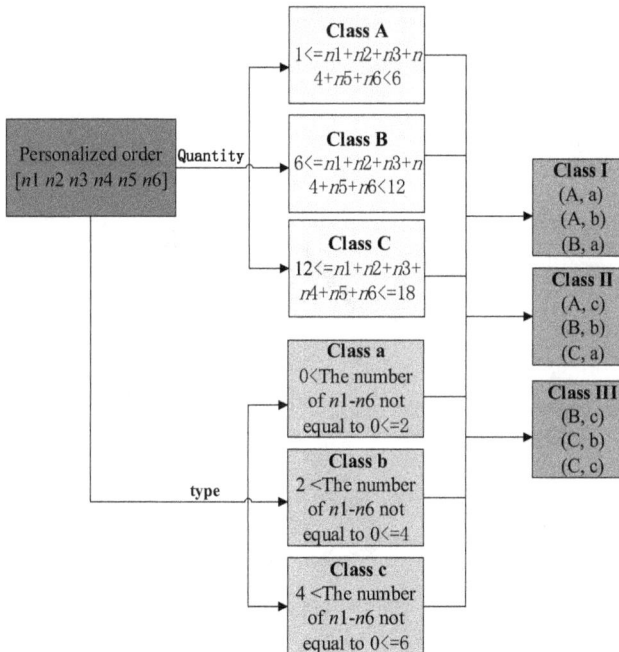

FIGURE 2.5 Diagram of representative order task division for customized production.

2.3.3.2.2 Thing–Edge–Cloud Collaborative Computing Mode   The thing–edge collaboration is mainly located at the bottom of the production site, where it can integrate the computing resources on the link between production equipment and edge computing nodes to fully utilize it, capitalize on the advantages of different equipment, and better enhance the capability of edge nodes. The TEC computing mode is shown in Figure 2.6. It is widely used in the IoT, especially in smart homes and smart manufacturing. Under the TEC computing mode, the thing is responsible for collecting data and sending it to the edge. Meanwhile, the calculation and control instructions of the edge are received for specific production operations. The edge is responsible for the centralized calculation of multi-channel data, issues instructions, and provides network, computing, and storage services. The thing can perform some simple calculations, and the edge, as the main body of the computing task and the core hub of the system, needs to undertake more computing tasks.

Collaborative computing between the edge-edge infrastructures is a current research hotspot, which can solve the contradiction between the resource requirements of intelligent algorithms and the limited resources and intelligent task requirements of edge devices and the single capability of edge devices. The EEC computing mode is shown in Figure 2.7. Specifically, the computing power of a single edge computing node is limited, and to improve the overall computing power of the system, time-sharing coordination between multiple edge computing nodes is required.

For example, when completing the training task of the deep neural network model, it is not feasible to train in a single edge computing node, which not only consumes a lot of time and computing power but is also easy to overfit the model due to the limitation of data volume to fail to obtain the optimal solution predicted by the model. Therefore, multiple

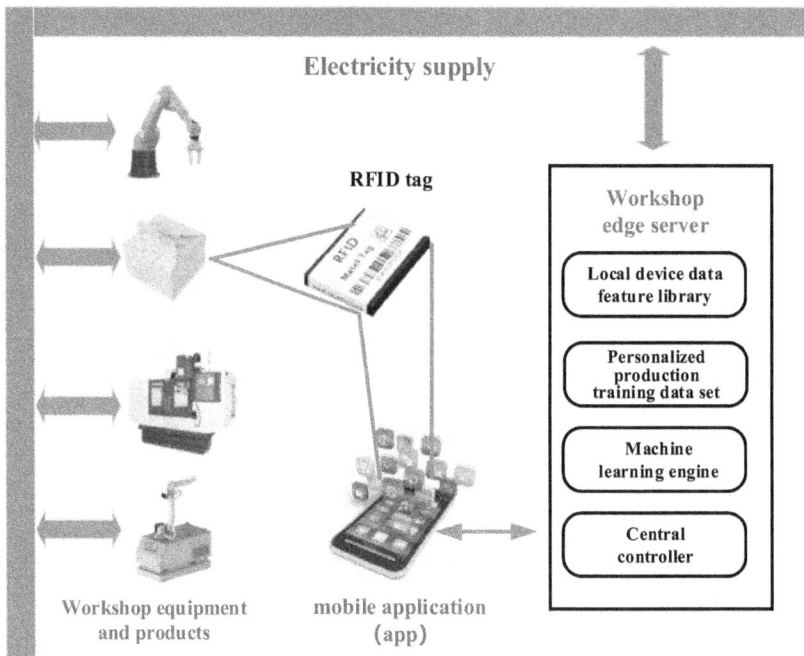

FIGURE 2.6   The thing–edge collaborative computing mode.

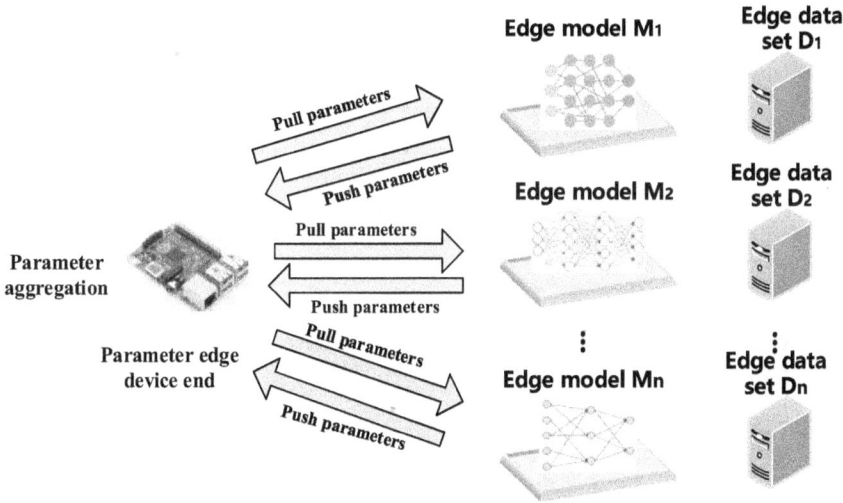

FIGURE 2.7   The edge–edge collaborative computing mode.

edge compute nodes are required to train the model together. The second is to solve the "data island" problem "data island" in production and manufacturing. The data source of a certain edge computing node has a strong locality and needs to cooperate with other edge computing nodes to complete a larger range of tasks. For example, in the operation monitoring of the customized production line of the whole factory, generally, one edge computing node can only obtain the operating status information of one workshop, and the cooperation among multiple edge computing nodes can be combined into an overview diagram of the workshop operating status of the whole intelligent factory.

In ECC computing, the edge is responsible for data computing and storage in the local area, while the cloud is responsible for big data analysis, mining, and algorithm training optimization. The ECC computing mode is shown in Figure 2.8. The collaboration of the edge-cloud can be divided into two parts. The first is functional collaboration. This kind of collaboration assumes different functions based on different geographic spaces and roles of different computing devices. For example, the edge is responsible for preprocessing, while the cloud is responsible for multi-channel data processing and service provision. The second is performance collaboration. This is due to the limitation of computing power, and computing devices of different levels undertake tasks with different computing power requirements, including longitudinal cutting and assignment of tasks.

2.3.3.2.3 Thing–Edge–Cloud Collaborative Computing Decision-Making Algorithm   In the thing–edge–cloud collaborative computing system, $C_i = \{J^c, B^c, Q^c\}$ represents cloud node C_i, where J^c represents the computing capacity of cloud node i, B^c represents the available bandwidth (in Mbps) between the cloud node and smart devices, and Q^c represents the cost of processing unit data on the cloud node. There are many edge computing nodes around smart devices to provide edge computing services. Moreover, $E_i = \{J^e, B^e, Q^e\}$ represents edge node E_i, where J^e represents the computing capacity of cloud node i, B^e represents the available bandwidth (in Mbps) between the edge node and smart devices, and

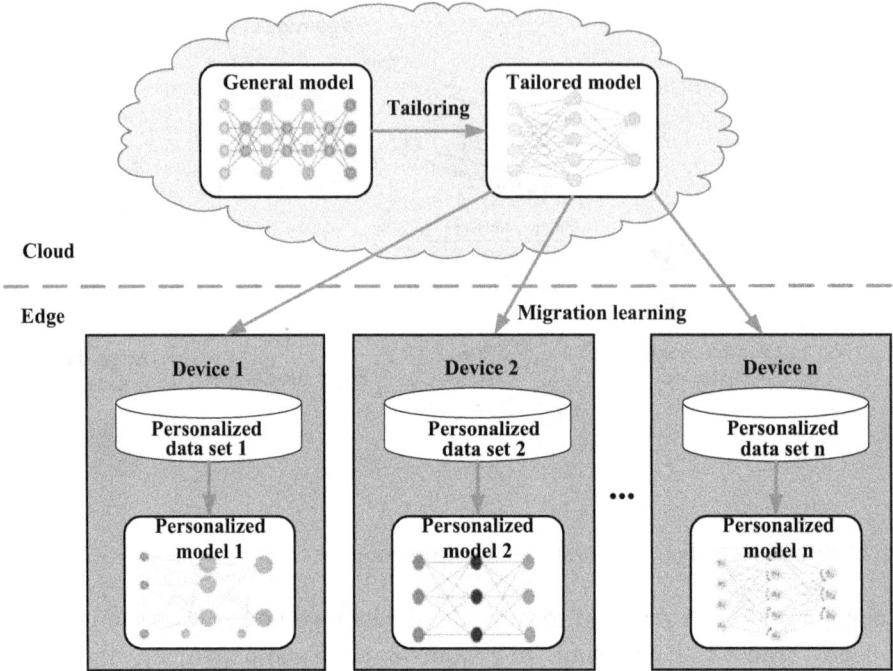

FIGURE 2.8   The edge-cloud collaborative computing mode.

Q^e represents the cost of processing unit data on the edge node. Smart device refers to any kind of equipment, apparatus, or machine with computational processing capabilities. S_i refers to a certain smart device, whose computing power is represented by J^s, and the task generated by the smart device can be represented as $T_j = \{U_j, D_j, R_j\}$, where U_j is the input data size uploaded by task j, D_j is the computational volume of the task processed on the cloud or edge node, and R_j is the data size returned by the computational results of the task. When choosing a collaborative computing mode for tasks, there are two main optimization goals for a customized production system, including increasing the utilization of edge computing nodes and reducing the average execution time of all tasks.

Task execution time is not only a factor of great concern in the personalized customization production process but also an important indicator of personalized customization production efficiency. Task execution time can be divided into four parts in the process of selecting edge collaborative computing mode:

1. The selection of decision stage t_d mainly includes information collection and making decisions based on the collected information and the necessary waiting time.

2. In the data transmission stage t_t, after making a decision, data needs to be transmitted from the source (thing) node to the edge computing node or cloud node. The transmission time can be expressed as:

$$t_t = U/B \qquad (2.32)$$

where U is the upload data size of the task and B (B^c or B^e) is the network bandwidth.

3. The task execution stage t_e refers to the time to execute tasks on smart devices, edge nodes, or cloud nodes. The time cost is as follows:

$$t_e = D / J \tag{2.33}$$

where D is the amount of task data that needs to be processed on smart devices, edge nodes, or cloud nodes; and J (J^s, J^e or J^c) is the computing capacity of smart devices, edge nodes, or cloud nodes.

4. The result return stage t_r refers to the time required to return the result data from the edge node or cloud node to the smart device. Here, we assume that the bandwidth of the two-way communication is stable and consistent. The time can be expressed as follows:

$$t_r = R / B \tag{2.34}$$

where R is the data size of the task result and B (B^c or B^e) is the network bandwidth.

Then, the execution time of task decision for collaborative computing is:

$$T_t = t_d + (U + R) / B + D / J \tag{2.35}$$

In general, the purpose of selecting the task computing method is to reduce the task execution time, that is, the execution time meets the following conditions:

$$T_s > t_d + (U + R) / B + D / J \tag{2.36}$$

where T_s is the time to execute the task on the smart device, and the other parameters are the same as in equation (2.35). If TEC, EEC, or ECC computing performs tasks faster than smart devices, then it is worthwhile to choose the edge collaborative computing mode.

Furthermore, a task priority sorting algorithm is proposed to optimize the waiting time of tasks in the queue. The task priority sorting algorithm is shown in Algorithm 2.4. Tasks are being prioritized based on the previous results. First, according to the task characteristic model established above, the computing amount, transmission amount, and amount of data returned by the computing result of each task are obtained. Then, the priority of the computing amount, transmission amount, and amount of data returned from the computing result is set to 0 or 1 because the amount of data returned from the computing result is usually relatively small, which is ignored. Finally, if the priority of the computing and transmission of the task are both 1, then the task is a high priority task; if the priority of the computing and transmission of the task are both 0, then the task is a low priority task; otherwise, the task is a medium priority task.

Equipment utilization is another important indicator that must be considered when making thing–edge–cloud collaborative computing decisions. In customized production, the improvement of equipment utilization can make full use of the computing resources

ALGORITHM 2.4   Pseudocode of Task Priority Sorting Algorithm

---

Initialization: Input task number N^t, task T_j, computing amount of task D, transmission amount of task U
Output: sorted task flow L
Begin

  Initialize task flow randomization
  Set the priority of D, U to 0 or 1
  for $j=1; j <= N^t; j ++$

    for $i=0; i<N^t; i ++$

      if the priority of D and U are 1 **then**

        $P(T_j)=0$ // T_j is a high priority task
        $L[i]=T_j$
        if the priority of D and U are 0 **then**

          $P(T_j)=1$ // T_j is a low priority task
          $L[i]=T_j$

        Else

          $P(T_j)=1/2$ // T_j is a medium priority task
          $L[i]=T_j$

        end if

      end if

    end for

  end for
  Reorder the elements in L using the insertion sort algorithm
  return L

End

---

of the equipment and improve the flexibility and intelligence of the production line. Equipment utilization refers to the ratio of the time taken to perform tasks on edge nodes to the time taken to complete tasks, which can be expressed as:

$$Z_t = D_e/(J*T_t) \qquad (2.37)$$

where Z_t represents equipment utilization, D_e represents the amount of task data to be processed on the edge node, and T_t represents the execution time of the task.

Based on the above model, the task execution time and equipment utilization are often related to the edge collaborative computing method of task selection, and thus, this chapter defines a decision variable $I_j \in \{-1,0,1\}$ to represent the edge collaborative computing method of a task, and this variable is also the decision amount of selection. This variable can be expressed as follows:

$$I_j = \begin{cases} -1, & \text{if task } j \text{ is TEC computing} \\ 0, & \text{if task } j \text{ is EEC computing} \\ 1, & \text{if task } j \text{ is ECC computing} \end{cases} \qquad (2.38)$$

where $I_j = -1$ if task j is the TEC computing mode; $I_j = 0$ if it is the EEC computing mode; and, otherwise, the task is the ECC computing mode, and then $I_j = 1$.

Next, we discuss the TCCD algorithms for optimizing execution time and device utilization. Choosing the TCCD method for tasks with a large amount of computing in personalized customization production should simultaneously reduce task execution time and improve equipment utilization. The optimization objectives are as follows:

$$\min\left[\left(\alpha * \overline{T_t} + \beta * 1/\overline{Z_t} + \gamma * N^t * Q\right) * I\right] \tag{2.39}$$

where $\overline{T_t}$ is the normalized value of the execution time of the task; $\overline{Z_t}$ is the normalized value of equipment utilization; N^t is the number of tasks to be processed on the edge node or cloud node; Q (Q^e or Q^c) is the cost of processing unit data on the edge node or cloud node; $N^t * Q$ represents the cost of TCCD task computing mode; α, β, and γ are the weights of execution time, equipment utilization, and cost of edge collaborative computing, and $\alpha+\beta+\gamma = 1$. These parameters can help the customized production system to adjust the bias of task execution time, equipment utilization, and cost. Besides, I represents the decision variable of the edge collaborative computing mode selection, that is, the edge collaborative computing mode selected for each task (whether to choose the TEC computing, the EEC computing, or the ECC computing mode. It satisfies $I = [I_1, I_2, ..., I_j, ..., I_{N^t}]$ and $I_j \in \{-1, 0, 1\}$). The other parameters have the same meaning as before.

Since there are three options for edge collaborative computing for each task, there are $3N^t$ possible solutions for TCCD for N^t tasks. Therefore, this is a nonlinear restricted programming problem, and the problem cannot be solved by a formula. If the enumeration method is used to traverse the entire solution set to find the optimal solution, then the computational time complexity is very high, and thus, this method is not suitable for solving this problem. Therefore, it is extremely necessary to design a fast and low time complexity heuristic algorithm to find the optimal solution and reduce the computational complexity. It can be observed that the TCCD variable I is a vector with only discrete values. This study used the discrete particle swarm algorithm to solve this nonlinearly restricted planning problem.

Based on this, the standard particle swarm optimization (PSO) is only suitable for searching for optimal solutions in a continuous space, and it cannot be used directly for discrete spaces. In the traditional PSO algorithm, all particles have their own positions and speeds. The meaning of particle position is a feasible solution in the solution space, while the particle speed represents the distance between the next position of a particle and the current position.

In the solution space of the J-dimensional optimization problem, the update formulas for the position and velocity of the jth dimension of the ith particle are as follows:

$$V_{ij}^{k+1} = wV_{ij}^k + c_1 r_1 (PBest_{ij}^k - X_{ij}^k) + c_2 r_2 (GBest_j^k - X_{ij}^k) \tag{2.40}$$

and

$$X_{ij}^{k+1} = X_{ij}^k + V_{ij}^{k+1} \tag{2.41}$$

To solve the problem of minimizing the task execution time and maximizing equipment utilization rate of intelligent equipment in the personalized customized production environment constructed in this paper, a discrete PSO (DPSO) algorithm based on the PSO algorithm is adopted.

In DPSO, each dimension X_{ij} of the particle position and each dimension $PBest_{ij}$ of the optimal historical position of the particle are set to discrete values −1, 0, or 1, and the particle velocity V_{ij} is the same as in the PSO algorithm. The function is defined as follows, where the parameter is the jth dimension of the current velocity of the ith particle:

$$S(V_{ij}^{k+1}) = \left[\left(V_{ij}^{k+1}\right)^2 - 1\right] \Big/ \left[\left(V_{ij}^{k+1}\right)^2 + V_{ij}^{k+1} + 1\right] \tag{2.42}$$

As shown, this function is a monotonically increasing function with a value range of (−1, 1). The function value of the particle's velocity can be regarded as the probability that the particle's position is −1, 0, or 1, which is consistent with the decision variables of the edge computing of personalized customized production. When the velocity value is large, the probability of particle position going to 1 is greater; when the velocity value is small, the probability of particle position going to −1 is greater; when the velocity value is small, the probability of particle position going to 0 is greater. Then, the jth dimension of the current position of the ith particle in the DPSO algorithm is updated to the following formula:

$$X_{ij}^{k+1} = \begin{cases} 1, & \text{if } R_{ij} < S(V_{ij}^{k+1}) \\ -1, & \text{if } R_{ij} > S(V_{ij}^{k+1}) \\ 0, & \text{otherwise} \end{cases} \tag{2.43}$$

where R_{ij} is a random number uniformly distributed in the interval $[-1,1]$.

In this paper, the position of the particle is the decision variable of the execution position of each task for the optimization of the execution time and equipment utilization of the TCCD method for personalized production. Using the function of equation (2.43), the speed can be used as a parameter to choose whether the position of the particle is −1, 0, or 1. According to DPSO, the first step is initialization, that is, to randomly assign decision variables to the particle swarm and then calculate the corresponding execution time of each particle and the equipment utilization value at this time through equations (2.35) and (2.37), and the historical optimal position of each particle and the historical optimal position of the population are obtained. The next step is a cyclic process in which the global optimal value of the problem is approached continuously. For each particle, the following operations are performed: update the particle velocity and position, calculate the equipment utilization rate and order task execution time corresponding to the new decision variable, and update the historical optimal position of each particle and the historical optimal position of the population according to the least principle of equation (2.39). When the set number of cycles is reached, a relatively optimal decision variable can be obtained to minimize the average execution time of all the order tasks and maximize equipment utilization. Therefore, the specific description of the TCCD method algorithm based on discrete PSO is shown in Algorithm 2.5. For this nonlinear constrained programming problem, the time complexity of DPSO is polynomial, while the time complexity of the exhaustive method is exponential. As shown, the edge cooperative decision algorithm based on discrete PSO has the advantages of low time complexity and reduced computational complexity.

ALGORITHM 2.5   Pseudocode of the TCCD Method Based on Discrete Particle Swarm Algorithm

---

Initialization: Input L obtained by Algorithm 1, task T_j // T_j belongs to L
Output: optimal task decision variable I, minimum task execution time T_t, maximum equipment utilization Z_t
Begin

> **for** each particle T_j
>
>> Initialize velocity V_{ij} and position X_{ij} for particle T_j
>> Calculate $PBest_{ij}$ and $GBest$
>> Evaluate particle T_j and set $PBest_{ij}=X_{ij}$
>
> **end for**
> $GBest=\min\{PBest_{ij}\}$ //$I=GBest$
> **while** did not reach the iteration termination number k
>
>> **for** $i=1$ to $L.size()$ // $L.size()$ is the scale of the particle
>>
>>> Update the velocity and position of particle T_j
>>> Calculate T_t and Z_t, and get $F(T_j)$ according to (2.39)
>>> Evaluate particle T_j
>>> **if** $F(T_j)>\min(F)$ **then**
>>>
>>>> Give up the current position of particle T_j
>>>
>>> **Else**
>>>
>>>> **if** fit$(X_{ij})<$fit$(PBest_{ij})$ **then**
>>>>
>>>>> $PBest_{ij}=X_{ij}$
>>>>
>>>> **end if**
>>>> **if** fit$(PBest_{ij})<$fit$(GBest)$ **then**
>>>>
>>>>> $GBest=PBest_{ij}$
>>>>
>>>> **end if**
>>>
>>> **end if**
>>
>> **end for**
>
> **end while**
> **return** I, T_t, Z_t

End

2.3.4 Resource Scheduling Strategy for Edge Computing in Smart Manufacturing

At present, smart manufacturing computing framework has faced many challenges such as the lack of an effective framework of fusing computing historical heritages and resource scheduling strategy [34] to guarantee the low latency requirement. In this subsection, we propose a hybrid computing framework and design an intelligent resource scheduling strategy to fulfill the real-time requirement in smart manufacturing with edge computing support [35]. First, a four-layer computing system in a smart manufacturing environment is provided to support the AI task operation with the network perspective. Then, a two-phase algorithm for scheduling the computing resources in the edge layer is designed based on greedy and threshold strategies with latency constraints. The proposed strategies have demonstrated the excellent real time, satisfaction degree, and energy consumption performance of computing services in smart manufacturing with edge computing.

2.3.4.1 System Architecture of Resource Scheduling Strategy for Edge Computing

This part presents the hybrid computing architecture in manufacturing and the working process of manufacturing computing.

2.3.4.1.1 Hybrid Computing Architecture in Manufacturing In the traditional framework, there are two layers to complete the computing tasks: fog/edge and cloud. All the tasks are transmitted into a cloud or edge on top of the traditional environment; the main drawback of this architecture lies in failing to fulfill the real-time requirement, especially with many tasks queuing in edge servers (ESs). After introducing smart networks nodes, agent devices with limited computing capability, cloud computing servers, and fog-computer servers, the traditional intelligent manufacturing system can be transferred to the hybrid computing architecture. Figure 2.9 shows the system architecture for manufacturing computing using different computing resources for the computational task.

Obviously, cloud servers have the strengths of data storage and computing power; edge servers are close to the industrial devices and equipment, thereby having benefits of real-time performance; device computing units can directly drive the mechanical structure; and software defined networking (SDN) can simply provide the cooperation of different network devices. Hence, from the network perspective, all computing resources are integrated into the hybrid computing architecture to meet the latency requirement. This hybrid architecture essentially contains four parts from the task-node perspective (such as manufacturing devices): (1) Device computing layer, (2) Edge computing layer, (3) Cloud server, and (4) SDN layer. All these elements are collected by the industrial networks (i.e., wired/wireless network). In the cloud layer, the servers are mainly used to resolve the computing-extensive tasks in which AI model is developed based on different information and big data. In edge computing layer, the edge computing servers are explored to finish the real time, AI works. Additionally, in the devices computing layer, the devices are mainly responsible for finishing the sensing and controlling works. Besides, the SDN

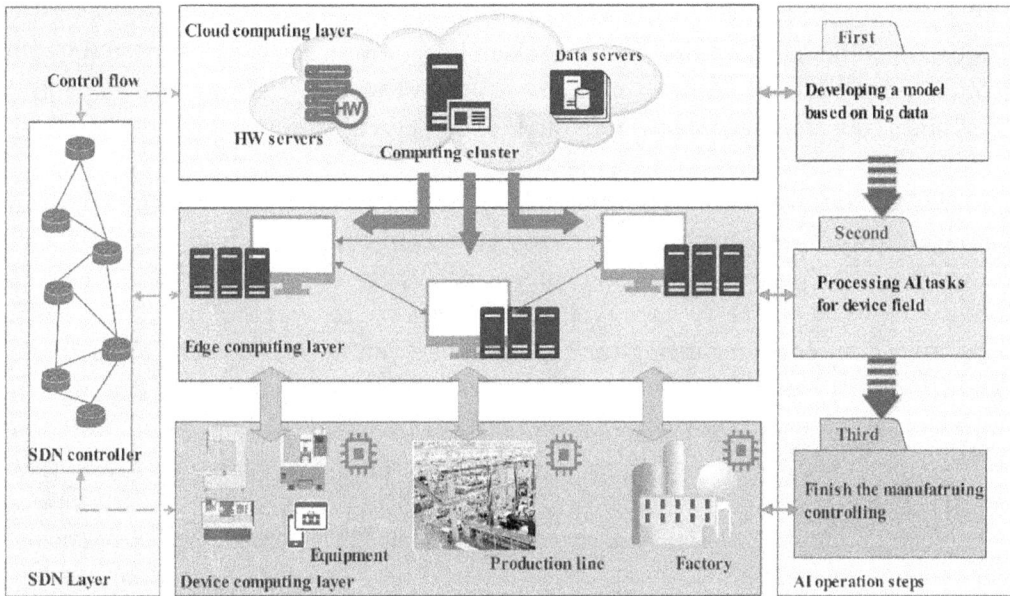

FIGURE 2.9   Architecture for hybrid computing system.

layer is used to control and coordinate different computing layers. There are differences between traditional maintenance and hybrid manufacturing computing architecture.

2.3.4.1.2 Working Process of Manufacturing Computing   In this part, we mainly focus on computing resource allocation in the proposed framework. The working process of manufacturing computing is briefly introduced. For the hybrid computing system, all computing tasks are created on the field devices, including producing machines, wireless network nodes, and mobile elements. Tasks are random events which should usually be processed in real-time manner. For scheduling a task, there are three factors to be considered: computation capability, queuing time, and data communication latency. While the latency of the task during the special time window can be computed according to the three factors, the AI task can be located at different computing layers in the hybrid computing system in accordance with the real-time requirement.

For edge computing layer, the computing capability and queuing time are the significant factors to determine the task completion time. It is obvious that there are differences for one edge computing layer from computing power, storage power. Since different servers may deal the task with different complexities, they may have different values of queuing time.

Actually, the historical legacy of computing resources and the low system latency are considered in the presented architecture. In particular, the computing framework can integrate different level computing resources in a smart factory, such as device computing, edge servers, and cloud servers. Therefore, our framework has outstanding performance in terms of low latency with the comprehensive utilization of various computing resources, especially for device layer.

2.3.4.2 Resources Scheduling in Edge Layer

This part mainly describes the resource scheduling in the edge layer. We first give the architecture of the edge layer, then present an algorithm for selecting edge computing server and a cooperation strategy for multiple edge servers.

2.3.4.2.1 Architectures of the Edge Layer   After the above analysis, the manufacturing edge layer (MEL), cloud, device computing resources of devices (LCRD) are constructed and connected to the manufacturing computing system via SDN wired /wireless networks. However, cloud servers are typically far from the devices and the system has to spend more time in transmitting the task data between devices and the cloud server. Meanwhile, LCRD is limited by computing capability and is responsible for dealing with the necessary tasks with the supporting of the local system normal operation. MEL that is close to the manufacturing equipment plays the most important role in processing the real-time tasks.

As shown in Figure 2.10, the MEL consists of multiple ESCs. Every ES is heterogeneous in the capacity of computing, storage, and task loads. In MEL, the ESs are connected via the high-bandwidth networks such as wired links and optical fiber. Therefore, ESs can form an ESC network with low delay. Therefore, every ES is deployed collocating with the devices to fulfill real-time computing tasks.

In order to achieve an efficient task process, ESs are placed in approximation to the devices. The tasks randomly are generated by the manufacturing equipment. They are then arranged and transmitted to the near and suitable ESs to ensure the real-time constraint. Obviously, there are two cases: (1) single ES can be qualified for the task, and (2) single ES cannot be qualified. Therefore, there are two strategies for computing resources scheduling: selection of algorithms for ES (SAE) and cooperation of edge computing cluster (CEC) to fulfill real-time requirement. The former way can be used to meet the low real-time requirement of computing tasks. SAE scheduling algorithm undertakes to choose the suitable ES from the edge server set (ESS) according to the task load, communication time, and computing power. Moreover, CEC is adopted for low latency requirement in which one ES cannot qualify to guarantee the low latency.

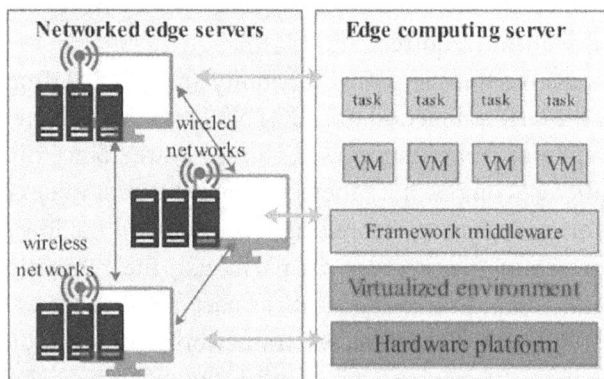

FIGURE 2.10   Mechanism of tasks scheduling for edge computing layer.

2.3.4.2.2 Algorithm for Selecting Edge Computing Server   MEL has the direct impact on the computing performances of the manufacturing task. It is indispensable to propose the scheduling algorithms for MEL and ESS. Scheduling for MEL contains two aspects: selection of algorithms for ES (SAE) and cooperation of edge computing for low latency task (CEC). Based on the requirement of specific application for manufacturing, the latency requirement of getting computing results depends on communication, computing, and queuing time.

In particular, the task x processing time in single edge server es T_{task} can be formulated as follows:

$$T_{task}(x,esc) = T_{trans}(x,es) + T_{que}(x,es)$$
$$+ T_{process}(x,es) + T_{re}(x,es) \tag{2.44}$$

where T_{trans}, T_{que}, $T_{process}$, and T_{re} are the times of transmitting task to edge server, queuing, processing, and receiving, respectively. Furthermore, assume that the data size of task and results are $\wp$, $\Im$ and the data rate is v, thus T_{trans} and T_{re} can be described as follows:

$$T_{trans}(x,es) = k_{x,esc} \frac{\wp(x)}{v(es)} \tag{2.45}$$

$$T_{re}(x,es) = k_{x,esc} \frac{\Im(x)}{v(es)}. \tag{2.46}$$

Meanwhile, let X be the set of tasks in edge server, namely $X = \{x_1, x_2, \cdots, x_{|X|}\}$. The set of computer instructions is denoted by $XN = \{xn_1, xn_2, \cdots, xn_{|X|}\}$, which is used to deal with X. For the new tasks, the queuing time can be formulated as follows:

$$T_{que}(x,es) = \sum_{i=1}^{|X|} \sum_{j=1}^{|xn_i|} \frac{IN_j}{V_{process}} \tag{2.47}$$

where IN_j and $V_{process}$ are the jth instruction of the ith task and the process speed of the edge server, respectively. In a similar way, we can get the equation for processing this task as follows:

$$T_{process}(x,es) = \sum_{j=1}^{|x|} \frac{IN_x}{V_{process}}. \tag{2.48}$$

According to formulations (2.44)–(2.46), the processing time of task x denoted by T_{task} can also be formulated as

$$T_{task}(x,es) = k_{x,esc} \frac{\wp + \Im}{v} + \sum_{i=1}^{|X|} \frac{IN \cdot xn_i}{V_{process}} + \frac{IN \cdot xn}{V_{process}}. \tag{2.49}$$

To ensure the real-time requirement of processing task x, the edge computing server must be subject to the following inequation:

$$T_{task}(x,es) \le T_{req}(x). \tag{2.50}$$

Assume that there are multiple edge computing servers close to the device which contains the task x. For easy understanding, let E be the set of ESs, namely $E = \{e_1, e_2, \cdots e_{|E|}\}$.

So, we propose Algorithm 2.6 to fulfill the strategy of SAE. In this algorithm, selecting a single ES strategy can mainly divide into three steps. First, the system searches all ESs and constructs the set of E. Then, according to equations (2.45)–(2.47), we can get the communication time T_{com} and queuing time $T_{process}$ of every ES in the set of E. Third, we evaluate the queuing time to determine whether it is larger than the deadline time of task x. Moreover, we update the candidate set Es of ES to processing the task. Fourth, in the light of the total time for resolving the task, we update Es. Finally, the device randomly selects the ES from Es for the task x.

2.3.4.2.3 Cooperation Strategy of Networked Edge Computing to Achieve Low Latency   In the previous part, the low real-time requirement of processing algorithm is given. It is obvious that Algorithm 2.6 may not deal with the computing-extensive task as there is only one ES assigned for this task. Therefore, we propose a method to cooperate multiple edge computing servers to create ESC to fulfill the latency constraints of single ES.

Indeed, once the edge servers are placed into smart factory, they are connected via industrial networks. Then, in the industrial system, the edge servers are clustered with cloud servers via SND controllers according to network distance between edge servers and cloud servers. Hence, to achieve low latency, the latter is adopted in the novel framework. The main idea of the method is explained as follows: (1) Selecting an edge server as the main server for dividing task and merging the results and (2) choosing other edge servers to cooperate to finish the task according to the latency.

Assume that the task x is be divided into N $(1 \le N \le |E|)$ subtasks, which are executed in parallel at an ESC to ensure the real-time demands. We denote the set of subtask by $x = \{sx_0, sx_1, sx_2, \cdots, sx_{N-1}\}$.

Let $Ec = \{ec_0, ec_1, ec_2, \cdots, ec_{N-1}\}$ be the set for cooperating to process the task x. For the subtask $sx_i \in x$ $(0 \le i \le N-1)$, the communication time can be given as follows:

$$T_{com}(ec_0, ec_i) = \begin{cases} \dfrac{D_{rough}(sx_i) + D_{result}(sx_i)}{V_{ec_0, ec_i}} & \text{if } i \ne 0 \\ 0 & \text{otherwise} \end{cases} \tag{2.51}$$

where $D_{rough}(sx_i)$ and $D_{result}(sx_i)$ are subtask rough and data size of results, respectively. The term V_{ec_0, ec_i} is the average data rate between ec_0 and ec_i. Furthermore, we can get the subtask processing time in the following equation:

ALGORITHM 2.6   Pseudocode of Selecting Single ES

Initialization: Input task x, xn, v, $v_{process}$, X, XN, E, $t_{requirement}$, $Es = \varnothing$

Begin:$Es \leftarrow E$
for $i \leftarrow 1$ to $|E|$

$\quad T_{com}(e_i) \leftarrow k_{e_i} \cdot \dfrac{\wp + \Im}{v}$

$\quad$ // computing the communication time;

$\quad t_{process} \leftarrow \dfrac{IN \cdot xn}{V_{process}}$

$\quad$ // computing the process time of task x;
$\quad$ **for** $j \leftarrow 1$ to $|X_{e_i}|$

$\qquad$ // computing the queuing and process time;

$\qquad t_{que}(ij) \leftarrow \dfrac{xn_{ij} IN}{V_{process}}$

$\qquad$ **if** $(t_{que}(ij) \geq t_{requirement})$

$\qquad\quad Es \leftarrow Es \setminus e_i$
$\qquad\quad$ //Selecting ES according with t_{que};
$\qquad$ **else** // Selecting ES with the total time of task x;

$\qquad\quad$ **for** $f \leftarrow 1$ to $|Es|$

$\qquad\qquad$ //computing the total time
$\qquad\qquad T_{task}(x, e_f) \leftarrow T_{com}(e_f) + T_{que}(e_f) + T_{process}(e_f)$
$\qquad\qquad$ **if** $(T_{task}(x, e_f) > t_{requirement})$

$\qquad\qquad\quad Es \leftarrow Es \setminus e_f$
$\qquad\qquad\quad$ //Selecting ES according with t_{que};

$\qquad\qquad$ **Else**
$\qquad\qquad$ **Break;**

$\qquad\quad$ **End for**

$\qquad$ **End if**

$\quad$ **End for**

End for

Return Es

$$T_{process}(sx_i, ec_i) = \frac{IN_{sx_i}}{V_{process}(ec_i)} \tag{2.52}$$

where IN_{sx_i} is the subtask instruction number and $V_{process}(ec_i)$ is the processing speed of the ith ES of Ec. It is worth mentioning that $|Ec| = N$, while we can get the formulation of $T_{sub\_task}(sx_i, ec_i)$ $(0 \leq i \leq N-1)$ as shown in equation (2.53), according to formulations (2.44) and (2.49):

$$T_{sub\_task}(sx_i, ec_i) = T_{com}(ec_0, ec_i) + T_{que}(sx_i, ec_i)$$
$$+ T_{process}(sx_i, ec_i) \tag{2.53}$$

where $T_{com}(ec_0, ec_i)$ is the communication time between ec_0 and ec_i, $T_{que}(sx_i, ec_i)$ and $T_{process}(sx_i, ec_i)$ are queuing time and the process time for subtask sx_i in edge computing server ec_i, respectively.

Recall the fact that the main ES is responsible for dividing task and merging the results. Therefore, the running task time in main ES is formulated as follows:

$$T_{main}(x, sx_0, Ec) = T_{divide}(x) + \sum_{i=1}^{N-1} T_{com}(ec_0, ec_i)$$

$$+ Max(T_{sub\_task}(sx_i, ec_i)) + T_{merge}(x, Ec) \qquad (2.54)$$

where $T_{divide}(x)$ and $T_{merge}(x, Ec)$ are the dividing time and the merging-result time for task x in edge computing servers set Es, respectively. Therefore, the total time for task x running at Ec is decided by the communication time between main edge computing server and the device of task x, as well as the running time $T_{main}(x, sx_0, Es)$ (as given in equation 2.54). It is formulated as follows:

$$T_{task}(x, Es) = T_{main}(x, sx_0, Ec) + T_{com}(ec_0, device_x) \qquad (2.55)$$

It is worth noting that T_{divide}, T_{merge}, $T_{com} \ll T_{process}$, hence according to equation (2.62), the equation can be simplified into the following:

$$T_{task}(x, Es) = Max(T_{sub\_task}(sx_i, ec_i)) + T_{com}(ec_0, device_x) \qquad (2.56)$$

Equation (2.50) gives the time constraints for processing the task. Therefore, we can get the following inequation:

$$Max(T_{sub\_task}(sx_i, ec_i)) + T_{com}(ec_0, device_x) \leq T_{req}(x)$$

$$Max(T_{sub\_task}(sx_i, ec_i)) \leq T_{req}(x) - T_{com}(ec_0, device_x) . \qquad (2.57)$$

$$T_{sub\_task}(sx_i, ec_i) \leq T_{req}(x) - T_{com}(ec_0, device_x)$$

According to the equations (2.52) and (2.57), it is easy to get the task instruction number of the ith ES. It is described as follows:

$$IN_{sx_i} \leq (T_{req}(x) - T_{com}(ec_0, device_x) - T_{que}(sx_i, ec_i)) V_{process}(ec_i). \qquad (2.58)$$

Furthermore, task time $T_{task}(x, Es)$ is determined by the maximum $T_{process}$. Therefore, in light of the above discussion, we propose the strategy of cooperating edge computing servers for the extensive task as shown in Algorithm 2.7.

In Algorithm 2.7, the steps can be mainly described as follows: First, according to the referenced equations, we compute subtask instruction number $IN(E)$ in the constraints of $T_{req}(x)$ for every ES in the set of E. Then, we sort $IN(E)$ in the descending order (i.e., from

ALGORITHM 2.7   Pseudocode of CES

Initialization

for $k \leftarrow 1$ to $|E|$

$\quad IN(e_k) \leftarrow (T_{req}(x) - T_{com}(ec_0, device_x) - T_{que}(e_k))V_{process}(e_k)$

$\quad$ //getting the subtask instruction number in the constraints of $T_{req}(x)$

$\quad IN(E) \leftarrow IN(e_k)$

$\quad Es' \leftarrow sort(IN(E))$

$\quad$ //sort () is the function for sorting the E according with the $IN(e_k)$;

$\quad main\_ESC \leftarrow Max(IN(E))$

$\quad$ //selecting the main ES

End for

for $i \leftarrow 1$ to $|ES'|$

$\quad Temp\_sum = IN(es_i') + Temp\_sum$

$\quad$ **if** $(Temp\_sum < xn)$

$\quad\quad Es \leftarrow Es / es_i'$

$\quad$ **else**

$\quad\quad$ **break**;

$\quad$ **End if**

End for

$divided\_task$ (x, Es) // divide the task x according with Es

$processing\_subtask$ () // processing subtask in selecting edge server

$Return\_subtask\_result$ ()

//returning the subtask result from different edge server

$RES \leftarrow merge\_subresult$ ()

//the main edge server merging the results

Return RES

---

largest to smallest) and create the sorted ES set *Es'*. Third, we sum the subtask instruction number, *temp_sum*, and evaluate whether the *temp_sum* meets the requirement of task *x*. Finally, the main ES divides the task *x*, finishes the processing task *x* by *Es*, and returns the result of the task (*RES*).

2.3.5 Cognitive Ability of Edge Computing in Customized Manufacturing

Computer-integrated manufacturing is a notable feature of Industry 4.0. Integrating machine learning (ML) into edge intelligent MIoT is a key enabling technology to achieve the intelligent MIoT. To realize novel intelligent applications of edge-enhanced MIoT, ML methods are proposed to improve the cognitive ability of edge intelligent MIoT in this subsection [36]. The ML methods are presented to enhance the cognitive ability [37] of MIoT including ML model of MIoT, data-driven learning and reasoning, and coordination with cognitive methods. The main purpose of this subsection is to point out the effects of ML-based optimization methods on the analysis of cognitive ability of edge computing in customized manufacturing from the macroscopic view.

Using the environmental perception, ubiquitous computing technology, and mobile communication technology, MIoT integrates many kinds of terminals into every phase of the industrial manufacturing. With the popularization of the low-cost sensors and

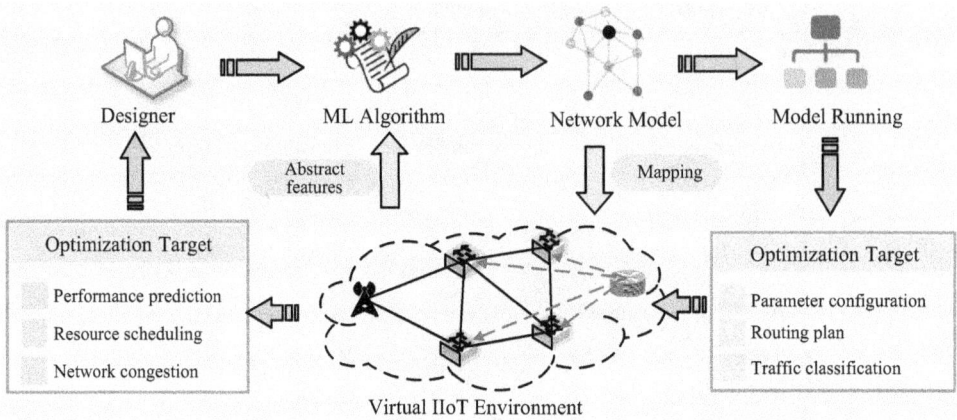

FIGURE 2.11   The ML-enabled network optimization method.

intelligent, distributed terminals, intelligent MIoT has gathered these devices together with edge computing to integrate physical information. Edge intelligent MIoT makes it easier to transfer cloud services with low latency, high bandwidth, and low jitter as shown in Figure 2.11. We can use the semantic association of information to perceive the changing industrial scenarios. The ML methods embedded into edge intelligent MIoT make the entire MIoT system have the ability of understanding, learning, and reasoning.

2.3.5.1 ML Model for Edge Intelligent MIoT

With the deeper integration of the industrial information, the real-time analysis of the industrial data or the offline training with ML methods is an important way to improve the decision-making accuracy. A proper ML model is the premise for the applications of intelligent MIoT. Notably, it is crucial to know the characteristics of the generic ML methods. Therefore, in the following, we briefly introduce three typical ML models, namely, deep learning (DL), reinforcement learning (RL), and deep reinforcement learning (DRL). The different types of ML/DL models applied in the MIoT are shown in Table 2.2.

DL is an ML method based on the data representations learning. DL aims to establish a neural network to mimic the functions of the human brain in analysis and learning. DL has significant advantages in feature extraction and model fitting. Particularly, it achieves

TABLE 2.2   Different Types of ML/DL Models Applied in the MIoT

Contrasts	Model Types	
	Deep Learning (DL)	Reinforcement Learning (RL)
Main features	Low-level features in a combined manner to form high-level properties or features	Take actions based on the environment to maximize the expected reward
Typical MIoT applications	Pattern recognition, Fault prediction, Service recommendation, Industrial network security and monitoring, etc.	Dynamic programming, Swarm intelligence, Multi-machine cooperation, Optimization scheduling for dynamic resources, etc.

a good approach capacity of a mathematical model based on a rich dataset. The character is helpful to build a high-precision simulation model for the intelligent MIoT services.

RL is an interactive learning method oriented to the decision goal. RL is also an interesting learning model more in line with the psychology of human actions. It learns how to map both single input and multi-input (e.g., the Markov decision process) to outputs with dependency relationship. In the context of possible operations for a given state, RL can be applied to change the environment state. During the learning process, the algorithm randomly explores the state-operation pairs (or construct a state-operation pairs table) in an observable environment. The learned information is utilized to accumulate the rewards of the state-operation pairs. The most awarded action is chosen to achieve the target state for the given environment. In terms of DRL, DL brings end-to-end convenience to RL and makes RL no longer confined to low-dimensional space. Thus, DRL is expected to be applied in the complex and high-dimensional environment of industrial IoT.

2.3.5.2 Data-Driven Learning and Reasoning

ML's essence is to train data following a specified target, therefore, dataset is crucial for the learning object of ML. The scale and variety of MIoT data are vitally important in improving the ML ability. ML is superior to the traditional methods because it uses the large-scale data. The typical features of MIoT data are high dimensionality, sparsity, and weak correlation. Therefore, ordinary IoT technologies are unable to provide the complete information representation for industrial environment in intelligent MIoT applications. ML does not mainly rely on expert experience, but rather on data that reflects the core problem. As a result, the mathematical model is more comprehensive than the expert system. Implementation of MIoT application with ML is commonly divided into two steps, namely, learning and reasoning. Learning is to obtain the weights and biases of the model through data training. Suitable mechanism is selected to evaluate the model. Reasoning can be regarded as data-based prediction, which is also an important way to acquire knowledge and discover the potential events. The quality of a model often depends on the accuracy of the reasoning results. The model building process of DL, which reflects the general process of building an ML model, is shown in Figure 2.12. By training the large-scale data, the candidate model with a higher fitting degree is selected through the verification set, and the cross-validation method is used to avoid overfitting. The test set enhances the model generalization ability. Similarly, the learning of prior knowledge is also needed in RL.

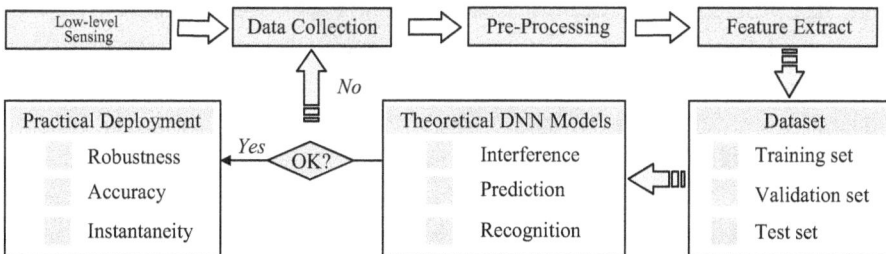

FIGURE 2.12 The data-driven process of a DL model building.

To enable intelligent applications of edge-enhanced MIoT, the cognitive MIoT provides rich datasets for ML to infer, predict, and make decisions when the external variables change. The cognitive MIoT acting on network transmission includes data marking, semantic, and feature abstraction. Moreover, the online learning or offline training is already promoting the typical MIoT applications (e.g., active operation and maintenance). By using the reasoning and prediction based on data, the intelligent decision-making is achieved according to the learning rules. Edge computing helps advanced data modeling and predictive analytics migrate to network edge for mainstream MIoT products. Data collection at network edge also helps the application model to perceive the environment, promote the improvement of the application model, and help the model adapt to the dynamic environmental change.

2.3.5.3 Coordination with Cognitive Methods

The collaboration between ML and cognitive methods is helpful to enhance the cognitive ability of edge intelligent MIoT. Semantic perception technology (e.g., ontology) helps to automatically search, discover, and access to the perceptible devices. The virtualization of physical entities and the description of perceptible devices can be realized through semantic technology. The semantic-based cognitive methods provide information sources for edge intelligent MIoT applications, which is the basic medium for interaction or collaboration. To utilize the resource in manufacturing cyber-physical system efficiently, we have proposed an ontology-based semantic modeling method for intelligent equipment. Our previous work realized the autonomous decision-making and reasoning for equipment in IoT-based manufacturing. Although the semantic-based cognitive methods can recognize the environment, it is difficult to realize the reasoning beyond the defined rules. The cognitive environment is still determined by the rule maker. The resource management is usually a difficult online decision-making task, and an appropriate solution depends on the understanding of the workload and environment. ML makes up for shortcomings in the data analysis for edge intelligent MIoT. Combined with the cognitive methods, ML methods can build an accurate semantic analysis model to deploy autonomous reasoning and take action to change the environment. In this way, the semantic-based cognitive technology provides an easy-to-understand perceptual environment for ML.

A narrow definition of cognitive methods can be regarded as activities related to information processing, including receiving and conversion of sensing information, concept formation, and problem determination. Cognitive methods are especially effective when faced with the data that is inaccurate, uncertain, and partially true in information statistics. ML is more about an abstraction method. It concerns mainly: (1) new knowledge discovery based on cognitive information; (2) high-order abstraction above underlying information; and (3) dynamic programming in a human manner. As the majority of the industrial IoT data is time series, higher-order ML algorithms can implement prediction based on historical data (e.g., predictive maintenance of equipment). Taking DL as an example, the application of DL in the industrial IoT system can be extended to the feature extraction of the network state, dynamic adaptive planning, system parameter configuration, and selection of optimization strategy. The DL algorithm helps to build a more comprehensive monitoring system for the MIoT, rather than relying on a sampling node or a specific domain knowledge. Accurate ML model is the premise to realize an intelligent service for edge MIoT. It is a challenge to get the real test sets that approximate the real network environment of the MIoT. As a result, an unsupervised ML model may be more suitable for solving the related problems. The embedded ML algorithms are helpful to provide more accurate models with the condition definition by domain experts.

2.4 VALIDATION OF RELEVANT KEY METHODS OF EDGE INTELLIGENCE IN SMART MANUFACTURING FACTORY

In this section, the validation of relevant key methods of edge intelligence in smart manufacturing factory is described. First, knowledge reasoning and sharing based on edge intelligence; second, adaptive transmission optimization in manufacturing with edge computing; and finally, reconfigurable for customized manufacturing based on edge intelligence.

2.4.1 Edge Intelligence-Based Knowledge Reasoning and Sharing

The development of multi-variety, mixed-flow manufacturing environments is hampered by a low degree of automation in information and empirical parameters' reuse among similar processing technologies. This subsection proposes a mechanism for knowledge reasoning and sharing between manufacturing resources that is based on edge intelligence [38,39]. The manufacturing process knowledge is coded using an ontological model [40], based on which the manufacturing task is refined and decomposed to the lowest-granularity concepts, i.e., knowledge primitives. On this basis, the learning process between devices is realized by effectively screening, matching, and combining the existing knowledge primitives contained in the knowledge base deployed on the cloud and the edge.

2.4.1.1 Experimental Prototype Platform

A prototype platform of a multi-variety, mixed-flow intelligent manufacturing production line is shown in Figure 2.13. The main system workflow is as follows. First, the user utilizes a web page or a mobile app to select the product's type, quantity, and personalized pattern. The user order is completed in the cloud. Once recorded, the order is issued directly to the manufacturing resource edge, generating the control flow related to the processing task. The control flow drives each manufacturing resource to perform the current processing task. The equipment utilization is optimized through a high degree of coordination and knowledge obtained in the previous processing operations. The manufacturing production line's efficiency is reflected in the multi-task collaborative production and dynamic production line reconfiguration. The mixed-flow manufacturing functions incorporated in the prototype platform to date include, for example, custom wood carving crafts processing, personalized U disk customization, and Bluetooth remote control assembly. In the prototype platform, Robot1 and CNC1 are matched, as well as Robot2 and CNC2, and are responsible for the clamping and engraving process of wood carving originals. The laser machine performs laser printing according to the pattern and text selected by the user

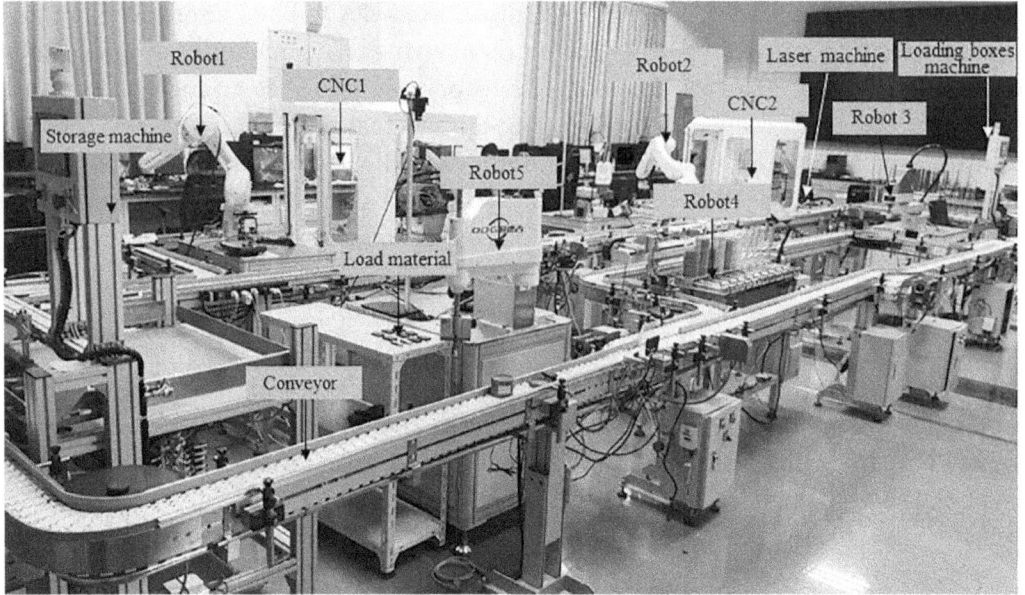

FIGURE 2.13   Prototype platform for mixed production line.

TABLE 2.3   Manufacturing Resource Function Description

Manufacturing Resource	Function Description
Robot1(2)	Pick up the raw materials into the CNC1(2) and put the finished products back on the conveyor belt
Robot3	Put the processed product in the box
Robot4	Select the box cover that matches the box and perform the capping operation
Robot5	Select raw materials and add to the conveyor belt
Laser machine	Laser printing personalized patterns
Loading boxes machine	Add boxes to the conveyor belt
Storage machine	Unload the packaged product
CNC1(2)	Carve the raw materials

when placing the order. Loading machine loads the product packaging/box. Robot3 and Robot4 cover the processed product packaging, whereas Robot5 uses machine vision to select the required raw materials for loading operations. The final products in the packaging box are uniformly unloaded. The specific functions of manufacturing resources are shown in Table 2.3. Because the prototype platform involved in this experiment is a multivariety mixed-flow platform, the order of each station in the processing process cannot satisfy a single product manufacturing process. To solve this problem, we have adopted a method of station matching. When the product passes through the station, the product manufacturing information will be read. If the product needs to be processed by this station, it will enter the corresponding production branch line, and if it is not needed, it will move directly to the next station.

2.4.1.2 Integration of Semantic Data and Knowledge Reasoning
Primitive data introduces difficulties during data analysis, especially when the volume of data significantly increases. Therefore, we need semantic data to conduct intelligent analysis and application. Figure 2.14 describes a combination of ontology-based knowledge modeling, OPC UA based sematic data generation, and semantic database for implementation of real-time processing of industrial big data. Information models for the machines, conveyors, products, and the system should be built previously. Based on the information model, the concepts in the domain and also the relationships that hold between those concepts can be defined with a software tool, like Protégé, to create an ontology model. According to the same information model, OPC UA address spaces are created for the machines, conveyors, and products, and the ontology is created for the system. At the running time, semantic data stored in OPC UA address spaces can be loaded up into the database and these data correspond to the properties of the ontology. The reasoning engine takes rules coming from the upper applications as targets to reason ontology with the real-time data from the database. After reasoning, results will be fed back to the application. The using of mature reasoning engines along with definable rules instead of programming contributes to flexibility, stability, and simplicity.

OPC UA expresses information in a similar way with ontology where objects and references (describing the relationship between objects) are commonly used. Therefore,

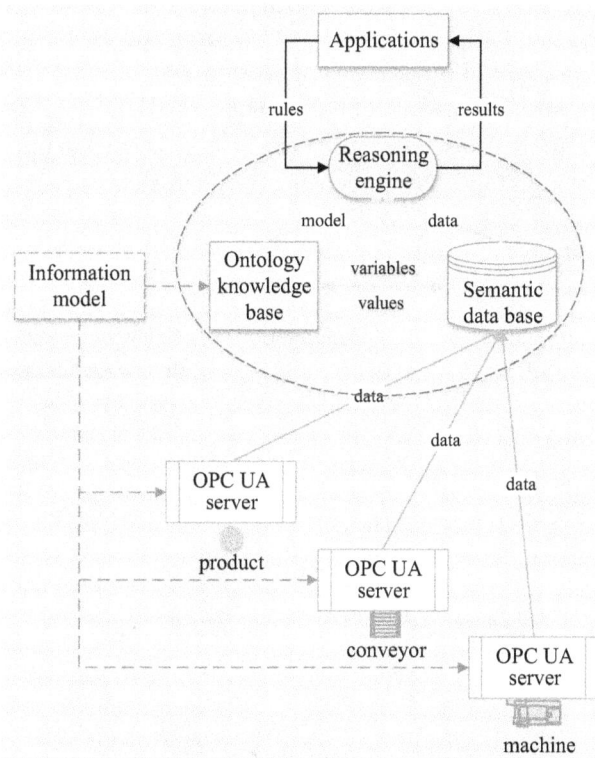

FIGURE 2.14 Integrated architecture of knowledge reasoning and semantic data.

developing an independent information model in an object-oriented way is necessary. However, any object can be viewed as a system that consists of other objects. This means we cannot define every objects that exist in our manufacturing system. One should focus on his application to simplify the information modeling. In our design, the self-organization and cloud assistance are two core ideas where machines, conveyors, and products organize themselves to process tasks, and the cloud assists this self-organized process to optimize system performance. Therefore, machines, conveyors, and products are identified as the first level objects, which can be subdivided as in Table 2.4. Note that the classification of machines and conveyors is based on functions instead of structures. For example, a robotic arm can be a processing machine if it is used for welding, but a conveyor if it is just used to move products from one place to another. Human being is always an indispensable factor for production; based on their duties, involved employee can be subdivided.

The recommended modeling language and related software tools are summarized in Table 2.5. The UML (Unified Modeling Language) is a powerful and application-independent modeling language to construct the information models. A number of software tools are available for UML modeling, some of which are free and even open source, such as JUDE-community, Argo UML, UMLet, Visual paradigm-community, and BOUml. To help the development of the OPC UA compliant programs, a few software tools have already been on the mark, among which the UaModeler from the Unified Automation Company is quite popular one.

The Protégé is a respective tool for ontology modeling which supports OWL (Web Ontology Language) 2.0. Finally, the Apache Jena is a free and open source Java framework for building semantic web and Linked data applications, supporting SWRL (Semantic Web Rule Language). Therefore, we have a set of tools where the JUDE-community uses UML to build the information model, based on which the Protégé is used to build ontology and the UaModeler is used to build OPC UA model. The OPC UA servers provide semantic data to

TABLE 2.4   Object Classification

Main Classes	Subclasses			
Machine	Processing machine	Testing machine	Assembling machine	Storing machine
Conveyor	Conveyor belt	AGV	Robotic arm that moves products	
Product	Bar code labeled product	QR code labeled product	RFID tag labeled product	Microcontroller labeled product
Staff	Operator	Maintenance technician	Team leader	Manager

TABLE 2.5   Recommended Language and Related Software Tools for Modeling

Model	Language	Recommended Tool
Information model	UML	JUDE-community
OPC UA model	OPC UA	UaModeler
Ontology model	OWL 2.0	Protégé
Application model	SWRL	Apache Jena

the cloud, combined with which the ontology is recurrently processed by the Jena taking application requirements that are expressed with SWRL rules as input. The original ontology built by the Protégé can be called file model, and this model becomes memory model after the file model is loaded into memory. After reasoning, the memory model turns into inference model. As the inference model reflects the change of the system, it may be different from the memory model. Fortunately, the inference model can be saved as a new file model so that the ontology model can evolve through reasoning.

2.4.1.3 Edge Intelligence-Based Knowledge Sharing

The use of ontologies enables abstraction, standardization, and formalization of equipment attributes and their relationships within the intelligent manufacturing field. Therefore, reusable knowledge extraction and storage can be achieved through the equipment's ontological modeling and processing. Due to its ease of use and maintainability, this research utilizes the open-source software Protégé with a graphical user interface to construct the devices' ontology. Compared with the existing research on ontology modeling in the manufacturing environment, the ontology model construction method proposed in this paper is developed around the manufacturing task modeling not only the physical resources involved in the manufacturing process but also the surrounding environment elements related to the resource, the evaluation of the execution effect, and the matching degree of the action and the task. In the description of the manufacturing process, this chapter also proposes a task decomposition scheme that allows it to adopt a more fine-grained representation method and also provides more knowledge materials for the reuse of knowledge.

The data flow of system operation is shown in Figure 2.15, including two parts: data uplink and data downlink. When data collection is performed, the edge server will first check the data collection node. If the local node does not have a mapping relationship with the mid-node of the ontology model, the mapping relationship will be established based on the characteristics of the data and the node type, and then the task decomposition steps will be performed and knowledge will be collected based on the decomposed action primitives. After completing the storage of the local knowledge base, upload and update the cloud knowledge base. When reusing knowledge, the local server will first decompose the target task and then perform similar task matching based on the action meta-model. If there is not enough knowledge available locally, it will perform a cloud knowledge query. After finding enough action primitives, the edge server will combine tasks and complete the configuration of local resources. As for the establishment of the ontology model and the task decomposition scheme, it will be described in the latter part of this section.

Intelligent manufacturing's physical resources include equipment used in a series of intermediate processes such as product processing, packaging, and transportation. The knowledge sharing realization requires organizing various links between manufacturing resources into a unified knowledge base to facilitate knowledge storage and recall. This section introduces the integrated ontology and semantic modeling of manufacturing resources

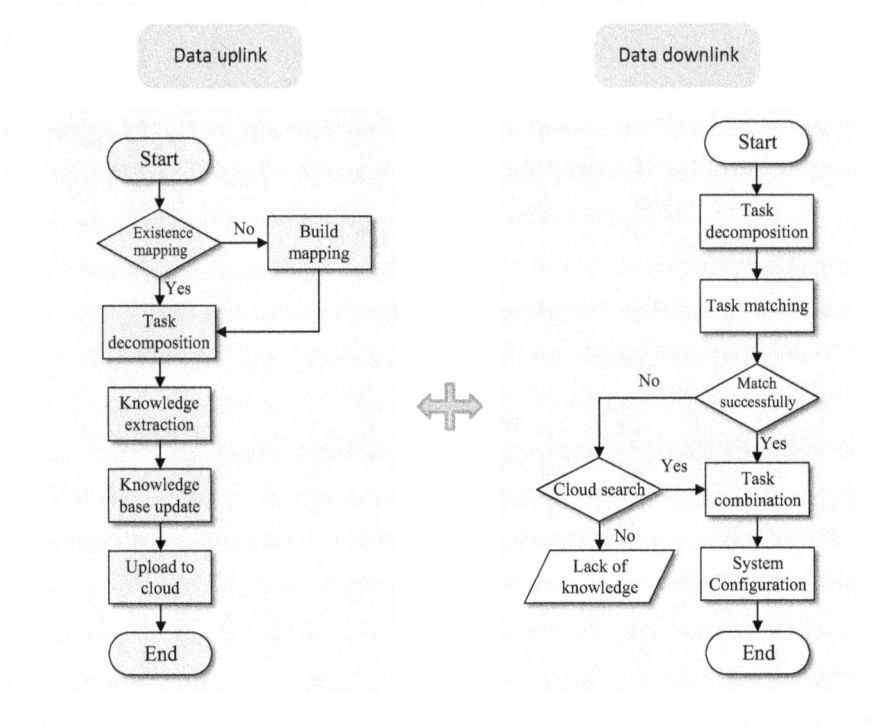

FIGURE 2.15 System operation data flow.

The manufacturing resources' ontology structure can be represented by a four-tuple {C, R, S, P}, where C is a set of ontological classes, R represents a set of relations between the ontological classes, S denotes a set of hierarchical relationships between classes (that is, the inheritance and realization of class relationships), and P is a set of other relationships between classes. Consider, for example, an ontology model $O_1 = \{(x_1, x_2, x_3, x_4), (r_1, r_2),$ $((x_1, x_2), (x_1, x_3), (x_3, x_4)), (r_1, (x_2, x_3); r_2, (x_2, x_4))\}$. Here, the classes involved are x_1, x_2, x_3, and x_4. According to the hierarchical relationship set, x_2 and x_3 are both subclasses of x_1, while x_4 is a subclass of x_3. In addition to inheritance and implementation between classes, there are two other relationships, r_1 and r_2, of which r_1 is the relationship between x_2 and x_3, and r_2 is the relationship between x_2 and x_4.

The central concept in the manufacturing resources' ontology is a manufacturing task, abstracted in accordance with the four-tuple definition. Based on this abstraction, the corresponding semantic model is constructed, as shown in Figure 2.16. The objects in the manufacturing process include raw materials and manufacturing resources. The attributes used include the equipment's name, number, physical size, capabilities, and the processing operations that can be performed. Environmental factors record the manufacturing task's processing environment, such as space, position, and the manufacturing process requirements regarding, for example, temperature, light, and air quality. The execution action module includes the action level classification, the action sequence combination, and the processing procedure record, thus covering the process's sequence information and state. Finally, the effect evaluation results from the previous executions using the current knowledge and includes each action's execution time cost, product quality, and time consumed.

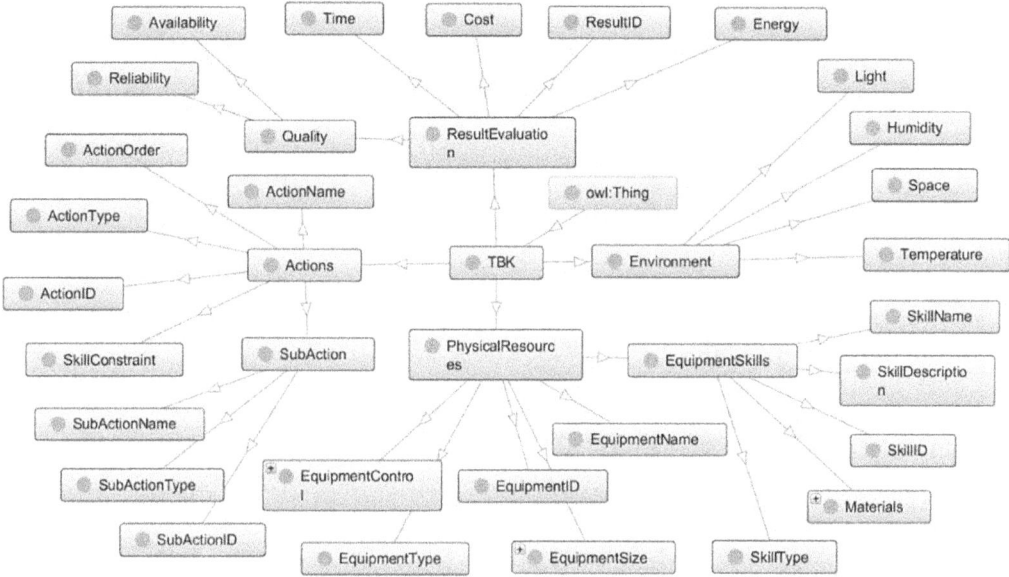

FIGURE 2.16   Task-based ontology model.

Each product's manufacturing process includes multiple manufacturing resources' action coordination. To extract and apply the manufacturing process experience, task and action models need to be decomposed, hierarchically divided, and described by a basic combination of action primitives. Additionally, the manufacturing process provides a clear hierarchical structure for subsequent knowledge reuse. This work embodies the hierarchical structure and logical relationship between the actions by ontologically modeling the manufacturing process actions.

Specifically, as shown in Figure 2.17, the execution actions' levels can be divided into the following: task-based actions (*ActionsOnTask*), station-based actions (*ActionsOnStation*), and the bottommost action primitives (*PrimitiveActions*). Task-based actions state the manufactured product's purpose or a function. Station-based actions are derived from the higher-level actions according to the processing procedure and are combined with the manufacturing process characteristics. The actions are defined for each manufacturing station, and the action primitives are used for processing. The process is refined and decomposed, and the actions are decomposed and combined from the manufacturing resources' level. Reusable action control information is obtained by decomposing different manufacturing processes to the primitive level and then reorganizing the action primitives to construct the new task.

2.4.1.4 Implementation Case

The experiment starts with the loading machine that reads the products' order information. The loading machine writes the packaging boxes' RFID tags, including the involved process and the corresponding processing actions. The pallets carrying the packaging boxes and materials pass through various stations. The RFID tags are read at the station, and the processing actions that can be performed at the current station are executed.

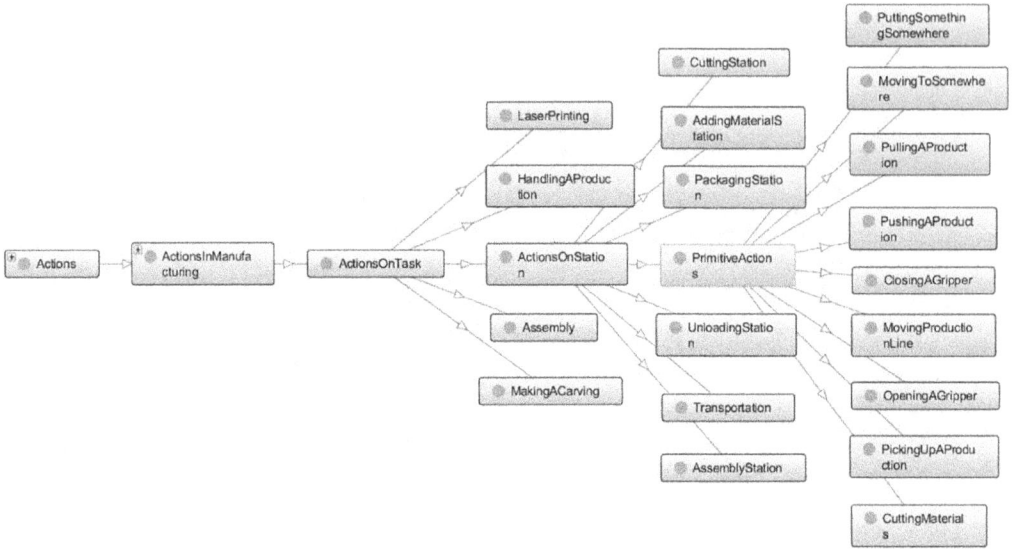

FIGURE 2.17   Action model hierarchy.

TABLE 2.6   Correspondence between Manufacturing Resources and Action Primitives ("+" means "contain")

Robot ID	Robot1	CNC1	Laser	Loading	Robot3	Robot5	Storage
Moving	+			+	+	+	+
Clamping	+	+		+	+	+	+
Placing	+			+	+	+	+
Cutting		+					
Printing			+				
Absorbing					+		

The relationship between the physical manufacturing resources in the prototype platform and the action primitives that can be executed is shown in Table 2.6. It can be seen from the table that each physical resource processing can be regarded as a collection of actions, and the action primitives it contains have a similar composition structure.

The presented multi-variety, mixed production line serves as an experimental platform. The equipment is manually configured and debugged to generate the equipment's initial processing experience (i.e., initial input). Once the ontology modeling on the manufacturing resources' edge is completed, the equipment that has not been manually debugged is configured by relying on the existing knowledge. More precisely, the existing process data model and control instructions are automatically configured. The proposed knowledge sharing mechanism's effectiveness is verified by the experimental results that compare the manually configured action cycle with the automatically configured result.

Since Robot1 and Robot2 have similar processing procedures, the two manipulators' action cycles can be read separately for comparison, as shown in Figure 2.18. Robot1 is manually configured and debugged, whereas Robot2 is automatically configured based on

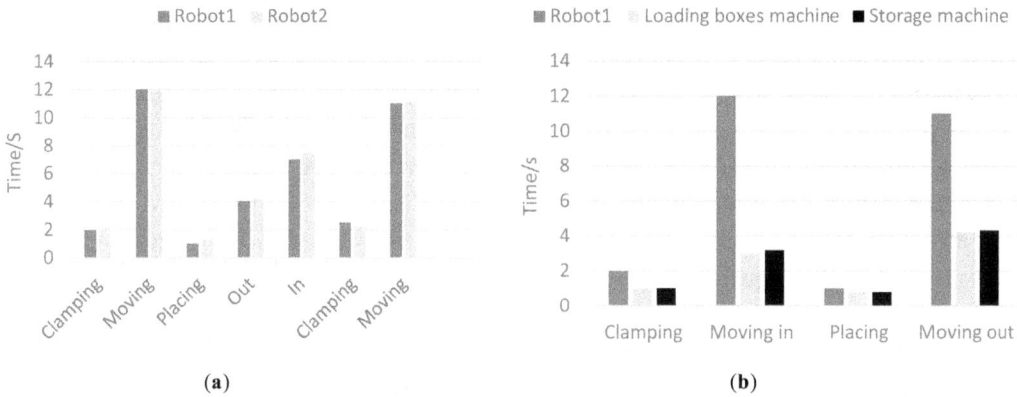

FIGURE 2.18 (a) Comparison of action cycles for similar manufacturing processes. (b) Multi-process action cycle comparison.

the knowledge sharing mechanism. The manipulators' main processing procedure can be detailed and decomposed into conveyor belt material clamping, moving the material to the processing equipment CNC, placing the material, exiting the CNC, waiting for the CNC to enter after the processing is completed, clamping the processed material, and moving the material. The raw materials arrive at the conveyor belt pallet. Figure 2.18a demonstrates that there is only a slight difference between the two manipulators' action cycles, indicating that Robot2 successfully learned Robot1's existing processing knowledge and experience. Nevertheless, certain differences in the configuration results can be found. In particular, Robot2 chose a more efficient execution plan by comparing the two actions' execution effect evaluations during the learning process. In order to reflect the reduction degree of knowledge reuse, we define the reduction degree $S_i = \dfrac{|T_i - T_i'|}{T_i}$ of action i, where T_i represents the action execution cycle of the knowledge source and T_i' represents the action execution cycle of the device after learning the existing knowledge. The reduction degree of the whole process adopts a weighted average $\sum S_i \cdot \dfrac{T_i}{T_a}$, where T_a represents the total processing time of the whole process. Combining the calculation of the specific execution cycle data of each action, it can be concluded that the reduction degree of Robot2 for the same task of Robot1 is 95.8%.

The action cycles of Robot1, loading machine, and storage machine are selected for comparison. The specific data is shown in Figure 2.18b. Robot1 and loading machine were manually configured and debugged, while the storage machine's process was learned through the local knowledge base. The figure shows data reading and comparison of the three devices' action primitives. The experimental results demonstrate that the storage machine's action cycle is very similar to that of the loading machine but significantly differs from Robot1's. The storage machine and the loading machine deal with the product's outer packaging. Thus, these machines' operation accuracy is lower than the value of Robot1 when gripping the engraving materials. Consequently, Robot1 has a lower processing speed. Regarding the equipment movement, the storage machine and the loading machine are responsible for the round-trip conveyor belt and the raw material warehousing or the finished product warehousing. These machines do not involve cooperation with other equipment. Robot1 needs to place the engraving materials into the CNC processing area accurately. The CNC's airtightness further restricts Robot1's movement. The three equipment are similar regarding the placement actions: the only difference being the processed object's size, the action cycles do not significantly differ. Following the analysis of the equipment learning process, it can be concluded that the knowledge reuse process is not a straightforward application, but rather a matching search for similar processing tasks to extract the knowledge that meets the similarity requirements.

2.4.2 Adaptive Transmission Optimization in Manufacturing with Edge Computing

In recent years, smart factory in the context of Industry 4.0 and MIoT has become a hot topic for both academia and industry. In MIoT system, there is an increasing requirement for exchange of data with different delay flows among different smart devices. However, there are few studies on this topic. To overcome the limitations of traditional methods and address the problem, we seriously consider the incorporation of global centralized SDN and Edge computing in MIoT with edge computing (EC) [41]. The performance of proposed strategy is evaluated by simulation. The results demonstrate that the proposed scheme outperforms the related methods in terms of average time delay, goodput, throughput, path difference degree, and download time. Thus, the proposed method provides better solution for MIoT data transmission.

2.4.2.1 MIoT Platform Structure with SDN and EC

2.4.2.1.1 Platform Structure Review   In manufacturing applications, the communication networks have been progressed greatly. Both clustered and grouped network topologies have been developed and proven as manageable structures for MIoT. Since the current structures are limited by the constraints of communication latencies, fixed bandwidth, coverage, and unbalanced deployment of computing resources, they are poorly adaptable to emerging MIoT demands. In order to increase the flexibility, scalability, and centralization MIoT, as well as to balance the reasonable deployment of computational resources, SDN and edge computing are integrated into MIoT. Therefore, in the proposed solution, the cloud, SDN, edge computing, and the other sub-systems constitute a novel framework, as shown in Figure 2.19, wherein all components are connected by communication infrastructures. To better understand the system, we simplify the framework into East–West flow (MIoT), North–South flow (SDN), and Computing plane (edge computing).

All objectives, workmen, users, and smart terminals are abstracted into different kinds of network nodes as follows: ordinary nodes, cluster heads, and sink nodes. In the clustered MIoT, different function nodes and data centers jointly construct the data exchange sub-system. Evidently, cluster heads and ordinary nodes establish a small sub-system for data gathering and delivering, while this system has basic functions in M2M communication. In industrial application, due to different services of the system, the data upload and offload need different time limits for different data flows. The solving of data transmission problem in the paper is focused on data transmission at data exchange sub-system.

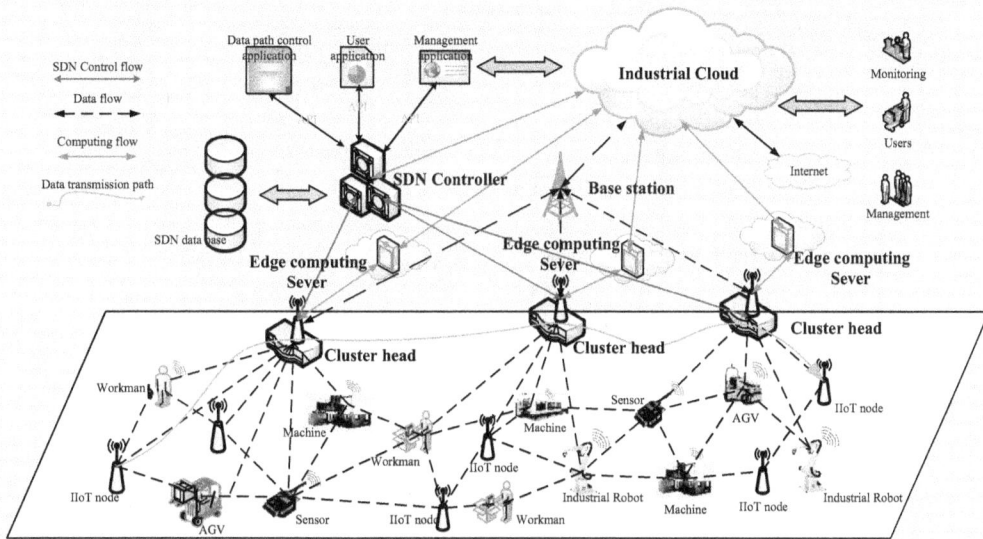

FIGURE 2.19   Data transmission structure based on SDN and edge computing in MIoT.

In SDN sub-system, all communication flows are divided into control flow and data flow. In the system, an open source SDN controller Open Mul is adopted in the framework. Open Mul is an Openflow/SDN controller platform. The SDN controller is connected with key network devices such as cluster heads. It is clear that these key nodes are amounted with controller interface, so they establish a link with SDN controller using OpenFlow protocol. MIoT nodes deliver the status parameter to SDN controller. Then, in the control layer, the controller makes decisions on data transmission control. In this research, SDN is adopted to control the transmission path and transmitted power of M2M.

To get a reasonable assignment of computing resources, the edge computing servers (ECSs) are deployed in MIoT. Downward, these ECSs establish the link between nodes and SDN controller; upward, ECSs are connected to the cloud. Every ECS is typically miniature data and computing center that is in the vicinity of nodes and SDN devices. In our scheme, edge computing is employed to derive the results in real time. The EC strategy reduces the time consumption and traffic load, compared to the traditional computing models.

2.4.2.1.2 Working Process   The main working process of this structure is as follows. First, every ordinary node forwards the data to the cluster heads or some other ordinary nodes, according to the real applications. After the completion of data gathering in the whole cluster, cluster heads transmit the corresponding information to sink node or base station. Then, sink node sends this information to the cloud. Second, to realize a flexible control of the entire network, the statuses of all devices are uploaded to SDN controller and mapping databases. Once the SDN application layer is modified and certain application function is adjusted, the SDN controller gathers the parameters and uploads the tasks to ECS. Third, when a computing task, such as changing of transmission path, is given, the ECS optimizes the network parameters such as power and hop path, and the optimization results are

TABLE 2.7   Simulation Parameters

Parameter	Value
Number of cluster heads	10
Number of ordinary nodes	100, 200
Communication amount	100–900 Mb
Requirement for communication time	500–3,000s
Maximal power communication range	50 m
Usual communication range	30 m
Custer head average channel rate	100 Mbps
Deadline Index I, II, and III	1,500, 2,000, 3,000 ms
Increasing power coefficient	1
Cost constant	0
Ordinary node average channel rate	1 Mbps

sent to SDN controller and MIoT devices. Lastly, after abortion of the control data, MIoT devices adjust their network conditions. Here, an adaptive data transmission path employs the same working process.

2.4.2.2 Evaluation Performances

In this part, the simulations are conducted to evaluate the performance of proposed method. First, we describe the simulation setup, performance metrics, reference schemes, and emulation scenarios. Then, the evaluation results are presented and discussed from various perspectives.

2.4.2.2.1 Simulation Setup   We developed the simulation structure and realized the proposed algorithm in MATLAB environment. We use multiple threads to simulate the related methods. One of the threads is used to simulate the SDN controller, which collects network parameters regularly from network nodes. Then, a state machine mechanism is adopted to emulate SDN controller actions by using the proposal involved in the paper such as routing path or increasing transmission power of cluster heads. The test cases were generated according to the MIoT node density and number of CHs. The main simulation settings are summarized in Table 2.7.

2.4.2.2.2 Performance Metrics and Reference Schemes   To evaluate the performance of the proposed methods, we introduce the following performance metrics:

1. **Average time delay**: The average time delay represents the time needed for data transmission from the source node to the destination node.

2. **Path difference degree (PDD)**: As shown in definition 1, this performance metric is used to measure the balance of transmission path. The energy consumption balance and the load balance are closely related with this performance metric.

3. **Goodput**: The goodput is the QoS performance metric. The Goodput is expressed as an amount of communication data successfully received at the destination within the required communication time.

4. **Throughput**: Throughput is the rate of successful message delivery over a communication wireless channel in industrial IoT.

5. **Download time from server (DLTS)**: Download time from server is measured in the emulations to reflect the time delay performance from server of different computing frameworks in MIoT.

We compared the proposed method ATOP with the following methods:

1. *SPND*, wherein the source node and the destination node choose the shortest path using a method such as Floyd or Dijkstra algorithm. The path is composed of MIoT NDs, and the node with maximal communication range is chosen.

2. *SPCN*, wherein as in SPND, the shortest path among NDs and CHs is chosen.

3. *CR*, wherein every node chooses its maximal rate node.

2.4.2.2.3 Evaluation Results

Average time delay: The average time delay of different methods for their best performance in terms of time delay and MIoT structure is presented in Figure 2.20, which demonstrates the average time delay increases with the increase of data amount for all methods. The performance metrics of average time delay for 100 and 200 NDs are presented in Figure 2.22a and b, respectively. Figure 2.20 shows that the average time delay increases with the raise of communication data amounts for the four methods. However, it is obvious that ATOP outperforms the other schemes in this metric with different data volumes. It is due to the optimization routing path of ATOP that selects the CHs as relay points with the higher communication rate and more communication resources. CR and SPCN achieve better performance than SPND, because CHs are in their communication path. Because NDs have the higher delay time during the communication, SPND method has the highest values for different data amounts as shown in Figure 2.20. When the data amount is equal to 700

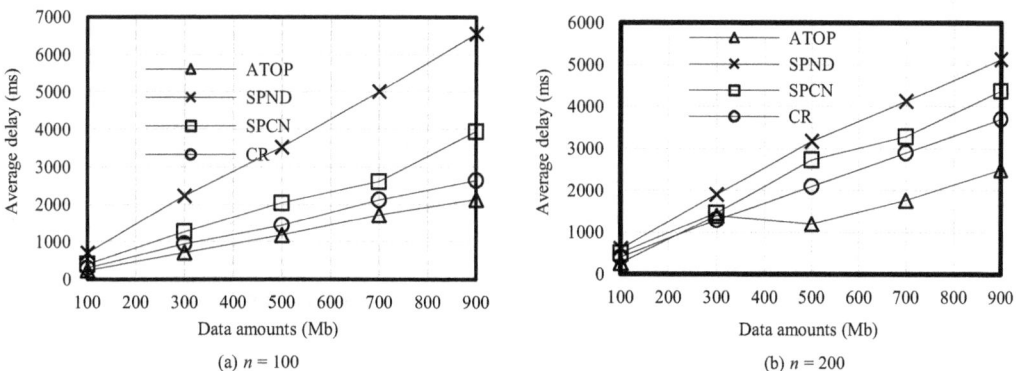

FIGURE 2.20   Comparison of average time delay: (a) $n=100$ and (b) $n=200$.

Mb, ATOP achieves the smallest average time delay; in other words, our methods have reduced the time consumption. In the same way, ATOP shows better performance than other methods. Compared to Figure 2.20a for 100 NDs, ATOP has a smaller value of average time delay than the others, and by increasing the number of nodes, more NDs could join in the communication among CHs.

PDD: The PDD performance in different ways for the same number of NDs and CHs is presented in Figure 2.21. In general, PDD of every strategy will increase with the increasing of deadline and ND number. In Figure 2.21a, with decreasing of deadline, PDD of ATOP also decreases; in other words, the smaller the value of deadline, the more routing paths will be deleted from the candidate path set. When the deadline is larger than 1,000 ms, the PDD is larger than 0.3. And if deadline is 1,500 ms, the PDD of ATOP will reach 0.5. However, the other methods achieve the value near zero, especially for SPND and SPCN, because they usually select the same path without considering the path difference. Smaller PDD indicates that more different paths are adopted to forward the information. Namely, ATOP provides more balanced load and energy consumption routing path; therefore, by employing the proposed method, it can reduce energy unbalance and prolong network life. In addition, PDD results are obtained for different numbers of NDs in Figure 2.21b. Similarly, the other methods have the PDD value near zero. The PDD of ATOP increases with the increase in the number of NDs due to more selected paths. Specifically, when ND number is 200, PDD of ATOP is 30 and 15 times of SPCN and CR, respectively.

Goodput: This performance metric is a useful index for evaluating how successful the data receiving from the source is with the deadline constraint. The goodput for different deadline levels and methods is presented in Figure 2.22a. In Figure 2.22a, it is obvious that when the deadline reaches different levels, goodput of the proposed method still has 100% successful receiving data rate, because the ATOP selects the most effective routing path and the method increases the transmitted power in order to get high communication rate. As presented in Figure 2.22a, CR and SPCN obtain

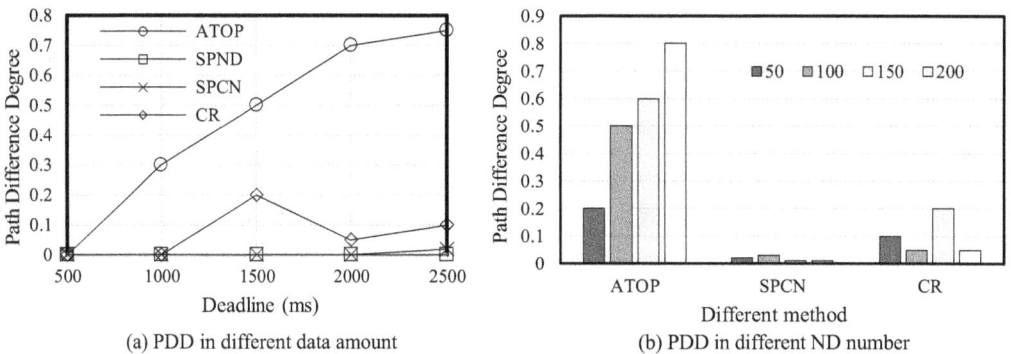

(a) PDD in different data amount (b) PDD in different ND number

FIGURE 2.21 Comparison of average PDD: (a) PDD in different data amount and (b) PDD in different ND numbers.

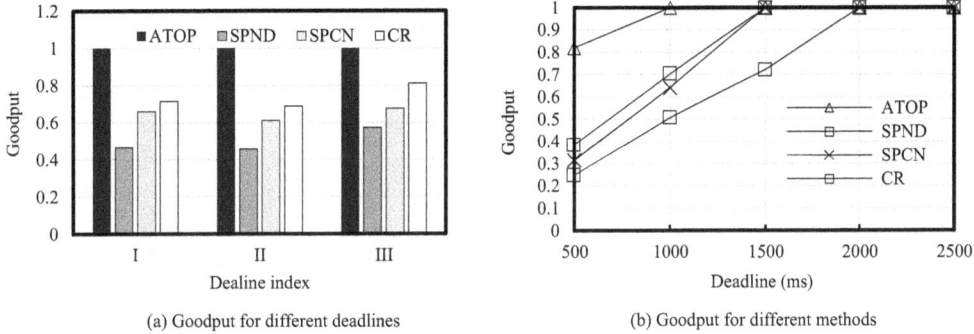

(a) Goodput for different deadlines

(b) Goodput for different methods

FIGURE 2.22 Comparison of goodput performance: (a) Goodput for different deadlines and (b) goodput for different methods.

good goodput as CH takes part in data transmission. Moreover, it is clear that SPND obtains the smallest value because it adopts the NDs to construct the routing path. The curve of goodput and different deadlines is presented in Figure 2.22b. With the increasing deadline, the performance of goodput raises for all methods. Moreover, by using ATOP, the success receiving data rate is higher than 80%, when the deadline is more than 500 ms. Hence, the ATOP can adapt to different deadline levels. The other methods are questionable in urgent communication.

Throughput: The performance of throughput in different data amounts with different methods is presented in Figure 2.23a, demonstrating that there are slight variations of throughput in different traffic load levels. In Figure 2.23a, we can observe that ATOP takes the highest throughput so as to choose the most effective path to forward the data. The average throughputs of ATOP, CR, SPCN, and SPND are 0.43, 0.35, 0.25, and 0.15 Mbps, respectively. In other words, ATOP throughput gets the max value in different data amounts. The proposed strategy provides the best performance of throughput and then followed by CR and SPCN. Due to the data rate limitation of NDs, the SPND shows the worst performance in terms of throughput.

DTLS: We use DTLS to evaluate the delay performance from server in different computing frameworks. DTLS in our proposed edge MIoT (EMIoT) framework and cloud computing architecture is presented in Figure 2.23b with different data amounts. We assume that the computational capacity of cloud computing server (CCS) is three times greater than EMIoT, and ND will spend three hops to download the data from cloud server. Further, CCS and our proposed framework have identical communication rate (1Mbps). According to Figure 2.23b, with the increase of data amount, the average delays increase in both frameworks (CCS and Edge MIoT). From another perspective, the EMIoT demonstrates better performance than CCP. Namely, NDs consume less time for data transmission in EMIoT. Specially, when traffic amount reaches 3,000 Mb, the ED MIoT framework reduces more than 40% of downloading time from server compared to CCS in light of direct connection between edge server and NDs. In addition, the EMIoT framework represents better solution for industrial applications. Lastly, the proposed framework achieves the best performance in terms of DTLS.

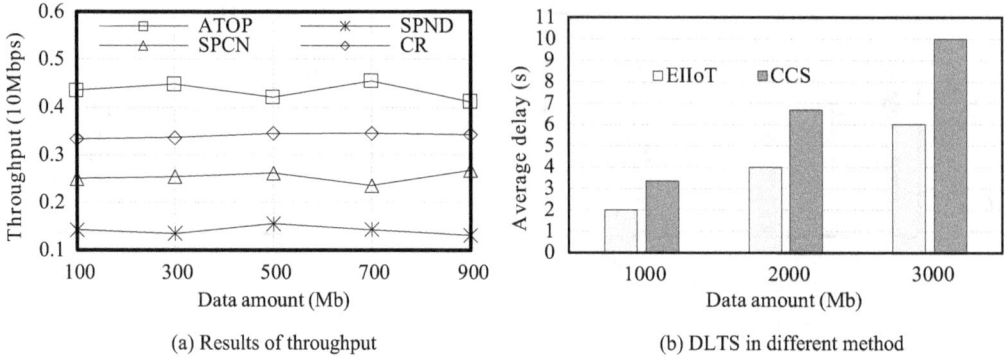

(a) Results of throughput

(b) DLTS in different method

FIGURE 2.23 Comparison of throughput and CODT in different levels: (a) results of throughput and (b) DLTS in different methods.

2.4.3 Intelligent Production Edges Design for Customized Manufacturing

The development of Industry 4.0 has provided the possibility to meet frequent changes in product type and batches, a sharp decline in the delivery cycle, constraints of quality cost, and other relevant parameters of customized production mode. Intelligent manufacturing, as a core of Industry 4.0, represents a deep integration of new IT technologies such as MIoT and Service-oriented architecture, as well as manufacturing process. To realize the intelligent manufacturing, an intelligent production edge is designed to provide the traditional devices with the abilities of data access and self-decision-making in this subsection [42].

2.4.3.1 IoT-Based Intelligent Production Edges Design

To meet the requirements for the presented architecture, we introduce the intelligent production edges (IPEs) into the system and equip them with the intelligence of edge decision-making and communication. Figure 2.24 shows the IPE framework. The IPE includes the data discoverer module, data converter module, instruction receiver module, and strategy maker module. The modules are described in the following.

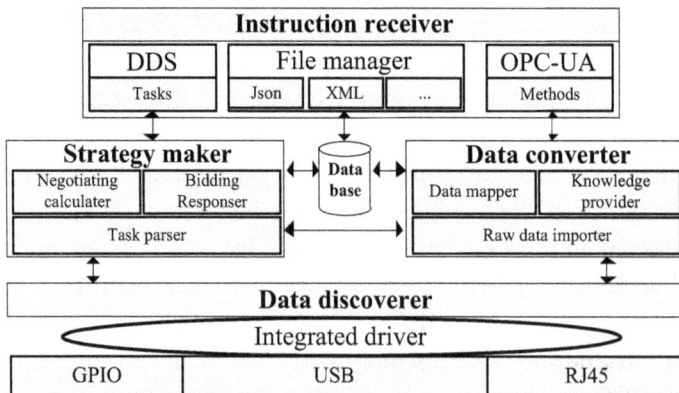

FIGURE 2.24 The IPE framework.

2.4.3.1.1 Hardware and Software Design   The data discoverer module is used to connect the IPE with the field devices so as to collect the data and publish the instructions. Moreover, the data discoverer module has three types of hardware driver interfaces, namely, GPIO, RJ45, and USB, which can communicate adaptively with the devices supporting different network types. The GPIO driver can connect various types of auxiliary devices that support only the GPIO pin communication, such as photoelectric switches and ultrasonic sensors. The RJ45 driver is used to connect the devices that support the TCP/IP communication, such as the robots and RFID readers. The USB driver is often used for communication after hardware conversion. In the wired-based communication, the USB is switched over to RS232/RS485 for serial communication to connect devices that support Modbus RTU and other protocols. As for wireless communication, USB is switched over to ZigBee/WiFi/Bluetooth to provide communication between devices in wireless sensor networks (WSNs). At the same time, the data discoverer module can match the relevant devices and realize network connections using the driver detection program. Because the hardware part has the reserved SPI and I2C interfaces, the devices supporting other third-party self-defined buses can develop their own drivers voluntarily for plug-and-play software and hardware adaptation. Based on such a design, the IPE, as the adapter of the architecture, can upload data from the field devices supporting different communication protocols and different hardware interfaces to the cloud. The cloud can publish the management and control instructions generated to various field devices for real-time control.

The OPC UA and DDS are used as communication methods between the cloud and the IPE for data acquisition and production task publication. Because the information model of the OPC UA provides data with the semantic meaning, the IPE needs to be able to interpret the information model. The data converter module is designed to map the raw data of the field devices to the variable values of the OPC UA information model objects. When uploading the data, the data converter module leaves the timestamps on the field device data collected by the data discoverer module. Then, it reads the information model and associates the data with the object variable values in the information model.

The DDS is chosen as a task publication method because it supports the publisher/subscriber mode, which coincides with the negotiation mechanism mentioned in Section 2.4. The strategy maker module is responsible for receiving task strategy files from the cloud through the DDS and parsing the task to generate the corresponding control instructions that are sent to the field devices for logical control. Moreover, the strategy maker module can also calculate the matching degree of the task information according to the field device capabilities.

The instruction receiver module, as an interface for the communication between the IPE and the cloud, is an OPC UA server and also a DDS node. It can respond to the calls of methods that are the service mechanism of the OPC UA and task queries that are the service mechanism of the DDS, and then return the results to the cloud after the service is completed. Therein, there are two types of services. One is updating information model mapping tables and the other is updating strategy files. Usually, because the two protocols cannot transfer files, when the above two services are called, it is necessary to set a URL address to accommodate the mapping table and the strategy file, and then the IPE can obtain the relevant files.

2.4.3.1.2 Interoperability between Cloud and IPE   In order to enable the manufacturing system to reconfigure the production logic by changing the interaction mode of different elements, it is necessary to integrate operation technology (OT) and information technology (IT) in the manufacturing process, that is, to solve the issue of interoperability between the field devices and cloud.

The OPC UA, as a current mainstream information middleware in the industry, is a platform-independent service-oriented communication architecture that can support the data transmission and semantic information modeling. Its information modeling framework supports the integration of information models and protocols, and it can model and transmit the semantic data directly by means of a user-specified data format. Ontology is a philosophical conception that systematically describes objective things, and it aims to capture the knowledge of relevant fields and define the objects, which forms the semantic basis for interoperability between field devices and cloud. The ISA95 standard is used to define the composition and the operation in different manufacturing levels, and its data model is compatible with IT and OT in the industrial field. Therefore, in this paper, the ISA95 device model is integrated with the OPC UA and ontology to form a common data model of the cloud and field devices and to establish a communication method that can satisfy the consistency of communication protocols in the IT layer and the heterogeneity of the communication protocols in the OT layer.

In general, traditional field devices in a production line are made by different device manufacturers and support various different communication protocols, including the TCP/Modbus, RTU/Bluetooth, and others; therefore, it is very difficult to integrate information obtained from different devices. The data discoverer module of the IPE can access third-party devices. As shown in Figure 2.25, the raw data importer acquires raw data such as an I/O signal in the field devices, and then, the data mapper maps the raw data into the information model presented in Figure 2.26 using the ISA95 device model mapping method. The knowledge provider uses the corresponding relationship between the ontology model and the information model to infer and obtain the information from a device by means of the semantic inference engine, and then exposes the device

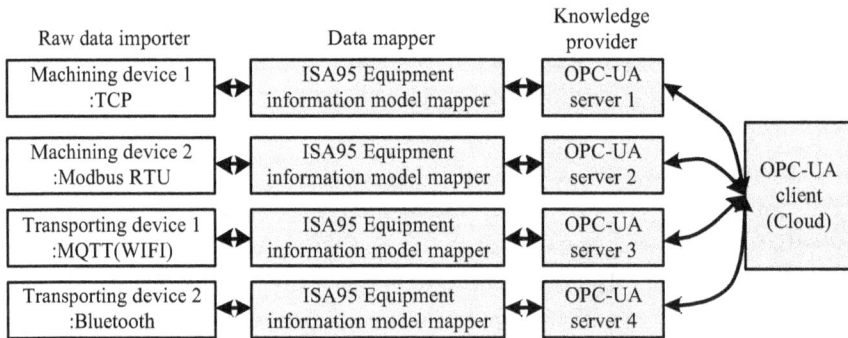

FIGURE 2.25   Data conversion from raw data to the OPC UA information model by mapping the model of the ISA95 device.

```
Information Model
Displayname
  ∨ 🗀 Objects
      ∨ 🗀 ISA95ObjectType
          ∨ 🗀 EquipmentType
              ▣ AssetAssignment
              ∨ 🍀 MachiningDevice1
                  > 🍀 ControlSystem
                  > 🍀 ElectricalSystem
                  ∨ 🍀 MechanicalSystem
                      > 🍀 Switch
                      > 🍀 Guiderail
                      > 🍀 Motor
                      ▣ DeviceStatus
                  > 🍀 ToolSystem
```

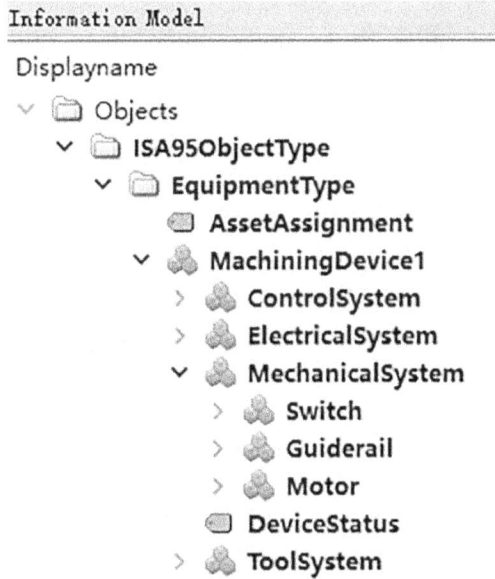

FIGURE 2.26   The information model of the OPC UA server based on ISA95 device model mapping method.

information model to the network via the OPC UA server. So the cloud can obtain the information of all the devices supporting different communication protocols from the OPC UA client.

The hierarchical structure of the information model is presented in Figure 2.26. The *ISA95ObjectType* is the basic model of the ISA95 from which all object types are derived; *EquipmentType* refers specifically to the type of device and third-party devices. For instance, machining device 1 is an instance of *EquipmentType*. Machining device 1 is composed of tool modules, electrical modules, control modules, mechanical modules, and other components. Additionally, it has the status property to indicate whether the current devices are available or faulty. These modules can be subdivided into circuit systems, various cutting tools, feedback systems, status properties, etc.

As for a production line, whether a product can be processed or not is determined by the device status in the process-related devices chain, not only by the failure of a certain device. So the information about all the devices should be combined to draw a reasonable conclusion. The CASOMA-IPE integrates the edge intelligent technology with the ontology technology to endow an IPE with the ability of decision-making and self-diagnosis based on the knowledge base. In the ontology model, because the upper object contains the lower component and property, the *Hascomponent* (Hc) relationship is established to represent the relationship between the upper and lower objects. Objects have not only the relational properties but also the status properties, and hence the *Hasproperty* (Hp) relationship is established to represent the relationship between upper objects and lower status properties.

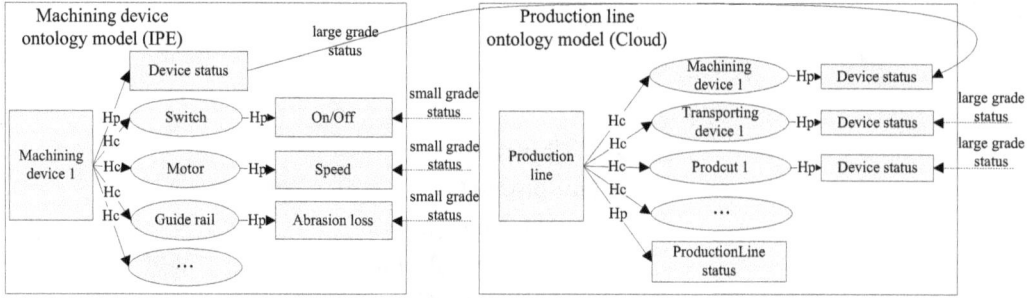

FIGURE 2.27 The mapping relationship between the machining device ontology model and the production line ontology model.

According to the ontology modeling principles, a machining device ontology model can be obtained as presented in Figure 2.27. The model consists of Switch, Motor, Guide rail, and so on. And these mechanical modules have their properties, including Speed, Abrasion loss, and On/Off. Before the machining device accepts a task, its status should be checked to judge whether it has the processing capacity or not. Thus, the following rules should be obeyed:

Rule 1: If Guide_rail && Abrasion_loss > "5%", Then Device_status = "*MillingFault.*"

Rule 2: If Machining_device_1 && Device_status = "*MillingFault*", Then Productionline_status = "*No accept Milling task*".

Rule 1 shows that, if the abrasion loss of guide rails is above 5%, the machining device 1 is in the condition of milling fault. Rule 2 shows that, if the device status of the machining device 1 is "*milling fault*," the production line will not accept the milling task. Therefore, the IPE can obtain the device status through Rule 1, and the cloud can obtain the production line status through Rule 2.

Figure 2.27 presents the mapping relationship between the machining device ontology model and the production line ontology model. The ontology model consists of two parts: the first part is the machining device model, whose input is a small grade status; the second part is a production line model, whose input is a large grade status of the production line. The real-time data uploaded from the field devices is input as a small grade status to update *Device status* in the machining device model and infer the current device status according to the device model rules. Moreover, in order to judge whether a product can be processed or not, the device status as a large grade status is input into the production line model to update *ProductionLine status* and obtain the production capacity of the current production line after the second round of the semantic inference. Accordingly, the judgment is the basis for producing the corresponding planning and scheduling strategies. For machining device 1, the ontology model and the information model are identical regarding the hierarchical structure and property status. Therefore, it is possible to infer device status from the device ontology model to update the device information model. Additionally, the data sources for the two models can be derived from the same table in the database.

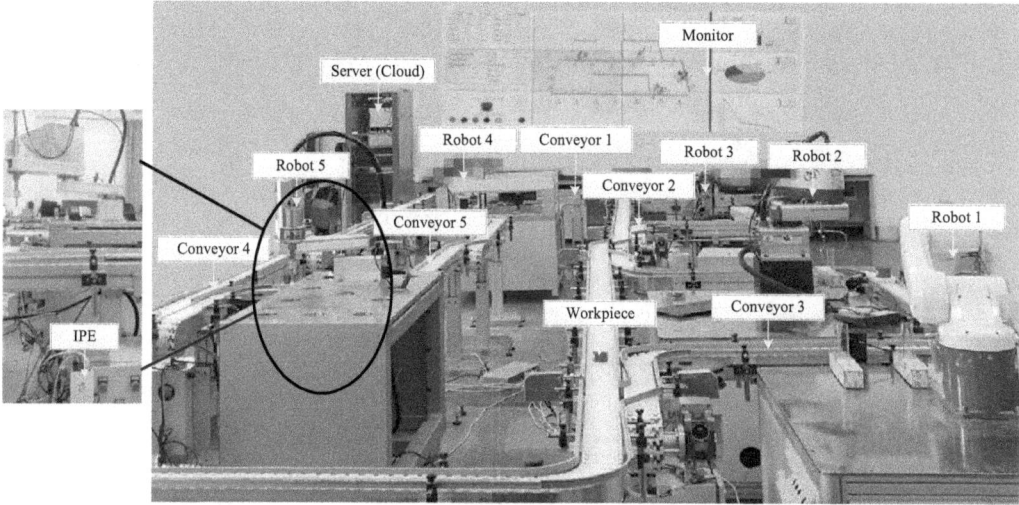

FIGURE 2.28   The experimental setup.

2.4.3.2 Experimental Verification

In order to verify the adaptability of the IPE and the reconfiguring function of the CASOMA-IPE production logic, two groups of experiments were conducted. The hardware used in the experiments included a server that served as the cloud, five robots used for machining simulation, five PLC-controlled conveyor belts used for transporting the workpieces, ten microcomputers (Raspberry Pi) that were used as MIoT-based IPEs, and a router. The IPEs were connected via Ethernet through the router. The experimental setup is shown in Figure 2.28.

For conveyor belts, robots, and sensors, the data was provided in different methods of traditional communication, including the Modbus RTU protocol, TCP/IP protocol, and GPIO. As a data bridge, the IPE provided the ability of browsing (reading and writing) the raw data from the cloud; thus, the interoperability was realized between the cloud and traditional devices. The data acquisition experiment was conducted to verify the IPE adaptability. The conveyor belts and robots were connected with the IPEs as OPC UA servers. Five sets of parameters {100, 200, 300, 400, 500} were the raw data quantity of each device. Besides, the cloud acted as an OPC UA client for data browsing.

In the experiment, the average data browsing time $T_{\text{totalexec}}$ was used as the evaluation criteria of the data acquisition performance:

$$T_{\text{totalexec}} = \frac{1}{n} \times \sum_{i=1}^{n} \sum_{j=1}^{m} T_{ij} \tag{2.59}$$

where T_{ij} represented the execution time when the ith IPE read/wrote the jth raw data to the device and responded to the OPC UA client, while $T_{\text{totalexec}}$ represented the average browsing time when the raw data of all devices were browsed by the OPC UA client.

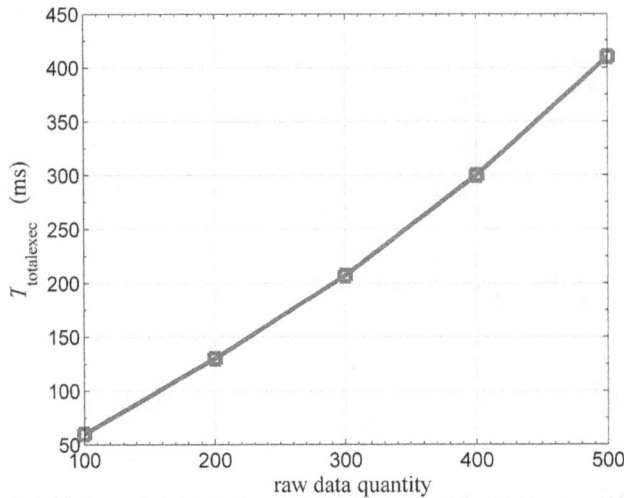

FIGURE 2.29 The data acquisition result.

Figure 2.29 shows the relationship between the raw data quantity and the average data browsing time of ten devices. With the increase in the raw data quantity, the average data browsing time tended to rise gradually. Moreover, the time spent on browsing a single data increased with the increase in the number of data. The reason was that when browsing more data, the memory consumption of the OPC UA protocol stack increased, resulting in the increase of the memory recovery time.

In the case of a device with 500 raw data, a browsing cycle of less than 450 ms could be maintained in the cloud. Such a speed of data browsing could ensure the timeliness of communication between the cloud and IPEs, and this data transmission architecture made the system compatible with the traditional devices.

2.5 SUMMARY

Edge computing, by bridging the gap between resource and network, clearly self-defines its characteristics promoting its usage and benefits into various areas. The MIoT which is potentially targeting an increase of information now grows to conquer and transform companies and organizations in the world around. Edge intelligence in customized manufacturing is of consideration which is very well discussed in this chapter taking the features of MIoT for enhancing Industrial operations and peripherals with advanced features.

REFERENCES

[1] Y. Zhang, C. Jiang, B. Yue, J. Wan and M. Guizani, "Information fusion for edge intelligence: A survey," *Information Fusion*, vol. 81, pp. 171–186, 2022.
[2] V. Hayyolalam, M. Aloqaily, Ö. Özkasap and M. Guizani, "Edge intelligence for empowering IoT-based healthcare systems," *IEEE Wireless Communications*, vol. 28, no. 3, pp. 6–14, 2021.
[3] V. Gupta and S. De, "Collaborative multi-sensing in energy harvesting wireless sensor networks," *IEEE Transactions on Signal and Information Processing over Networks*, vol. 6, pp. 426–441, 2020.

[4] H. Dai, H. Wang, G. Xu, J. Wan and M. Imran, "Big data analytics for manufacturing internet of things: Opportunities, challenges and enabling technologies," *Enterprise Information Systems*, vol. 14, no. 9–10, pp. 1279–1303, 2020.

[5] S. M. Noghabaei, R. L. Radin, Y. Savaria and M. Sawan, "A high-sensitivity wide input-power-range ultra-low-power RF energy harvester for IoT applications," *IEEE Transactions on Circuits and Systems*, vol. 69, no. 1, pp. 440–451, 2022.

[6] A. Aalerud, J. Dybedal and D. Subedi, "Reshaping field of view and resolution with segmented reflectors: Bridging the gap between rotating and solid-state LiDARs," *Sensors*, vol. 20, no. 12, p. 3388, 2020.

[7] R. I. Ansari, C. Chrysostomou, S. A. Hassan et al., "5G D2D networks: Techniques, challenges, and future prospects," *IEEE Systems Journal*, vol. 12, no. 4, pp. 3970–3984, 2018.

[8] M. Fu, J. Liu, H. Zhang and S. Lu, "Multisensor fusion for magnetic flux leakage defect characterization under information incompletion," *IEEE Transactions on Industrial Electronics*, vol. 68, no. 5, pp. 4382–4392, 2021.

[9] X. Wang, Y. Han, C. Wang, Q. Zhao, X. Chen and M. Chen, "In-edge AI: Intelligentizing mobile edge computing, caching and communication by federated learning," *IEEE Network*, vol. 33, no. 5, pp. 156–165, 2019.

[10] M. Satyanarayanan, "The emergence of edge computing," *Computer*, vol. 50, no. 1, pp. 30–39, 2017.

[11] M. D. Hssayeni and B. Ghoraani, "Multi-modal physiological data fusion for affect estimation using deep learning," *IEEE Access*, vol. 9, pp. 21642–21652, 2021.

[12] W. Shi, J. Cao, Q. Zhang, Y. Li and L. Xu, "Edge computing: Vision and challenges," *IEEE Internet of Things Journal*, vol. 3, no. 5, pp. 637–646, 2016.

[13] H. M. Shakir and J. Karimpour, "Systematic study of load balancing in fog computing in IOT healthcare system," *2021 International Conference on Advanced Computer Applications (ACA)*, Maysan, Iraq, 2021, pp. 132–137.

[14] Y. He, N. Zhao and H. Yin, "Integrated networking, caching, and computing for connected vehicles: A deep reinforcement learning approach," *IEEE Transactions on Vehicular Technology*, vol. 67, no. 1, pp. 44–55, 2018.

[15] H. Lu, Q. Liu, D. Tian, Y. Li, H. Kim and S. Serikawa, "The Cognitive internet of vehicles for autonomous driving," *IEEE Network*, vol. 33, no. 3, pp. 65–73, 2019.

[16] N. Abbas, Y. Zhang, A. Taherkordi and T. Skeie, "Mobile edge computing: A survey," *IEEE Internet of Things Journal*, vol. 5, no. 1, pp. 450–465, 2018.

[17] H. Li, K. Ota and M. Dong, "Learning IoT in edge: Deep learning for the Internet of Things with edge computing," *IEEE Network*, vol. 32, no. 1, pp. 96–101, 2018.

[18] M. Simsek, A. Aijaz, M. Dohler, J. Sachs and G. Fettweis, "5G-enabled tactile internet," *IEEE Journal on Selected Areas in Communications*, vol. 34, no. 3, pp. 460–473, 2016.

[19] E. Bastug, M. Bennis and M. Debbah, "Living on the edge: The role of proactive caching in 5G wireless networks," *IEEE Communications Magazine*, vol. 52, no. 8, pp. 82–89, 2014.

[20] L. Liu, X. Zhang, M. Qiao and W. Shi, "SafeShareRide: Edge-based attack detection in ride-sharing services," *2018 IEEE/ACM Symposium on Edge Computing (SEC)*, Seattle, WA, USA, 2018, pp. 17–29.

[21] R. Yang, F. R. Yu, P. Si, Z. Yang and Y. Zhang, "Integrated blockchain and edge computing systems: A survey, some research issues and challenges," *IEEE Communications Surveys & Tutorials*, vol. 21, no. 2, pp. 1508–1532, 2019.

[22] G. Li, G. Kou and Y. Peng, "A group decision making model for integrating heterogeneous information," *IEEE Transactions on Systems, Man, and Cybernetics: Systems*, vol. 48, no. 6, pp. 982–992, 2018.

[23] T. Rausch and S. Dustdar, "Edge intelligence: The convergence of humans, things, and AI," *2019 IEEE International Conference on Cloud Engineering (IC2E)*, Prague, Czech Republic, 2019, pp. 86–96.

[24] S. Qiu, L. Liu, Z. Wang et al., "Body sensor network-based gait quality assessment for clinical decision-support via multi-sensor fusion," *IEEE Access*, vol. 7, pp. 59884–59894, 2019.

[25] D. Zhang, N. Vance and D. Wang, "When social sensing meets edge computing: Vision and challenges," *2019 28th International Conference on Computer Communication and Networks (ICCCN)*, Valencia, Spain, 2019, pp. 1–9.

[26] L. Bragilevsky and I. V. Bajić, "Tensor completion methods for collaborative intelligence," *IEEE Access*, vol. 8, pp. 41162–41174, 2020.

[27] Y. Kang, J. Hauswald, C. Gao et al., "Neurosurgeon: Collaborative intelligence between the cloud and mobile edge," *ACM SIGARCH Computer Architecture News*, vol. 45, no. 1, pp. 615–629, 2017.

[28] B. Chen, J. Wan, A. Celesti, D. Li, H. Abbas and Q. Zhang, "Edge computing in IoT-based manufacturing," *IEEE Communications Magazine*, vol. 56, no. 9, pp. 103–109, 2018.

[29] D. Xu, T. Li, Y. Li et al., "Edge intelligence: Empowering intelligence to the edge of network," *Proceedings of the IEEE*, vol. 109, no. 11, pp. 1778–1837, 2021.

[30] J. Wan, B. Chen, M. Imran et al., "Toward dynamic resources management for IoT-based manufacturing," *IEEE Communications Magazine*, vol. 56, no. 2, pp. 52–59, 2018.

[31] C. Jiang, J. Wan and H. Abbas, "An edge computing node deployment method based on improved k-means clustering algorithm for smart manufacturing," *IEEE Systems Journal*, vol. 15, no. 2, pp. 2230–2240, 2021.

[32] X. Li and J. Wan, "Proactive caching for edge computing-enabled industrial mobile wireless networks," *Future Generation Computer Systems*, vol. 89, pp. 89–97, 2018.

[33] C. Jiang and J. Wan, "A thing–edge–cloud collaborative computing decision-making method for personalized customization production," *IEEE Access*, vol. 9, pp. 10962–10973, 2021.

[34] X. Li, J. Wan, H.-N. Dai, M. Imran, M. Xia and A. Celesti, "A hybrid computing solution and resource scheduling strategy for edge computing in smart manufacturing," *IEEE Transactions on Industrial Informatics*, vol. 15, no. 7, pp. 4225–4234, 2019.

[35] J. Wan, B. Chen, S. Wang, M. Xia, D. Li and C. Liu, "Fog computing for energy-aware load balancing and scheduling in smart factory," *IEEE Transactions on Industrial Informatics*, vol. 14, no. 10, pp. 4548–4556, 2018.

[36] B. Chen, J. Wan, Y. Lan, M. Imran, D. Li and N. Guizani, "Improving cognitive ability of edge intelligent IIoT through machine learning," *IEEE Network*, vol. 33, no. 5, pp. 61–67, 2019.

[37] J. Wan, J. Li, Q. Hua, A. Celesti and Z. Wang, "Intelligent equipment design assisted by Cognitive Internet of Things and industrial big data," *Neural Computing and Applications*, vol. 32, pp. 4463–4472, 2020.

[38] S. Wang, J. Wan, D. Li and C. Liu, "Knowledge reasoning with semantic data for real-time data processing in smart factory," *Sensors*, vol. 18, no. 2, p. 471, 2018.

[39] X. Wang and J. Wan, "Cloud-edge collaboration-based knowledge sharing mechanism for manufacturing resources," *Applied Sciences*, vol. 11, no. 7, p. 3188, 2021.

[40] H. Yan, J. Yang and J. Wan, "KnowIME: A system to construct a knowledge graph for intelligent manufacturing equipment," *IEEE Access*, vol. 8, pp. 41805–41813, 2020.

[41] X. Li, D. Li, J. Wan, C. Liu and M. Imran, "Adaptive transmission optimization in SDN-based industrial internet of things with edge computing," *IEEE Internet of Things Journal*, vol. 5, no. 3, pp. 1351–1360, 2018.

[42] H. Tang, D. Li, J. Wan, M. Imran and M. Shoaib, "A reconfigurable method for intelligent manufacturing based on industrial cloud and edge intelligence," *IEEE Internet of Things Journal*, vol. 7, no. 5, pp. 4248–4259, 2020.

CHAPTER 3

Heterogeneous Networks in Smart Manufacturing Factory

INDUSTRIAL NETWORKS PROVIDE INFRASTRUCTURE for the full interconnection of human, machine, material, and other production factors and promote the full flow and seamless integration of industrial data. However, the diverse quality of service (QoS) requirements, massive intelligent devices, and complex communication means lead to the coexistence of wired and wireless heterogeneous networks in smart factories, which brings great challenges to the realization of high interconnection, deep integration, and dynamic reconfiguration of smart factories. This chapter analyzes the cause, classification, and current solution framework of smart factory heterogeneous network, with an emphasis on network QoS optimization objectives, including low latency and high reliability, network load balancing, and high security and privacy protection.

3.1 OVERVIEW OF HETEROGENEOUS NETWORKS IN SMART MANUFACTURING FACTORY

With the deep integration of the new-generation information and communication technologies with advanced manufacturing technologies, such as the IoT, industrial Internet, cloud computing, and AI technology, the phenomenon of large-scale intelligent device access, wired/wireless network coexistence, diversified network QoS requirements, and complex network topology is emerging in smart factories, which seriously limits the improvement speed of digitalization, information, and intelligence in smart factories. The details are described as follows [1]:

- Large-scale intelligent devices in the factory mainly include intelligent production equipment (such as intelligent processing centers and additive manufacturing equipment), intelligent detection equipment (like machine vision inspection equipment and industrial endoscopes), intelligent logistics equipment (automatic guided vehicle, truss manipulators, and suspended conveyor chains), VR/AR virtual assembly

DOI: 10.1201/9781003460992-3

equipment, industrial robots, intelligent sensors, and network infrastructure. These devices from different manufacturers and suppliers follow different communication protocols, and their openness and intelligence levels are also different. Industrial wired networks typically utilize twisted pairs, coaxial cables, and optical fibers. Twisted pairs are commonly used in Ethernet networks: 10-Mbps Ethernet, 100-Mbps Ethernet, 10-Gbps Ethernet, and 100-Gbps Ethernet. The transmission rate of optical fiber has reached several to dozens of Gbps, and the transmission band of coaxial cables can reach up to 1 GHz. In comparison, industrial wireless networks work in the frequency range from 3 Hz to 3,000 GHz, where cellular networks generally adopt the frequency ranges of 800–900 or 1,700–2,600 MHz. Addition to differences in bandwidth, transmission rate, and stability, wired and wireless heterogeneous networks are also different from the physical layer and data link layer. Specifically, the physical layer includes Gauss Frequency Shift Keying (GFSK), Differential Quadrature Phase Shift Keying (DQPSK), Orthogonal Frequency Division Multiplexing (OFDM), Quadrature Amplitude Modulation (QAM), and other technical means. And the MAC layer includes multiple access methods, such as TDMA, CDMA, Frequency Division Multiple Access (FDMA), Orthogonal Frequency Division Multiple Access (OFDMA), Non-Orthogonal Multiple Access (NOMA), Carrier Sense Multiple Access with Collision Avoidance (CSMA/CA), and Carrier Sense Multiple Access with Collision Detection (CSMA/CD).

- The industrial network is an important infrastructure for the high correlation between factory equipment and the interconnection and internetwork between the equipment and the industrial cloud. Industrial wired networks present the advantages of high reliability and fast transmission speeds. Fieldbus technologies, such as Modbus and PROFInet, have been widely used in factory automation and in multiple industrial processes. Industrial wireless networks, with the advantages of low energy consumption and self-organized network, can reduce costs and simplify the factory infrastructure, which have been rapidly developed and generally applied within the industrial field. However, the topology of wireless network is vulnerable to power loss, mobility, and channel fading of wireless nodes, as well as radio frequency interference, high humidity, vibration, and dust in industrial environment. Therefore, wireless communication networks cannot fully replace wired networks in every industrial field; wired and wireless heterogeneous networks coexist in smart factories.

- As shown in Table 3.1, the requirements of various business applications in smart factories are far different in terms of latency, bandwidth, reliability, security, mobility, and the number of connections, and each business application has complex requirements. For example, industrial automation control requires about 10 ms end-to-end latency and extremely high safety and reliability. High-precision real-time positioning in a complex industrial environment requires the fusion of varieties of different positioning technologies in a short time. Applications such as online monitoring of equipment health and performance are required to process massive data and generate

TABLE 3.1   Diversified Network QoS Requirements in Smart Factories

	Data Size (bytes)	Periodicity (ms)	Network Synchronization	Transmission Requirements	Jitter Tolerance	Packet Loss Tolerance	Importance of Message Timeliness
Real-time and synchronous data	20~100	0.1~2	Yes	Deterministic	0	—	High
Periodic data	50~1,000	2~20	No	Latency	<latency	1~4 frames	High
Events	100~1,500	—	No	Latency	—	Yes	High
Network control	50~500	—	No	Bandwidth	—	Yes	Medium
Configuration diagnosis	500~1,500	50~1,000	No	Bandwidth	—	Yes	Low
Audio/video	1,000~1,500	—	No	Latency	—	Yes	Low

active preventive maintenance strategies in a short time. AR services need to use ultra-high bandwidth networks, which are more than 1,600 Mbps, and the energy meter reading business needs to provide massive connections through the network. Therefore, it is impossible to meet all current or future application needs through only one single network.

- Each network communication product is supported by a group of specific suppliers or organizations, and different manufacturers have different certifications and authorizations, network compatibility and interoperability, and access control for the products they operate. Taking TSN as an example, Beckhoff released the TSN bridge communication module EK1000; National Instruments (NI) has produced several controllers integrating the TSN technology; SERCOS released Rexroth motion control system based on TSN switch bridging; SIEMENS released PROFInet; and Mitsubishi released CC-link IE TSN; Manufacturers such as Huawei, Cisco, Moxa, and Hirschmann have also released TSN switch products. Therefore, with the gradual elimination and transformation of products and the continuous optimization of market costs, it will be unrealistic to assume a universal and unified industrial environment from a technical perspective.

3.2 CLASSIFICATION AND KEY TECHNOLOGIES OF HETEROGENEOUS NETWORKS IN SMART MANUFACTURING FACTORY

For smart factories, it is widely understood that wired and wireless heterogeneous networks must permeate the smart factory for transmitting data, commands, and other information between the cloud and the equipment, including both machines and products. The development of industrial networks is as shown in Figure 3.1 [2]. In all cases, networks play an important role in the implementation of Industry 4.0 [3]. In other words, networks play a role similar to the human body's nervous system, including industrial wired networks, industrial wireless networks (IWNs), and power line carrier (PLC) communication.

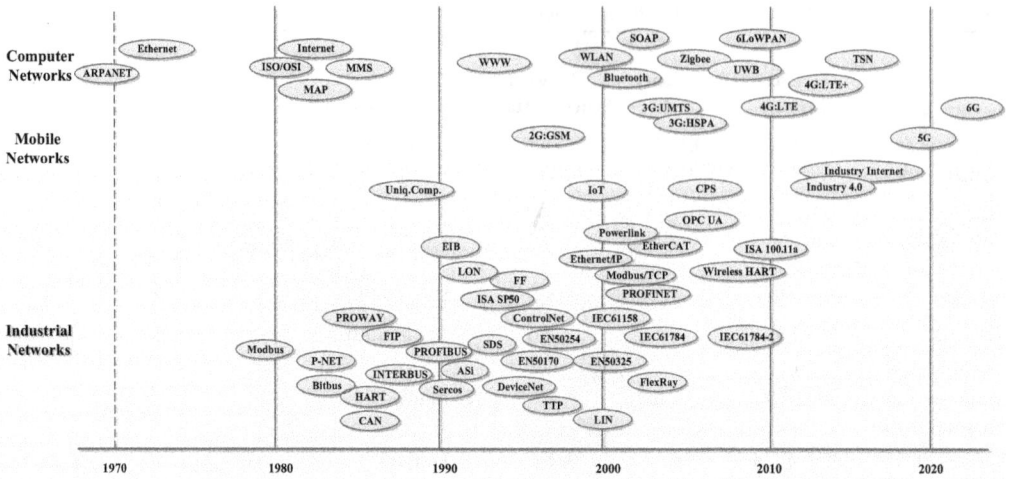

FIGURE 3.1   Milestones achieved toward industrial network development [2].

3.2.1 Industrial Wired Networks

3.2.1.1 Fieldbuses

Fieldbus technology has been developing for nearly 50 years, there are more than twenty bus standards in the International Electrotechnical Commission (IEC), including PROFIBUS, CC-Link, Modbus-RTU, and INTERBUS, and their communication rate varies from several kbps to tens of kbps. As the most mature field-level system network, Fieldbus networks mainly provide data communication support between field sensor devices and controllers, controllers and actuators, or controllers and input and output control substations. Thomesse et al. [4] and Sauter et al. [5] reviewed in detail the Fieldbus network, including its origin and development, standardization process, service capability, and further analyzed its feasibility in the fields of factory automation, process automation, building automation, and transportation system.

However, there are many problems with Fieldbus technology, such as low communication ability, short distance, and poor anti-interference ability. Moreover, more than 250 manufacturers worldwide have more than 1,500 products supporting the PROFIBUS standard alone. The technical routes, protocol specifications, and technical indicators adopted by different manufacturers are different. The openness and compatibility of the bus technology is insufficient, which increasingly affects the interconnection and intercommunication between the associated devices and systems.

3.2.1.2 RTEs

RTE networks can meet the requirements of clock synchronization and time deterministic between drivers and other field devices by modifying or adding time scheduling and clock synchronization protocol mechanisms based on the IEEE 802.3 Ethernet standard. Up to now, most RTE solutions have been standardized by the IEC. They can be broadly classified into three categories: Ethernet based on the standard TCP/IP protocol stack, Ethernet based on standard protocols, and Ethernet based on real-time improvement. Industrial Ethernet methods based on the TCP/IP protocol stack can deal with non-deterministic communication factors through reasonable control. The typical protocol includes Modbus/ TCP and Ethernet/IP. Industrial Ethernet based on the standard Ethernet protocol, which can be considered a soft RTE network, utilizes standard Ethernet hardware and specific Ethernet frame to transmit data. The typical protocols include EtherCAT and PROFINET RT. Real-time improved Ethernet is a hard real-time solution that uses hardware to control the communication between real-time and non-real-time channels. Real-time MAC communication is used in real-time channels to avoid message conflict. At the same time, the original protocol is carried out in non-real-time media, which can achieve real-time performance with delays more minor than one microsecond. The typical protocols include SErial Realtime COmmunications System III (SERCOS III), and PROFInet IRT.

With high-efficiency communication ability and flexible network topology, the market share of RTE networks has exceeded that of the Fieldbus networks, becoming the preferred communication protocol for automation and control systems. Surveys [6–8] have compared and analyzed several industrial real-time communication solutions, such as PowerLink, SERCOS III, and EtherCAT, by considering their QoS performance and hardware requirements and by examining their relative advantages and disadvantages. Felser et al. [9] summarized multiple solutions for Ethernet standardization and described typical network protocols in detail, including protocol architectures, application models, and network topologies. However, despite various RTEs being unified at the physical layer, the technical routes to solve real-time performance are different. Each RTEs have formed network systems with their own independent configuration, forming automation islands with each other, which has the problem of interconnection and interoperability. In addition, it also leads to the fact that the standard Ethernet cannot be directly connected with the RTEs of various automation systems, which limits the expansion and general applicability of RTE technology.

3.2.1.3 TSN

In the traditional network, the data flow is transmitted in a best-effort service mode, which inherently lacks determinism and real-time and cannot meet the factory's requirements for data reliability. Moreover, when time-sensitive and non-time-sensitive data flow are transmitted in the same network, the non-time-sensitive data flow will affect the scheduling and transmission of the time-sensitive data flow. Thus, the transmission delay of time-sensitive data streams cannot be guaranteed. The traditional methods modify the Ethernet network protocol or deploy independent dedicated Ethernet network devices in the critical production processes to meet the real-time communication requirements of industrial applications (e.g., machine control, process control, and robot control). However, these approaches cannot guarantee network interoperability, scalability, and compatibility, making them unsuitable for future smart factories. Based on the traditional Ethernet network, TSN establishes network clock synchronization, traffic scheduling, and other mechanisms to make it have strong interconnection and high reliable low-latency transmission capability, which has become the evolution direction of the next-generation industrial network.

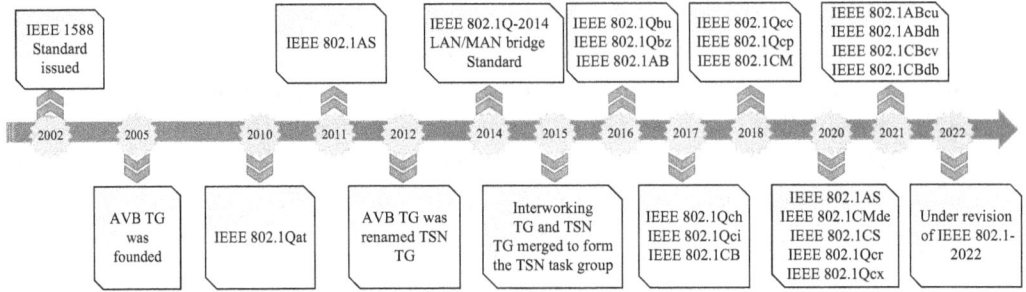

FIGURE 3.2   Timeline describing the release of the various TSN-related standards.

The TSN is a technology that represents an extension of Audio and Video Bridging (AVB) to automotive, mobile communication, and general industry. The formulation, testing, promotion, and product certification of TSN standards are jointly promoted by organizations, including the AVnu Alliance, IIC, and OPC UA Foundation. A timeline describing the release by the IEEE organization of the TSN-related standards is illustrated in Figure 3.2. Finn et al. [10] and Messenger et al. [11] described the latest progress toward TSN with bounded low latency and zero congestion loss and pointed out the upcoming challenges in the TSN standardization process. Bello and Steiner [12] comprehensively introduced the specific standards and projects of the TSN, discussed the applicability of each criterion in various industrial applications, and provided essential opinions and directions of TSN development. To further improve and expand the application scale of TSN technology, multiple organizations such as the Deterministic Network (DetNet), the IEC/IEEE 60802, and the Field-Level Communication (FLC) working groups have actively participated in constructing general architecture, formulating interoperability standards and specifications, and the integration of OPC UA and TSN. Organizations such as the Labs Network Industrie 4.0 (LNI4.0), the Industrial Internet Consortium (IIC), the Edge Computing Consortium (ECC), and the Alliance of Industrial Internet (AII) Industry have actively promoted the construction of a TSN testbed. At present, the mainstream automation manufacturers have released their TSN products or test products, including Beckhoff's EK1000 module, SIEMENS's PROFInet over TSN, and Mitsubishi's CC-Link IE TSN. Moreover, Cisco, Huawei, Moxa, and Hirschmann vendors have also issued their TSN switch products.

TSN defines the transmission mechanism of time delay sensitive data, including clock synchronization, data flow scheduling, load shaping, and resource reservation. It can provide deterministic transmission services with bounded low delay, low jitter, and extremely low data loss rate for applications, and provide a more unified and universal architecture and standard for heterogeneous real-time Ethernet. The IEEE 802.1AS and IEEE 802.1 AS-Rev define the precision time synchronization mechanism, where the precision error is microsecond or even nanosecond. By inserting a timestamp containing local clock information into the data frame, the clock synchronization of each network node is ensured to meet the deterministic transmission of time-sensitive flows in complex application scenarios. The IEEE 802.1Qav standards define the queuing and forwarding mechanism, which forwards and queues time-sensitive data flow. The IEEE 802.1Qbv classifies Ethernet network data flow into different types. Each type is a constraint for accessing the network at a specific time, granting a priority policy in transmitting real-time data. The IEEE 802.1Qbu+IEEE 802.3br standards define a preemptive mechanism that can interrupt the transmission of standard Ethernet or jumbo frames, to transmit high-priority frames without discarding the messages that were previously interrupted. In TSN, the peak traffic is cached by switch devices, and the stable transmission and end-to-end transmission delay of the time-sensitive data flow are ensured through traffic shaping and other mechanisms.

TSN provides a set of MAC layer protocol standards for the Ethernet to ensure the reliability and real-time transmission of data in complex network environments. At present, RTEs have a large number of installed nodes around the world, and their compatibility and interworking with TSN is of great practical significance. Mainstream RTE technology organizations have put forward integration schemes with TSN, such as PROFInet over TSN [13], CC-link IE TSN [14], and EtherCAT TSN communication profile [15]. These integration schemes are compatible with TSN at the data link layer and retain their own application rules, enabling seamless integration with the corresponding automation systems. Therefore, it can be predicted that with the improvement of integration technologies between control protocols of various manufacturers and TSN, various real-time Ethernet controllers and I/O network segments can be physically connected in a common TSN network, and several independent real-time streams can be mapped to different RTEs' application layers through the configuration of TSN switch.

3.2.2 IWNs

Compared with industrial wired networks, IWNs are an important foundation for realizing intelligent manufacturing. They support the mobility requirements of industrial field equipment (such as mobile robots, AGV, and UAV), improve the flexibility and management efficiency of the network, and ease to carry out information transformation and upgrading of traditional manufacturing enterprises. At present, wireless technologies such as Industrial Wireless Sensor Networks (IWSNs), Radio Frequency Identification (RFID), Bluetooth Low Energy (BLE), ZigBee, and WiFi HaLow have been applied and popularized in factory information, industrial non-real-time control, and data acquisition. In addition, wireless technologies are also penetrating into the field of industrial real-time control and becoming a powerful supplement and alternative to the traditional industrial wired control network. For example, the fifth-generation communication (5G) has clearly regarded industrial control as an important application scenario with low delay and high reliability.

3.2.2.1 IWSNs

IWSNs are composed of sensors, wireless communication modules, middleware, and network management. Sensors can collect data from various devices and items in factories, wireless communication modules can achieve wireless transmission of data, middleware can process and analyze data, and network management can achieve management and control of the entire industrial wireless sensor network.

IWSNs are suitable for industrial applications such as industrial intelligent monitoring with the advantages of self-organization, rapid deployment, low cost, and inherent intelligent processing. Currently, there are four approved international standards for IWSNs based on the IEEE 802.15.4 standard, including the ZigBee developed by the Connectivity Standards Alliance, the WirelessHART standard developed by the HART Communication Foundation, the ISA100.11a wireless standard developed by the International Association of Automation (ISA), and the Wireless network for Industrial Automation-Process Automation (WIA-PA) developed by China. Since the four IWSN standards differ in network architecture, protocol characteristics, network security, and application scenarios, several surveys/review literatures have comprehensively analyzed and compared them [16–20].

3.2.2.2 RFID

RFID technology is complementary to the IWSNs and be applied to logistics management, production process monitoring, quality control, and maintenance, among other areas. The RFID system consists of tags, readers, and middleware. The tag is the core of the RFID system, containing a chip and an antenna that can store and transmit information about the item. The reader communicates wirelessly with the tag to obtain the stored information. Middleware is the core software of the RFID system, processing and analyzing the information obtained by the reader to achieve logistics management, production process monitoring, and quality control.

In smart factories, RFID technology can achieve real-time monitoring and management of the logistics process. By identifying and tracking items through RFID tags, item tracking and traceability can be achieved, improving logistics efficiency and accuracy. At the same time, RFID technology can achieve real-time monitoring and management of the production process. By identifying and tracking items in the production process, real-time monitoring and control can be achieved, improving production efficiency and product quality [21]. In addition, RFID technology can also achieve functions such as quality control and maintenance, improving the reliability and stability of the production process. Overall, RFID technology can help enterprises achieve intelligent production and management, improving production efficiency and product quality.

However, RFID technology also has some disadvantages in smart factories. First, the cost of RFID tags is relatively high, and further cost reduction is needed for large-scale applications. Second, the coverage of RFID technology is limited, and many readers need to be installed within the factory, increasing the deployment and maintenance costs of the system. In addition, RFID technology also has some security and privacy issues that need to be addressed.

3.2.2.3 BLE

BLE technology is a wireless communication technology designed for low power consumption and short-range communication. It presents the advantages of centralized low power consumption, media access control, strong anti-interference ability, low cost, and ease of deployment [22,23]. BLE-based wireless sensor networks can be used to monitor and collect data on various parameters, such as temperature, humidity, and vibration, enabling real-time monitoring of production processes and equipment. BLE-based smart tags can also be attached to goods and materials, enabling real-time tracking and tracing of goods throughout the production process and supply chain.

The Bluetooth mesh network is a BLE network topology for many-to-many device communication, which meets the network reliability, scalability, and security in smart factory but does not support real-time communication. To overcome this limitation, Patti et al. [24] proposed a real-time protocol for industrial wireless mesh networks based on BLE. Using the TDMA method and optimizing transmission allocation, this protocol provided real-time support for data packets while ensuring a good balance between the maximum guaranteed latency and effective throughput. Leonardi et al. [25] developed a multi-hop real-time BLE (MRT-BLE) protocol that allowed bounded latency and priority support in mesh networks. In addition, the communication distance of BLE is relatively short, making it unsuitable for large-scale production facilities. And BLE security is also a concern, as it can be vulnerable to hacking and data breaches if not properly secured.

3.2.2.4 ZigBee

ZigBee is a wireless communication technology that is specifically designed for low-power, low-data-rate applications. ZigBee devices operate in the 2.4 GHz unlicensed frequency band, enabling multiple devices to share the same wireless medium without interference. ZigBee also supports mesh networking, enabling devices to communicate with each other through intermediate devices, which improves the overall network coverage and reliability. Additionally, ZigBee also has a relatively low cost, making it an attractive option for industrial applications.

However, there are also some challenges and limitations associated with ZigBee technology. For example, the communication distance of ZigBee is relatively short, making it unsuitable for large-scale production facilities. Additionally, ZigBee security is also a concern, as it can be vulnerable to hacking and data breaches if not properly secured.

3.2.2.5 WiFi HaLow

In smart factories, the WiFi network, a wireless technology based on the IEEE 802.11 standard, supports wireless interconnection of up to 256 devices and enables data transmission and Internet access. However, factors such as metal structures and device operation in factories can interfere with WiFi signals, leading to decreased signal quality and limited coverage range. Additionally, a large number of devices connecting to the network may also result in insufficient network capacity. WiFi HaLow, adopting the IEEE 802.11ah technology, is considered a long-distance, low-power WiFi suitable for industrial IoT applications, supporting low-power connections required by applications such as sensor networks and wearable devices. Compared with WiFi, WiFi HaLow uses the 900 MHz frequency band, which is more suitable for devices with small data loads and low power consumption.

WiFi HaLow can cover a maximum range of 1 km when the antenna is suitable, and the signal is stronger and not easy to be interfered. In addition, each node in the WiFi HaLow network can communicate with two or more neighboring nodes with higher reliability. In [26], the authors conducted a comprehensive review and analysis of the existing research on WiFi HaLow from various perspectives such as target characteristics, research objectives, applicable scenarios, advantages, and disadvantages, and introduced related hardware prototypes, products, and simulators. Lee et al. [27] developed the world's first full system on chip (SoC) including a digital baseband and an analog/radio frequency transceiver and verified the feasibility of WiFi HaLow in long-distance, high-throughput, and low-power IoT solutions. However, although WiFi HaLow can provide long-distance, low power consumption, and high-security wireless connections, WiFi HaLow has a weak networking ability and few network nodes, which cannot directly connect to the cloud and are not suitable for multi-point control.

3.2.2.6 5G and Beyond

The 5G network, as the key enabling technology of industrial Internet, becomes an important part of the infrastructure of the factories. 5G provides ultra-high data rate, ultra-large terminal connections, and ultra-low latency for industrial control applications, such as machine vision and remote control, and meet the demands of flexible mobility, wireless connection, accurate control, and efficient cooperation of terminal equipment in the industrial control.

The Third Generation Partnership Project (3GPP), an umbrella term for multiple standards organizations that develop mobile telecommunication protocols, has developed and maintained the 5G New Radio (NR) and related 5G standards, including 5G-Advanced. According to the 5G specification released or planned by 3GPP, Rel-18 is moving into focus, and Rel-15, Rel-16, and Rel-17 have been frozen. The 5G Alliance for Connected Industries and Automation (5G-ACIA) is the center and global forum for solving, discussing, and evaluating the related technical, regulatory, and business issues of the 5G in the industrial field. The 5G-ACIA has been working to integrate industrial 5G systems into factories according to the I4.0 principles, and has released the white paper: 5G for industrial Internet of things (IIoT): capabilities, features, and potential, to further promote the application of 5G in industrial automation. The white paper described 5G capabilities related to IIoT applications, including support for Ethernet integration, TSN, and security in nonpublic networks, and discussed how the vertical industry accesses 5G technology to boost the development of IIoT deployment and spectrum access.

The certified 5G core technologies contain SDN/NFV, network slicing, NR, low latency, edge/cloud computing, and service orchestration. SDN is a new network architecture that decouples the network control plane and forwarding plane to simplify network management and control. It has the advantages of centralized management, agile expansion, and fine-grained control. It is a potential solution for smart factory applications such as cross-network fusion and scheduling, resource dynamic management and control, routing optimization and configuration, energy-saving, and secure transmission.

To meet the needs of differentiated business processing capabilities and diverse network functions in smart factories, network slicing, supported by SDN and NFV, is considered a reasonable solution. Network slicing virtualizes multiple end-to-end networks based on the common hardware through slicing technology, dynamically deploy network services of the end-to-end network on-demand, and ensure that the slices do not affect each other. Network resources are allocated on-demand through a network slicing method, and solutions for different time delay, energy consumption, network mobility, connection density, and connection cost are provided. To provide low-latency and highly reliable network services, it is necessary to construct critical transaction slices, mobile broadband slices, and large connection slices. The crucial transaction slice deploys the user data function module in the local data center and optimizes the network connection with local traffic diversion, which reduces the delay as much as possible, simultaneously ensuring the real-time control and response to the production. Moreover, the optimization of 5G for business requirements is reflected in different functional characteristics of network slicing and the flexible deployment of the slicing scheme. Based on distributed cloud computing and NFV, network function modules inside the slice are flexibly deployed in multiple distributed data centers according to the business requirements, providing customers with shared or isolated access resources, transmission resources, cloud resources, and other infrastructure resources.

With the wide application and deployment of 5G in industrial fields such as factory automation, smart grid, video detection, cloud-based robots, and intelligent logistics, communication technology will be further integrated with cloud computing, big data, and artificial intelligence to promote the evolution and development of mobile communication technology in the direction of the next-generation mobile communication system (6G), which is becoming the next focus of science and technology strategic competition among large countries in the future. Table 3.2 summarizes the strategic layout of 6G R&D carried out by governments around the world. There are also a large number of scholars in academia who have reviewed and explored 6G network specifications, network architecture, network requirements, and enabling technologies [28–32]. Compared with 5G,

TABLE 3.2   The Strategic Layout of 6G R&D by Governments around the World

Country	Time	Contributions
Finland	2018	Launch the 6G flagship project of "6genesis—supporting 6G wireless intelligent society and ecosystem"
	2019	The world's first 6G summit was held and the world's first 6G white paper was released, which promoted the start of 6G research
United States	2020	The American Telecommunications Industry Solutions Alliance (ATIS) released the 6G action proposal
	2020	The Federal Communications Commission (FCC) of the United States opened the 95 GHz–300 GHz terahertz bands as the experimental spectrum and officially launched the research and development of 6G technology
South Korea	2019	The science and information and communication technology of Korean Ministry (MSIT) held a 6G forum and officially announced the start of 6G research
	2020	MSIT released the future mobile communication R&D strategy leading the 6G era
Japan	2020	The world's first "6G Technology Comprehensive Strategic Plan Outline and Roadmap" with 6G as the national development goal and initiative was released
The European Union	2020	Several related strategies such as "Shaping Europe's Digital Future," "European New Industrial Strategy," and "2030 Digitalization Guide" have been launched successively
	2021	The 6G flagship project Hexa-X was launched
China	2019	The China IMT-2030 (6G) Promotion Group was established
	2021	The "White Paper on the Overall Vision and Potential Key Technologies of 6G" was released

6G provides a greater connection, higher rate and reliability, lower latency, wider coverage, and more intelligence. It is used to realize various applications, including intelligent machines communicating with each other and their digital twins, huge sensor networks, high trust, and security. However, 5G and 6G are still in the research and experimental stage in automation control and machine-to-machine (M2M) communication, and there is still a long way to apply to smart factories.

3.2.3 PLC

Smart grid includes all the intelligent infrastructure in smart factory energy supply, which is essential to achieve energy-saving production. Although the traditional Fieldbus, the Ethernet, and the wireless Fieldbus are available for smart grids, they are not cost-effective. PLC utilizes the existing power line channel to transmit data, which has the advantages of no rewiring, easy maintenance, and no blocking by metals and walls. The PLC network is used to combine a multitude of sensors and IoT devices, representing a feasible solution for smart factories [33–35].

The PLC network uses the power line as an information transmission medium to load the modulated high-frequency carrier signals for voice or data transmission. According to the voltage level, PLC is split into three categories: 35 kV and above high-voltage PLC, 10 kV medium-voltage PLC, and 380/220 V low-voltage PLC. The high-voltage PLC has a relatively single carrier channel, and the technology is mature. While the low-voltage PLC has a complicated network structure and a poor carrier channel, which is difficult to study and apply. Additionally, PLC is also separated from the carrier signal frequency and bandwidth. Specifically, it is divided into the narrowband PLC (NBPLC, conforming to IEEE P.1901.2 standard) and the broadband PLC (BPLC, conforming to IEEE p.1901.1 standard). The NBPLC has a frequency bandwidth from 10 kHz to 500 kHz. It is comparable to the Fieldbus such as CAN, PROFIBUS DP, PROFIBUS PA, Fieldbus Foundation, and Modbus, as well as the wireless Fieldbus like WirelessHART in terms of performance, which is used as the communication infrastructure for non-real-time industrial applications, e.g., intelligent instrument, automatic meter reading, intelligent measurement, remote monitoring, and alarm. Moreover, the combination of the NBPLC network and wireless communication improves the diversity, availability, and reliability of communication. The BPLC has a frequency bandwidth from 2 to 20 MHz and a communication rate of more than 1 Mbps, providing reliable data transmission in a short time and applies to such real-time industrial applications as real-time meter reading, remote real-time monitoring, and channel monitoring and management.

However, the power line, different from the ordinary data communication line, was originally used for the transmission of power rather than data. The data transmission channel of PLC is unstable, characterized by significant noise and severe signal attenuation. Free and insulator discharge in high-voltage circuits will generate considerable noise, and the low-voltage PLC channels are usually disturbed by impulse noise, synchronous or asynchronous noise, radio broadcasting, and other noise. Therefore, it is critically necessary to adopt reasonable modulation and demodulation technology to realize reliable, stable, and high-speed data communication. As the network technologies such as spread spectrum communication, OFDM, network self-organization, network reconfiguration, and network routing are gradually used in the PLC, the channel coding, modulation algorithm, and power control of PLC have been improved, while the multipath propagation, frequency fading, and noise interference have also been effectively weakened. G3-PLC and PRIME are typical PLC technical schemes based on OFDM modulation, effectively improving the anti-interference ability of PLC and achieving reliable data transmission in a harsh industrial communication environment. In addition, line impedance is another challenge for PLC. When the channel of high-voltage PLC has defects or capacitive load, the impedance value of the carrier channel will change, resulting in communication interruption. While the load switch on the low-voltage PLC is random, the extensive line impedance difference leads to the instability of the channel. Consequently, the carrier devices with a fixed impedance are impossible to be used in smart factories.

PLC technology, the mainstay of the smart grid, enables to collect and transmit the power equipment signals accurately and efficiently and analyze the operation state of the power grid accurately. Nevertheless, with the extensive use of motors, robots, and other equipment in the industrial environment, the reliability of PLC communication is affected by the power grid topology and channel noise. It is still unclear how to further decrease the bit error rate of channel coding and improve signal transmission stability and anti-interference ability. The dual-mode communication combined with PLC technology and micro power wireless communication is expected to solve the carrier signal attenuation and signal island and will become the next development direction. The standard of dual-mode communication technology is currently being developed. Hence, the primary development trend of PLC technology is from NBPLC to BPLC, and then to dual-mode communication at present and in the next few years.

3.3 SDN- AND EC-BASED HETEROGENEOUS NETWORKS FRAMEWORK FOR SMART FACTORIES

The massive data, deep integration, and diverse QoS requirements of smart factories have put forward higher requirements for the flexibility, real-time, and scalability of factory network systems, and the gateway-based smart factory network architecture is also undergoing revolutionary changes. To achieve low latency, high flexibility, and configurable network environment in smart factories, an SDN-based heterogeneous network architecture, which separates manufacturing data forwarding and network control, is shown in Figure 3.3. The system, with a central SDN network controller, can be seen as a north–south trend. The southbound interface is connected to the underlying intelligent devices through the edge intelligent gateway (EIG), which implements flexible forwarding of manufacturing data. The northbound interface connects to the cloud platform via the specific APIs to upload and download the data. It can also be used for deployment of various network application services [36].

The EIG plays an important role as a middleware that connects intelligent devices with the SDN controllers and switches. First, the EIG provides access to the smart devices and plays the role of shielding heterogeneity of the system, which is a prerequisite for

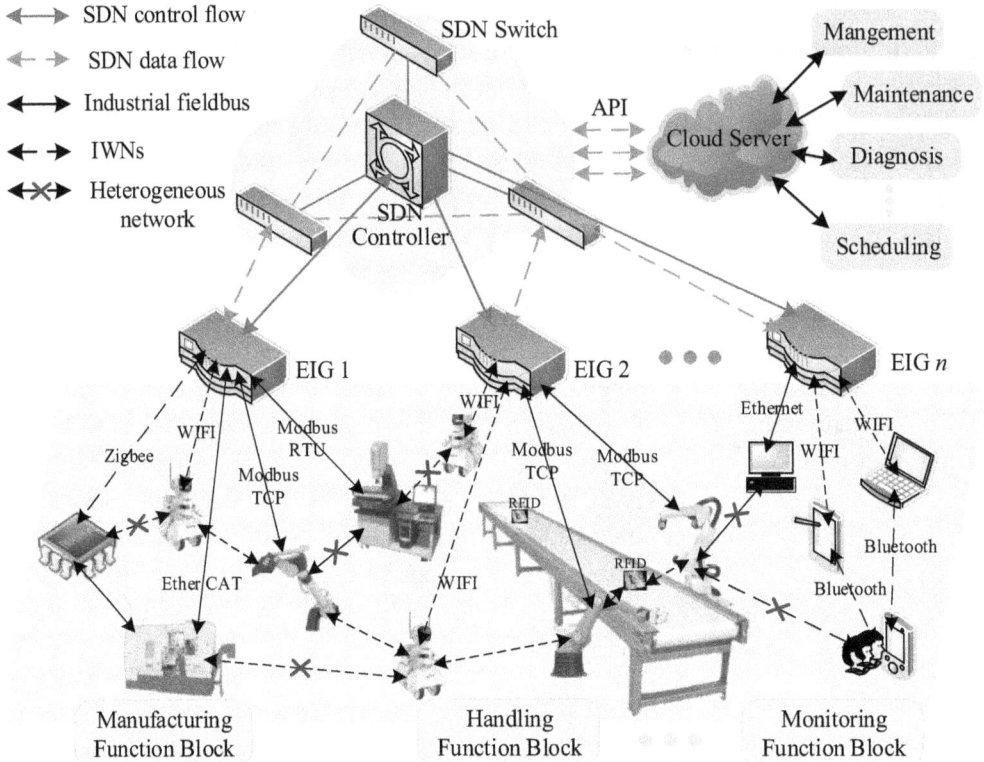

FIGURE 3.3   Heterogeneous network architecture based on SDN and EC.

centralized control and unified forwarding of multi-source manufacturing data. Second, the EIG has computational and storage capabilities, which provide a basis for complex network operations such as data processing, data caching, and bandwidth allocation for different data flows. In addition, the EIG role as a multi-protocol interface makes it compatible and scalable, which is of great significance for complex and heterogeneous manufacturing systems [3,37].

Traditional manufacturing systems are highly coupled and heterogeneous, which is detrimental to efficient cross-network fusion of manufacturing data and network reconstruction. The concept of modularity is utilized for dividing the manufacturing workshop into different functional blocks, where each functional block is made up of several underlying devices to form a relatively functional manufacturing subsystem [2]. Each functional block is connected to the SDN controller via an EIG to achieve information interaction between the subsystem and the global system.

Data transmission between devices in the system is a highly dynamic information interaction process, which must follow flexible and unified data forwarding rules. The main operation process of this system is as follows: (1) The devices in each function block upload data to the EIG, which collects and integrates the subnet device data and transmission parameters, including source device ID, target device ID, data transfer volume, and latency requirement; (2) the SDN controller obtains the data transmission information from the EIG and forwards the corresponding data through the SDN switches to the target devices. Of course, an information and state mapping table for each function block device and EIG needs to be created and maintained in advance in the SDN controller; and (3) the SDN controller schedules network resources according to the actual operating conditions and real-time requirements of data flows to realize an efficient cross-network fusion of multi-source manufacturing data. To better understand the system, the framework is simplified into East–West flow (IIoT), North–South flow (SDN), and Computing plane (edge computing).

3.3.1 East–West Flow Plane

In a smart factory, the east–west data flow usually refers to the communication between various devices, robots, sensors, etc. within the factory. These devices are usually intelligent and can be connected and communicated through the network to achieve collaborative and optimized production lines. Specifically, the east–west data flow in a smart factory has the following characteristics:

1. **Large amount and variety of data**: Smart factories involve a large number of devices, robots, sensors, etc., and each device generates a large amount of data, including device status, production data, and quality data. These diverse types of data need to be transmitted and processed through the network.

2. **High speed and efficiency**: The devices in a smart factory are usually high speed and efficient, capable of achieving fast data transmission and processing. For example, communication between robots needs to respond in real-time, and the data collected by sensors needs to be processed in real-time.

3. **Various network connection methods**: The devices in a smart factory are usually connected and communicated through wired or wireless networks, including Ethernet, WiFi, Bluetooth, and other methods. These connection methods need to be selected and configured according to the characteristics and requirements of the devices.

4. **High availability and reliability requirements**: In a smart factory, the east–west data flow is usually a critical component supporting the production line, requiring high availability and reliability to ensure the continuity and stability of the production line.

5. **Security requirements**: The east–west data flow in a smart factory usually involves sensitive information such as production data and device status, requiring a high level of security to ensure data confidentiality and integrity.

In summary, the east–west data flow in a smart factory is an important component in achieving intelligent manufacturing and smart factories, requiring consideration of a large amount and variety of data, high speed and efficiency, various network connection methods, high availability and reliability, and security requirements.

3.3.2 North–South Flow Plane

In a smart factory, the north–south data flow usually refers to the data flow from the factory to the enterprise's internal system or cloud platform, such as production data, equipment status, and quality data that need to be transmitted to the enterprise's internal system or cloud platform for processing and analysis. These data usually need to be transmitted through the internet or dedicated lines and need to consider data security and service quality factors. Specifically, the north–south data flow in a smart factory has the following characteristics:

1. **Large amount and variety of data**: Smart factories involve a large amount of production data, equipment status, quality data, etc. These data are diverse and need to be collected, transmitted, and processed.

2. **High speed and efficiency required**: The north–south data flow usually needs to transmit and process data in real-time or near real-time to support real-time monitoring and decision-making in the enterprise's internal system and cloud platform.

3. **High availability and reliability required**: The north–south data flow is usually a critical component supporting decision-making in the enterprise's internal system and cloud platform, requiring high availability and reliability to ensure data continuity and stability.

4. **Network performance needs to be considered**: Because the north–south data flow usually needs to go through multiple network devices for forwarding and processing, network performance such as bandwidth, delay, and other indicators need to be considered.

5. **Data security needs to be considered**: Because the north–south data flow usually involves sensitive data such as production data and equipment status, data security measures such as encrypted transmission and firewalls need to be considered.

In summary, the north–south data flow in a smart factory is an important component in achieving intelligent manufacturing and smart factories. In SDN subsystem, an open source SDN controller, such as NOX and POX controller, Ryu controller, Floodlight controller, SDN Open Network Operating System (ONOS), OpenContrail controller, Helium controller, and Open Mul controller, is able to be adopted in the smart factory. The SDN controller relates to key network devices, such as cluster heads. Network nodes deliver the status parameter to the SDN controller. It is clear that these key nodes are amounted with controller interface, thus they establish a link with SDN controller using the southbound interface protocol, such as the OpenFlow protocol, OpenFlow Configuration Protocol (OF-Config), NETCONF, Open vSwitch Database Management Protocol (OVSDB), Path Computation Element Protocol (PCEP), Interface to the Routing System (I2RS), eXtensible Messaging and Presence Protocol (XMPP), and OpFlex protocol. At the same time, the SDN controller may provide information (e.g., equipment utilization rate) with the application layer through the northbound interface and API.

3.3.3 Computing Plane

To get a reasonable assignment of computing resources, the edge computing servers (ECSs) are deployed in smart factories. Downward, these ECSs establish the link between nodes and SDN controller; upward, ECSs are connected to the cloud. Every ECS is typically miniature data and computing center that is in the vicinity of nodes and SDN devices to derive the results in real-time and reduce the time consumption and traffic load. In an industrial environment, the smart factory architecture based on SDN and EC in heterogeneous networks may provide three kinds of services, i.e., data collection, data transmission, and data processing, as follows:

1. **Data collection service**: In this layer, the applications are designed according to the provided API, and the data format (e.g., data type, and data attributes) may be re-customized, thus allowing for adaptation to different application requirements. For example, for the same equipment, different users may focus on different parameters, which may be accomplished by dynamically configuring the data format.

2. **Data transmission service**: Wireless or wired networks are used to forward the perception data to the industrial cloud or transmit data from one control node to other control nodes. The interaction among these devices forms the foundation of a new intelligence design approach, which allows, for example, designs of interactive mechanisms for avoiding deadlocks or deployment the system resources according to real-time requirements of different applications. All these problems may be approached from the perspective of configuring the data transmission service.

3. **Data processing service**: Figure 3.4 shows the data transmission and data processing. For software-defined data processing, the data attributes and application requirements should be considered to determine how to deal with the data. For example, an industrial robot with laser navigation carries out the path planning, which involves a large amount of data processing. If these data are uploaded to the cloud and then the

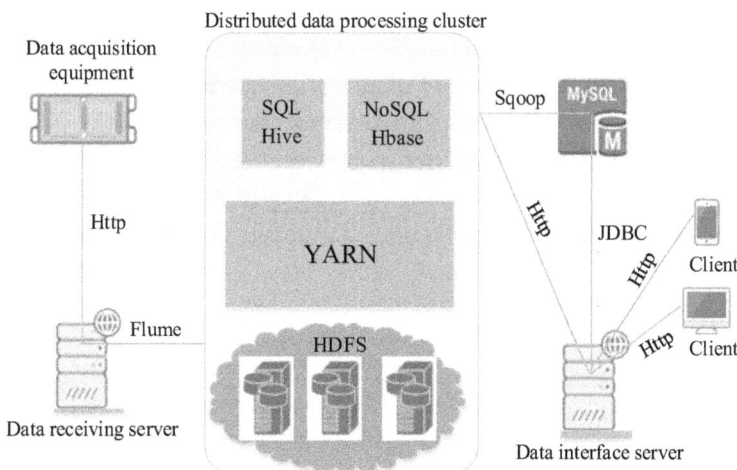

FIGURE 3.4 Information processing of heterogeneous networks in smart factories.

path information is computed, considerable computing resources may be saved and hardware costs can be correspondingly reduced. Therefore, software-defined data processing may provide more flexibility.

3.4 AI-ENABLED QOS OPTIMIZATION OF HETEROGENEOUS NETWORKS IN SMART MANUFACTURING FACTORY

It is known that information and communication technologies are the basements of smart manufacturing. Intelligent service and optimization driven by AI are the core force of smart factories. In a broad sense, machine learning, computer vision, knowledge graph, natural language processing, speech recognition, intelligent robot, and other richer researches together constitute AI. Thus, taking full advantage of AI-driven software-defined industrial network (SDIN) and related technologies has become one of the significant methods for information intelligent interaction and network QoS optimization for realizing the smart factories.

3.4.1 Cloud-Assisted Ant Colony-Based Low Latency of Mobile Handover

In a smart factory, timeliness is often characterized by determinism and real-time. Determinism accounts for the ability to carry out a specific action at a precise instant. Examples are the periodic transmission of a message from a sensor carrying a measurement value or the delivery of a set-point value from a controller. Real-time refers to the ability of a station to correctly deliver a message within a specified deadline. Table 3.3 gives the latency requirements for different applications in smart factories. In addition, the moving wireless nodes and the harsh industrial environment can increase the wireless interference and bit error rate (BER), and therefore reliability is also an important aspect in heterogeneous network performance.

Reliability and latency are the core requirements of industrial network communication. The wireless network technology has gradually permeated into the industrial field including data acquisition and production control. In IMWNs, connectivity restoration and maintenance directly determine the performance of the whole network in which the handover strategies are research hotspots. However, current handover strategies have many disadvantages, such as frequent handover requests, latency, and the lack of consideration of the network performance, which are harmful to smart factory networks. Therefore, to optimize for industrial wireless node handover sequences, execution, and handover schemes in different cases and decrease the processing capacity of mobile WNs, a Cloud-assisted Ant colony-based Fixed-Path (CAFP) mobile node handover strategy is proposed [38].

TABLE 3.3   Real-Time Requirements in Smart Factories

Industrial Applications	Main Features of the Application	Delay Requirement
Machine control	Low latency, high reliability, and accurate positioning	<20 ms
Machine vision	High bandwidth and precise positioning	<20 ms
Industrial AR and monitoring	High bandwidth and low latency	<20 ms
Mobile robot	High bandwidth, high reliability, and precise positioning	<20 ms

3.4.1.1 Mobile Handover Model

There are many differences between IWNs and wireless networks. As shown in Figure 3.5, industrial MNs must follow a certain or a fixed path to avoid mechanical collisions, such as moving AGVs, or products being processed in an assembly line. And the second difference is that MNs focus on the QoS performance like real-time, stability, and security. To achieve stability within networks, many heterogeneous APs with different job capacities, communication qualities, and different functions are installed in the same place. Mobile nodes face many challenges, such as a complex network topology and a large mobility range. In such situations, it is critical that the network can adopt a novel handover scheme to meet these challenges.

To better illustrate mobile node handover issues, based on a fixed path, the following assumptions and definitions are made:

Assumption 1: In IWNs, APs have fixed positions with a certain communication range. The capacity of APs is heterogeneous, and the overall network is a hybrid network, composed of static and mobile nodes.

Assumption 2: Mobile nodes move along a fixed path. V (m/s) is the speed of the mobile node, and r_m (m) is the communication range. Let $P=\{p_1, p_2,..., p_m\}$ represent the set of m discrete points in its trajectory, where $p_i(x_i, y_i)$ are the coordinates of the ith discrete point of the path.

Assumption 3: In an industrial domain, there are n APs, and A is the set of these APs, $A=\{a_1, a_2,..., a_n\}$, with a communication range set $R=\{R_{a1}, R_{a2},..., R_{an}\}$. Additionally, a_j has coordinates (X_j, Y_j).

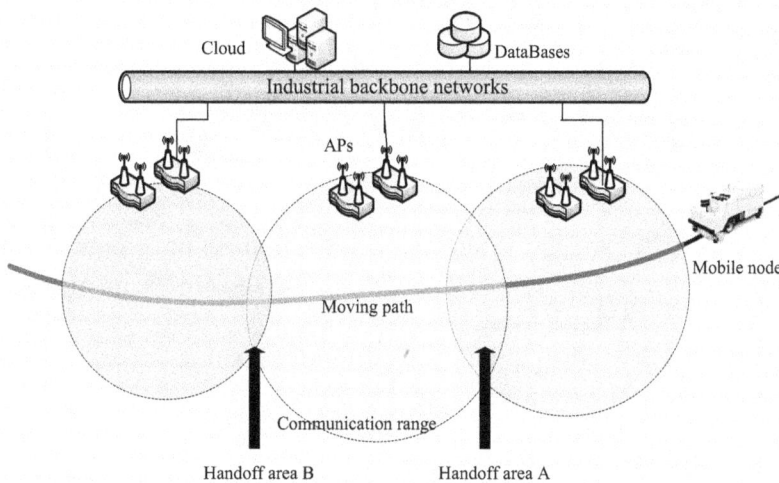

FIGURE 3.5   Mobile nodes in the new industrial communication framework.

Definition 3.1

In order to facilitate and simplify the problem analysis, we create an affiliation matrix H between p and A (where P is the column and A is the row), i.e.

$$H = \begin{pmatrix} h_{11} & \cdots & h_{1n} \\ \vdots & \ddots & \vdots \\ h_{m1} & \cdots & h_{mn} \end{pmatrix} \tag{3.1}$$

where H_{ij} is as follows:

$$h_{ij} = f((x_i, y_i),(X_j,Y_j)) = \begin{cases} 1 & \text{if } d_{ij} \leq d \\ 0 & \text{if } d_{ij} > d \end{cases} \tag{3.2}$$

where d_{ij} is the distance between p_i and a_j, $d_{ij} = \sqrt{(x_i - X_j)^2 + (y_i - Y_j)^2}$, having units m, and d is the effective communication distance, also having units m. Based on equation (3.1), H can be simplified as a 0–1 matrix.

Definition 3.2

Let $\Omega = \{\omega_1, \omega_2,\ldots, \omega_m\}$ be the payload set of A. $\omega_i = q_i/max(\Omega) \times \%$, where q_i is a_i communication data, having units bytes.

In industrial mobile wireless networks, along a fixed path, each mobile node i tries to minimize its handover number and payload, which in turn maximizes the network performance. So, the objective of this work is to minimize the optimization function, which can be formulated as

$$\min F$$

$$F = \partial \cdot \sum_{j=1}^{n} C_N \cdot \Omega + \wp\, T_{\text{handover}} \cdot C_{\text{Handover}} \tag{3.3}$$

subject to $h_{ij} > 0$

where ∂ and $\wp$ are the weighting factors, C_N is the number of communication paths between the ith ant and AP a_j, and T_{handover} and C_{Handover} are the handover time and the handover number, respectively.

Based on the above assumptions and definitions, the following theorems and properties can easily be obtained:

Theorem 3.1

Given an access point a_j, if its communication range is not within p_i, then $\sum_{i=0}^{m} h_{ij} = 0$.

Proof: Based on Definitions 3.1 and 3.2 and equation (3.2), without loss of generality, let p_i with (x_i, y_i) denote the communication range of the access point a_j. The following equation can be obtained:

$$d_{ij} = \sqrt{(x_i - x_j)^2 + (y_i - y_j)^2} > d; \qquad (3.4)$$

i.e., since any $h_{ij} = 0$, namely $h_{1j} + h_{2j} + \ldots + h_{nj} = 0$, $\sum_{i=0}^{m} hij = 0$. This completes the proof.

Theorem 3.2

For a moving path discrete point P_i within the handover area, for the matrix H, $\sum_{i=0}^{n} h_{ij} \geq 2$.

Proof: Since a discrete point P_i is within the handover area, based on the wireless handover concept, there are more than two APs that can cover P_i. Without loss of generality, we assume that there are k access points that can cover P_i, the covering access points set $A_i = \{a_j, a_{j+1}, \ldots, a_{j+k}\}$. Based on Definition 3.1 and 3.2, the following equation can be obtained:

$$\begin{cases} d_{ij} = \sqrt{(x_i - X_j)^2 + (y_i - Y_j)^2} \leq d \\ d_{ij+1} = \sqrt{(x_i - X_{j+1})^2 + (y_i - Y_{j+1})^2} \leq d \\ d_{ij+2} = \sqrt{(x_i - X_{j+2})^2 + (y_i - Y_{j+2})^2} \leq d \\ \vdots \\ d_{ik} = \sqrt{(x_i - X_k)^2 + (y_i - Y_k)^2} \leq d \end{cases} \qquad (3.5)$$

i.e., $h_{ij} = h_{ij+1} = h_{ij+2} = \ldots = h_{ik} = 1$, namely $h_{ij} + h_{ij+1} + h_{ij+2} + \ldots + h_{ik} = k$. Combining these equations, the following formula is obtained:

$$\begin{cases} \sum_{i=0}^{n} h_{ij} = h_{i1} + h_{i2} + h_{i3} + \cdots + h_{ij} + h_{ij+1} + h_{ij+2} + \cdots + h_{in} \geq k \\ k \geq 2 \end{cases} \qquad ; \qquad (3.6)$$

Therefore, $\sum_{i=0}^{n} h_{ij} \geq 2$. This completes the proof.

Property 1: Given an AP a_j that covers most of the points P_i, $\sum_{i=0}^{m} h_{ij}$ $(j = 1, 2, 3 \ldots n)$ provides the maximum value.

Property 2: Based on the dynamic payload of APs, the search for the optimal AP series can be NP-Hard.

When there are many heterogeneous nodes in an industrial network, this may result in APs providing different QoS to mobile nodes. Specifically, for different APs, mobile nodes may have different performance, considering latency and payloads. Therefore, we create the payload matrix of fixed moving paths l and loads Ω, and obtain the following equation:

$$H' = H\Omega^{T}. \tag{3.7}$$

After analysis, the handover problem is transformed into a search task for the best AP series. It is clear that this is a combinatorial optimization problem, i.e., the problem becomes to find $\min f(h_c, \Omega)$, where h_c is the handover time. The problem may be considered as a Hamilton loop problem, with reference to the graph theory. Therefore, Property 2 can be proved, since the Hamilton loop problem is an NP-hard problem. Accordingly, an intelligent optimization algorithm is adopted.

3.4.1.2 CAFP Handover Optimization Strategy

The procedure of the CAFP handover optimization mechanism is as follows: (1) **Step 1:** Using an industrial cloud with an intelligent optimization algorithm (Ant Colony), compute the best series of APs following the moving path sequences. The overall network performance of every AP, QoS for APs, and each mobile node's behavior (such as its velocity) are considered during this step. Additionally, the payloads of the APs, the handover number, and the latency are used as evaluation criteria. (2) **Step 2:** The IWNs are responsible for issuing optimal series of APs to mobile nodes. At the same time, information about the node and the performance of the AP is uploaded to the cloud for other events. Finally, the APs in the optimal series of APs reserve resources for nodes to access within the scheduled time window. (3) **Step 3:** Each industrial wireless mobile node follows a fixed path, and the best series of APs finish the handover and create an effective wireless link. When there are no APs in the best series, the mobile node adopts a contingency handover scheme.

Because ant colony algorithms are simple and have fast convergence characteristics, they are an effective method of solving NP-hard problems. Thus, such an algorithm is employed to search for the optimal series of APs for industrial wireless mobile node handover. Because the ant colony algorithm relies on a pheromone distribution to find optima, the initial value of the pheromone and the update strategy directly determine the quality of the search results. In order to obtain the global optimal solution, local and global methods are used to update the pheromone. The local pheromone update formula can be defined as follows:

$$\gamma_{ij}^{k+1} = (1-\xi) * \gamma_{ij}^{k} \tag{3.8}$$

where γ_{ij}^{k+1}, γ_{ij}^{k}, and ξ are the updated pheromone at points (i, j), the current pheromone, and the local pheromone evaporation coefficient, respectively. A locally updated pheromone is aimed at enabling the ants to increase the probability that points are being passed, thus achieving the purpose of the global search. Correspondingly, a global pheromone increases the pheromone of optimal results. The global pheromone update formula is given in the following equation:

$$
\left\{
\begin{aligned}
\gamma_{ij}^{k+1} &= (1-\theta)*\gamma_{ij}^{k}+\theta\Delta\gamma_{ij}^{k} \\
\Delta\gamma_{ij}^{k} &= \frac{Z}{\min(fit(ant_i))}
\end{aligned}
\right. \tag{3.9}
$$

where θ, Z, and $fit(ant_i)$ are the global pheromone evaporation coefficient, a constant, and the fitness function value of the ith ant, respectively.

A heuristic function directly determines the ant selection probability between the current position and the next position. In this paper, we use the heuristic function $Q(ij)$ of point (i, j) based on the relationship between APs and mobile nodes to obtain the following equation:

$$
\left\{
\begin{aligned}
Q(ij) &= \Omega(ij)^{\alpha_1} \cdot Cov(ij)^{\alpha_2} \cdot S(ij)^{\alpha_3} \\
Cov(ij) &= \sum_{i=0}^{m} h_{ij} \\
S(ij) &= H(ij)
\end{aligned}
\right. \tag{3.10}
$$

Let α_1, α_2, and α_3 be the weighting factors, $Cov(ij)$ the number of access nodes that cover the pathway points, which can decrease the handover times, and $S(ij)$ a security function to prevent the ant from choosing APs with no coverage.

$fit(ant_i)$ contains the metrics of the search results. Taking payloads, handover times, and the communication time into account, the following fitness function can be created:

$$
fit(ant_i) = \partial \cdot \sum_{j=1}^{f} C_N(j)\cdot\Omega(j)+\wp\, T_{handover}\cdot C_{Handover}(ant_i). \tag{3.11}
$$

where ∂ and $\wp$ are the weighting factors, $C_N(j)$ is the number of communication paths between the ith ant and AP a_j, and $T_{handover}$ and $C_{Handover}(ant_i)$ are the handover time and the handover number, respectively.

From the above analysis of the ant colony algorithm, formally, the mobile node optimization is formulated as follows:

$$
\min fit(ant_i)
$$

$$
\text{subject to } h_{ij} > 0. \tag{3.12}
$$

Using the above assumptions and definitions, the algorithm can be divided into three phases. The specific algorithm design is now described.

Phase 1: Search for an optimal series of handover APs for the cloud, using the global network information. The wireless mobile nodes are deployed in a working network with APs. The purpose of this step is to obtain optimal results based on the whole QoS of the mobile node along its trajectory, considering the number of handovers and the AP payload. This phase is composed of three parts. In the first part, the industrial cloud creates an affiliation matrix H based on the APs and the fixed-point coordinates (X_j, Y_j) and (x_i, y_i) obtained from equations (3.2) and (3.3). The ant colony is adopted to search for the best handover series of APs, based on equations (3.7)–(3.10), and outputs the best AP set A' and positions (x'_i, y'_i). Finally, the handover APs and their context are issued for the IWNs. The proposed Phase 1 is shown in Algorithm 3.1.

Phase 2: This phase of the algorithm is implemented locally in mobile nodes. The industrial wireless mobile nodes obtain A'_{min} and the handover position at the starting position, as shown in Algorithm 3.2. The mobile node moves at speed V, and after time t, the node determines whether it reaches the handover position k. If the above conditions are met, the mobile node picks up the corresponding access node from A'_{min}. After obtaining the access node's permission, the mobile node establishes wireless connection with A_k and completes the handover. Finally, the updated mobile node A_k's handover sequence and the QoS of the subnet are uploaded to the cloud.

ALGORITHM 3.1   Search for Optimal Series of Handover APs

1	**Input**: $P=\{p_1, p_2,..., p_m\}$, $\Omega=\{\omega_1, \omega_2,..., \omega_m\}$, $A=\{a_1, a_2,..., a_n\}$, and the position of x_i, (x_i, y_i), and a_j, (X_j, Y_j).
2	**Output**: $A' = \cup a_j$, and the corresponding set of handover positions $P'_{handover}(x'_i, y'_i)$
3	**Begin**
4	Create the matrices H, H', and $H_{A'}$ based on equations (3.2) and (3.3)
5	**for** $f=1$ to m // implement the algorithm m times to get the minimum value
6	**for** each ant k
7	**if** $(h_{1j}>0)$ // judge whether the APs cover the fixed path
8	Randomly choose an access point node at the initial point
9	**for** $i=1$ to n // from the tracing point (x_j, y_j) to (x_n, y_n)
10	Handover to the next access point node j with probability P
11	Update the local pheromone based on equation (3.4)
12	**end for**
13	Compute the total load $fit(ant_m)_k$ of the tour constructed by the kth ant
14	Record the kth handover series.
15	Update the global pheromone value based on equation (3.5)
16	**end if**
17	**end for**
18	$Min(fit(ant_m)_k)$, and recoding handover series A_{min} and $P'_{handover}(x'_i, y'_i)$.
19	**end for**
20	$Min(Min(fit(ant_m)_k))$, and the corresponding handover series A'_{min} and $P'_{handover}(x'_i, y'_i)$
21	**End Begin**

ALGORITHM 3.2   Procedure of the Mobile Node for Local Handover

1	**Input**: the minimum load access point nodes, and the corresponding handover series $A'_{min} = \{A_1, A_2,..., A_\Lambda\}$, $P'_{handover}(x'_i, y'_i)$ and V.
2	**Output**: the link to access points according to $A'_{min} = \{A_1, A_2,..., A_\Lambda\}$.
3	**Begin**
4	**for** the kth handover
5	**if** the mobile node arrives at a handover point $P^k_{handover}$ based on Vt.
6	Search for the kth access point
7	Link to kth access point
8	**end if**
9	Update the $A'_{min} = \{A_1, A_2,..., A_\Lambda\}$.
10	**end for**
11	**End Begin**

ALGORITHM 3.3   Procedure of the Mobile Node for Handover in Unpredicted Cases

1	**Input**: local $RSSI$, PLR, and $A_{min} = \{A_1, A_2,..., A_\Lambda\}$.
2	**Output**: the candidate access point.
3	Begin
4	**if** $A_{search} \notin A_{min} = \{A_1, A_2, \cdots, A_\Lambda\}$
5	**if** the mobile node requires handover
6	Select the best access point A_k from A_{search}, based on $RSSI$, PLR.
7	Create a wireless link with A_k
8	**end if**
9	**end if**
10	**End Begin**

Phase 3: This phase is the recovery procedure of the mobile node in unpredicted cases. Connectivity restoration and maintenance are very important for industrial mobile nodes. While moving between APs, the mobile node may not complete the handover if an AP breaks down. To avoid this drawback, an unpredicted back-off procedure is provided. This algorithm phase is aimed at connectivity restoration and maintenance and further at achieving better survivability of the whole network. The specific process is as follows. When the mobile node does not search for an AP from A_{min} after several cycles, it should choose a new AP based on the $RSSI$ and PLR.

3.4.2 Data Transmission Strategies with Different Delay Constraints

In smart factories, there is an increasing requirement for exchange of data with different delay flows among different smart devices. To overcome the limitations of traditional methods and address the problem, the requirements can be divided into two groups according to data streams with different latency constraints, that is ordinary and emergent stream. In the low-deadline situation, a coarse-grained transmission path algorithm provided by finding all paths that meet the time constrains in hierarchical Internet of Things.

After that, by employing the path difference degree, an optimum routing path is selected considering the aggregation of time deadline, traffic load balances, and energy consumption. In the high-deadline situation, if the coarse-grained strategy is beyond the situation, a fine-grained scheme is adopted to establish an effective transmission path by an adaptive power method for getting low latency [39].

3.4.2.1 System Model

The smart factory network with V vertices and E edges is denoted as $G=\{V, E\}$ representing the set of nodes and links, where V represents the factory network nodes (ND) and E represents the links. The communication range of nodes in the case of low real-time performance is R. Furthermore, if e_{ij} exists in set E, then nodes v_i and v_j can directly communicate with each other. Every node has its energy and network resources. The smart factory cluster can be denoted by $C_i=\{c_i, m_i\}$, where c_i and m_i represent the cluster head (CH) and its members, respectively. Therefore, setting $\Omega=\{C, L\}$ to simplify set G, with C vertices and L edges representing the set of cluster heads and links. The link value l_{ij} between c_i and c_j, $c_i, c_j \in C$, is given as follows:

$$l_{ij} = \begin{cases} 1, c_i \cap c_j \neq \emptyset \text{ or } d_{CiCj} \leq R \\ 0, \qquad\qquad \text{other}. \end{cases} \tag{3.13}$$

where d_{CiCj} is the Euclidean distance between c_i and c_j.

According to equation (3.13), the adjacency matrix is represented by A. It is assumed that in a cluster, every ND can directly communicate with CH; in other words, there is one hop between CH and ND. The data transmission rate between CH and ND is expressed as TR_{CN}, and the data transmission rate between CH and CH expressed as TP_{CC} (in bps). To simply the problem, it is assumed that ND has the same communicate rate. In addition, cluster heads have high capacities of data rate and energy. Then, the data rate between the pair of nodes $PV=\{v_i, v_j\}$ can be formulated as:

$$TR_{ij} = \begin{cases} TR_{CN} & \exists\, PV \in ND \\ TR_{CC} & \forall\, PV \in C \end{cases} \tag{3.14}$$

Define ξ_{ij} as the weight of transmission between cluster heads c_i and c_j; if $i=j$, $\xi_{ij}=0$; otherwise, the weight value is formulated by:

$$\xi_{ij} = l_{ij} \cdot \frac{P_{ij}}{H_{ij}} \cdot \frac{TR_{CC}}{TR_{ij}} \tag{3.15}$$

where P_{ij} and H_{ij} are the numbers of paths and hops between v_i and v_j, respectively. Apparently, the greater the value of ξ_{ij} is, the better the communication between v_i and v_j will be. The value of P_{ij} is formulated as:

$$P = \begin{cases} 0 & \text{if } l_{ij} = 0 \\ 1, l_{ij} = 1, d_{C_iC_j} < R \\ \|C_i \cap C_j\|, & \text{other} \end{cases} \quad (3.16)$$

where $d_{C_iC_j}$ is the distance between c_i and c_j. If $P_{ij} = 1$, then $H_{ij} = 1$, and if $P_{ij} = 0$, then $H_{ij} = \infty$; otherwise, $H_{ij} = 2$.

According to equation (3.15), it is easy to get the transmission weight matrix W of cluster heads, which is expressed as:

$$W = \begin{bmatrix} \xi_{11} & \cdots & \xi_{1n} \\ \vdots & \ddots & \vdots \\ \xi_{n1} & \cdots & \xi_{nn} \end{bmatrix} \quad (3.17)$$

Hence, according to equations (3.13) and (3.14), the transmission time $f(t_{ij})$ between c_i and c_j is defined by:

$$f(t_{ij}) = \begin{cases} \dfrac{\theta}{TR_{CC}}, j \in C \cap i \in C, \\ \displaystyle\sum_{p_{ij}} \dfrac{\theta}{TR_{CN} \cdot H_{ij}}, \text{other} \end{cases} \quad (3.18)$$

where θ is the transmission data volume, and p is the number of all paths with two hops between c_i and c_j.

For the path $p = \{v_i \ldots, c_k, c_{k+1}, \ldots, v_j\}$ where v_i and v_j represent start and end nodes, respectively, the forwarding time is defined by:

$$T_p = \sum_{k \in p} f(t_{kk+1}) \quad (3.19)$$

The coarse-grain level problem is stated as follows: when a multiple cluster $C = \{c_1, c_2, c_3, \ldots\}$ with the transmission data volume θ is given, our objective is to find a routing path between v_i and v_j which meets the deadline time constraints.

Definition 3.3

Let *PrePath* and *CurPath* be the sequence of previous and current communication paths between v_i and v_j, respectively. Then, the Path Difference Degree (PDD) is defined by:

$$PDD_{curpath, prepath} = \frac{Setdif(CurPath, PrePath)}{\varphi_{CurPath} \cdot |CurPath|} \quad (3.20)$$

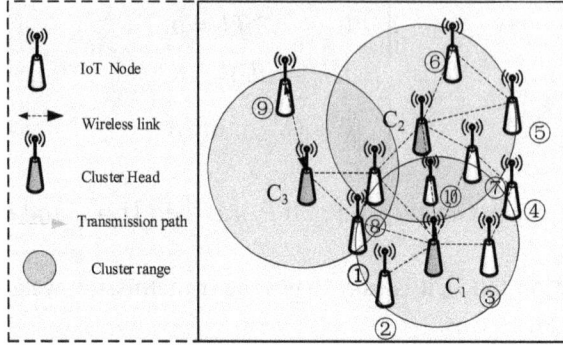

FIGURE 3.6   Illustration of representative system model.

where the function of *setdif* (A, B) returns the number of different members of set A from set B, and φ is the path adopted frequency for communication between v_i and v_j. φ is the member of path adopted frequency set ϕ. Actually, multiple paths may exist between v_i and v_j. However, if one path is too frequently used, load and energy unbalance is caused. Given the delay time constraint Tc, we further formulate the current problem as:

$$\text{Max } PDD$$

$$\text{S.T.} t_{ij}^{cur} \leq T_C \tag{3.21}$$

To understand the problem better, an example is given to illustrate the above model and network structure. As shown in Figure 3.6, there are three CHs and ten NDs. Since, the member set of C_1 is $M_1=\{1, 2, 3, 4, 8, 7, 10\}$ and the member set of C_2 is $M_2=\{5, 6, 7, 8, 10\}$, the intersection of M_1 and M_2 is $\{7, 8, 10\}$. Moreover, we get the value of P_{12} which is equal to 3, and the hop number H_{12} which is equal to 2. According to (3), $TR_{CC=}$ $100TR_{ND}$, hence it is easy to get the weight value, $\xi_{12}=150$. Furthermore, the weight value is $\xi_{23}=50$. As $\xi_{12}>0$, for a given pair of NDs (v_2, v_6), the multiple paths exist between them (C_1, C_2). Consequently, there are three routing paths that meet deadline, $Path_1=\{2, C_1, 7, C_2, 6\}$, $Path_2=\{2, C_1, 8, C_2, 6\}$, and $Path_3=\{2, C_1, 10, C_2, 6\}$ given $\varphi_1=\varphi_2=\varphi_3=1$, thus $PDD_{12}=0.2$.

For given IIoT nodes v_i and v_j, the minimum hop number is $H_{ij} \geq \lfloor D_{ij}/2R \rfloor$. For any vertices v_k, its one-hop neighbor vertex is v_f. So, the distance of two vertices must meet the condition of $D_{kf} \leq \min\{R_f, R_k\}$. We assume that the communication range has the same value R; so $D_{kf} \leq R$. Hence, for the nodes with the same communication range, the maximizing one-hop distance is R. For given IIoT nodes v_i and v_j, the straight line between them is LS_{ij}, and LS_{ij} can be covered by $\lfloor D_{ij}/2R \rfloor$ circles with radius R, and the minimal hop number is $\lfloor D_{ij}/2R \rfloor$.

According to the above description, the minimal transmission time $T_{\min}$ for current network structure and state can be obtained. For any link between v_i and v_j, $T_{\min}$ is formulated by:

$$T_{\min}^{ij} = 2\frac{\theta}{TR_{ND}} + \left\lfloor \frac{D_{ij}}{2R} \right\rfloor \frac{\theta}{TR_{CH}}. \tag{3.22}$$

Assuming that the ith node transmit power is P_t, the received power P_r at the jth node is defined by:

$$P_r = P_t - \alpha(d_0) - 10\beta \lg(d_{ij}/d_0). \tag{3.23}$$

where $\alpha(d_0)$ is the attenuation at reference distance d_0, and the path loss exponent β varies depending on the network deployment scenario. For simplification, define $d_0 = 1$ m; thus, the formulation can be rewritten by:

$$P_r = P_t - \alpha - 10\beta \lg(d_{ij}). \tag{3.24}$$

Definition 3.4

The Maximal Transmission Radius (MTR) $R_{\max}$ of node is the maximal transmission range obtained with the maximal transmission power $P_{\max}$.

Furthermore, only when P_r is greater than the definite power P_C, the communication link defined by equation (3.25) can be created:

$$R_{\max} = 10^{(Pt-Pc-\alpha)/10\beta} = \gamma \cdot 10^{P_t/10\beta} \tag{3.25}$$

The maximal communication range $R_{\max}$ is non-decreasing function. If the maximal communication range for a given function is f, then $f = R_{\max}$, thus f is derived as follows: $\frac{\partial f}{\partial P_t} = \frac{\gamma}{10\beta} \cdot \ln 10 \cdot 10^{P_t/10\beta}$. Furthermore, $\beta > 0$, $\gamma > 0$, $P_t > 0$, so $\frac{\partial f}{\partial P_t} > 0$. In other words, f is an increasing function. Therefore, the maximal communication range $R_{\max}$ is a non-decreasing function. It is obvious that when P_t increases, $R_{\max}$ also increases. Therefore, by increasing the transmission power, the number of hops and transmission time can be reduced.

It is well known that to provide a communication to the larger distance (i.e. to increase the communication radius), the transmission power has to be increased. On the other hand, to evaluate the cost of increasing transmission power, the cost function is defined as follows:

$$Cp = \lambda e^\tau + \varepsilon. \tag{3.26}$$

where τ is the increasing transmission power, λ is the increasing power coefficient, and ε is the cost constant.

The problem of adaptive power control is stated as follows: for a given delay time constraint T_c and set of all paths, the minimal power control cost is defined as follows:

$$\text{Min} \, Cp$$

$$\text{s.t.:} t_{ij} \leq T_c$$

$$0 < P_j \leq P_{\max}.$$ (3.27)

$$T_c \leq T_{\min}^{ij}$$

where t_j is the communication time after the transmission power changes and P_j is the new transmitted power. According to equation (3.20), the problem finds the minimal power that meets the related constraints.

3.4.2.2 Coarse-Grain Optimal Path Algorithm

Based on the initial network, weight adjacency matrix A and weight matrix W, and by using the function of transmission time $f(t_{ij})$ between v_i and v_j, all paths between cluster heads c_i and c_j need to be calculated. Namely, this is a classical generation tree problem. All paths of cluster head pair (c_i, c_j) are found by some algorithms, such as Depth First Search (DFS), and stored in ECS or industrial cloud. The variable $Path_{ij}$ represents all paths between c_i and c_j, whereas all paths of every cluster head pair are represented with the set $Path\_AllCH$. Summing up all the techniques, an online solution is designed. The adaptive transmission optimization (ATO) based on the coarse-grain optimal path algorithm is shown in Algorithm 3.4.

The key steps of the above algorithm are explained as follows. Line 1 represents the initialization of current network state, such as source and destination ND, Cluster head node set, etc. As shown in Lines 2–3, when SDN meets the transmission requirement of nodes v_i and v_j, according to the set C, the algorithm finds the corresponding cluster heads and constructs communication cluster head pair $CHPair <c_i, c_j>$. Then, Lines 4–10 derive all transmission paths between c_i and c_j from the set of $Path\_AllCH$. Moreover, by calculating the transmission time, all paths with threshold less than T_c are found and stored into set $pathSS$. Once a set of paths that meet time constraint is determined, the maximal path difference degree of these paths is found by Lines 11–17. Lastly, algorithm returns the path with the maximal path difference degree. Time complexity of the main algorithm loop is $O(|Path_{ij}|)$, and the worst time complexity is $O(|CH|)$.

Using the example in situation, Figure 3.6 explains Algorithm 3.4, which is used for finishing the communication between (v_2, v_6). According to the above analyses, if the forwarding times of paths are less than Tc, there are three paths. $Path_1 = \{2, C_1, 7, C_2, 6\}$, $Path_2 = \{2, C_1, 8, C_2, 6\}$, and $Path_3 = \{2, C_1, 10, C_2, 6\}$. Assume that path1 is $PrePath$. So, given $\varphi_1 = 4$, $\varphi_2 = 2$, $\varphi_3 = 1$, PDD (1, $PrePath$) = 0.05, PDD (2, $PrePath$) = 0.1, and PDD (3, $PrePath$) = 0.2. As PDD (3, $PrePath$) is the maximal value, $Path\_selection = Path_3$. In other words, path 3 finishes the communication between (v_2, v_6).

ALGORITHM 3.4   Pseudocode of Coarse-Grain Optimal Path Algorithm

---

Require: vi, vj, C, $Path\_AllCH$, T_C, TR, W, ϕ

Ensure: $Path\_selection$

1    $V \leftarrow \langle vi, vj \rangle$, $Path_{ij} \leftarrow \emptyset$, $\varpi_0 \leftarrow 0$

2   From the set C, search the cluster head pair (c_i, c_j)
       of NDs v_i and v_j

3   Construct the communication cluster head pairs corresponding $CHPair\langle c_i, c_j \rangle$ of v_i and v_j

4   Select all paths of $Path_{ij}$ between cluster head c_i and c_j, from $Path\_AllCH$, according with W.

5   **for** $i = 0$ to $|Path_{ij}|$ **do**

6           $T_i = \Sigma t$ //calculating every path time by equation (3.18)

7   **if** $T_i < T_C$

8                  $PathSS \leftarrow path_i$ //gathering the path

9   **end if**

10   **end for**

11   **for** $j = 0$ to $|PathSS|$

12   calculate $\tilde{\omega}_j$ //calculating the path difference degree

13   **if** $\tilde{\omega}_j > \tilde{\omega}_0$ //determining whether $\tilde{\omega}_j$ is larger than the current maximum $\tilde{\omega}_0$

14                 $\tilde{\omega} \leftarrow \tilde{\omega}_j$

15                 $Path\_selection \leftarrow PathSS_j$ //find the path with maximal path difference degree

16   **end if**

17   **end for**

18   $\varphi_{Path\_selection} = \varphi_{Path\_selection+1}$

19   Update ϕ

20   **return** $Path\_selection$.

3.4.2.3 Adaptive Transmission Power for Fine Grain Algorithm

Algorithm 3.4 is designed for a common situation, in other words for a high transmission delay. However, the network usually faces a low transmission delay. Therefore, the power is increased to create a strong link so cluster heads can directly communicate with each other providing better performance, especially data rate. Therefore, adaptive transmission optimization is designed for urgent situation, as shown in Algorithm 3.5.

The main steps are described as follows. First, SDN controller gains the constraint and related parameters of network, and then according to Algorithm 3.5, designs a new transmission path. The direct communication between cluster heads with increasing the transmission power $P(d(c_m, c_{m+1}))$ is found for all paths by Lines 2–9, according to cluster heads distance $d(c_m, c_{m+1})$. Then, by recalculating the transmission time, the paths that meet the time constraints are found by Lines 10–13. In Lines 14–18, the minimal cost path is obtained. Lastly, algorithm returns a new path for the networks. Similar to Algorithm 3.4, the worst time complexity of main algorithm loop in is $O(|CH|^2)$.

Similarly, employing the example in situation, Figure 3.6 shows Algorithm 3.5. The communication time between (v_1, v_6) by adopting Algorithm 3.4 cannot meet the time deadline constraints T_C. For addressing the high-deadline situation, Algorithm 3.5 is used. First, calculate the corresponding cluster head pairs of (v_1, v_6) with results $\langle C_1, C_2 \rangle$,

ALGORITHM 3.5   Pseudocode of Adaptive Transmission Power Optimization

---

Require: v_i, v_j, $Path_{ij}$, Tc, T_{min}

Ensure:, $Path\_new$

1	**for** $i=0$ to $|Path\_all|$
2	Construct the communication all cluster head pairs corresponding of v_i and v_j, stored in $CP(v_i, v_j)$ from the $Path_{ij}$
3	**for** $m=0$ to $|CP|$
4	Calculate distance pair of cluster head $d(c_m, c_{m+1})=|c_m, c_{m+1}|$
5	**if** $R<d(c_m, c_{m+1})<R_{max}$
6	$P_t(c_m) \leftarrow P(d(c_m, c_{m+1}))$ //increase the current transmission power
7	$TR_{m, m+1} = TR_C$ // update the transmission rate
8	**end if**
9	Calculate the T_m by equation (3.18)
10	**if** $T_i \leq T_c \leq T_{min}$ // determining whether a new path meets the requirement
11	$NewPathSet \leftarrow Path\_all[m]$
12	**end if**
13	**end for**
14	**end for**
15	**for** $k=0$ to $|NewPathSet|$
16	Calculate Cp_k //calculating a new power changed cost
17	**if** $Cp_k < Cp_0$
18	$Cp_0 \leftarrow Cp_k$; //find the path with the minimal cost
19	$Path\_new \leftarrow NewpathSet[k]$
20	**end if**
21	**end for**
22	**return** $Path\_New$.

$<C_3, C_2>$. Obviously, $R<d(C_1, C_2)<d(C_3, C_2)<R_{max}$, and $T(C_1, C_2)$, $T(C_3, C_2) \leq T_C \leq T_{min}$, so $CP(C_1, C_2)<CP(C_3, C_2)$. Then, select $\{v_1, C_1, C_2, v_6\}$ as the new transmission path with C_2 increasing the transmission power $P(d(C_m, C_{m+1}))$.

3.4.3 Load-Balanced Packet Broadcast Scheme Based on Neighbor Information

Flooding is one of the major data dissemination schemes to broadcast packets in the WSNs. In basic flooding, there are too many redundancies in the network, which consume much energy. Consequently, a load-balanced packet broadcast scheme that explores flooding based on neighbor information is proposed, namely neighbor-based load-balanced packet dissemination (NLPD). In NLPD, the nodes receiving the packets decide whether to rebroadcast the packet based on neighbors' information [40].

3.4.3.1 Problem Statement

3.4.3.1.1 System Model   For the network model in the WSNs, some assumptions are described as follows:

- **All sensor nodes use Boolean sensing model**: In this model, the communication zone of each node is a circle with radius R_c, which is the communication distance of sensor nodes. Each node only can communicate with the nodes in its communication zone.

- **All sensor nodes are nearly static**: After deployment, the nodes do not change their position any more. Also, there are no nodes added to the network and no nodes leaving the network except that the nodes die naturally. One example of broadcasting in WSN applications is shown in Figure 3.7.

In Figure 3.7, the network is deployed to monitor a forest environment. The sink node is interested in if there is any part of the forests on fire. The sink node broadcasts the packet containing the question that are you on fire? to all nodes in the network. As the communication radius of sink nodes is limited, only a few sensor nodes in the network can receive the query packet through one hop. The packet should be retransmitted by the intermediate nodes to all sensor nodes in the area. One of the main packet dissemination strategies is the basic flooding.

3.4.3.1.2 Analysis of Basic Flooding   In basic flooding, every node that receives the packet rebroadcasts the packet automatically. Let's call nodes that have processed the packet infective node (IN), while nodes that have not processed the packet susceptive node (SN).

The spreading process is shown in Figure 3.8. There is a circle region of nodes that have processed the packet centered at the source node which grows outwards with time, ultimately reaching the whole network. Only the nodes that lie at the edge of the circle can move the message forward, such as *Node j*, *Node h*, and *Node i* in Figure 3.8. *Node k*'s rebroadcast cannot affect any more nodes. Only the edge nodes can infect nodes that lie in their communication area. Only nodes in the shaded region can benefit from edge nodes' rebroadcast. So, a rebroadcast can only provide about 0% ~ 61% additional area. In Figure 3.8, *Node j*'s rebroadcast only provides about 28.6% additional area; *Node h*'s rebroadcast provides 0% additional area and *Node i*'s rebroadcast can provide 66.7%

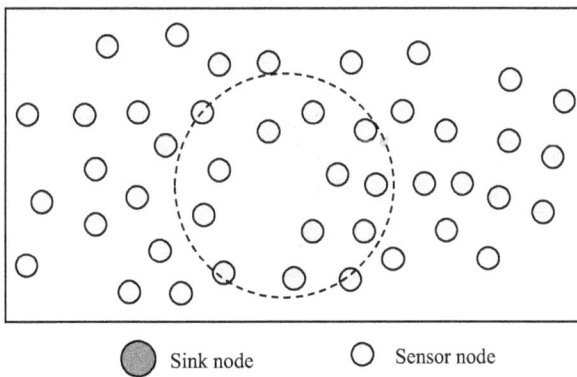

Sink node   Sensor node

FIGURE 3.7   Broadcasting example for environment monitoring.

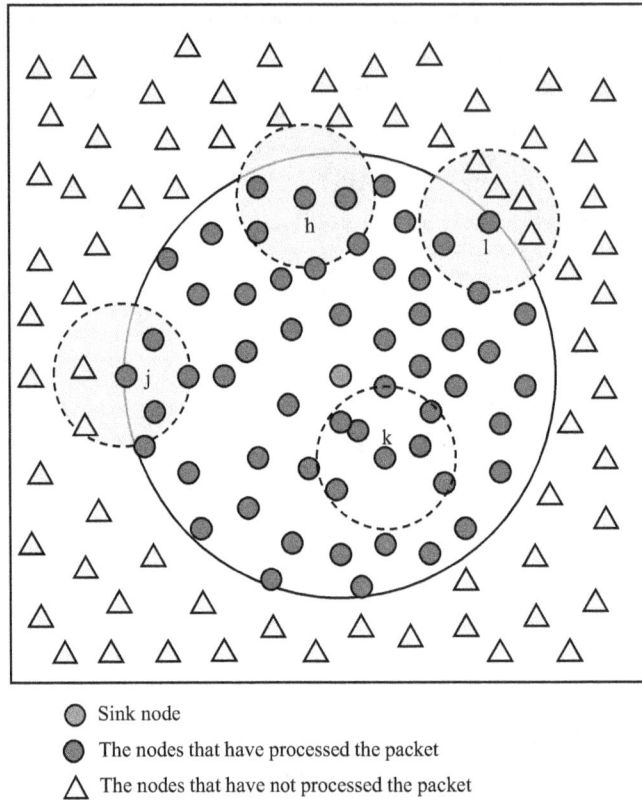

○ Sink node
● The nodes that have processed the packet
△ The nodes that have not processed the packet

FIGURE 3.8 Packet's spreading process in sensor networks with basic flooding.

additional area. So, the rebroadcast of *Node i* is encouraged and the rebroadcast of *Node h* and *Node j* is limited. The redundant rebroadcast times must be reduced as much as possible. In the spread process of packet, each node receives the packet duplicates from its all neighbors. The number of duplicates that every node receives depends on the node's degree. If the node has more neighbors, it receives more duplicates and consumes energy more quickly. If the node is in the area with high density, the effect of its rebroadcast is worse. Thus, consider using a threshold to limit the nodes with high degree.

3.4.3.2 Proposed Solution

3.4.3.2.1 Overview   In basic flooding scheme, when the source node wants to search for the data that it is interested in or provide software update, it distributes a packet to other nodes. When the nodes receive the packet, they rebroadcast the packet automatically, which results in serious redundancy. One approach to alleviate the above problem is to inhibit some nodes from rebroadcasting the packet to reduce the redundancy. As the nodes are randomly distributed in the monitoring areas, the node density is different in different parts of the area. If all nodes in dense area rebroadcast the packet, all nodes receive duplicates and consume energy quickly.

As shown in Figures 3.9 and 3.10, the area of the two monitored regions is S, which is equal to $1.5S_d$, where S_d is the monitored area of a node. In Figure 3.9, there are 7 nodes in

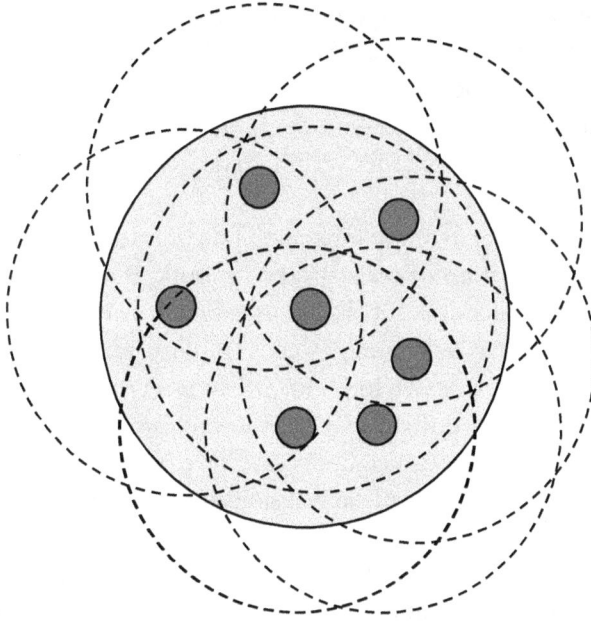

FIGURE 3.9   Broadcast efficiency of nodes in dense area.

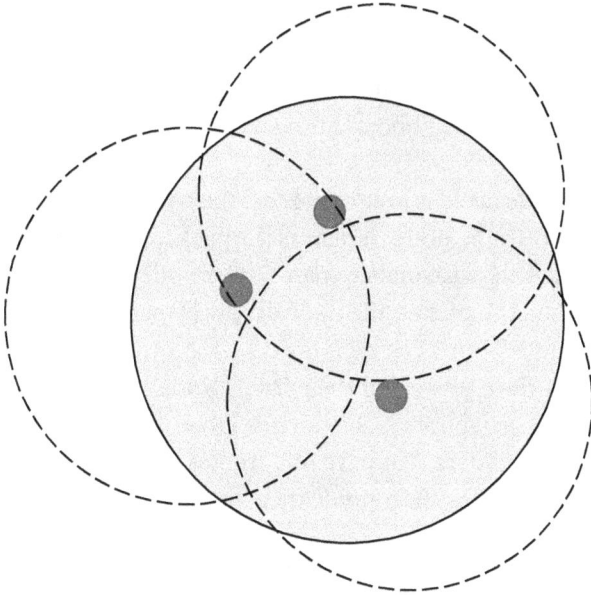

FIGURE 3.10   Broadcast efficiency of nodes in sparse area.

the area; if all nodes rebroadcast the packet, the total cover area of 7 nodes is $7S_d$, but the useful cover area is S, and therefore the efficiency is about 21.4%. But in Figure 3.10, there are 3 nodes in the area; if all nodes rebroadcast the packet, the total cover area of 3 nodes is $3S_d$, the useful cover area is S, and the efficiency is about 50%. So, consider taking measures to limit the nodes in dense area rebroadcasting the packet.

Nodes in dense areas must have high node degree, while the nodes in spares area are with low degree. The broadcasting effect of nodes in dense area is worse than that of nodes in sparse area. The nodes in dense areas consume much more energy than the nodes in sparse areas. The energy consumption of nodes in dense area must be reduced. Therefore, to reduce the rebroadcast times of nodes with high degree, a threshold Q which stands for nodes' degree is defined.

When the sink node wants to query for some node or interested events, it broadcasts a query packet to all or apportion of nodes in the network. But the communication of the sink node is limited. The query packet should be rebroadcasted by intermediate nodes to the whole network. In order to reduce energy consumption, intermediate nodes should not rebroadcast the packet automatically. In NLPD, when receiving previously unseen packet, the intermediate node can have three measures to process it: rebroadcasting the packet, transmitting the packet to one of its neighbors that have not processed the packet, or dropping the packet. The nodes make localized decisions on the above-mentioned parameter— the total number of neighbors.

Each node holds a variable q, which stands for the total number of neighbors of the node. It is computed in deployment of network through HELLO packet. When deploying the network, each node broadcasts a HELLO packet. Each node's q is initialized to zero. After receiving a HELLO packet, the nodes update their q plus one. Consequently, every node gains knowledge of all neighbors within one-hop transmission radius. When a new node is added to the network or leaves the network, it broadcasts a HELLO packet to let its neighbors update their neighbor information tables. The total number of neighbors can reflect the area's density. The neighbors' number is higher, therefore the area's density is larger.

After getting q, the node decides how to process the packet. If q is larger than or equal to the predefined threshold Q, the node transmits the packet to one of its neighbors that have not processed the packet. If q is smaller than Q, the node rebroadcasts the packet. If all neighbors have processed the packet, the node drops the packet directly. The three process situations can be seen in Figure 3.11. In Figure 3.11, the threshold 7 is predefined. Because all neighbors of Node 1 have processed the packet, Node 1 drops the packet directly. The total number of Node 3's neighbors is 8, which is larger than threshold, thus it transmits the packet to one of its neighbors that have not processed the packet. As for Node 2, it has 5 neighbors, which are smaller than predefined threshold, and thus it rebroadcasts the packet.

3.4.3.2.2 Algorithm Design   In NLPD, each node has three measures to process the received unseen packet: dropping it, transmitting it to one of its neighbors that have not processed the packet, or rebroadcasting it. Every node makes decision based on its neighbor information. The NLPD algorithm is shown in Algorithm 3.6. Before running the algorithm, each node has gotten its neighbor's total number. The parameter Q has been defined before. Much of the algorithm's running time is spent on searching for the cached information. When the node decides to send the packet to one of its neighbors, it needs to search for the next hop in the cached information. Compared to basic flooding,

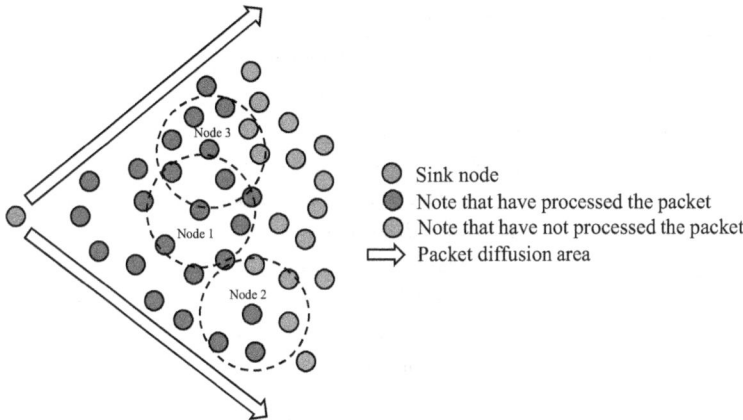

FIGURE 3.11   Three measures to process the packet.

ALGORITHM 3.6   NLPD
___

On hearing a new broadcast packet *m* at Node *x*

Begin

1:        Decide how to process the packet.

2:        **if** all neighbors have processed the packet **then**

3:            Drop directly.

4:        **end if**

5:        **if** $x_{ncitotalnum} < Q$ **then**

6:            Rebroadcast the packet.

7:        **Else**

8:            Transmit the packet to one of neighbors.

9:        **end if**

10:       Initial the counter $c_x = 0$.

11:       **if** *Node n* has processed the packet **then**

12:           **if** $c_x < 3$ **then**

13:               Select a *Node n* from the neighbors randomly, c_x++.

14:           **Else**

15:               Search the $x_{neighbor}$ sequentially,

16:               Get the first *Node n* that has not process the packet,

17:               Send the packet to *Node n*.

18:           **end if**

19:       **end if**

20:   **End**
___

the searching process of NLPD algorithm may spend more time. Consequently, the diffusion time of our strategy may be more than that of basic flooding. If the cached information is little, the node can process the received packets quickly. Sensor nodes do not need to spend much time on processing the packets, but only need to spend time on transmitting the packet.

TABLE 3.4   Notation Definition

Notation	Definition
d_k	Data flow request between node pairs
P	All reachable paths of the d_k
p_k	The transmission path of the d_k at time t
w_k	The probability that data flow d_k passes through the path p_k
L_k	The transmission delay of the d_k at time t
U_E	The network bandwidth utilization at time t
CV_{load}	Degree of network load balancing
L_J	The network average latency at time t
Thp_t	The network throughput
Jit_t	The average network jitter time

3.4.4 Network Load Balancing and Routing Optimization Based on Deep Reinforcement Learning

The routing algorithm based on a single routing metric parameter is difficult to meet the QoS requirements of multi-source data flows in the software-defined factory heterogeneous network, resulting in link congestion and waste of network resources. To solve the problem, a network load balancing and routing optimization method for software-defined factory heterogeneous networks based on the double deep Q network (DDQN) is proposed [41].

3.4.4.1 Load Balancing and Routing Optimization Problem

To simultaneously meet the network delay, network load balancing, and intelligent traffic scheduling requirements of multi-source data flows, the load balancing and routing optimization problem in software-defined factory network will be described as follows.

It should be noted that the network topology model under the software-defined factory networks is an undirected graph $G = (V, E)$, where V represents the set of all SDN switch nodes, which can be expressed as $V = \{v_1, v_2, ..., v_M\}$, where M is the number of nodes. E represents physical links, which can be expressed as $E = \{e_1, e_2, ..., e_N\}$, where N is the number of links, and the network link bandwidth between any two nodes is represented by B_{en}. Assuming that the starting node and the destination node of any data flow d_k request are d_{src} and d_{dst}, respectively, the set of K reachable paths is $P = \{p_1, p_2, ..., p_K\}$, where the probability of passing the path p_k is w_k. Table 3.4 summarizes the symbols used in this chapter.

Definition 3.5

Network average latency L_J: It is expressed by the average transmission delay of J data flows at a certain time. For any data flow d_k, if its transmission path goes through h ($h < N$) hops from its d_{src} to the d_{dst} and the transmission delay of per hop is L_i, then the transmission delay of the data flow is $L_k = \Sigma_i \in_h L_i$. Therefore, the network average transmission delay of J data flows at time t is calculated using the root mean square, which can be expressed as follows:

$$L_J = \sqrt{\frac{\sum_{k \in J} L_k^2}{|J|}} = \sqrt{\frac{\sum_{k \in J} \left(\sum_{i \in h} L_i \right)^2}{|J|}} \qquad (3.28)$$

Definition 3.6

Network utilization U_E: The SDN controller periodically sends statistical packets through the OpenFlow protocol to obtain statistics about SDN switch ports, including the number of bytes received ΔRx_{bytes} and the number of bytes sent ΔTx_{bytes} within the statistical duration Δt. Therefore, the used bandwidth of any link *en* at time *t* can be expressed as follows:

$$U_{en}^{used} = \frac{\Delta Rx_{bytes} + \Delta Tx_{bytes}}{|\Delta t|} \qquad (3.29)$$

Further, link utilization is represented by the ratio of the occupied bandwidth to the total link bandwidth, that is:

$$U_{en} = U_{en}^{used} / B_{en} \qquad (3.30)$$

Therefore, the network utilization at any time can be expressed as follows:

$$U_E = \sqrt{\frac{\left(\sum_{k \in J} \left(\sum_{en \in p_k} U_{en} \right)^2 \right)}{J}} \qquad (3.31)$$

Definition 3.7

Network load balancing degree CV_{load}: The traditional OSPF routing algorithm always selects the path with the shortest path for data transmission, resulting in heavy load on some links and unbalanced link load. The coefficient of variation is able to measure the degree of data dispersion. Therefore, the coefficient of variation to characterize the degree of network load balancing is adopted, that is:

$$CV_{load} = \frac{\sigma\left(\forall U_{en}^{used} \right)}{\mu\left(\forall U_{en}^{used} \right)} = \frac{\sqrt{\left(\sum \left(U_{en}^{used} - \mu\left(\forall U_{en}^{used} \right) \right)^2 \right) / N}}{\mu\left(\forall U_{en}^{used} \right)}$$

$$= \frac{\sqrt{\left(\sum \left(U_{ei}^{used} - \left(\left(\sum_{i=1}^{N} U_{ei}^{used} \right) / N \right) \right)^2 \right) / N}}{\left(\sum_{i=1}^{N} U_{ei}^{used} \right) / N} \qquad (3.32)$$

To improve the performance of the whole network and achieve network load balancing, it is necessary to avoid selecting nodes with heavy load and links with high link utilization for data transmission. Therefore, the optimization problem of the software-defined factory networks is modeled as follows:

$$\text{Min}\left(\mu U_E + v L_J\right) \tag{3.33}$$

$$\text{s.t.} \sum_{p_k \in P} w_k = 1, \forall k \in J \tag{3.33a}$$

$$w_k \geq 0, \forall p_k \in P, \forall k \in J \tag{3.33b}$$

$$L_k < L_{\max} \tag{3.33c}$$

$$U_{en} < B_{en} \tag{3.33d}$$

$$\mu + v = 1, 0 \leq \mu, v \leq 1 \tag{3.33e}$$

where equation (3.33) is the optimization goal, which meets different network latency or transmission bandwidth requirements of data flows. Equation (3.33a) indicates that d_k is on all possible forwarding paths between the source node and the destination node. Equation (3.33b) indicates that the probability of each possible forwarding path being selected is not negative. Eq. (3.33c) indicates that the transmission delay of the data flow request d_k does not exceed its threshold. Equation (3.33d) indicates that the link utilization rate of the data flow request d_k cannot exceed the link bandwidth. μ and v in equation (3.33e) are the weight values of network load balancing and network delay, respectively, which meet the requirements of μ, $v \in [0,1]$ and $\mu + v = 1$.

3.4.4.2 Design of DDQN-Based Network Load Balancing and Routing Optimization Algorithm

3.4.4.2.1 DDQN   In general, reinforcement learning (RL) can be modeled as a Markov decision process (MDP), which is described by a quintuple as follows:

$$\xi = \langle S, A, P, R, \gamma \rangle \tag{3.34}$$

Among them, ξ represents the environment interacting with the agent, S represents the state space of the environment, and each $s_t \in S$ represents the observed value of the state of the environment at a certain moment. A represents the action space of the environment, including all the action a_t under state s_t. P represents the state transition function, $p(s_t, a_t, s_{t+1})$ is the probability of taking action a_t from state s_t and then transitioning to s_{t+1}. R represents the reward function, and $r(s_t, a_t)$ is the reward value returned by the environment when the agent performs the action a_t under state s_t. γ represents the discount factor of the reward, which is used to give weights to the rewards at different times, $\gamma \in (0,1]$.

The value function is used to quantify the value of each state, and the cumulative decay reward at time t is represented by G_t as follows:

$$G_t = r_{t+1} + \gamma r_{t+2} + \cdots + \gamma^{T-1} r_{t+T} = \sum_{l=t}^{T} \gamma^{l-t} r_{l+1} \tag{3.35}$$

where T is the number of iteration termination steps. Policy π is the probability distribution of all actions under state s_t, which determines which action to choose. Defining the action state value function, $Q_\pi(s_t, a_t)$ is the expectation of the cumulative decay reward obtained by the agent in the state s_t following the policy π to select the action a_t, namely:

$$Q_\pi(s,a) = E_\pi[G_t | s_t, a_t] = E_\pi[r_{t+1} + \gamma Q_\pi(s_{t+1}, a_{t+1}) | s_t, a_t]; \tag{3.36}$$

The function that maximizes the action state value function in all policies is the optimal action-value function, that is, $Q^*(s, a) = \max_\pi Q_\pi(s, a)$. When the action-value function is known, the ϵ-greedy strategy is used to select the action that maximizes the action-value function, namely:

$$\pi^*(a|s) \leftarrow \begin{cases} 1 - \varepsilon + \dfrac{\varepsilon}{|A(s)|}, & a = \arg\max_{a \in A} Q^*(s,a) \\ \dfrac{\varepsilon}{|A(s)|}, & other \end{cases} \tag{3.37}$$

In addition, DDQN uses different value functions for action selection and action evaluation to solve the problem of value function overestimation, hence its time difference target formula is:

$$Y_t^{DDQN} = r_{t+1} + \gamma Q(s_{t+1}, \arg\max_a Q(s_{t+1}, a_t; \theta); \theta') \tag{3.38}$$

where the action selection is performed by the action-value function: $a_{max} = \arg\max Q(s_{t+1}, a_t; \theta)$, and its neural network parameter is θ. After the maximum action is selected, the action evaluation is performed by the action-value function: $Y_t = r_t + \gamma Q'(s_{t+1}, a_{max}; \theta')$, and its neural network parameter is θ'.

3.4.4.2.2 Optimization Problem Transformation
3.4.4.2.2.1 *State* Most of the existing RL-based load balancing and routing optimization solutions adopt the traffic matrix to define the state space, but they rarely consider the dynamic change of resources (computing, storage, and bandwidth resources). Therefore, the network bandwidth measured by the SDN controller is fully utilized, and the state s_t at time t is defined as follows:

$$s_t = [D_t, P_t, U_t] = \begin{bmatrix} d_1, d_2, \cdots, d_J \\ p_{d_1}, p_{d_2}, \cdots, p_{d_J} \\ U_{e_1}, U_{e_2}, \cdots, U_{e_N} \end{bmatrix} \tag{3.39}$$

where J data flow requests D_t (represented by the starting node and destination node of each data flow) and their currently reachable paths P_t constitute one feature. The link bandwidth U_t is another feature, and the number of links equals its number of features, so the input features of DDQN-based optimization algorithm are $N+1$.

3.4.4.2.2.2 Action After sensing the network state information, the DDQN agent further learns the real-time status of each link in the network and quickly calculates the optimal forwarding path of the data flow, that is, action a. If the QoS optimization function is expressed by variable π, then the action at time t can be expressed as follows:

$$a_t = \pi(s_t) = \begin{bmatrix} d_1, d_2, \cdots, d_J \\ p_{d_1}, p_{d_2}, \cdots, p_{d_J} \end{bmatrix} \tag{3.40}$$

The transmission path action of each data flow may remain unchanged (that is, the transmission path used last time), or a new path may be selected from P.

3.4.4.2.2.3 Reward The rewards obtained by the DDQN agent are related to network QoS performance indicators, such as network delay, packet loss rate, and network throughput. Since the network throughput increases linearly with the increase of network load, it is prone to network congestion when the network load increases to a certain extent, which increases network transmission delay and data packet loss rate. Considering the two QoS indicators of network average transmission delay and network load balancing degree, the reward function is expressed by vector r_t as follows:

$$r_t = r(i \rightarrow j | s_t, a_t) = -\mu CV_{load} - \nu L_J \tag{3.41}$$

where μ, $\nu \in [0,1]$ is the adjustable weight, and $\mu + \nu = 1$. When $\nu = 1$, the agent only considers optimizing the transmission delay of the data flows.

3.4.4.2.3 Optimization Algorithm Based on DDQN After obtaining the network topology, link status, and data flow request information perceived by the SDN controller, as well as the network delay and link bandwidth measured by the SDN controller, the DDQN-based QoS optimization algorithm uses the DDQN algorithm to calculate the best forwarding path of each data flow. The specific process is shown in Figure 3.12, which is mainly divided into two processes: data collection and sampling, and the learning and training process of optimal path selection.

3.4.4.2.3.1 Data Collection and Sampling To reduce the correlation of interaction sequence data (composed of the current state s_t, transmission path a_t, reward r_t, next state s_{t+1}, that is (s_t, a_t, r_t, s_{t+1})) between DDQN agent and software-defined industrial environment, and to ensure that the data used for batch training are independent and identically distributed (i.i.d.). The DDQN adopts a replay buffer to store the interactive sequence data into the

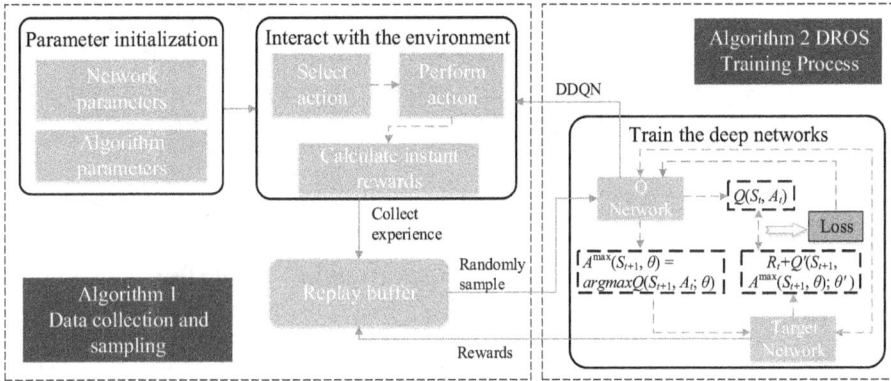

FIGURE 3.12   The process of DDQN-based optimization algorithm.

historical experience database, and then uniformly and randomly samples a certain size of data from the database to train the target network and Q- network; the specific data collection and sampling process is shown in Algorithm 3.7.

The DDQN collects and stores the continuously generated data tuple (s_t, a_t, r_t, s_{t+1}) into the replay buffer in chronological order. If the replay buffer is full of samples, the initially inserted data will be deleted. When the data stored in the replay buffer is sufficient, a certain size of data to learn is evenly and randomly extracted by the DDQN agent. Therefore, the introduction of the replay buffer not only reduces the correlation distribution of interactive sequence data but also improves the use efficiency of interactive sequence data.

ALGORITHM 3.7   Data Collection and Sampling Process

1:	The SDN controller perceives information such as network topology, link status and d_k, calculates P, L_k, and U_{en}^{used}, and sends the above information to the DDQN agent
2:	The DDQN agent initializes the replay memory D to capacity N
3:	The DDQN agent initializes the action-value function Q with random weights θ
4:	The DDQN agent initializes the target action-value function Q' with weights $\theta'=\theta$
5:	The DDQN agent interacts with the software-defined industrial environment to obtain interactive sequence data with certain correlations
6:	for episode$=1$, N do
7:	Initialize observation state s_t
8:	for $t=1$, T do
9:	With probability ε, select a random action A_t
10:	Otherwise, select $a_t = argmax_A Q(s_t, a_t; \theta)$
11:	Execute action a_t and observe reward r_t
12:	Set observation new state s_{t+1}
13:	Store transition (s_t, a_t, r_t, s_{t+1}) in D
14:	$s_t = s_{t+1}$
15:	end for
16:	end for

3.4.4.2.3.2 Process of Optimal Path Selection Based on DDQN DDQN uses experience replay to randomly sample and adopts two independent neural networks with the same structure, namely, Q network and target network. The specific learning and training process of the optimization algorithm is as follows:

- First, the maximum hop limit is used to eliminate the excessively long paths in P, then Q network uses the ϵ-greedy strategy to select the action a_{max} corresponding to the maximum Q value, thus completing the extraction of state features and the learning process from the state to the Q value.

- Then, the target network uses the selected action a_{max} to calculate the target Q value Y_t^{DDQN} by the Q-learning method.

- The loss function is calculated by the mean square error, that is: $L(\theta) = E[(Y_t^{DDQN} - Q(s_t, a_t; \theta))^2]$, and all the Q network parameters θ are updated by the gradient time series difference method of a neural network, that is, $\theta = \theta + \alpha L(\theta) \nabla Q(s_t, a_t; \theta)$, where α is the learning rate.

- The Q network parameters are copied to the target network at intervals of a certain number of pieces of training, that is, $\theta' = \theta$. This reduces the correlation between the output values of the two networks and realizes the iterative learning from the Q value to the final target Q value.

- Combined with the calculated optimal forwarding path and network topology of each data flow, the SDN controller determines the forwarding port of each SDN switch, encapsulates the routing information into flow tables and sends them to the corresponding SDN switches.

Therefore, the optimization algorithm based on DDQN not only utilizes the feature extraction characteristics of deep learning and the back-propagation optimization characteristics of neural networks based on a reward function, but also makes full use of the centralized control characteristics of SDINs. Algorithm 3.8 presents the learning and training process of the proposed QoS optimization and traffic scheduling algorithm; the trained model is able to determine the transmission path of data flow quickly according to the network state and ϵ-greedy policy.

3.4.5 Blockchain for Network Security and Privacy Protection

It is apparent that network security has become an important topic in recent years. Indeed, modern industrial networks have to ensure the seamless remote access to segments, or even single nodes, located up to the lowest levels of factory and process automation systems. This is, nowadays, feasible via the Internet: industrial automation devices (e.g., programmable controllers or even simpler components such as sensors/actuators) may be equipped with Web server modules. In these new scenarios, security issues must be addressed. Also, the classical countermeasures against threats adopted by general-purpose networks (e.g., the use of firewalls, cryptography techniques, and intrusion detection systems) might

ALGORITHM 3.8   Learning and Training Process of the Proposed Algorithm

1:	Initialize algorithm parameters, including iterations T, discount factor γ, learning rate α, exploration rate ϵ, number of samples for batch gradient descent m, target network parameter update frequency C.
2:	Initialize Q network Q and its parameters θ, target network Q' and its parameters θ'.
3:	Initialize target action-value function Q with weights $\theta'=\theta$
4:	for episode$=1$, T do
5:	for $j=1$, m do
6:	Sample randomly m samples (s_j, a_j, r_j, s_{j+1}) from D, $j=1, 2,..., m$
7:	Set $Y_j = r_j$ if episode terminates at step $j+1$
8:	Otherwise, $Y_j = r_j + \gamma Q'(s_{j+1}, argmaxQ(s_{j+1}, a_j; \theta); \theta')$
9:	Update all parameters of Q network θ by minimizing the loss function: $L(\theta) = E[(Y_j - Q(s_j, a_j; \theta))^2]$
10:	if T % C$=1$:
11:	Reset $\theta' = \theta$
12:	end for
13:	end for
14:	Obtain the best forwarding path at of the data flow, generate flow tables, and send them to the SDN switches

negatively impact the performance of industrial networks. Hence, their possible adoption is not straightforward and has to be investigated. Defense mechanisms are briefly surveyed and listed in Table 3.5.

However, due to the differences between wireless networks and wired networks as well as the special requirements of industrial application, a large amount of critical security and private data is very vulnerable to the threats and attacks, which would cause not only sensitive data leakage but also would bring huge social and economic problems, even endanger the national security. Therefore, the Blockchain-based architecture, which is an emerging scheme for constructing the distributed networks, is introduced to form a new, multi-center, and partially decentralized IIoT architecture [42].

3.4.5.1 Blockchain-Based IIoT Architecture for Smart Factory

Currently, most of the smart factories are based on the Cloud-Based Manufacturing (CBM) architecture. Such an architecture enables users to access the shared pool of manufacturing resources anytime and anywhere on demand, and rapid configuration and management of resources can be realized with the minimal work and third-party interaction. However, the centralized architecture is very fragile. Namely, as long as the central node is damaged, all services will be suspended. Therefore, to build a decentralized system with nodes supervising each other mutually, the Blockchain-based IIoT architecture for a smart factory is shown in Figure 3.13, including the sensing layer, the management hub layer, the storage layer, the firmware layer, and the application layer.

The sensing layer includes various types of sensors and at least one microcomputer with a certain computing power, which can obtain information on various equipment, and preprocesses the collected data. On the one hand, the management hub layer parses the

TABLE 3.5 Attacks and Coding Methods

Attacks	Methods
Eavesdropping	Encryption and decryption technique
Denial of Service	Random back-offs
Node Compromise	Code testing schemes
Selective Forwarding	Multi-hop acknowledgments
Sybil Attack	Link layer encryption and authentication
Sinkhole and Wormhole Attacks	Unique symmetric shared key
Physical Attack	Tamper-proof hardware

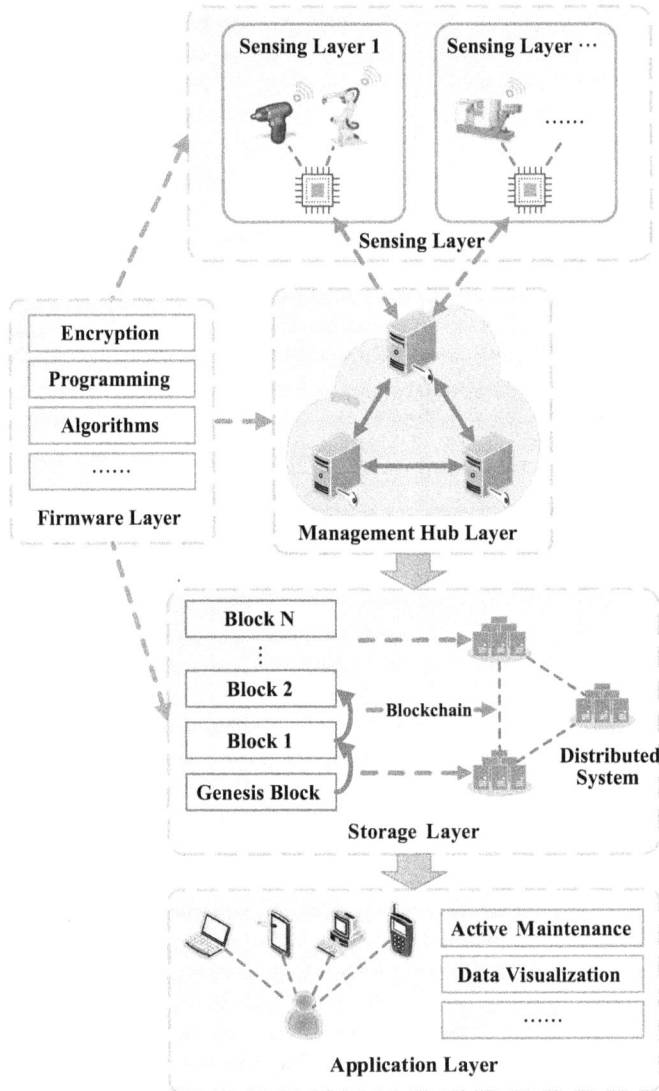

FIGURE 3.13 The Blockchain-based IIoT architecture for a smart factory.

uploaded data, encrypts the data, packages the data to generate blocks, and stores it in the database. On the other hand, the management hub layer needs to integrate and manipulate different equipment based on the production scheduling strategy and respond to the users' requests in real-time to provide the customized services. The storage layer plays the role of a data center, keeping the encrypted, tamper-resistant data and blockchain records, which are stored in a distributed form and synchronized at a certain interval. To make the sensing layer, the management hub layer and the storage layer complement each other effectively, the firmware layer is proposed, which involves the underlying implementation technologies to connect each layer, including the data acquisition, distributed algorithms, and data storage technology. The application layer provides users with different kinds of services such as real-time monitoring and failure prediction.

3.4.5.1.1 Division of the Blockchain-Based Architecture   The Blockchain-based architecture is divided into the intranet and the extranet. The intranet aims to collect and store data, while the extranet aims to utilize the data to provide different services to different users.

The intranet consists of the sensing layer, the management hub layer, and the storage layer. Due to the limitation on the computing power, a peer-to-peer (P2P) network is not considered. The data of each equipment node need to be uploaded to the designated management hub and managed by the management hub. Every equipment node needs to request permission from the management hub for different operations. The Bitcoin uses the Unspent Transaction Output (UTXO) format to ensure good anonymity and security. Since the number and authority of participating equipment nodes in the private blockchain are strictly limited, instead of the UTXO, the state of sensors is recorded directly. Moreover, considering the diversity and complexity in the IIoT system, the sensing layer is equipped with microcomputers, such as STM32 and Raspberry Pi, to preprocess the data, which help effectively reduce the overhead of the upper system.

The extranet consists of the management hub layer and the application layer. The main difference between the intranet and the extranet is that the intranet is oriented to the equipment, while the extranet is oriented to the users. Therefore, the extranet needs to connect to the Internet and consider connection, algorithms, tools, etc. to utilize the data and create a reasonable access method to provide services for the users. That means users can customize diverse management hubs according to their own needs, and high QoS requirements can be ensured.

In order to ensure data security and privacy, both intranet and extranet have the whitelist and dynamic authentication mechanisms to restrict nodes. In the whitelist mechanism, a whitelist is usually used together with a blacklist, which determines the right to access or deny. Such a mechanism can quickly verify the access traffic and filter the malicious traffic, providing fast and convenient security and privacy. Considering the equipment nodes and user nodes in the architecture, the whitelists and blacklists should be created, respectively, for the intranet and the extranet. Furthermore, the dynamic verification mechanism is also an effective way to guarantee good security and privacy of data. The permission acquired by equipment nodes and user nodes is time-limited. When the time limit is reached, the permission and the Proof of Work (PoW) need to be re-verified.

What's more, the management hub layer and the user layer are also provided with a self-running algorithm to generate a paired verification code. Users need to provide the code to maintain access permission when re-verifying.

3.4.5.1.2 Management Hub Layer   In the blockchain system, besides the transaction node, there is a special node responsible for recording blocks. In the Blockchain-based architecture, that is the management hub, constituting the management hub layer.

In order to ensure that all management hubs are trusted, some consensus algorithms must be applied to the architecture, such as PoW, Proof of Stake (PoS), and Practical Byzantine Fault Tolerance (PBFT). By raising a mathematical question with a reward as the PoW, the Bitcoin system encourages the recording nodes to compete. The reward guarantees the participation of the recording nodes, and the computational cost brought by the mathematical problem solving makes the malicious operation costly. However, this solution not only wastes a lot of computational resources but also brings the problem of scalability.

In the IIoT system, more attention should be paid to the resource utilization and data interaction efficiency. Moreover, unlike the untrusted system in the Bitcoin, the proposed system is a private blockchain system with all nodes trusted initially. Therefore, the competition and reward mechanism are abandoned. Each workshop in a smart factory is equipped with one or more specialized management hubs for data management. Then, the Statistical Process Control (SPC) or other comparison algorithms are introduced to complete the PoW. For SPC, according to the specific equipment, eigenvalues such as control limits and average values are first set. Then, some statistical analysis of the uploaded data is carried out to compare it with the set values and authenticate the transaction. Each time a new block record is generated, the PoW should be carried out once. Such a design can improve the fault tolerance, scalability, and real-time capability of the architecture.

Instead of setting up the cloud, multiple management hubs are considered to form a private cloud, which allows users to connect and access the data. It can be considered that the management hub is equivalent to the fusion of an intelligent edge gateway and a cloud, which is a simplification and improvement of the edge computing architecture. Although the sensing layer and the management hub layer are a centralized LAN system, from the global perspective, multiple management hubs constitute a partially decentralized system. Multiple distributed nodes can reduce the pressure of a single central node and avoid system failure caused by a single central node crash. Such design is a suitable solution for the Blockchain technology in combination with the actual industrial environment.

3.4.5.1.3 Private Blockchain   As shown in Figure 3.14, a private and unique block structure is designed to record data in the storage layer, which contains two parts: block header and block body. Structured data are stored in the block header, recording multiple characteristic values of the current block, including the hash value of the previous block, the timestamp of the current block, and the Merkel root generated by specific algorithm on the collected data. Generally speaking, a block is more like an index file. The specific equipment data is still stored by the database, and the access record (storage, reading, and

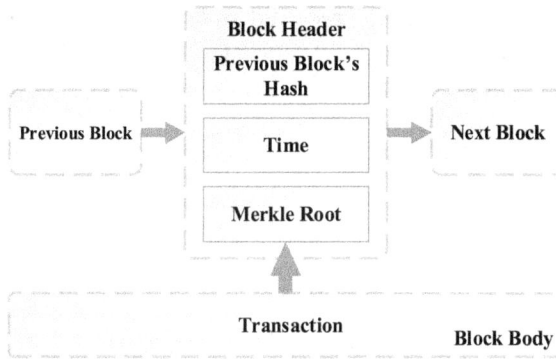

FIGURE 3.14   Block structure.

TABLE 3.6   Transaction in the Block Body

Address	Request	Merkle Root	Action
1	Store	1	Deny
2	Read	2	Deny
3	Control	3	Deny

control) of each node is stored in the block. Through this setting, each operation will be strictly supervised via blocks, while the advantages of traditional databases can be maintained. Therefore, various operations of each node can be traced and the data interaction can be highly protected.

In Table 3.6, the examples of the information recorded in the block body, including the request address, the request content, the Merkel root, and the response from the management hub, are given. The block limits the number of data records, which means a new block will be generated after a certain number of insertions. In fact, the storage layer is the physical storage device in the management hub, such as Solid State Drive (SSD) and Hard Disk Drive (HDD), which constitutes a distributed storage system.

In order to achieve good privacy as well, the hash algorithm and the asymmetric encryption algorithm are introduced. For the Blockchain-based architecture, the SHA256 and elliptic curve cryptography algorithm are chosen. By using the SHA256, the Merkle root can generate, which effectively compresses the volume of data to link each block. When using the encrypted data, users can perform the same hash calculation and compare the hash codes to verify the data. The elliptic curve cryptography algorithm is used to generate the public and private keys to encrypt the data. Data in the database will be encrypted into ciphertext via the public key. Users should provide the private key to decrypt the encrypted message for customized services. These two algorithms can ensure good privacy of the data and prevent illegal manipulation.

3.4.5.2 Data Security and Privacy Model

The Blockchain-based architecture needs to address three major requirements: confidentiality, integrity, and availability (CIA), which can be referred to as the CIA requirement. The confidentiality ensures that only authorized users can read the message; the integrity ensures that no changes are made to the received and sent messages; the availability means that each service and data are available. The Bell-La Padula (BLP) model and Biba model are suitable for CIA requirements. Based on these two models, some simplifications are made and some attributes are introduced:

$$S = \{s_1, s_2, s_3, \ldots, s_n\} \tag{3.42}$$

$$O = \{o_1, o_2, o_3, \ldots, o_n\} \tag{3.43}$$

$$\mu = \{M_1, M_2, M_3, \ldots, M_n\} \tag{3.44}$$

$$A = \{w, r, c\} \tag{3.45}$$

$$L = \{l_1, l_2\} \tag{3.46}$$

where S is the set of subjects. O is the set of objects. μ is the set of access matrixes, which indicates the access privileges that the subjects have to the objects. A is the set of access attributes in which w stands for storing, r stands for reading, and c stands for controlling. L means different privilege levels in which $l_1 < l_2$. In the Blockchain-based architecture, the security level can be easily combined with the integrity level to get comprehensive privilege levels of different subjects or objects. Based on the definition above, an Access Control List (ACL) is designed as shown in Table 3.7.

It can be concluded that the same level of subjects have all permissions to objects, high-level subjects have read and control access to low-level objects, and low-level subjects have no permissions to objects above their own level. On the one hand, data should be restricted to flow from low level to high level; on the other hand, data from low level cannot be tampered by high level. These two rules help maintain fine CIA requirements. Furthermore, there are three entities in the architecture: equipment nodes, management hubs, and user nodes. Equipment nodes and management hubs should belong to l_2, while user nodes should belong to l_1. The data can just flow from equipment nodes and management hubs to user nodes; and user nodes have no authorization to write or modify the data. However, equipment nodes and management hubs have all permissions to each other, so that effective data interaction can be implemented.

Nevertheless, a formula is also defined to determine if the current state is safe:

$$V = S \times O \times A \times \mu \times L \tag{3.47}$$

TABLE 3.7 Access Control List

Subject	Object	
	l_1	l_2
l_1	$\{w, r, c\}$	$\{r, c\}$
l_2	$\varnothing$	$\{w, r, c\}$

TABLE 3.8 Defensive Mechanisms

Subject	Object		
	Equipment Nodes	Management Hubs	User Nodes
Equipment nodes		Whitelist, PoW, Dynamic verification, Merkle Root	
Management hubs			
User nodes		Whitelist, Dynamic verification, Asymmetric encryption, Merkle Root	

In this formula, $S \times O \times A$ indicates that a subject uses some method to access an object; μ indicates the access matrix; and L indicates the privilege levels. Once all the elements are secure and trusted, a secure state can be ensured. Since we have divided the architecture into different levels and made some definitions, which means $S \times O \times A$ and L have been strictly restricted and abided by, μ is the last element that should be considered. Therefore, as shown in Table 3.8, some defensive mechanisms have been designed to help control access matrix.

Actually, the management hubs play the role of a data transmission intermediary. Except for the rigorous hierarchy, different means of verification are set to help keep the architecture secure. Together with Blockchain technology, not only can they provide effective methods to control all kinds of access which help prevent from malicious activities, but also can enhance the security and privacy of the architecture, forming a more mature and stable system.

3.4.5.3 Data Interaction Process Design

The process of data interaction in the Blockchain-based architecture is designed to prevent the possible attacks and threats: leakage of permissions, DoS or DDoS attacks, network sniffer, compromised-key attack and invasion. Taking the temperature collection as an example, the Blockchain-based architecture should achieve data acquisition first, which depends on the microcomputers. Generally, a microcomputer can manage one or more sensors and connect to one management hub. After the data is acquired, the microcomputers are required to register a unique ID, which will be put in the whitelist in the connected management hub. The whitelist has backups in each management hub. If one management hub crashes, the connected microcomputer can be standby or choose to change its network settings to switch to other management hubs. The flowchart of preparation is shown in Figure 3.15.

In the data interaction process, the attacks of stealing and abusing node permissions (primarily the sensing layer and the management hub layer) are considered. Therefore, two defense mechanisms are designed. On the one hand, in the sensing layer, the whitelist mechanism, the dynamic verification mechanism, and the PoW consensus algorithm are integrated to prevent malicious traffic and injection of erroneous data. On the other hand, as a multi-center system is created, under the supervision of the other management hubs,

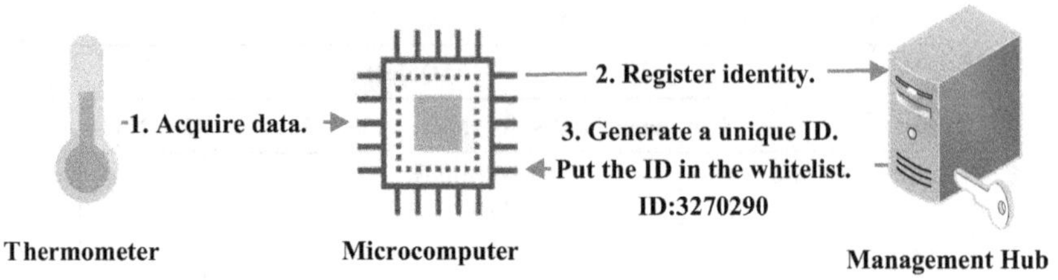

FIGURE 3.15 Data preparation flowchart.

the invaded management hub can be discovered, excluded, and replaced quickly. These two defense mechanisms can guarantee the stable operation of the underlying system.

The detailed process of applying for the storage permission by the equipment node is shown in Figure 3.16, and it can be concluded as Algorithm 3.9. After obtaining the permission through the whitelist verification, the data is first put into the buffer pool. When reaching a certain amount of data, the management hub will use built-in comparison algorithms to calculate the characteristic values and compare them with the set values through which PoW is accomplished. If the requirements are satisfied, the data in the buffer pool will be put into the database; meanwhile, the uploaded data can be directly transmitted to the database for a period, and all operations within the permission of the equipment node will be allowed; otherwise, the data collected in the buffer pool will be discarded and the permission request will be rejected. It should be noted that all data being transmitted to the

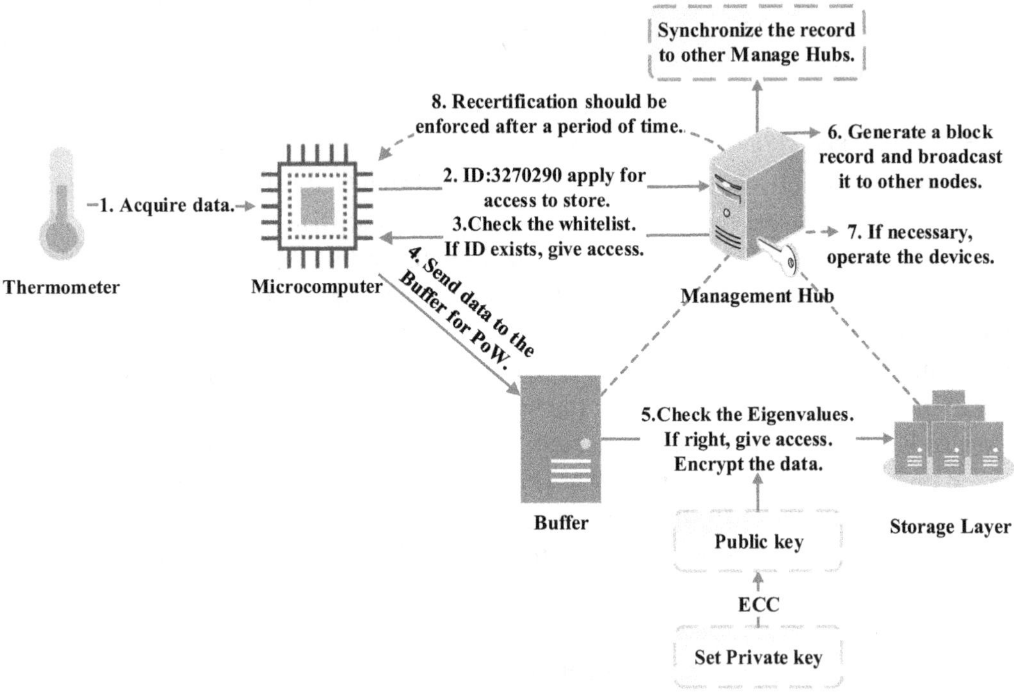

FIGURE 3.16 The intranet flowchart (equipment interaction).

ALGORITHM 3.9   Data Interaction in the Intranet

```
01 Begin
02    for i←1 to mComputer[1...a]
03       find the connected mComputer[j] for mComputer[i]
04       register ID
05    end for
06    wait()   //wait for application
07    if(requestReceived == true)
08       if(compare the mComputer with whitelist[1...a] == true)
09          tick()   //record the running time
10          wait for enough insertions in the buffer for the PoW
11          if(execute PoW == true)
12             generate and broadcast a block record
13             subsequent data is uploaded to the database directly
14          else
15             deny and generate a block record
16             discard the data in the buffer
17          end if
18          if(time == set value) close the connection
19       end if
20       else
21          deny, generate and broadcast a block record
22       end if
23    end if
24 End
```

database need to be converted into ciphertext by the public key. The management hub will generate a new block record for each permission request. The block record is then broadcasted to the other management hubs, which will verify the block record again and record it. However, after a certain period, the dynamic verification mechanism requires the re-authentication, and therefore the system needs to repeat the process of 1–6 in Figure 3.16. The flowchart is an example of applying for the storage permission, and the process of reading and control request are the same.

As discussed above, the data interaction is implemented in the intranet. As shown in Figure 3.17, the flowchart of the extranet interaction is similar to that of the intranet. Since the management hub is connected to the Internet, the Dos attacks or DDoS attacks will be very frequent. Therefore, the whitelist mechanism, the dynamic verification mechanism, and the asymmetric encryption mechanism are set up for the extranet. The whitelist mechanism and the dynamic verification mechanism are the same as that of the intranet, performing the screening and eliminating malicious traffic on the Internet. On the other hand, the asymmetric encryption technology is designed specifically for the extranet to prevent unauthorized access. The process is also concluded as Algorithm 3.10.

In the Blockchain-based architecture, the management hub is also a record node. Each block and equipment data have a copy in each management hub. When block records reach the limit, the management hub with a low overhead will generate a block to record all the access applications during that period. As shown in Figure 3.18, after the block is generated, it will be placed in the storage layer and synchronized with the other management hubs. In addition to the above defense technologies, the double Merkel roots are introduced in the

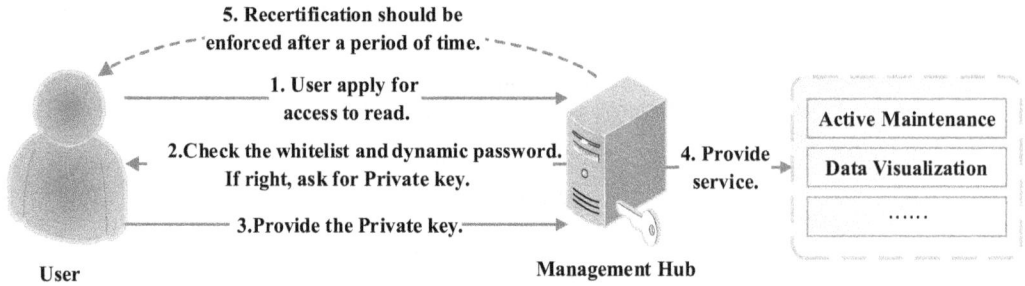

FIGURE 3.17   The extranet flowchart (users apply for service).

ALGORITHM 3.10   Data Interaction in the Extranet

```
01 Begin
02    for i←1 to user[1…d]
03       find the connected mComputer[j] for users[i]
04       register ID
05    end for
06    wait()   //wait for application
07    if(requestReceived == true)
08      if(compare with the whitelist[1…a] and password == true)
09          tick()   //record the running time
10          user verify the data, provide private key to get service
11          generate and broadcast a block record
12             if(time == set value) close the connection
13             end if
14      else deny, generate and broadcast a block record
15      end if
16    end if
17 End
```

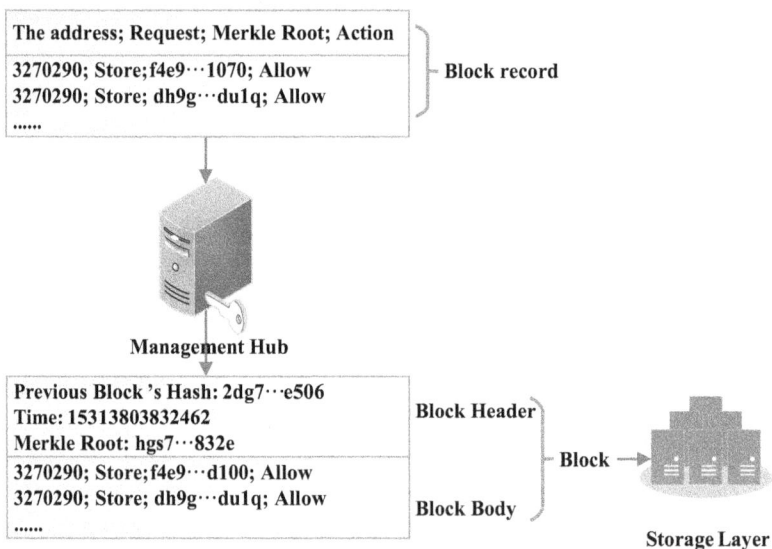

FIGURE 3.18   Block generation.

block record to protect the data. The first one is applied on the data in the buffer pool existing in the block record; the second one is applied on the data in the block body existing in the block header. Such kind of nesting guarantees that the data will not be sniffed and the malicious invasion is difficult to achieve.

3.5 VALIDATION OF QOS OPTIMIZATION METHODS OF HETEROGENEOUS NETWORKS IN SMART FACTORIES

In this section, the validation of relevant QoS optimization methods of heterogeneous networks in smart manufacturing factory was described. First, the low latency assisted by proactive caching of EC was verified. Second, the cloud-assisted ant colony-based low latency of mobile handover method was verified. Finally, the network load balancing and routing optimization based on deep RL was verified.

3.5.1 Validation of EC Proactive Caching for Low Latency

In the simulation, ten edge clusters are randomly deployed in an industrial domain. Therefore, five edge clusters cover the MN path L with an edge server caching (ESC) capacity of $Se=500$ Mbits. Each edge cluster has 5 SNs covered paths and meets the minimum sojourn time with a caching capacity of $Ss=100$ Mbits. The big data F contains 100, 200, 300, or 500 Mbits. The total communication rate mr_i is 0.5, 1, 1.5, 2, 3, 3.5, 4, or 4.5 Mbps. In addition, we consider that the length of each edge cluster is a covered path $l_i=l_j=100$ m and that every SN covers a path length of 20 m. The MNs ensure an equivalent speed, where the moving speed vi is 1 m/s. The configuration of the simulation parameters is shown in Table 3.9.

Four performance metrics are used to evaluate the above method: *goodput, time delay, throughput,* and *energy consumption.* The details are explained as follows. The **goodput** is the amount of useful information delivered to the MN with an imposed deadline. The **time delay** is the standard metric to measure the time consumption; it is also the difference in time between when the MN sends the required large amount of data F and when F is completely received. The **throughput** is the rate of successful message delivery over a wireless

TABLE 3.9   Parameter Values

Parameters	Values
Number of covered path edge clusters E	5
Number of neighbor SNs in each edge cluster $N(e_i)$	5
Big data size F	100–500 Mbits
Portion of big data size of fi D_f	5 Mbits
ESC capacity Se	500 Mbits
Covering path SN caching capacity Ss	100 Mbits
MN moving speed Vi	1 m/s
Each edge cluster covered path li	100 m
Communication rate between MN with SN mr_i	0.5–4.5 Mbits/s
Communication rate between MN with edge server cr_i	0.5–4.5 Mbits/s
Average time delay of caching node T_{delay}	1 s
Handover time of MN for fetching data $T_{Handover}$	1 s
Communication efficiency of MN in edge server γ_i	1

channel of MNs. The ***energy consumption*** is the total energy consumption using wireless communication while receiving data. For the performance comparison, three schemes are compared to the above method. For the sake of convenience, the above method is simple as "PCE." *Cloud server download* (CSD): In this type of architecture, mobile network nodes fetch data from cloud servers using multiple hops. In the simulation, it assumed that four hops exist between the cloud server and the MN. *Edge server cache* (ESC): By adopting CES (edge computing server) caching, MNs directly download large amounts of data from the edge server. *Multiple communication paths* (MCP): Under this strategy, downloading the data from the TCS is provided via MCP.

Goodput: Figure 3.19 plots the average goodput obtained by various simulation settings in edge computing mobile networks. Figure 3.19a shows the goodput values of the MN and the mobile trajectories for various transmission data sizes. The results of the PCE outperform the reference schemes that increase the average goodput using the same communication rate. Moreover, by increasing the transmission data size, the average goodput of the PCE is kept at a higher level, while the goodput values are more than 0.9 when the data size changes from 100 to 500 Mbps. This result occurs because our proposal considers MN properties and neighboring node influences and adopts optimization strategies. Figure 3.19b plots the goodput value for various communication rates and under different schemes. The PCE has a higher goodput value than the CSD, MCP, and ESC. Furthermore, our proposal can ensure successfully retrieved data from MNs to download large amounts of data from TCSs.

Time delay: Figure 3.20 shows the relationship between the time delay, data size, and communication rate. In Figure 3.20a, we assume that the complete network has the same communication rate (0.5 Mbps). We can predict that the time delay increases as the data size increases. However, PCE has the best performance in terms of the time delay. When the data arrives at a rate of 500 Mbps, PCE can save 5%, 30%, or 45% of transmission time in MNs relative to the ESC, MCP, and CSD methods, respectively.

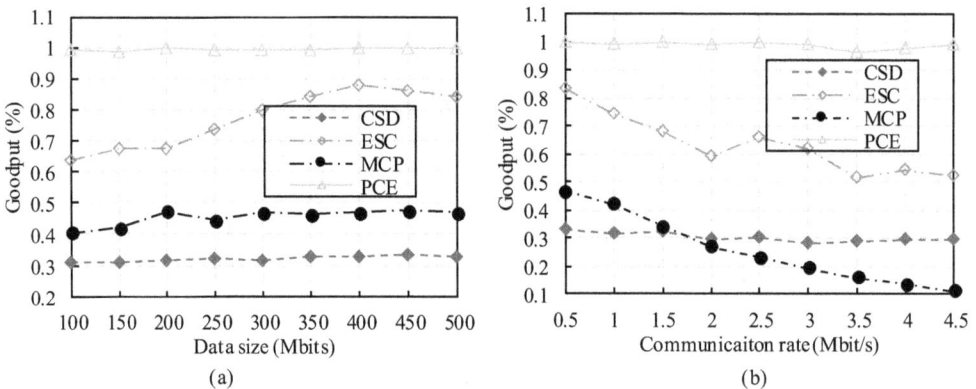

FIGURE 3.19 Comparison of goodput performance: (a) various data sizes and (b) various communication rates.

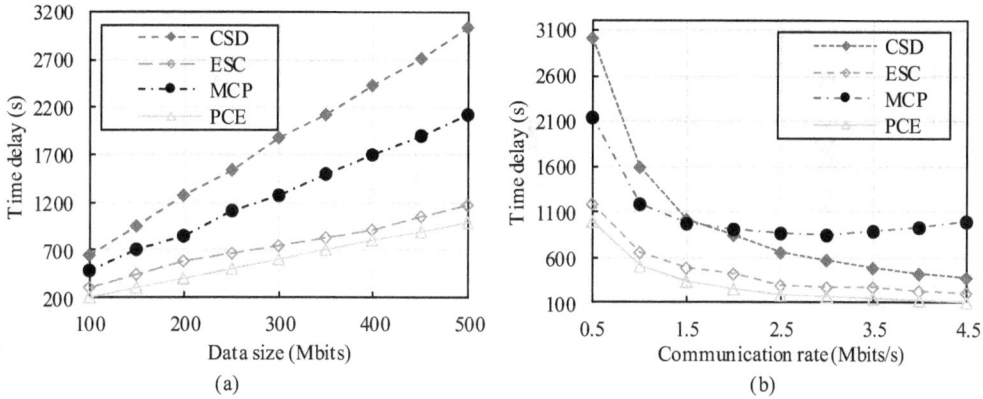

FIGURE 3.20 Comparison of time delay performance: (a) various data sizes and (b) various communication rates.

Similarly, the time delay values of the various schemes are shown in Figure 3.20b. As the communication rate increases, the time delay of all the strategies drops swiftly. The PCE still has a lower time delay for MNs in many other ways. In summary, our proposal leverages the edge server and neighbor nodes to cache large amounts of data. PCE helps to increase the real-time performance of complete IWNs and guarantees time determinism.

Throughput: The throughput is a well-suited evaluation indicator for networks. We evaluate throughput performance from an MN perspective in simulation. Figure 3.21a shows the throughput value of three data size situations for the reference scheme with a communication rate of 0.5 Mbps. The results demonstrate that our proposal outperforms the reference schemes for throughput. Moreover, regardless of the data size situation, the throughput values of PCE are more than 0.45. Furthermore, in the PCE method, MNs spend less time waiting. Figure 3.21b shows the throughput values of various communication rates. PCE has the highest throughput values in many ways. When the communication rate is 4.5, the PCE can increase the throughput by 9, 3, and 1.5 times in MNs relative to MCP, CSD, and ESC, respectively. In other words, PCE offers superior throughput performance.

Energy consumption: The average energy consumption of the four competing schemes with respect to various traffic loads is shown in Figure 3.22a. The resulting pattern is shown in Figure 3.22a. As the MN spends less time fetching data, PCE achieves a significantly lower energy consumption than other methods. As the data size increases, most methods consume more energy for commutation purposes. To establish a relationship between energy consumption and communication rate, we plot the energy consumption with respect to the communication rates of various strategies in Figure 3.22b. PCE guarantees the least energy consumption. When the communication rate is 4 Mbps, PCE can save 38%, 64%, and 83% energy consumption for ESC, CSD, and MCP, respectively. These results indicate a significant energy efficiency and longer lifetime of industrial MNs and its network.

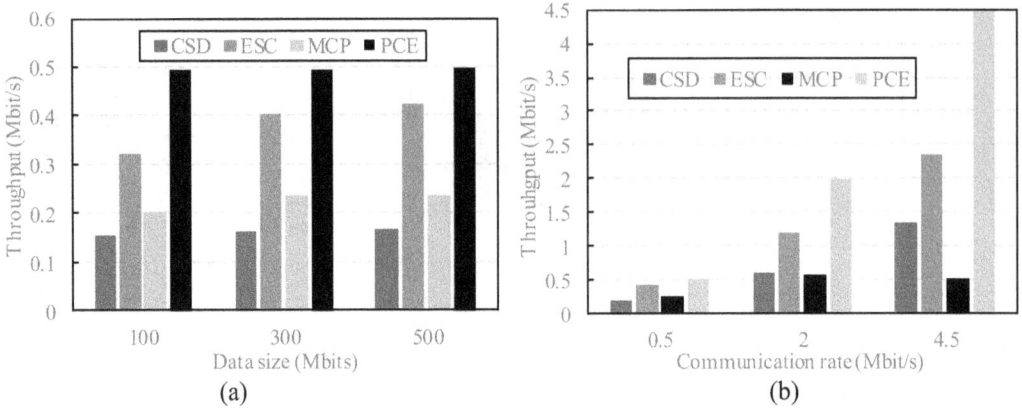

FIGURE 3.21 Comparison of throughput performance: (a) various data sizes and (b) various communication rates.

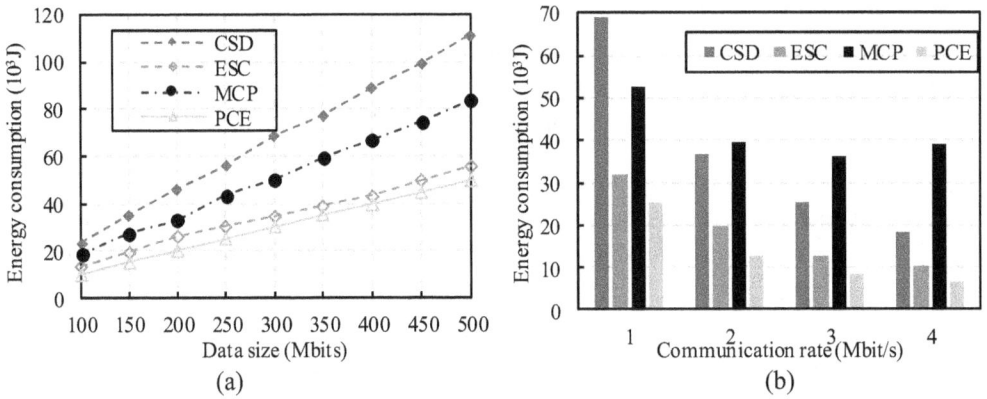

FIGURE 3.22 Comparison of energy consumption: (a) various traffic loads and (b) various communication rates.

3.5.2 Validation of Mobile Handover Latency Optimization

To evaluate the mobile handover latency optimization method, the simulator generates an evaluation environment and performs related actions. The analysis is based on a random fixed path of a mobile node and the payload of APs. However, due to space limitation, the results are presented for a single fixed path and determined payload. The following industrial domain network size is considered: $1,000*1,000\,\mathrm{m}^2$. Without loss of generality, we consider the movement trajectory ($y = 1,000 - 2*x, 0 \leq x \leq 500; y = 2*x - 1,000, 500 \leq x \leq 100$) as a simulation example. At the same time, for simplification, in the experiments, the even discrete points are selected for the expression of the trajectory. As depicted in Figure 3.23, the APs are deployed in the simulation domain randomly (using the normal distribution). The solid circle denotes the AP positions covering the trajectory, while the hollow ones are off the trajectory. The dotted line is the travel path of the mobile nodes. The big circles are the communication ranges of the APs.

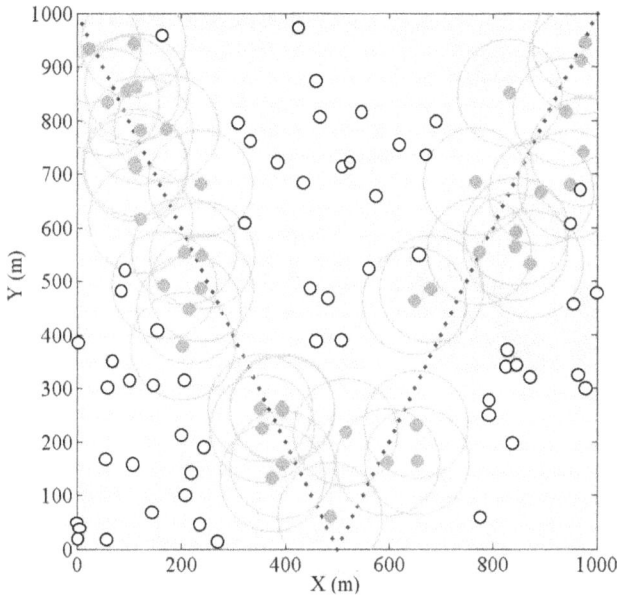

FIGURE 3.23 Simulation environment.

TABLE 3.10 Simulation Parameters

Parameter	Value	Parameter	Value
ξ	0.9	α_3	1
θ	0.2	Z	1,000
α_1	1	∂	1
α_2	1	$\wp$	1

For the wireless mobile nodes, the transmission ranges R_i are set as 100 m and the moving speed is set as 1 m/s. The handover time is between 1 and 15 ms. For the ant colony, the number of ants is set to be between 10 and 50. The other parameters of the proposed approach are provided in Table 3.10. In the experiments, it assumed that $Cov(ij)$, $S(ij)$, and $\Omega(ij)$ have the same weight, so in equation (3.9), the values of these coefficients were set as 1. The $fit(ant_i)$ contains the metrics of the search results. Due to similar considerations, ∂ and $\wp$ are set as 1.

Convergence of the optimal solution in the fitness function and handover number: First, the convergence of a different number of loops and ants is studied by measuring the fitness value and the handover number. The results are reported in Figure 3.24. As shown in Figure 3.24a, when the number of ants is 10, the fitness value gradually decreases with the number of loops. However, for the same number of loops, as the number of ants increases from 10 to 50, the fitness function value decreases from 4,700 to approximately 3,100. In short, the results show that as the number of cycles and the number of ants increase, the fitness function value has an overall downward trend. This shows that the CAFP mobile node handover strategy has a convergent fitness function value.

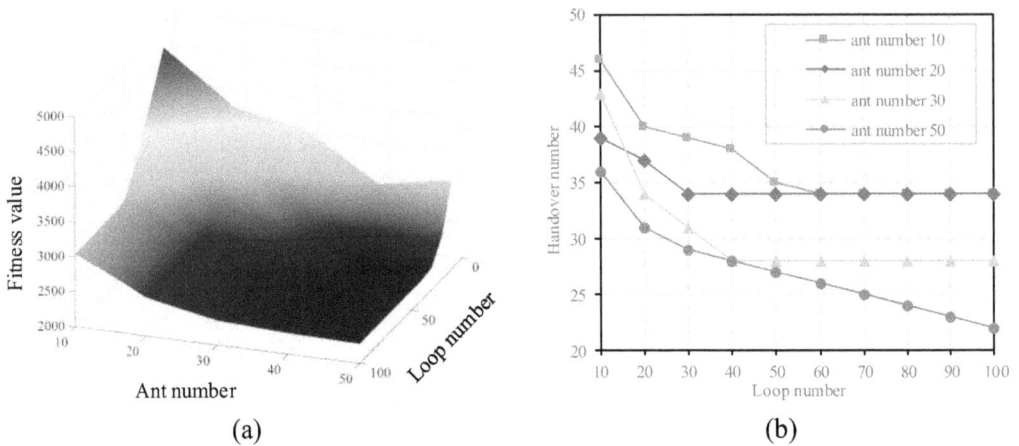

FIGURE 3.24   Convergence of the optimal solution: (a) fitness value and (b) handover number.

Figure 3.24b further investigates the handover number convergence of the optimal solution. It can be easily noticed that all the curves of the handover number show a downward trend. Once the number of loops reaches 50, the handover numbers have smaller changes. In addition, as the number of ants increases, the handover numbers also have a downward trend. Overall, in industrial wireless networks, there is convergence of the CAFP method in terms of the mobile node handover number.

Handover number comparison: A metric of the total number of handovers can be obtained as the sum of the mobile node handover numbers in the network in order to finish all tasks from start to finish. It is known that as the number of handovers increases, mobile nodes take more time to complete handover between APs, and the network devices waste more resources on management. Therefore, the total number of handovers is an important metric for the evaluation of mobile node handover. The number of handovers is a basic standard used to measure the merits of a handover algorithm. The handover number results for the CAFP approach in comparison with the Greedy-load, RSSI-threshold, and Random algorithm are shown in Figure 3.25.

Figure 3.25a shows the total number of handovers for a mobile node to finish the fixed path, with every handover taking 20 ms. The simulation results indicate that the CAFP method has better performance than the Greedy handover scheme. Compared to the RSSI and Random approach, the CAFP has little differences in terms of the evaluation index of the handover number. In the other words, the CAFP does not increase the number of handover events. The results in Figure 3.25a show that the Greedy-load method has the largest number of handovers. The results are plotted in Figure 3.25b to obtain a complete picture of the number of handovers for different handover times. From this graph, it can be observed that the Greedy-load method is constant, and the RSSI-threshold and Random methods have only small changes. However, the CAFP reduces the number of handovers as the handover time gradually increases. For instance, for the handover period of 18 ms, the CAFP method has

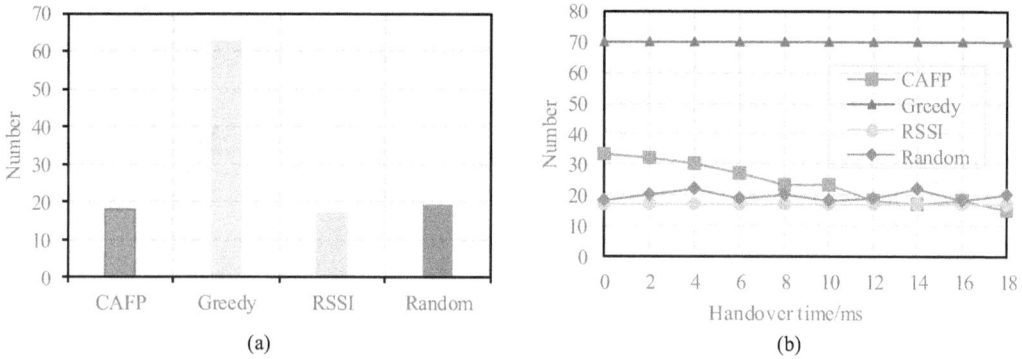

FIGURE 3.25   Comparison of handover numbers: (a) handover number and (b) handover number using different approaches

a smaller number of handover requests compared to the other schemes, reducing the number from 30 to approximately 20. The result varies, because with an increase in the handover time, the weight of handover in the optimization function increases. In addition, it can be observed that our approach has a decreasing trend for the handover number. In other words, the CAFP handover number has good adaptability to different handover times.

Load and time consumption in different approaches: The payload of APs, the handover time, and the number of handovers should all be considered across the whole course of the mobile node. The total time consumption is the best standard metric for handover strategies. For easy comparison and analysis, the sum of total handover times, communication times, and load latency is used to evaluate the algorithms. Figure 3.26 presents the final values for time consumption. Similar performance results can be observed in Figure 3.26a. The Random method and RSSI method have a higher load, while the CAFP approach and the Greedy algorithm have smaller values. The results in Figure 3.26b indicate that the CAFP has the lowest average time consumption, saving approximately 5%, 14%, and 42% of time compared to the Greedy-load, RSSI-threshold, and Random method, respectively. This is because the payload of a mobile node acts as an evaluation index in the optimization function. Figure 3.26c shows the final overall time consumption in the case when the handover time is changed. These results show that for all handover algorithms, as the handover time increases, the time consumption also has an overall upward trend. However, when the handover time is larger than 8 ms, the CAFP delivers the best performance in terms of time consumption. Furthermore, the graph shows that the Greedy-load method has good performance with small mobile nodes and APs. In addition, the CAFP shows strong robustness in terms of the handover time and AP payload. On the one hand, the cloud can provide the best handover solution based on the global factors. On the other hand, the CAFP maintains better connectivity restoration performance, owing to the back-off strategy for malfunctioning APs.

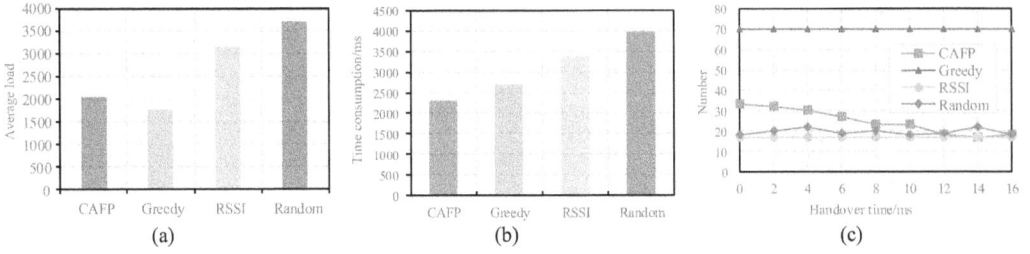

FIGURE 3.26 Load and time consumption for different approaches: (a) average load, (b) average time consumption, and (c) time consumption.

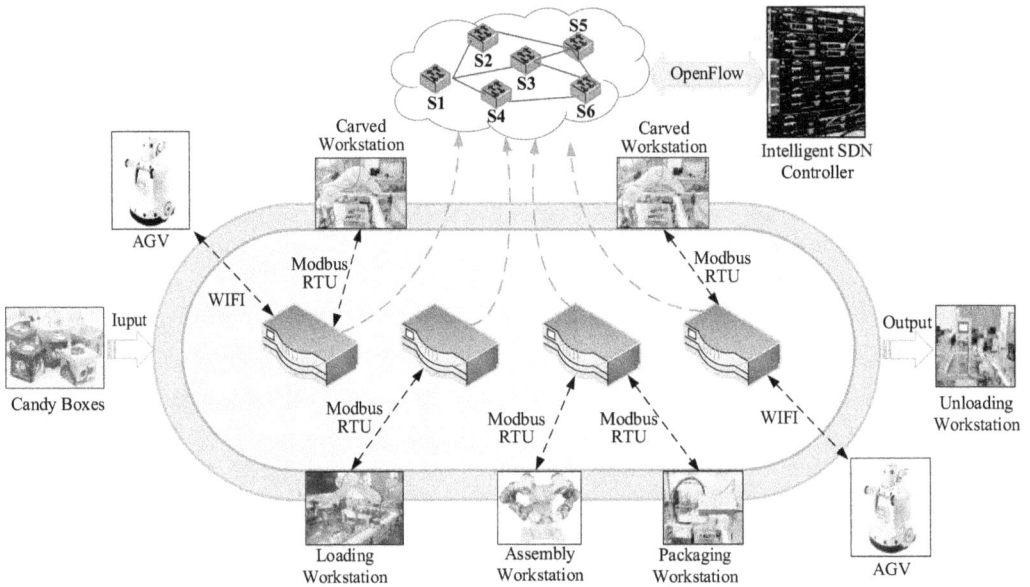

FIGURE 3.27 Production prototype of customized candy packaging for smart factories.

3.5.3 Validation of Load Balancing and Routing Optimization

3.5.3.1 Experimental Environment and Scheme

To reduce the cost of deployment, upgrade, and testing caused by directly transforming the traditional smart factory network system, a network simulation environment is constructed using the Mininet network emulator and Ryu controller, and the DDQN-based load balancing and routing optimization method is verified based on the network simulation environment. The experimental topology is based on the production prototype of customized candy packaging shown in Figure 3.27 (abbreviated as production prototype). The production prototype comprises bottom-level processing equipment, heterogeneous network equipment, private cloud servers, and remote cloud servers. The bottom-level equipment includes nine functional modules including box loading, material loading, product packaging, product assembly, product lid cover, laser printing, fine carving, and unloading. Taking the box loading module as an example, the equipment associated with the manipulator included Raspberry Pi, RFID reader, photoelectric switch, conveyor belt, candy box (with an RFID tag), and other equipment. The interconnection between devices is realized through wired/wireless heterogeneous networks. The heterogeneous network communication protocols involved include Modbus, TCP/IP, EtherCAT, RS-232, and WiFi. Therefore, the production prototype which is various equipment and heterogeneous networks is suitable as the verification platform of the DDQN-based load balancing and routing optimization method for smart factories.

The network topology contains four edge intelligent gateways and six SDN switch nodes in the production prototype network system. To verify the performance of the proposed algorithm such as convergence and generalization, two network topologies are defined:

- **Topo-1:** Based on the network node and link connection in Figure 3.27, all links have the same bandwidth, and the link bandwidth is greater than the maximum data flow size.

- **Topo-2:** Based on the network node and link connection in Figure 3.27, all links have the same bandwidth, and the link bandwidth is less than the minimum data flow size.

And this paper designed two sets of experimental schemes, namely: (1) In Topo-1, the convergence and effectiveness of the DDQN-based QoS optimization algorithm are verified under different μ, v, and α; (2) The generalizability of the DDQN-based QoS optimization algorithm is verified, and the advantages of the DDQN-based QoS optimization algorithm with OSPF-based and DQN-based QoS optimization algorithm under the two network topologies are compared and evaluated.

3.5.3.2 Experimental Setup and Details

The DDQN-based QoS optimization algorithm runs on the following hardware environment: NVIDIA Geforce GTX 1060 GPU, 16GB memory, i7-8700 CPU, and Ubuntu 18.04 operating system. The DDQN-based QoS optimization algorithm uses TensorFlow as the code framework and designs a two-layer fully connected network in which the hidden layer includes 32 neurons. The ReLU is used as the activation function, and the Adam optimizer is used for optimization. The ϵ-greedy algorithm is used for exploration, and its initial value equals zero. It linearly increases from 0.01 to 0.9 in the process of interacting with the environment and then keeps 0.9 for exploration. Other hyperparameters are listed in the Table 3.11.

TABLE 3.11   Simulation Parameters

Parameters	Value
Topology	Topo-1, Topo-2
Number of episodes	1,200
Epsilon	Linearly increases from 0 to 0.9 in 0.01 increments
Buffer size	10,000
Target Q update frequency	100
Batch size	32
(μ, ν)	(0, 1), (0.2, 0.8), (0.4, 0.6), (0.6, 0.4), (0.8, 0.2), (1, 0)
Learning rate α	0.5, 0.05, 0.005, 0.0005
Discount factor γ	0.9

Each experiment uses 1 random seed and interacts with the environment in 1,200 steps. To verify the convergence and effectiveness of the DDQN-based QoS optimization algorithm, testing the proposed method with different μ, ν, and α in Topo-1, and taking the reward, the network average latency, and the degree of network load balancing as the evaluation indicators.

- The reward is calculated according to equation (3.41), the larger the value is, the more rewards are obtained.

- The average network delay is calculated according to equation (3.28). The smaller the delay, the lower the possibility of network congestion.

- The degree of network load balance is calculated according to equation (3.32). The smaller the value, the more balanced the network load.

Figure 3.28 shows the convergence effect of the DDQN-based QoS optimization algorithm under the same μ, ν and different learning rates α. It can be seen that when $\alpha=0.0005$, it can converge quickly, and the optimization effect of network average latency and degree of network load balancing is more obvious. Figure 3.28 depicts the optimization and convergence of average reward value, network average latency, and network load balancing under different μ and ν with the same learning rate α ($\alpha=0.0005$). As expected, the DDQN-based QoS optimization algorithm can converge quickly under different μ and ν, especially when $\mu=0.6$ and $\nu=0.4$ to achieve good optimization. At the beginning of training (about 50 steps), a large amount of data transmission leads to a rapid increase in network delay and unbalanced network load. Because the DDQN-based QoS optimization algorithm gives corresponding penalties, the reward value decreases. As the training steps increase, the experience of the DDQN-based QoS optimization algorithm is gradually enriched, and it is able to respond to network changes and make dynamic adjustments. The average network delay is gradually reduced, the network load is gradually balanced, and the reward value is slowly increased to convergence.

3.5.3.3 Performance Evaluation

To further evaluate the advantages and generalization of the DDQN-based QoS optimization algorithm, three QoS optimization algorithms were performed on Topo-1 and Topo-2, respectively, including the DDQN-based QoS optimization method (which reward function only considers the network latency), the OSPF-based, and DQN-based. And the average network delay, network jitter, and network throughput were compared when the assembly station transmits a 5M size data flow to the engraving station. Figure 3.29 shows the test results of data flow transmission and Table 3.12 summarizes the details of data flow transmission, including average delay and average jitter.

It can be seen from Figure 3.29a and d that the average transmission latency on Topo-1 and Topo-2 of the DDQN-based QoS optimization algorithm is all lower than that of OSPF-based and DQN-based. According to Table 3.12, the average latency of the DDQN-based QoS optimization method in Topo-1 is 111.2 ms, while OSPF and DQN are 202.6 and 112.2 ms, respectively, and therefore the network delay is optimized by 1.82× and 1.01×, respectively. The DDQN-based QoS optimization method in Topo-2 is able to respond quickly to

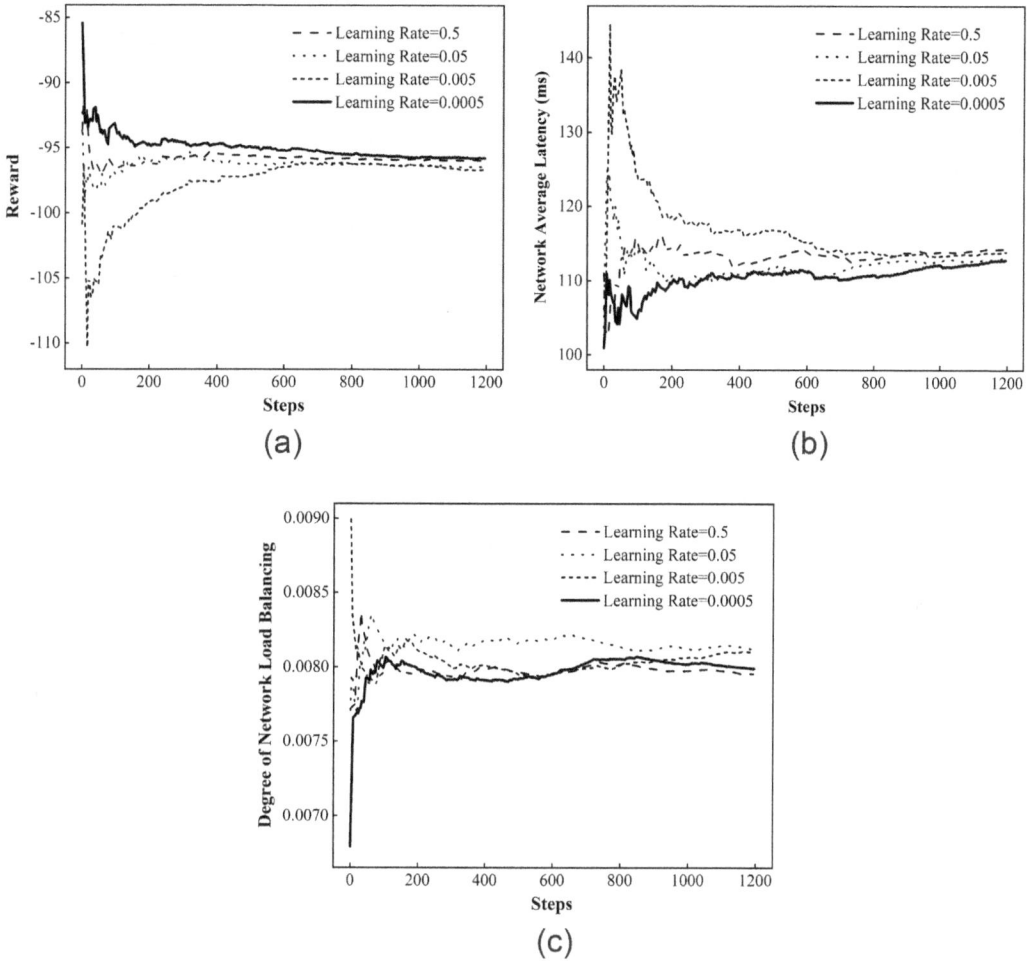

FIGURE 3.28   Performance of the proposed algorithm under same μ, ν and different learning rates α. (a), (b), and (c) represent the reward, the average network delay, and the degree of network load balancing, respectively.

the network latency, and its average latency is 350.1 ms, while OSPF and DQN are 576.8 and 354.2 ms, respectively, thus the network delay is optimized by 1.65× and 1.01×, respectively.

Figure 3.29b and e shows the jitter time when the assembly station transmits 5M data flow to the engraving station for 10 tests in Topo-1 and Topo-2, respectively. The jitter value of the DDQN-based QoS optimization algorithm in Topo-1 is lower than that of the other two methods in 9 of the 10 times and its winning rate is 90%. The jitter value of the DDQN-based QoS optimization algorithm in Topo-2 is lower than that of the other two methods in 7 of the 10 times, and its winning rate is 70%. Table 3.12 is the average value of 10 tests. Compared with OSPF-based and DQN-based, the jitter time of the DDQN-based QoS optimization algorithm in Topo-1 is optimized by 1.22× and 1.09×, respectively, while in Topo-2 is 1.08× and 1.10×, respectively.

Figure 3.29c and f shows the network throughput when the assembly station transmits 5M data flow to the engraving station for 10 tests in Topo-1 and Topo-2, respectively, and Table 3.12

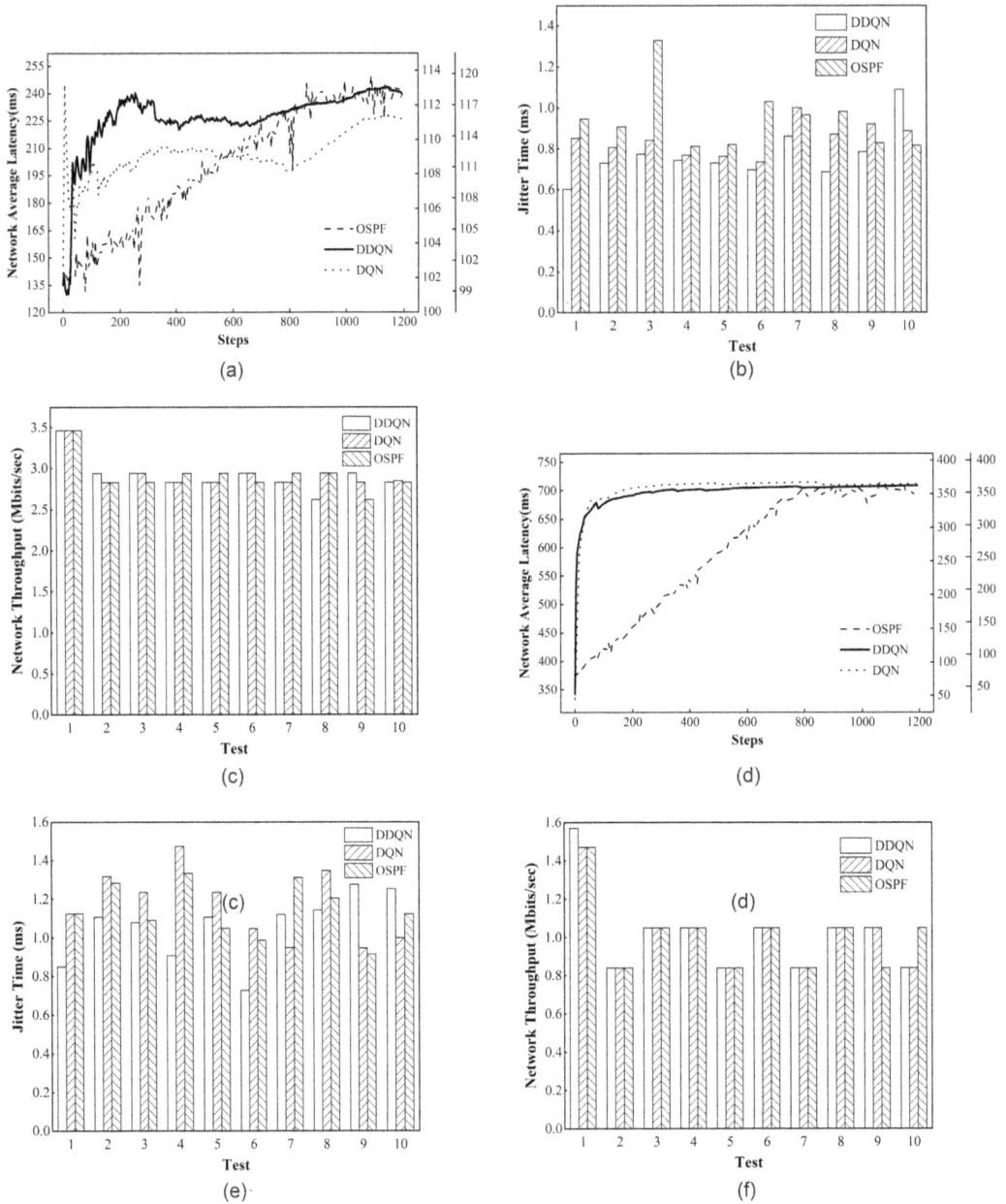

FIGURE 3.29 Performance comparison of three methods for data flow transmission on two network topologies.

shows the average value of 10 tests. In Topo-1, since the link bandwidth is greater than the maximum data flow size, the network throughput of the three algorithms is not much different, but in Topo-2, the DDQN-based QoS optimization algorithm, compared to OSPF-based and DQN-based, optimizes the network throughput by 1.22× and 1.09×, respectively.

TABLE 3.12 Comparison of Three Algorithms in Different Topologies

	Method	Average Latency (ms)	Average Jitter (ms)	Average Throughput (Mbits/s)
Topo-1	DDQN	111.2	0.769	2.916
	DQN	112.2	0.844	2.928
	OSPF	202.6	0.943	2.916
Topo-2	DDQN	350.1	1.057	1.018
	DQN	354.2	1.167	1.008
	OSPF	202.6	0.943	2.916

Therefore, the DDQN-based QoS optimization algorithm is applicable to the network environment not only with sufficient link bandwidth (Topo-1) but also with the limited (Topo-2), which has a certain generalization. At the same time, in the two topologies, the QoS optimization of the DDQN-based QoS optimization algorithm is better than the other two algorithms, including network delay, network jitter, and network throughput, which helps to reduce network delay and promote the network load balance.

3.6 SUMMARY

This book chapter significantly discussed the heterogeneous network in smart manufacturing factories and its background and development process, and later described various heterogeneous networks and discussed how heterogeneous networks help smart manufacturing. Coming to technologies SDN, EC, and cloud computing with heterogeneous networks and how these technologies are enhancing the revolutions in the smart manufacturing factory are discussed. Based on network QoS requirements, we discussed how AI technology is helping in the QoS optimization of heterogeneous networks.

REFERENCES

[1] D. Xia, C. Jiang, J. Wan, J. Jin, V. C. Leung and M. Martínez-García, "Heterogeneous network access and fusion in smart factory: A survey," *ACM Computing Surveys*, vol. 55, no. 6, pp. 1–31, 2023.

[3] J. Wan, S. Tang, Z. Shu et al., "Software-defined industrial internet of things in the context of industry 4.0," *IEEE Sensors Journal*, vol. 16, no. 20, pp. 7373–7380, 2016.

[2] M. Wollschlaeger, T. Sauter and J. Jasperneite, "The future of industrial communication: Automation networks in the era of the internet of things and industry 4.0," *IEEE Industrial Electronics Magazine*, vol. 11, no. 1, pp. 17–27, 2017.

[4] J.-P. Thomesse, "Fieldbus technology in industrial automation," *Proceedings of the IEEE*, vol. 93, no. 6, pp. 1073–1101, 2005.

[5] T. Sauter, "The three generations of field-level networks-evolution and compatibility issues," *IEEE Transactions on Industrial Electronics*, vol. 57, no. 11, pp. 3585–3595, 2010.

[6] J.-D. Decotignie, "The many faces of industrial ethernet [past and present]," *IEEE Industrial Electronics Magazine*, vol. 3, no. 1, pp. 8–19, 2009.

[7] J.-D. Decotignie, "Ethernet-based real-time and industrial communications," *Proceedings of the IEEE*, vol. 93, no. 6, pp. 1102–1117, 2005.

[8] P. Danielis, J. Skodzik, V. Altmann et al., "Survey on real-time communication via ethernet in industrial automation environments." *Proceedings of the 2014 IEEE Emerging Technology and Factory Automation*, 2014, pp. 1–8.

[9] M. Felser, "Real-time ethernet-industry prospective," *Proceedings of the IEEE*, vol. 93, no. 6, pp. 1118–1129, 2005.

[10] N. Finn, "Introduction to time-sensitive networking," *IEEE Communications Standards Magazine*, vol. 2, no. 2, pp. 22–28, 2018.

[11] J. L. Messenger, "Time-sensitive networking: An introduction," *IEEE Communications Standards Magazine*, vol. 2, no. 2, pp. 29–33, 2018.

[12] L. L. Bello and W. Steiner, "A perspective on IEEE time-sensitive networking for industrial communication and automation systems," *Proceedings of the IEEE*, vol. 107, no. 6, pp. 1094–1120, 2019.

[13] "PROFINET over TSN," https://www.profibus.com/technology/industrie-40/profinet-over-tsn.

[14] "CC-Link IE TSN," https://www.cc-link.org/en/cclink/cclinkie/cclinkie_tsn.html.

[15] "EtherCAT and TSN," https://www.ethercat.org/en/ethercat_and_tsn.htm.

[16] D. Raposo, A. Rodrigues, S. Sinche, J. Sá Silva and F. Boavida, "Industrial IoT monitoring: Technologies and architecture proposal," *Sensors*, vol. 18, no. 10, pp. 3568, 2018.

[17] I. Tomić and J. A. McCann, "A survey of potential security issues in existing wireless sensor network protocols," *IEEE Internet of Things Journal*, vol. 4, no. 6, pp. 1910–1923, 2017.

[18] D. V. Queiroz, M. S. Alencar, R. D. Gomes, I. E. Fonseca and C. Benavente-Peces, "Survey and systematic mapping of industrial Wireless Sensor Networks," *Journal of Network and Computer Applications*, vol. 97, pp. 96–125, 2017.

[19] R. E. Mohamed, A. I. Saleh, M. Abdelrazzak and A. S. Samra, "Survey on wireless sensor network applications and energy efficient routing protocols," *Wireless Personal Communications*, vol. 101, pp. 1019–1055, 2018.

[20] B. Bhushan and G. Sahoo, "Requirements, protocols, and security challenges in wireless sensor networks: An industrial perspective," in *Handbook of Computer Networks and Cyber Security: Principles and Paradigms*, eds: B. Gupta, G. Perez, D. Agrawal and D. Gupta. Springer, Cham, 2020, pp. 683–713.

[21] W. Chen, "Intelligent manufacturing production line data monitoring system for industrial internet of things," *Computer Communications*, vol. 151, pp. 31–41, 2020.

[22] C. Gomez, J. Oller and J. Paradells, "Overview and evaluation of bluetooth low energy: An emerging low-power wireless technology," *Sensors*, vol. 12, no. 9, pp. 11734–11753, 2012.

[23] K. Cho, W. Park, M. Hong et al., "Analysis of latency performance of Bluetooth low energy (BLE) networks," *Sensors*, vol. 15, no. 1, pp. 59–78, 2015.

[24] G. Patti, L. Leonardi and L. L. Bello, "A Bluetooth low energy real-time protocol for industrial wireless mesh networks." *Proceedings of the IECON 2016-42nd Annual Conference of the IEEE Industrial Electronics Society*, 2016, pp. 4627–4632.

[25] L. Leonardi, G. Patti and L. L. Bello, "Multi-hop real-time communications over bluetooth low energy industrial wireless mesh networks," *IEEE Access*, vol. 6, pp. 26505–26519, 2018.

[26] L. Tian, S. Santi, A. Seferagić, J. Lan and J. Famaey, "Wi-Fi HaLow for the Internet of Things: An up-to-date survey on IEEE 802.11 ah research," *Journal of Network and Computer Applications*, vol. 182, pp. 103036, 2021.

[27] I.-G. Lee, D. B. Kim, J. Choi et al., "WiFi HaLow for long-range and low-power Internet of Things: System on chip development and performance evaluation," *IEEE Communications Magazine*, vol. 59, no. 7, pp. 101–107, 2021.

[28] M. Giordani, M. Polese, M. Mezzavilla, S. Rangan and M. Zorzi, "Toward 6G networks: Use cases and technologies," *IEEE Communications Magazine*, vol. 58, no. 3, pp. 55–61, 2020.

[29] Z. Zhang, Y. Xiao, Z. Ma et al., "6G wireless networks: Vision, requirements, architecture, and key technologies," *IEEE Vehicular Technology Magazine*, vol. 14, no. 3, pp. 28–41, 2019.

[30] T. Huang, W. Yang, J. Wu, J. Ma, X. Zhang and D. Zhang, "A survey on green 6G network: Architecture and technologies," *IEEE Access*, vol. 7, pp. 175758–175768, 2019.

[31] P. Yang, Y. Xiao, M. Xiao and S. Li, "6G wireless communications: Vision and potential techniques," *IEEE Network,* vol. 33, no. 4, pp. 70–75, 2019.

[32] W. Saad, M. Bennis and M. Chen, "A vision of 6G wireless systems: Applications, trends, technologies, and open research problems," *IEEE Network,* vol. 34, no. 3, pp. 134–142, 2019.

[33] S. Rinaldi, P. Ferrari, A. Flammini, M. Rizzi, E. Sisinni and A. Vezzoli, "Performance analysis of power line communication in industrial power distribution network," *Computer Standards & Interfaces,* vol. 42, pp. 9–16, 2015.

[34] M. Yigit, V. C. Gungor, G. Tuna, M. Rangoussi and E. Fadel, "Power line communication technologies for smart grid applications: A review of advances and challenges," *Computer Networks,* vol. 70, pp. 366–383, 2014.

[35] A. Verl, S. Schmitz, D. Yang and K.-H. Wurst, "Industrial powerline communication for machine tools and robotics," *Production Engineering,* vol. 4, pp. 295–305, 2010.

[36] J. Wan, J. Yang, S. Wang, D. Li, P. Li and M. Xia, "Cross-network fusion and scheduling for heterogeneous networks in smart factory," *IEEE Transactions on Industrial Informatics,* vol. 16, no. 9, pp. 6059–6068, 2020.

[37] R. S. Peres, A. D. Rocha, A. Coelho and J. Barata Oliveira, "A highly flexible, distributed data analysis framework for industry 4.0 manufacturing systems." in *Service Orientation in Holonic and Multi-Agent Manufacturing,* eds: T. Borangiu, D. Trentesaux, A. Thomas, P. Leitão and J. Oliveira. Springer, Cham, 2016, pp. 373–381.

[38] D. Li, X. Li and J. Wan, "A cloud-assisted handover optimization strategy for mobile nodes in industrial wireless networks," *Computer Networks,* vol. 128, pp. 133–141, 2017.

[39] X. Li, D. Li, J. Wan, C. Liu and M. Imran, "Adaptive transmission optimization in SDN-based industrial Internet of Things with edge computing," *IEEE Internet of Things Journal,* vol. 5, no. 3, pp. 1351–1360, 2018.

[40] T. Qiu, F. Xia, Y. Ding, L. Liu and J. Wan, "A neighbour-based load-balanced packet dissemination scheme for wireless sensor networks," *International Journal of Sensor Networks,* vol. 22, no. 4, pp. 220–228, 2016.

[41] D. Xia, J. Wan, P. Xu and J. Tan, "Deep reinforcement learning-based QoS optimization for software-defined factory heterogeneous networks," *IEEE Transactions on Network and Service Management,* vol. 19, no. 4, pp. 4058–4068, 2022.

[42] J. Wan, J. Li, M. Imran and D. Li, "A blockchain-based solution for enhancing security and privacy in smart factory," *IEEE Transactions on Industrial Informatics,* vol. 15, no. 6, pp. 3652–3660, 2019.

CHAPTER 4

Intelligent Fault Diagnosis and Maintenance in Smart Manufacturing Factory

FAULT DIAGNOSIS AND MAINTENANCE is an important topic both in practice and in research. There is intense pressure on industrial systems to continue reducing unscheduled downtime, performance degradation, and safety hazards, which requires detecting and recovering from potential faults as early as possible. With the exponential growth of monitoring data, fault diagnosis and maintenance face enormous challenges dealing with industrial big data. It is like an iceberg where only a small part of fault information floats on the surface. It is hard to use the previous diagnostic methods to explore the true hidden value. At this time, the problem of transforming the growing volumes of data into the value is a considerable issue. The problem mainly includes two aspects. The first one is how to diagnose and predict failures rapidly or even in real time using novel processing systems. The second one is how to deeply dig out the "big" value of big data by improving the existing methods or leveraging new ones. This section contains three main contents: a method of dimensionless fault eigenvalue extraction, methods for fault diagnosis and fault prediction, and intelligent maintenance under Industry 4.0.

4.1 OVERVIEW

In modern industry, production equipment develops toward being extremely precise, efficient, and intelligent. Small performance degradation or security risks may bring serious consequences. It is vitally important to have a valid diagnosis approach to ensure the safe operation of the equipment. Before the arrival of big data, the previous research on fault diagnosis mainly depended on the richness of domain knowledge, the accuracy of diagnostic models, and the completeness of data samples [1]. These methods have the advantages of simplicity, interpretability, and ease of development. But they are susceptible to

DOI: 10.1201/9781003460992-4

disturbances in the environment and produce a tremendous computation pressure when facing large-scale complex systems.

In the context of Industry 4.0, related technologies, such as the Internet of Things [2], wireless sensor networks [3], and cloud computing [4], have developed rapidly. At the same time, data acquisition and storage become easier and easier, promoting the arrival of the era of industrial big data. Intelligent fault diagnosis is supported by artificial intelligence technology. By mining and utilizing the laws and values hidden in industrial big data, it can judge the online operation status of equipment, detect early failures, evaluate the degree of failure qualitatively or quantitatively, reveal the decline law of equipment performance, and predict the remaining service life of equipment. Intelligent failure based on big data breaks the deadlock between the large amount of diagnostic data for mechanical equipment and the relative scarcity of diagnostic experts. It is a key component of intelligent manufacturing and has become an important content of "Made in China 2025" [5,6].

Vibration analysis is still the most popular approach to the health monitoring of rotating machinery [7–12]. Although the conventional intelligent diagnosis framework based on shallow learning models and vibration analysis has been studied for decades, it cannot avoid the tedious feature extraction and selection that is relied on rich domain experience and knowledge [13,14].

In recent years, thanks to the rapid development of artificial intelligence technology, deep learning models have been developed and successfully applied in the field of end-to-end machinery intelligent fault diagnosis, which have shown decent results for dealing with different challenges in anomaly detection and condition monitoring of equipment, including stacked autoencoder (SAE), deep belief network (DBN), convolutional neural network (CNN), and long short-term memory (LSTM) [15–21]. However, intelligent fault diagnosis and maintenance still faces many challenges as follows:

- Multiple concurrent faults are a common problem in fault diagnosis. Complexity is the basic characteristic of multiple concurrent failures. The technologies of the multiple concurrent fault diagnosis depend on the complexity of the mechanical system and their relevance. Because of the complexity of the mechanical structure, the same kind of characteristic can often accompany different faults and the same kind of fault can cause a variety of characteristics. A fault may correspond to several characteristics and a characteristic may correspond to several faults, which is why the fault diagnosis is difficult. Uncertainty is an important characteristic of faults in complex systems. There are subjective and objective factors for uncertainty such as instability of voltage, load disturbance, and current frequency. When the working conditions change, it affects the fault diagnosis model performance.

- The effective training of various deep learning models heavily depends on sufficient, labeled, and identically distributed samples, which is quite difficult in practical fault diagnosis. Therefore, it is necessary to study high-precision fault diagnosis methods of machinery under limited labeled samples.

- Most of the existing methods focus on vibration analysis of the rotor bearing. However, vibration analysis has the problem of affecting the equipment structures and the difficulty in installing sensors. Besides, the processing of vibration signals is very complicated due to the long transfer path of the signal, changeable working conditions, and strong noise in real applications.

4.2 DIMENSIONLESS EIGENVALUE EXTRACTION AND FAULT DIAGNOSIS

In fault diagnosis, using dimensionless parameter method to detect and predict the equipment failure is effective. Dimensionless parameters have many categories such as the waveform, the peak, the margin, and pulse and kurtosis indices. They are not easily affected by mechanical conditions, such as changes in the load and speed. Therefore, they are widely used in mechanical fault diagnosis. Moreover, they are sensitive to parameters particularly in operating voltage and speed.

This section proposes a new fault diagnosis method for the petrochemical rotating machinery fusing the dimensionless indices and the Pearson's product moment correlation coefficient (PPMCC) to enhance the efficiency and accuracy of fault diagnosis [22]. The order statistic correlation coefficient and Pearson's correlation coefficient are used to calculate the correlation coefficients of dimensionless indices, which are given by dimensionless algorithms after preprocessing the raw data. Different fault types are recognized by comparing the correlation coefficient and each dimensionless indicator. The numerical results revealed that the proposed fault diagnosis method has the highest accuracy.

4.2.1 System Architecture

In this section, we fuse dimensionless indices using OSCC and PPMCC. Regarding the project background, there is a considerable noise introduced during data collection, which will impact the signal processing. The complex structure and process of large rotating machinery operation usually results in multiple concurrent failures, which makes it very difficult to diagnose faults. Different fault characteristics present complex signals due to coupling and ambiguity, which are often the simple linear superposition of multiple single faults.

We combined the waveform indices, pulse indices, peak indices, margin indices, and kurtosis indices into a dimensionless index, which is not sensitive to disturbance. There are still difficulties in seeking the uncertain characteristic for some fault. Therefore, we propose a method fusing the dimensionless indices using the correlation coefficient, as shown in Figures 4.1 and 4.2. The point of the measurement setup is the white dots as shown in the Figure 4.1. In the experiment, we only need to set the point collecting the data. At the same time, from the point collecting the data, we can obtain the waveform indicator, the pulse indicator, the margin indicator, the peak indicator, and the kurtosis measures. Its algorithm model is as shown in Figure 4.2.

Our laboratory developed a detection and diagnosis platform for petrochemical rotary equipment failure based on an immune dimensionless detector. Xiong et al. [23] proposed a rotating machinery diagnosis technology for concurrent faults based on dimensionless indices to analyze the characteristics of numbers and concurrent faults, and explore the

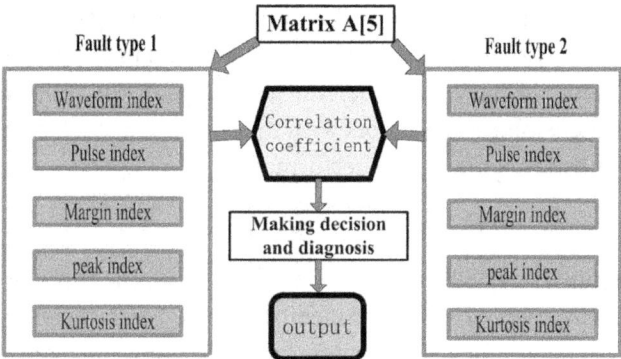

FIGURE 4.1 The algorithm model and the architecture of measurement data.

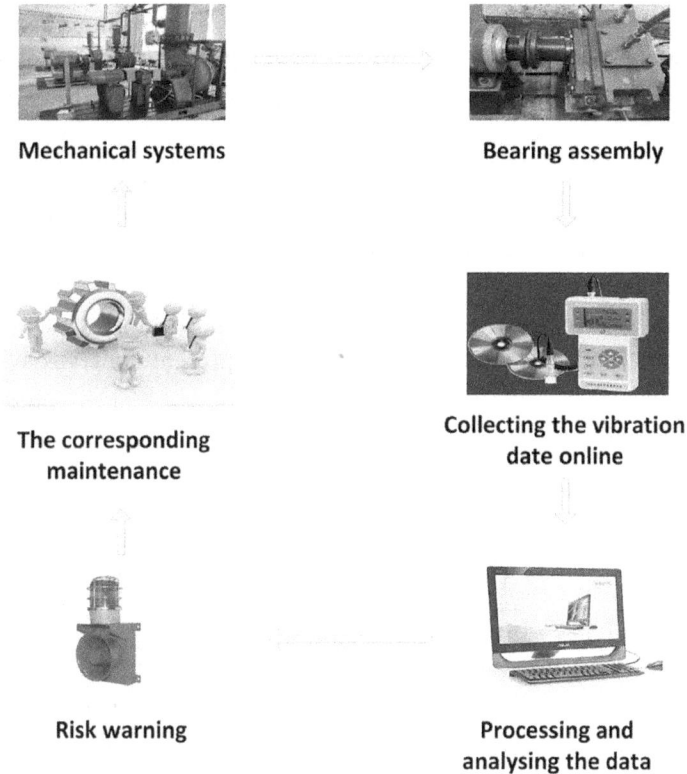

Mechanical systems

Bearing assembly

The corresponding
maintenance

Collecting the vibration
date online

Risk warning

Processing and
analysing the data

FIGURE 4.2 The system model.

relationship between concurrent failures and single failures. At the same time, he formulated methods using evidence theory into the fault diagnosis, which used the immune system to perform data analysis and processing from which a five dimensionless indices-based diagnosis ability factor in the time-domain signal can be obtained. Finally, each index of the fault diagnosis ability is considered as a probability distribution function and coupled with information fusion rules of evidence theory to diagnose faults [24].

Modern industrial applications involve a multitude of rotating machinery types, such as steam turbines, blowers, generators, and motors. The model proposed in this paper is validated on a petrochemical rotary machine. The advantage of the proposed method is that we combined the dimensionless indicators with the correlation coefficient, which allows us to increase the diagnostic accuracy and performance.

4.2.2 Definitions and Basic Theories

Dimensionless diagnosis is a fault diagnosis method using the dimensionless equipment parameters. Some examples of dimensionless diagnostic parameters are the waveform index, pulse index, margin index, peak index, kurtosis index, and so on. The corresponding expressions are provided below.

This section of the paper presents three fundamental definitions. The first one is the definition of the dimensionless index, including the waveform indices, pulse indices, margin indices, peak indices, and kurtosis indices, all represented using a general expression. Then, the two correlation coefficients, OSCC and PPMCC, are defined. Finally, the fusion of the dimensionless index with the correlation coefficients is given at the end of this section.

4.2.2.1 Definitions of Dimensionless Index

A dimensionless index is formed by the ratio of two physical quantities having the same dimension. It is an effective fault diagnosis function, which is less sensitive to the changes in amplitude and frequency characteristics, and is not affected by the conditions of the mechanical motion; it depends on the shape of the probability density function. The dimensionless indices used for fault diagnosis in the present context have the following definition:

$$\Delta\Psi_x = \frac{\left[\int_{-\infty}^{+\infty}|x|^T M(x)dx\right]^{\frac{1}{T}}}{\left[\int_{-\infty}^{+\infty}|x|^m M(x)dx\right]^{\frac{1}{m}}} = \frac{\sqrt[T]{E(|x|^T)}}{\sqrt[m]{E(|x|^m)}} \tag{4.1}$$

where x is the vibration amplitude and $M(x)$ is the probability density function of vibration amplitude. Also, we measured the acceleration of gear. Assigning values to T and m, we obtain five dimensionless indices as follows:

Waveform indicator: $T = 2, m = 1$:

$$W_f = \frac{\left[\int_{-\infty}^{+\infty}|x|^2 M(x)dx\right]^{\frac{1}{2}}}{\left[\int_{-\infty}^{+\infty}|x| M(x)dx\right]} = \frac{\sqrt{E(|x|^2)}}{\sqrt{E(|x|)}} \tag{4.2}$$

Pulse indicator: $T \to \infty, m = 1$:

$$P_f = \frac{\lim\limits_{x \to \infty} \left[\int_{-\infty}^{+\infty} |x|^T M(x) dx \right]^{\frac{1}{T}}}{\left[\int_{-\infty}^{+\infty} |x| M(x) dx \right]} = \frac{\lim\limits_{T \to \infty} \sqrt[T]{E(|x|^T)}}{\sqrt{E(|x|)}} \qquad (4.3)$$

Margin indicator: $T \to \infty, m = 1/2$:

$$CL_f = \frac{\lim\limits_{x \to \infty} \left[\int_{-\infty}^{+\infty} |x|^T M(x) dx \right]^{\frac{1}{T}}}{\left[\int_{-\infty}^{+\infty} |x|^{\frac{1}{2}} M(x) dx \right]^2} = \frac{\lim\limits_{T \to \infty} \sqrt[T]{E(|x|^T)}}{[E(\sqrt{|x|})]^2} \qquad (4.4)$$

Peak indicator: $T \to \infty, m = 2$:

$$F_f = \frac{\lim_{T \to \infty} \left[\int_{-\infty}^{+\infty} |x|^T M(x) dx \right]^{\frac{T}{T}}}{\left[\int_{-\infty}^{+\infty} |x|^2 M(x) dx \right]^{\frac{1}{2}}} = \frac{\lim_{T \to \infty} \sqrt[T]{E(|x|^T)}}{\sqrt{E(|x|^2)}} \qquad (4.5)$$

Kurtosis measures: $T \to \infty, m = 4$:

$$K_v = \frac{\int_{-\infty}^{+\infty} |x|^4 M(x) dx}{\left(\int_{-\infty}^{+\infty} |x|^2 M(x) dx \right)^2} = \frac{E(|x|^4)}{[E(|x|^2)]^2} \qquad (4.6)$$

From the above definitions, it is clear that the dimensionless indices are based on the probability density function of the monitored signal, which is not affected by the signal's absolute level and is less sensitive to the vibration detector's sensitivity to the signal's amplifier. Therefore, the calculation of the dimensionless indices does not require a monitoring system calibration, which is convenient for the fault diagnosis of practical equipment.

On a practical level, fault diagnosis using dimensionless indices is performed as follows. First, in a petrochemical machine unit, the dimensionless parameters are obtained using data collected online, i.e., when the machine is operating under normal (or faulty) condition.

The range of the dimensionless indices is determined by the maximum and the minimum of the calculation results.

Under a single fault vibration, there are N data samples.

If N is very large, from the definitions presented above, the expectation of a dimensionless index can be approximated as follows:

$$X^{[T]} = E\left(|x|^T\right) = \frac{1}{N} \sum_{i=1}^{N} |x_i|^T \tag{4.7}$$

Dimensionless index can be approximate as follows:

$$\Delta\Psi_x = \frac{\sqrt[T]{x^{-T}}}{\sqrt[m]{x^{-m}}} \tag{4.8}$$

When $T = \infty$, have

$$\sqrt[T]{x^{-T}} \approx \max_{j=1,2\ldots n} |x_j| \tag{4.9}$$

From the above equation, we get $x^{-T} \in [\alpha_T, \beta_T]$ by the multiple historical sets of vibration monitoring data for a single fault. Fault interval of dimensionless index is given as follows:

$$\Delta\Psi_x = \frac{\left[\int_{-\infty}^{+\infty} |x|^T M(x) dx\right]^{\frac{1}{T}}}{\left[\int_{-\infty}^{+\infty} |x|^m M(x) dx\right]^{\frac{1}{m}}} = \frac{\sqrt[T]{x^{-T}}}{\sqrt[m]{x^{-m}}} \in [c_x, d_x] \tag{4.10}$$

$$= \frac{\left[\sqrt[T]{\alpha_T}, \sqrt[T]{\beta_T}\right]}{\left[\left[\sqrt[m]{\alpha_m}, \sqrt[m]{\beta_m}\right]\right]} \in \left[\frac{\sqrt[T]{\alpha_T}}{\sqrt[m]{\beta_m}}, \frac{\sqrt[T]{\beta_T}}{\sqrt[m]{\alpha_m}}\right]$$

4.2.2.2 Theory of the Correlation Coefficient

Order Statistic Correlation Coefficient (OSCC):

Definition: Let (X_i, Y_i) be data pairs from the sample n of continuous joint cumulative distribution function of bivariate normal distribution. X_n and Y_n are independent of each other. Ascending order of data pairs (X_i, Y_i) is new data pairs $(X_i, Y_{[i]})$, among them the arrangement of $Y_{[1]} \ldots Y_{[n]}$ is following with $X_1 \ldots X_n$. The OSCC is defined as

$$T_n(X, Y) = \frac{\sum_{i=1}^{N} \left[X_{(i)} - X_{(n-i-1)}\right] Y_{[i]}}{\sum_{i=1}^{N} \left[X_{(i)} - X_{(n-i-1)}\right] Y_{(i)}} \tag{4.11}$$

Pearson Product Moment Correlation Coefficient $(PPMCC)$: The well-known PPMCC is defined as

$$L_n(X,Y) = \frac{\sum_{i=1}^{N} \left(X_i - \bar{X} \right)\left(Y_i - \bar{Y} \right)}{\sqrt{\sum_{i=1}^{N} \left(X_i - \bar{X} \right)^2 \sum_{i=1}^{N} \left(Y_i - \bar{Y} \right)^2}} \qquad (4.12)$$

The OSCC and PPMCC display the following fundamental characteristics: (1) The correlation coefficient remains constant under a monotonically increasing transformation; (2) when two variables are independent of each other, the correlation coefficient is zero; (3) when two variables are positively related, the correlation coefficient is 1, whereas the correlation coefficient is −1 when two variables are negatively correlated. This is a very ideal linear measurement.

Fusing Dimensionless Index with OSCC and PPMCC: The sensitivity of all dimensionless indexes is different from the disturbance detection of machine. Therefore, for the purpose of this experiment, we propose a fault diagnosis method fusing the dimensionless indices using the OSCC and PPMCC. Fusing the dimensionless index with the OSCC or PPMCC, we have $X = \left(W_f, P_f, CL_f, F_f, K_v \right)_1$ and $Y = \left(W_f, P_f, CL_f, F_f, K_v \right)_2$

4.2.3 Experiments

4.2.3.1 The Environment and Conditions of Experiment

The experiment described in this paper was based on the large rotating machinery experimental platform in Guangdong province's main laboratory for petrochemical equipment fault diagnosis, as shown in Figures 4.3–4.5.

The bearings used for collecting experimental data are shown in Figure 4.4, and Figure 4.5 is the lab of level 2 units multistage centrifugal fan. In this study, an EMT390 sensor was used for data acquisition. The motors used were three-phase asynchronous motors, whose model number was JW5624. In this experiment, we collected the data under the condition that the speed of the spindle was 1,200 revolutions per minute and the sampling frequency is 20 k Hz. The data collected were stored in MATLAB, which was also used for the algorithm's implementation. Experimental data were collected for five types of simulated bearing faults, including lack of bearing ball, wear of bearing's inner ring, wear of bearing's outer ring, missing teeth of the large gear, and missing teeth of large gear with outer ring wear of left bearing. Vibration acceleration signals were recorded while the rotating machinery was running, and the waveform index W_f, peak index F_f, pulse index P_f, margin index CL_f, kurtosis index K_v were obtained through the linear calculation.

To obtain more accurate and reliable experimental data, we collect two data groups, A and B, each with 49 datasets and each dataset with 1,024 data points. Getting five dimensionless index datasets by pretreatment of the dimensionless index, the value range of the indices was determined through the minimum and maximum of each dimensionless index. Finally, the correlation coefficients of the dimensionless indices of each training sample were obtained through the OSCC and PPMCC algorithms using the relevant equations.

FIGURE 4.3 The experimental environment.

FIGURE 4.4 The experimental bearing.

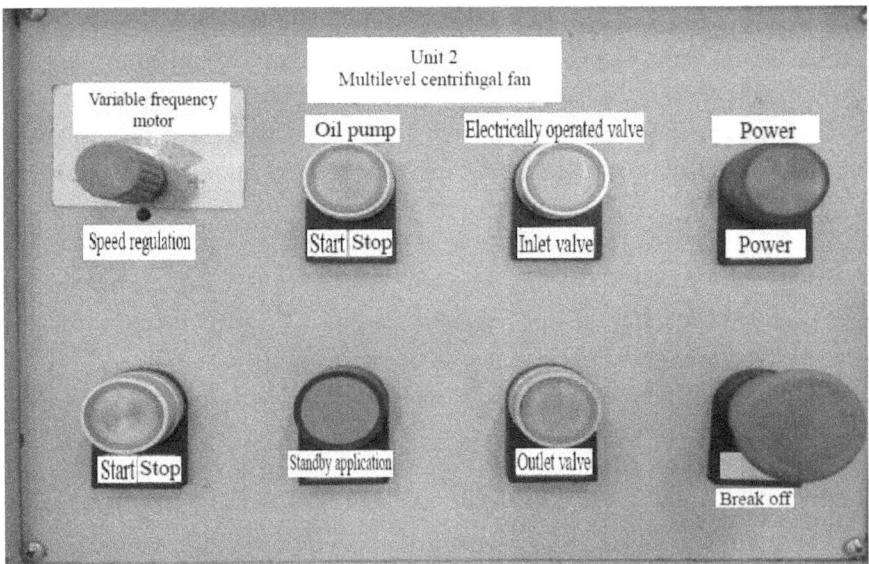

FIGURE 4.5 The experimental unit.

4.2.3.2 Implementation Steps

We collected six types of online datasets for petrochemical units, including a normal one, and sets in the presence of faults, specifically: lack of bearing ball, wear of bearing's inner ring, wear of bearing's outer ring, missing teeth of the large gear, and bend shaft. We obtained the dimensionless index by feeding the original data to the dimensionless index algorithm, and the correlation coefficients between the dimensionless indices were obtained using the OSCC and PPMCC equations.

By comparing the experimental data between the groups and finding the relationships between various data, we can obtain a reasonable assessment method for various kinds of fault diagnosis. As shown in Figure 4.6, the specific experimental steps are as follows:

Step 1: Replace the bearing of large mechanical rotating equipment with a bearing that has known fault.

Step 2: Set site to collect data, as shown in Figure 4.3.

Step 3: Collect online datasets for the faulty bearings for all fault types.

Step 4: Use the EMT390 sensor to gather the data and enter it into the database.

Step 5: Use the original data to obtain the waveform index W_f, peak index F_f, pulse index P_f, margin index CL_f, and kurtosis index K_v through linear calculations and determine the range of each fault index.

Step 6: Calculate the datasets of the dimensionless index through the OSCC and PPMCC equations to obtain the correlation coefficients $T_n(X,Y)$ and $L_n(X,Y)$.

Step 7: Contrast experimental datasets, and determine the relationships within the data, from which a reasonable judgment for fault diagnosis can be determined.

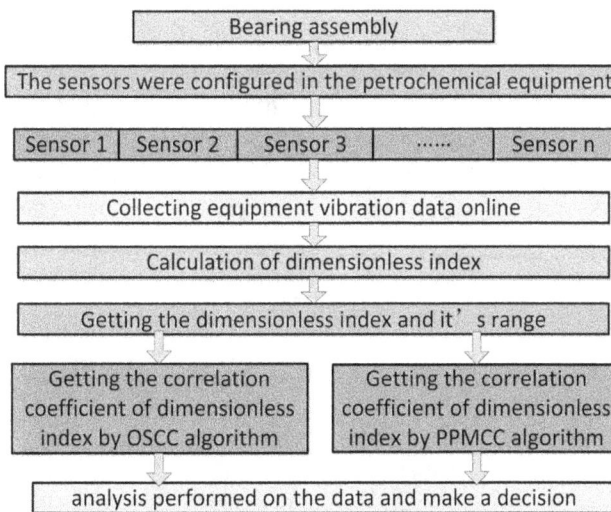

FIGURE 4.6   The method of experimental research steps.

4.2.3.3 Calculating of Dimensionless Indices

We obtained the ranges of dimensionless indices using the input samples, as shown in Table 4.1: Group A data and group B data were collected at different times. As shown in Tables 4.1 and 4.2, "zc," "zcqgz," "wqms," "nqms," "dclqc," and "dzw" represent the "normal bearing," "Lacking ball bearing," "Outer ring crack bearing," "Inner ring crack bearing," "the bearing missing teeth of the large gear," and "Bend shaft," respectively, as shown in Figure 4.7. Comparing Tables 4.1 and 4.2, it is easy to see that the ranges are similar for the two time periods.

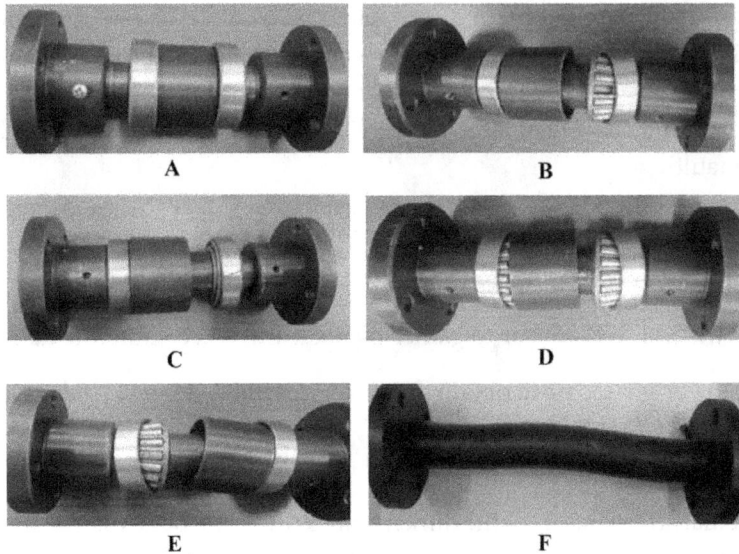

FIGURE 4.7 The all types of the bearings: (a) Normal bearing, (b) lacking ball bearing, (c) outer ring crack bearing, (d) inner ring crack bearing, (e) the bearing missing teeth of the large gear, and (f) bend shaft.

TABLE 4.1 Ranges of the Dimensionless Indices from the First 24 Datasets of Group A

Fault index	Waveform	Pulse	Margin	Peak	Kurtosis
zc	[1.367,1.506]	[4.827,7.729]	[6.028,9.979]	[3.461,5.315]	[4.522,6.656]
zcqgz	[1.284,1.401]	[4.164,8.820]	[5.016,11.03]	[3.199,6.293]	[3.428,7.412]
wqms	[1.238,1.322]	[2.136,5.032]	[2.496,5.927]	[1.726,3.968]	[2.995,4.227]
nqms	[1.248,1.314]	[3.045,4.441]	[3.574,5.358]	[2.441,3.456]	[2.963,3.840]
dclqc	[1.306,1.439]	[4.164,7.258]	[5.083,9.097]	[3.140,5.164]	[3.796,6.176]
dzw	[1.300,1.379]	[3.800,7.531]	[4.651,9.261]	[2.883,5.552]	[3.578,5.589]

TABLE 4.2 Ranges of the Dimensionless Indices from the First 24 Datasets of Group B

Fault Index	Waveform	Pulse	Margin	Peak	Kurtosis
zc	[1.390,1.500]	[4.724,7.990]	[5.979,10.53]	[3.383,5.327]	[4.647,6.904]
zcqgz	[1.286,1.364]	[4.054,9.454]	[4.908,11.59]	[3.110,6.930]	[3.545,6.511]
wqms	[1.267,1.331]	[3.507,6.603]	[4.192,7.862]	[2.716,5.130]	[3.261,4.245]
nqms	[1.241,1.285]	[2.941,4.330]	[3.493,5.108]	[2.344,3.453]	[2.765,3.465]
dclqc	[1.281,1.359]	[3.927,5.902]	[4.686,7.175]	[3.060,4.414]	[3.240,4.838]
dzw	[1.210,1.372]	[2.740,7.266]	[3.138,8.923]	[2.264,5.409]	[3.558,6.669]

However, there is a considerable overlap between the ranges of dimensionless indices among different faults, which is difficult to distinguish between different fault cases. In the following, we fuse the dimensionless indices with OSCC and PPMCC using two different combination methods.

4.2.3.4 Correlation Coefficient Calculation

1. **Combination method 1:** From Tables 4.3 and 4.4, we obtain the correlation coefficient values of the normal index by fusing normal data from group A with normal data in group B. The four correlation coefficient values in each Table 4.4 cell come from fusing the normal data of group A with the fault data of group A, the normal data of group A with the fault data of group B, the normal data of group B with the fault data of group A, and the normal data of group B with the fault data of group B. The coefficient values between the dimensionless indices in Table 4.3 were obtained using the OSCC algorithm, whereas the values in Table 4.4 were obtained using the PPMCC algorithm.

   From Tables 4.3 and 4.4, by comparing the OSCC with the PPMCC algorithms and the corresponding correlation coefficients of the fault indices, we see that the change trend of the corresponding datasets is similar, including the change

TABLE 4.3   The Order Statistic Correlation Coefficient

Fault Index	Waveform	Pulse	Margin	Peak	Kurtosis
zc	0.1046	0.1200	0.1059	0.1372	0.1058
zcqgz	−0.0564	−0.2411	−0.2312	−0.1957	−0.0790
	−0.2608	0.1207	0.0582	0.1754	0.0730
	0.1233	−0.1019	0.0135	−0.0166	−0.0519
	0.1165	0.1864	0.1788	0.1844	0.0790
wqms	−0.2601	0.0881	0.0930	0.1119	−0.0190
	−0.0033	0.2503	0.2157	0.3016	−0.1202
	−0.2120	−0.0147	−0.0245	−0.0209	−0.0199
	0.1757	0.1568	0.1762	0.0885	0.2426
nqms	0.1115	0.1205	0.1386	0.0908	0.1533
	−0.0825	0.1619	0.1626	0.1395	0.1717
	0.1094	0.0026	0.0079	−0.0023	−0.1664
	0.0945	0.1607	0.1617	0.1465	0.2156
dclqc	0.1343	−0.1389	−0.0993	−0.1030	−0.1035
	−0.0664	0.1093	0.1108	0.1034	−0.0344
	−0.1042	0.1547	0.1538	0.1531	−0.2384
	−0.0180	−0.2684	−0.2721	−0.2874	−0.1688
dzw	0.0480	−0.2891	−0.3002	−0.2973	0.0783
	−0.0064	−0.0788	−0.0543	−0.0800	0.2215
	0.1871	0.2288	0.2295	0.2560	0.1804
	0.1604	−0.2296	−0.2225	−0.2296	0.0825

TABLE 4.4 Pearson's Product Moment Correlation Coefficient

Fault Index	Waveform	Pulse	Margin	Peak	Kurtosis
zc	0.0831	0.1187	0.0986	0.1364	0.0988
zcqgz	−0.0609	−0.2010	−0.2041	−0.1870	−0.1280
	−0.2481	0.1004	0.0598	0.1295	0.0476
	0.1255	0.0028	0.0095	−0.0046	−0.0489
	0.1490	0.1629	0.1611	0.1528	0.0974
wqms	−0.2528	0.0918	0.0931	0.0914	−0.0483
	−0.0202	0.2340	0.2065	0.2695	−0.1305
	−0.1769	0.0148	0.1788	0.0228	−0.0275
	0.2033	0.1551	0.1796	0.0981	0.2350
nqms	0.1198	0.1274	0.1344	0.1096	0.1640
	−0.0053	0.1379	0.1444	0.1054	0.1243
	0.0812	−0.0452	−0.0349	−0.0494	−0.1930
	0.1099	0.1826	0.1891	0.1730	0.2338
dclqc	0.1347	−0.1264	−0.0955	−0.1335	−0.0972
	−0.0901	0.1054	0.0987	0.1053	−0.0472
	−0.1414	0.1314	0.1349	0.1469	−0.2179
	−0.0594	−0.2337	−0.2381	−0.2283	−0.1480
dzw	0.0376	−0.2944	−0.2958	−0.2971	0.0628
	−0.0026	−0.0766	−0.0622	−0.1083	0.1857
	0.1687	0.2104	0.2052	0.2177	0.1813
	0.1824	−0.1986	−0.1798	−0.2224	0.0999

of symbols and numerical data. However, there are no rules for the experimental results of this data combination method.

2. **Combination method 2:** The first 29 datasets of each group are used as training data to obtain the order statistic correlation coefficient and Pearson's product moment correlation coefficient, as shown in Tables 4.5 and 4.6.

The data in Tables 4.5 and 4.6 show that OSCC is not suitable for method 2 data combination. Each kind of sensitivity of the dimensionless index of mechanical vibration is certain, which means that the size of the order is certain when five dimensionless parameter values are collected. Moreover, the characteristics of OSCC

TABLE 4.5 The Order Statistic Correlation Coefficient

Group A Group B	zc	zcqgz	wqms	nqms	dclqc	Dzw
zc	1	1	1	1	1	1
zcqgz	1	1	/	/	/	/
wqms	1	/	/	/	/	/
nqms	1	/	/	1	/	/
dclqc	1	/	/	/	1	/
dzw	1	/	/	/	/	1

TABLE 4.6 Pearson's Product Moment Correlation Coefficient

Group A / Group B	zc	zcqgz	wqms	nqms	dclqc	Dzw
zc	0.9997	0.9917	0.9947	0.9963	0.9970	0.9963
zcqgz	0.9863	0.9999	/	/	/	/
wqms	0.9848	/	0.9976	/	/	/
nqms	0.9874	/	/	0.9982	/	/
dclqc	0.9897	/	/	/	0.9991	/
dzw	0.9934	/	/	/	/	0.9999

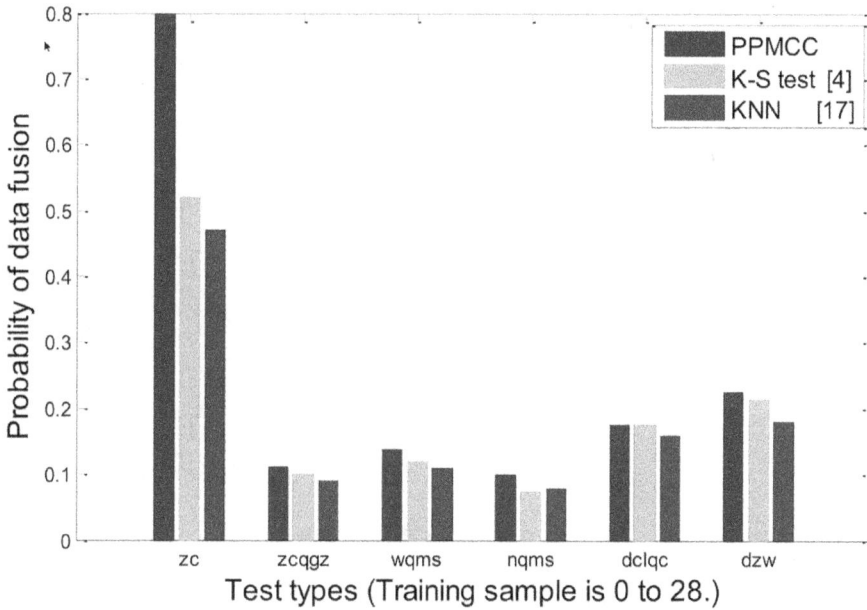

FIGURE 4.8 The probability of each type of failure data fusion compared to the conventional method.

are calculated using the data of the permutation order; this is the reason why the calculated data are all 1. However, this method of data combination is more appropriate for data fusion using the PPMCC and allows determination of the malfunction. Therefore, in the following, we shall only refer to data fusion using the PPMCC for the rest of this experimental process.

We then use the datasets 29 to 48 of each group as test data, fusing them with the first 29 datasets of group A and group B using the PPMCC.

The method to define the fault is that fusing the data collected online and the data of known fault type in the database. The fault type of the maximum degree of fusion was found between the data collected online and the data of known fault types, from that the fault type is decided.

As shown in Figure 4.8, the experimental results also show that this method has a stronger ability to diagnose faults when compared with the K-S test method and the K-nearest neighbor (KNN) method, with an average accuracy improvement of 5.8% and 7.7%, respectively.

4.2.3.5 Discussion

In Figures 4.9–4.12, "zc," "zcqgz," "wqms," "nqms," "dclqc," and "dzw" represent the "normal," "lack of ball bearing," "wear of bearing's outer ring," "wear of bearing's inner ring," "missing teeth of the large gear," and "bend shaft," respectively. All experiments were repeated 80 times. The normal test experiments yielded a correct diagnosis 64 times, which

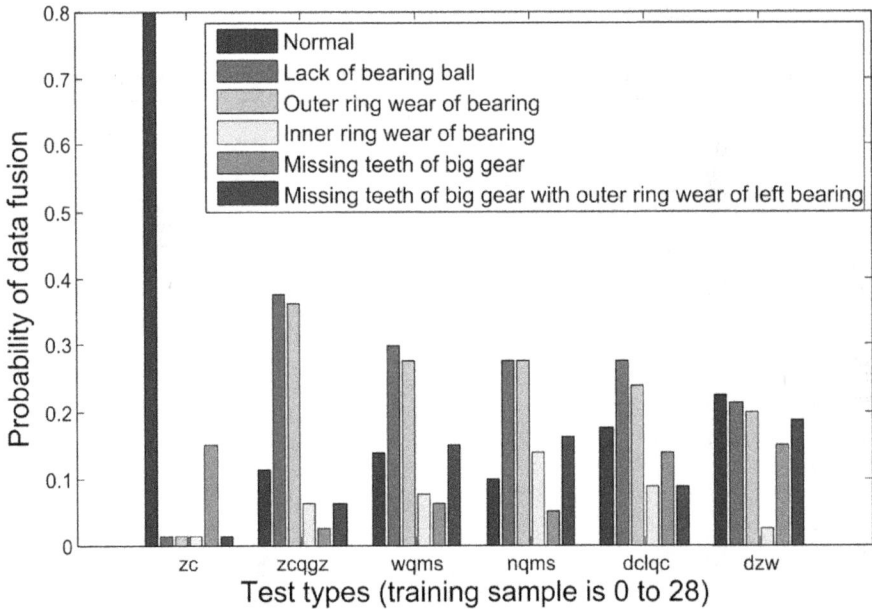

FIGURE 4.9 The probability of each type of failure data fusion (Training sample is 0 to 28).

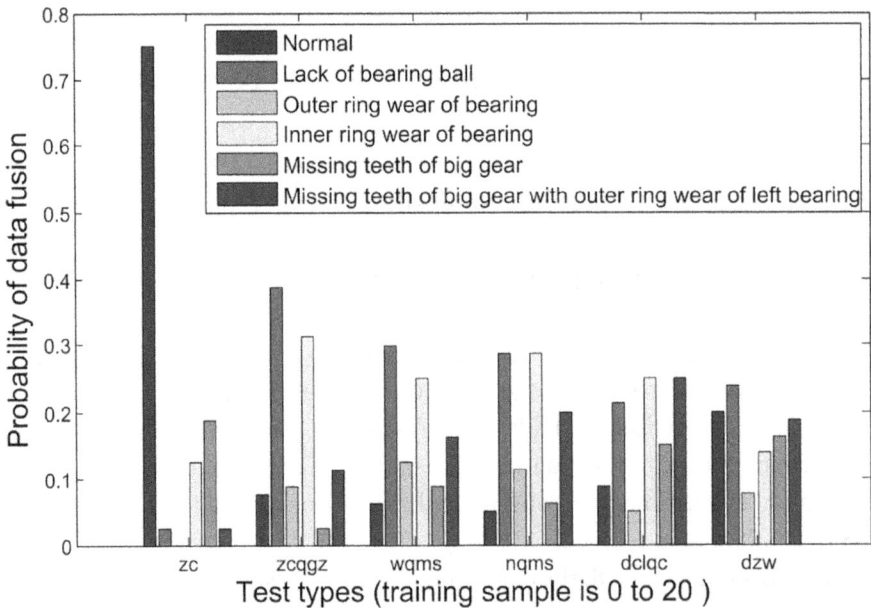

FIGURE 4.10 The probability of each type of failure data fusion (Training sample is 0 to 20).

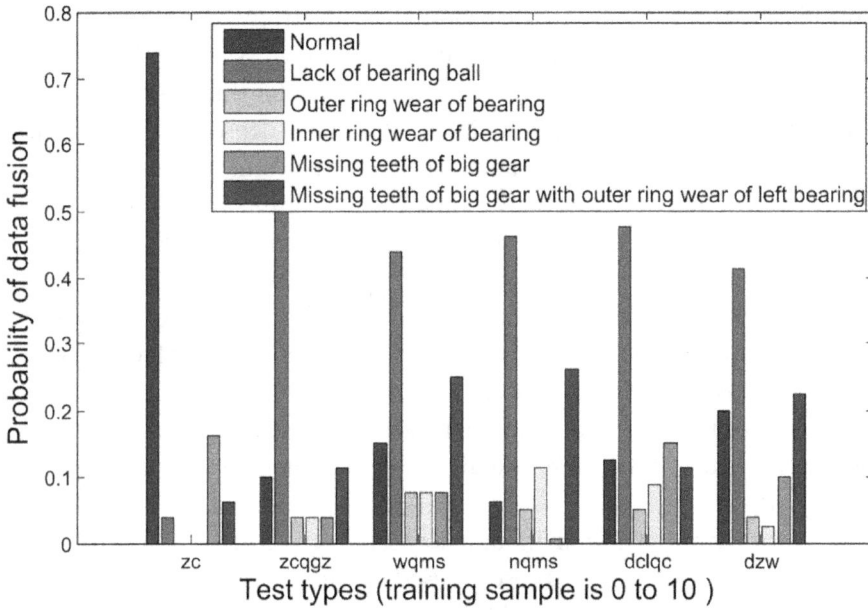

FIGURE 4.11   The probability of each type of failure data fusion (Training sample is 0 to 10).

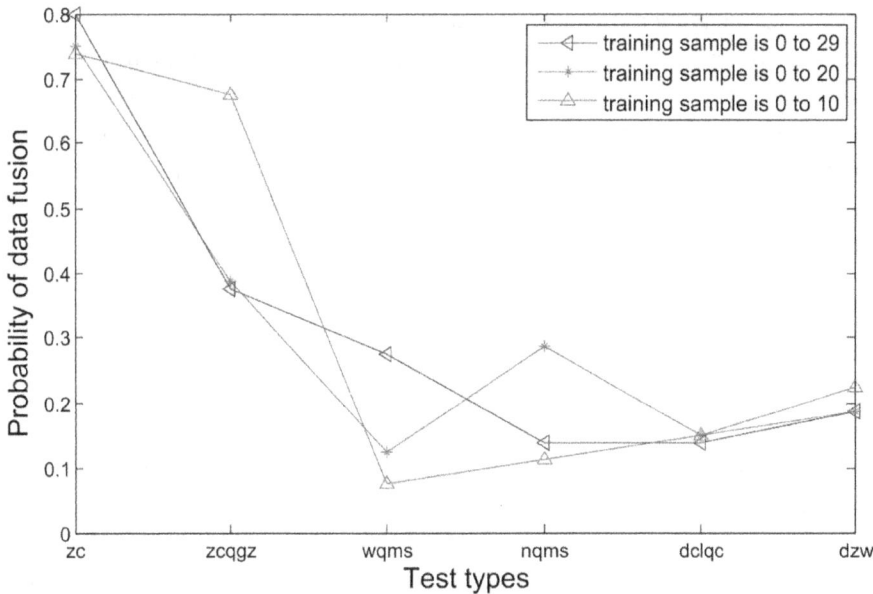

FIGURE 4.12   The probability of each type of failure data fusion.

corresponds to an accuracy of 80%. In the case of the outer ring wear, the experiments gave 22 correct diagnoses, i.e., an accuracy of 27.5%, while the outer ring wear experiments yielded the same result. The inner ring wear experiments gave 11 correct diagnoses, i.e., an accuracy of 13.75%. The large gear tooth deficiency experiments gave successful diagnoses 15 times, which means an accuracy of 18.75%. The above results correspond to an overall accuracy of 50%.

As shown in Figures 4.9 and 4.10, we conducted comparative experiments by sampling the dataset samples from 0 to 20 and 0 to 10, respectively. As shown in Figure 4.11, the fused probability obtained were similar for each fault type for three different sample sizes.

In the experimental results of combination method 1, the degree of interference is larger, making it difficult to distinguish the fault types. For the combination method 2, clearly the OSCC is not a suitable approach. The experimental results from Table 4.6, where the online acquired datasets are from the first 29 sets of each group, show that the degrees of data fusion between the same working states are higher than different working states, with the values being closer to 100% between the same working states. Combination degree is higher. In general, this method appears to be more applicable for fault diagnosis. This method is not limited to fault diagnosis in petrochemical units of rotating machines, but can also be extended to all kinds of industrial machinery and equipment, including the key equipment in the field of petroleum, chemical industry, metallurgy, machinery manufacturing, and aerospace.

The characteristics of the new method: It is based on the correlation coefficient of fused data and thus avoids unilateral diagnoses caused by information on a single fault characteristic. The method cannot be affected by conflicting evidence from different sources of information. The dimensionless indices' fusion coefficient of correlation was applied for large petrochemical machinery fault diagnosis.

The advantages of using dimensionless indices are as follows. The dimensionless indices reflect the fault status based on the probability density function $p(x)$. Dimensionless indices are not affected by the absolute level of the vibration signal and have nothing to do with the vibration detector, the sensitivity of the amplifier, and the magnification of the testing system, so it cannot be affected by errors of measurements, even in cases where the sensor's or the amplifier's sensitivity was changed. The sensitive degree of various non-dimensional indexes to different faults is different. Many faults have no obvious effects on the dimensionless indices. Changes in conditions, load, and speed have nearly no influence on the dimensionless indices.

The main limitations of correlation coefficients are that the correlation coefficient can only measure the linear relationships between the normal and fault indicators, and it cannot measure the other relationships between the normal and fault indicators. The method not only depends on the edge of the index data distribution but also relies on the joint distribution of the index data. Under a monotonic transformation, the value of the correlation coefficient often changes.

In order to overcome these defects, this paper introduces the dimensionless correlation coefficient fusion index and fault interval method.

4.3 INTELLIGENT FAULT DIAGNOSIS OF INDUSTRIAL MACHINERY

Due to the development of sensors and internet technologies, data-driven approaches are developing rapidly. With massive monitoring data of machinery being collected, how to mine the effective information is significant for its operation and maintenance. Advanced machine learning methods, such as deep learning and transfer learning, have been widely applied to fault feature mining and intelligent diagnosis of machinery. In the context of customized manufacturing, industrial machinery operates for a long time under conditions of high dynamics and short-term strong overload, and key rotating parts such as bearings and gears are often prone to various failures. The occurrence of failure will directly affect manufacturing cost, product quality, and safe production.

This section introduces four intelligent fault diagnosis methods for industrial machinery.

4.3.1 Supervised Fault Diagnosis Method Based on Modified Stacked Autoencoder Using Adaptive Morlet Wavelet

SAE possesses the properties of unsupervised learning, high-efficiency training, and easy implementation, and it has wide application in different fields, such as fault diagnosis, image classification, data denoising, and feature reduction.

However, the challenges exist when applying the basic SAE to the practical fault diagnosis task of rotating machinery. On one hand, various fault categories and severities may reduce the distinguishable characteristic differences hidden in the raw vibration data [25]. On the other hand, for some rotating machines such as planetary gearboxes, the coupled vibration of multiple components and complicated transmission paths will cause stronger nonstationarity and increased interference of the collected vibration signals [26]. Neural networks designed with general activation functions may have limitations in establishing accurate mappings between a nonlinear and nonstationary input data and various output patterns [27,28]. Unlike the popular transformation functions, wavelet function has a special attribute of time-frequency localization. Inspired by the successful application of wavelet neural networks in fault diagnosis [29], there exists a strong motivation to modify the basic SAE by using a wavelet function. Because of the strong similarity to periodic impulse components of mechanical vibration signals [30,31], Morlet wavelet has been successfully utilized to modify the basic SAE in recent work [27,32]. However, the two waveform parameters of Morlet wavelet activation function are manually selected and fixed, which cannot flexibly match the characteristics of the analyzed data. Besides, Morlet wavelets with different parameters probably show different performance when dealing with the analyzed data [31]. Thus, there exists a strong motivation to design an adaptive Morlet wavelet by flexibly adjusting the two parameters to achieve the best match with the characteristics of the analyzed signals. The fruit fly optimization algorithm (FOA) can effectively carry out a global optimization with many advantages including faster convergence, stronger stability, higher precision, and easier implementation than many other optimization algorithms [33]. In the last 2 years, FOA has been gradually used to optimize the intelligent diagnosis model of rotating machinery and has shown good performance [34]. Thus, modified stacked auto-encoder (MSAE) with adaptive Morlet wavelet using FOA has the potential to properly match the characteristics of the analyzed data in different diagnosis tasks.

The other challenge is due to the increase of the width and the depth of the SAE model, which would require updating of a larger amount of weights, making it harder to train. Even though a weight decay term is usually added to the cost function of SAE to avoid overfitting, numerous nonzero connection weights will lead to a reduction in sparsity and affect reconstruction quality [35,36]. Thus, more effective weight decay strategies are crucial.

To address these challenges, in this section, MSAE that uses an adaptive Morlet wavelet and improved training algorithm is proposed to automatically diagnose various faults of rotating machinery [37].

4.3.1.1 Main Algorithms

4.3.1.1.1 Principle of Autoencoder   AE belongs to unsupervised deep-learning-based models, and its architecture is shown in Figure 4.13. The training goal of AE is to achieve the reconstruction of the inputs as accurately as possible by adjusting the model parameters. The main formulas of AE are as follows:

$$h = s_g(Wx + b) \tag{4.13}$$

$$z = s_f(W'h + b') \tag{4.14}$$

$$C_1 = \frac{1}{2}\sum_{i=1}^{m}(z_i - x_i)^2 + \beta\left(\sum_{j=1}^{p} r\log\frac{r}{\hat{r}_j} + (1-r)\log\frac{1-r}{1-\hat{r}_j}\right) \tag{4.15}$$

where C_1 is the cost function of AE, β is a sparse penalty coefficient, r is a sparse constant, $x = [x_1, x_2, ..., x_m]$ is an unlabeled input sample whose feature vector and the reconstruction vector are $h = [h_1, h_2, ..., h_p]$ and $z = [z_1, z_2, ...z_m]$, respectively, s_g and s_f are the activation functions in the hidden layer and the output layer, respectively, which are generally selected as Sigmoid (Sigm) or rectified linear unit (ReLU) activation functions, W, W' are weights, and b, b' are biases.

4.3.1.1.2 MSAE Construction   Based on the idea of wavelet transform, wavelet function has been a new choice of the activation functions applied in the neural network, which can make full use of the time-frequency localization characteristics. To establish an accurate nonlinear mapping between the collected nonstationary vibration data and various working states, this chapter uses Morlet wavelet as the activation function of the hidden layer of

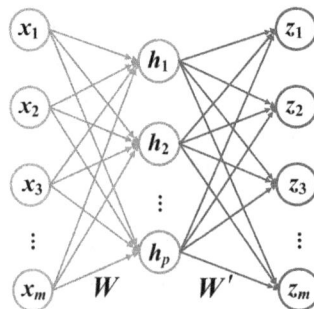

FIGURE 4.13   Architecture of the AE model.

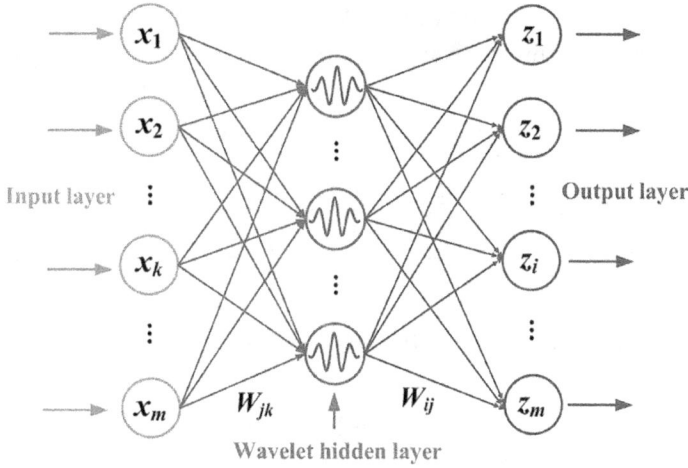

FIGURE 4.14   MAE based on Morlet wavelet function.

the basic AE due to its greater similarity to the fault characteristic components hidden in the collected vibration signal than other wavelets.

Morlet wavelet is expressed as

$$\psi(t) = \frac{1}{\sqrt{f_b \pi}} \cos(2\pi f_c t) \exp\left(-\frac{t^2}{f_b}\right) \tag{4.16}$$

in which f_b is the bandwidth and f_c is the central frequency, and these two parameters influence the performance of Morlet wavelet function. Based on the Morlet wavelet activation function, a new version of AE model called modified AE (MAE) is constructed as shown in Figure 4.14.

For the input sample $x = [x_1, x_2, \ldots, x_m]$, the expression of the hidden layer output is given by

$$h_j = \frac{1}{\sqrt{f_b \pi}} \cos\left(\frac{2\pi f_c\left(\frac{\left(\sum_{k=1}^{m} W_{jk} x_k - c_j\right)}{d_j}\right)}{d_j}\right)$$

$$\cdot \exp\left(-\frac{\left(\frac{\left(\sum_{k=1}^{m} W_{jk} x_k - c_j\right)}{d_j}\right)^2}{f_b}\right) \tag{4.17}$$

where h_j is the output of the hidden node j, d_j and c_j are the scale factor and the shift factor, respectively, W_{jk} is the weight between the hidden node j and the input node k, and W_{ij} is the weight between the hidden node j and the output node i. Set the nonlinear transformation of the output layer as a tanh function. Then, the final reconstructed output is

$$z_i = s_g \left(\sum_{j=1}^{p} W_{ij} h_j \right) = \tanh \left(\sum_{j=1}^{p} W_{ij} h_j \right) \tag{4.18}$$

To avoid overfitting, a weight decay term can be added to the cost function as

$$C^{\mathrm{T}} = \frac{1}{2} \sum_{i=1}^{m} (z_i - x_i)^2 + \frac{\lambda}{2} \sum_{i,k=1}^{m} \sum_{j=1}^{p} \left((W_{ij})^2 + (W_{jk})^2 \right)$$

$$+ \beta \left(\sum_{j=1}^{p} r \log \frac{r}{\hat{r}_j} + (1-r) \log \frac{1-r}{1-\hat{r}_j} \right) \tag{4.19}$$

where λ is a weight decay factor and C^{T} is the traditional cost function. However, a large number of connecting weights in this decay strategy will lead to a reduction in sparsity. The nonnegative constraint of the connecting weights was first proposed in 2016 to further improve sparsity and reconstruction quality by reducing the negative weights. In the last 2 years, SAEs integrated with the nonnegative constraint have been gradually used for the fault diagnosis of rotating machinery, and their diagnosis results are better than those using a conventional weight decay term. Here, to achieve higher quality reconstruction, the nonnegative constraint is introduced into the cost function of the MAE as

$$C^{\mathrm{E}} = \frac{1}{2} \sum_{i=1}^{m} (z_i - x_i)^2 + \frac{\delta}{2} \sum_{L=1}^{2} \sum_{I=1}^{s_L} \sum_{J=1}^{s_{L+1}} G\left(W_{JI}^{(L)} \right)$$

$$+ \beta \left(\sum_{j=1}^{p} r \log \frac{r}{\hat{r}_j} + (1-r) \log \frac{1-r}{1-\hat{r}_j} \right) \tag{4.20}$$

$$G\left(W_{JI}^{(L)} \right) = \begin{cases} \left(W_{JI}^{(L)} \right)^2, & \text{if } W_{JI}^{(L)} < 0 \\ 0, & \text{if } W_{JI}^{(L)} \geq 0 \end{cases} \tag{4.21}$$

where the second term represents the nonnegative constraint, δ represents a penalty coefficient, C^{E} represents the enhanced cost function, and s_L represents the node dimension in layer L. The training task of the MAE is also to adjust the weights $w_{JI}^{(L)}$ so as to make C^{E} a minimum. Gradient descent with back propagation is a simple and fast way to update the weights as follows:

$$W_{JI}^{(L)} = W_{JI}^{(L)} - \eta \frac{\partial C^M}{\partial W_{JI}^{(L)}} \quad L = 1, 2 \tag{4.22}$$

$$\frac{\partial C^M}{\partial W_{JI}^{(L)}} = \frac{\partial C_1}{\partial W_{JI}^{(L)}} + \delta g\left(W_{JI}^{(L)}\right) \tag{4.23}$$

$$g\left(W_{JI}^{(L)}\right) = \begin{cases} W_{JI}^{(L)}, & W_{JI}^{(L)} < 0 \\ 0, & W_{JI}^{(L)} \geq 0 \end{cases} \tag{4.24}$$

where η denotes the learning rate, $W_{JI}^{(1)} = W_{jk}$, $W_{JI}^{(2)} = W_{ij}$.

MSAE with stacked trained MAEs can further capture the valuable features hidden in the input samples, as shown in Figure 4.15. After that, the learned deep features are used as the input of the Softmax classifier for fault classification.

4.3.1.1.3 Adaptive Morlet Wavelet Design   According to equation (4.16), the performance of Morlet wavelet relies on the parameters f_b and f_c, as shown in Figure 4.16. Different time-frequency resolutions can be acquired by adjusting these two parameters. Thus, it is important to design an adaptive Morlet wavelet by flexibly adjusting the two parameters to achieve the best match with the characteristics of the analyzed signals.

FOA can effectively search global optimization with many advantages, and it is adopted for flexibly determining the adjustable parameters of Morlet wavelet in the present chapter.

More details of FOA can be found in Ref. [3]. The flowchart of the scheme is given in Figure 4.17 and summarized as follows (the fitness is misclassification rate on validation samples):

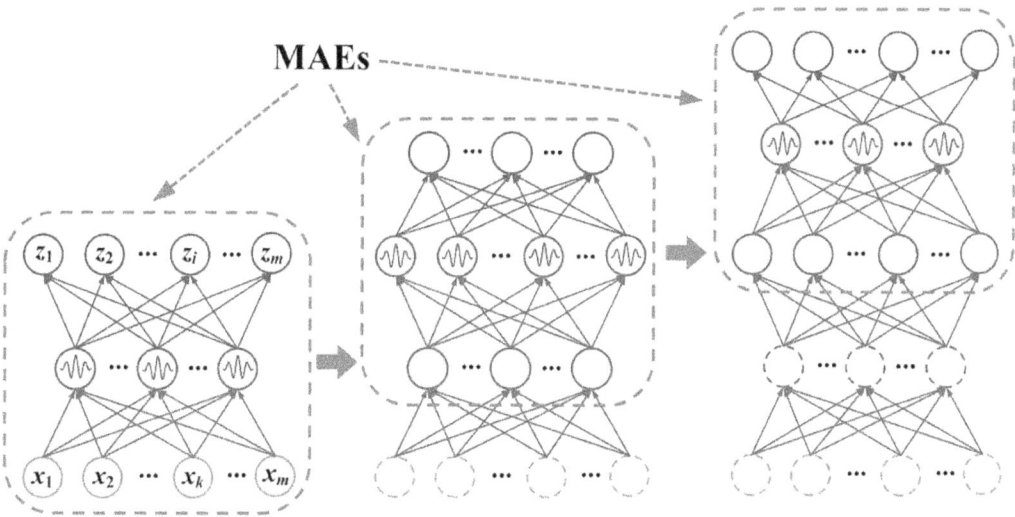

FIGURE 4.15   Construction of an MSAE.

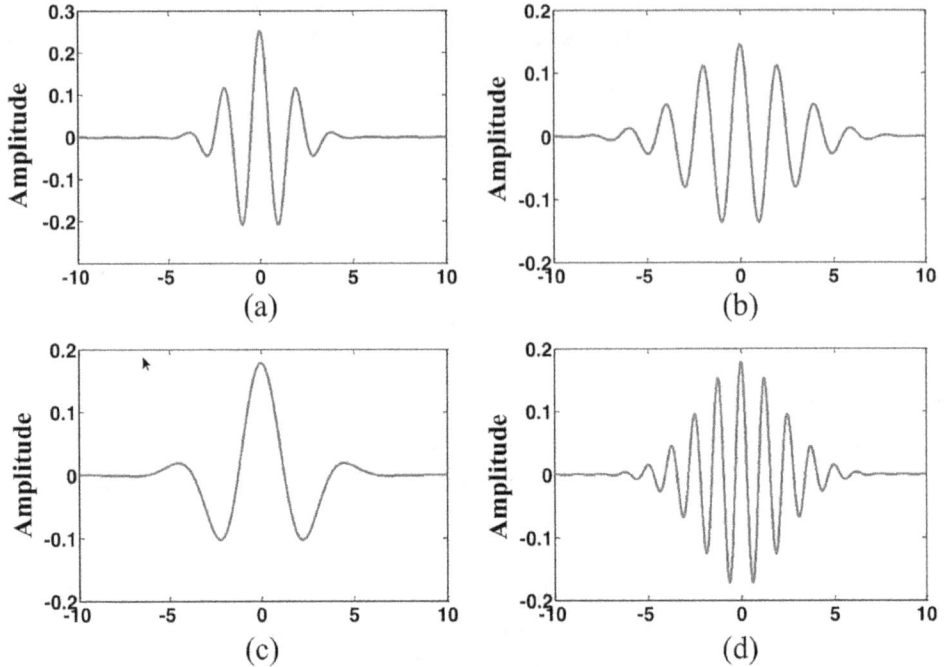

FIGURE 4.16 Morlet wavelets with different parameters f_b and f_c: (a) $f_b = 5$ and $f_c = 0.5$; (b) $f_b = 15$ and $f_c = 0.5$; (c) $f_b = 10$ and $f_c = 0.2$; and (d) $f_b = 10$ and $f_c = 0.8$.

1. Prepare an MSAE model with an initial Morlet wavelet that has already been trained by using the training samples.

2. Input the validation samples. Determine the maximum epoch number, population size, and initial location of the fruit fly swarm.

3. Give each fruit fly a random search direction and distance for foraging the food based on the smell.

4. Calculate the smell concentration judgment value using the distance between each fruit fly and the origin.

5. Search the smell concentration of each location of the fruit fly by substituting the smell concentration judgment value into the fitness function and then look for the minimum smell concentration among the fruit fly swarm.

6. The fruit fly swarm saves the best smell concentration value and will fly toward the best location using vision. Repeat Steps 3–5, and continue the optimization until reaching the maximum epoch number.

7. The designed MSAE with adaptive Morlet wavelet is used to analyze the testing samples.

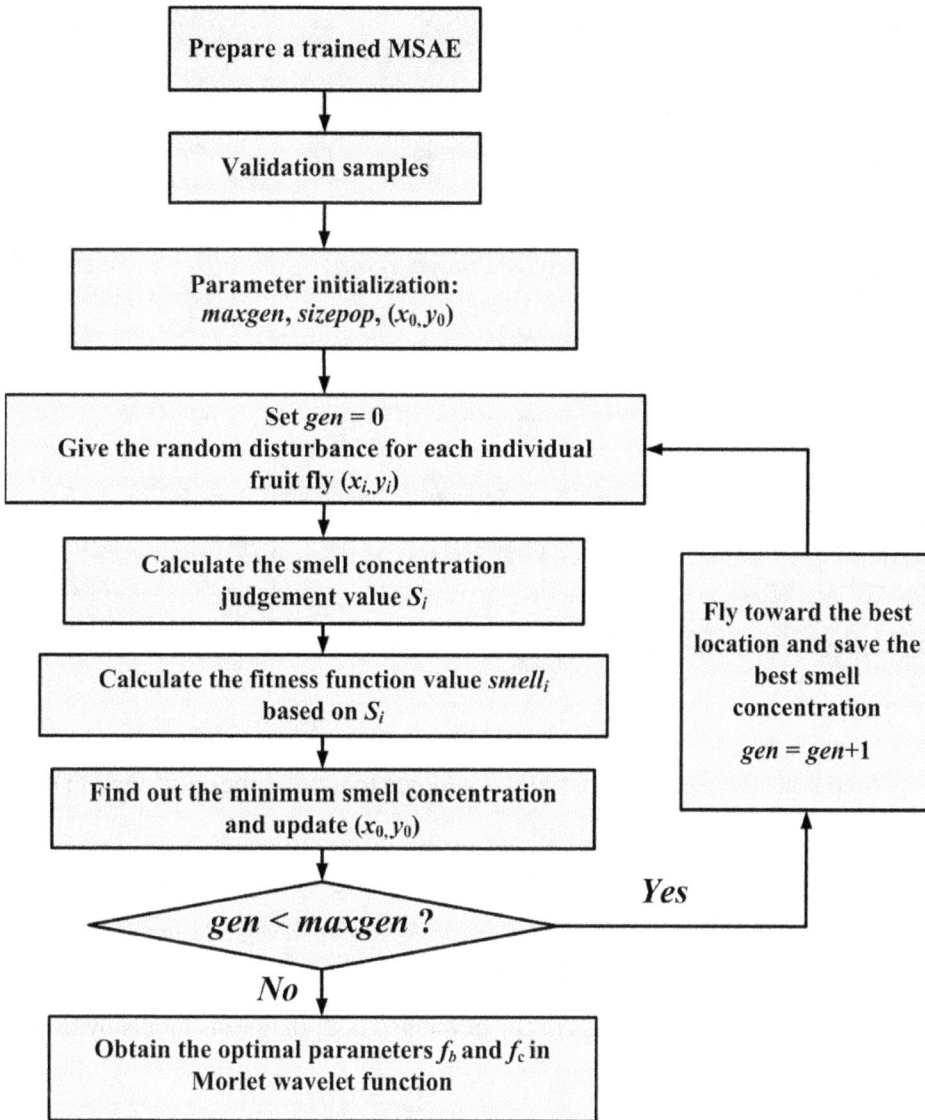

FIGURE 4.17   Adaptive Morlet wavelet design of the MSAE using FOA.

4.3.1.2 Overall Framework of the Proposed Method

Figure 4.18 gives the overall framework of the proposed method.

The following are its main steps:

Step 1: Collect the raw vibration data of the key parts of the rotating machine, which are divided into training, validation, and testing samples.

Step 2: Design MSAE with an adaptive Morlet wavelet.

Step 2.1: Morlet wavelet is employed as the activation function to design MSAE based on equations (4.16)–(4.18).

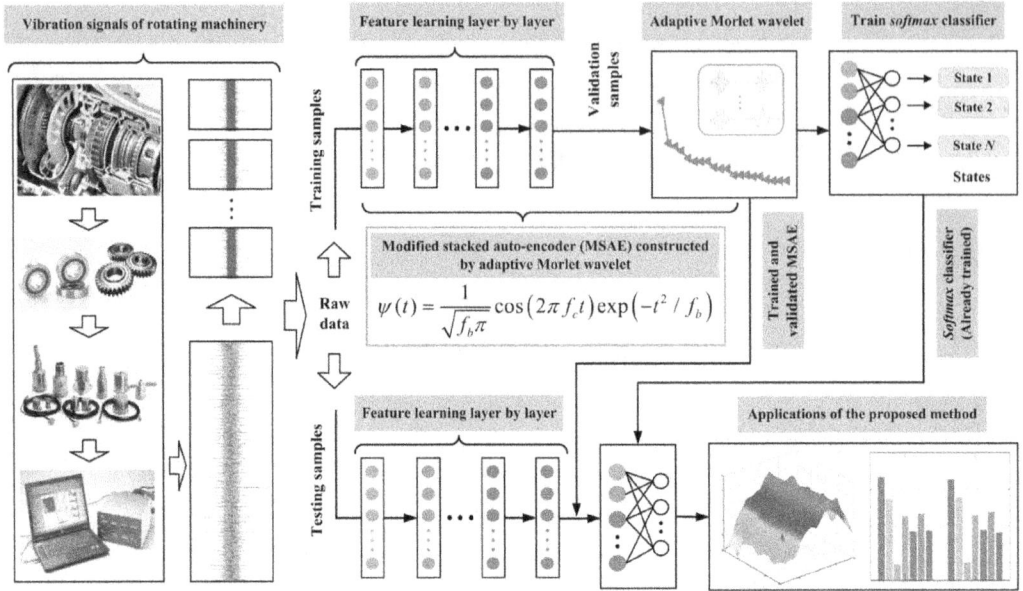

FIGURE 4.18   Overall framework of the proposed method.

Step 2.2: Nonnegative constraint is adopted to modify the cost function based on equations (4.8) and (4.9).

Step 2.3: Update the weights of the MSAE using equations (4.22)–(4.24).

Step 2.4: FOA is adopted for the adaptive Morlet wavelet design of the MSAE to minimize the misclassification rate of the validation samples.

Step 3: Verify the effectiveness of the developed fault diagnosis model by the testing samples.

4.3.1.3 Experimental validation

4.3.1.3.1 Case 1: Fault Diagnosis of a Sun Gear Unit   In Case 1, the effectiveness of the proposed approach is tested using the measured vibration data of sun gears from the Drivetrain Dynamics Simulator (DDS), University of Connecticut. As shown in Figure 4.19, the DDS mainly includes motor, motor controller, brake, parallel gearbox, and planetary gearbox. An accelerometer (model: PCB608A11) is installed to collect vibration signals at a sampling frequency of 20 kHz under the stable running of the monitored components. The specifications of the accelerometer including frequency range, measure range, and sensitivity are 0.5–10 kHz, ±50 g, and 100 mV/g, respectively.

A total of nine working conditions of sun gears in a planetary gearbox are employed as shown in Figure 4.20. Specifically, the fault types include missing tooth, root crack, spalling, and chipping tip, which are common fault modes of gear. For chipping tip, there are five damage degrees.

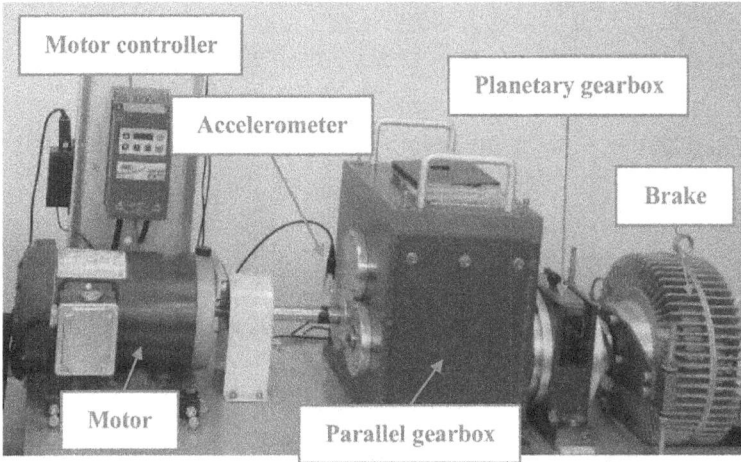

FIGURE 4.19   Experimental setup of the planetary gearbox.

FIGURE 4.20   Pictures of the nine types of sun gears.

Detailed sample distributions of the nine states of sun gears are listed in Table 4.7. Every state has 360 data samples, and each sample is a vibration signal segment with 1,024 data points. Randomly selected 200 out of 360 samples are employed for training, while the rest 60 and 100 samples are used for validation and testing, respectively. Time-domain waveforms of the nine working states of sun gears are plotted in Figure 4.21.

The proposed method is compared to popular methods that use deep learning, including four kinds of SAEs, two kinds of DBNs, and a classical CNN model called LeNet-5. Their inputs are all selected as the 1,024-dimensional raw vibration data.

To reduce the impact of contingency on the diagnosis results, ten trials for each method are run and the detailed results are shown in Figure 4.22. For each run, the overall classification accuracy is the ratio of the total number of correctly classified samples to the total

TABLE 4.7   Nine Working States of Sun Gears

Working States of SUN Gears	Number of the Training/Validation/Testing Samples	Labels of the States
Healthy	200/60/100	1
Missing tooth	200/60/100	2
Root crack	200/60/100	3
Spalling	200/60/100	4
Chipping tip (Severe 1)	200/60/100	5
Chipping tip (Severe 2)	200/60/100	6
Chipping tip (Severe 3)	200/60/100	7
Chipping tip (Severe 4)	200/60/100	8
Chipping tip (Severe 5)	200/60/100	9

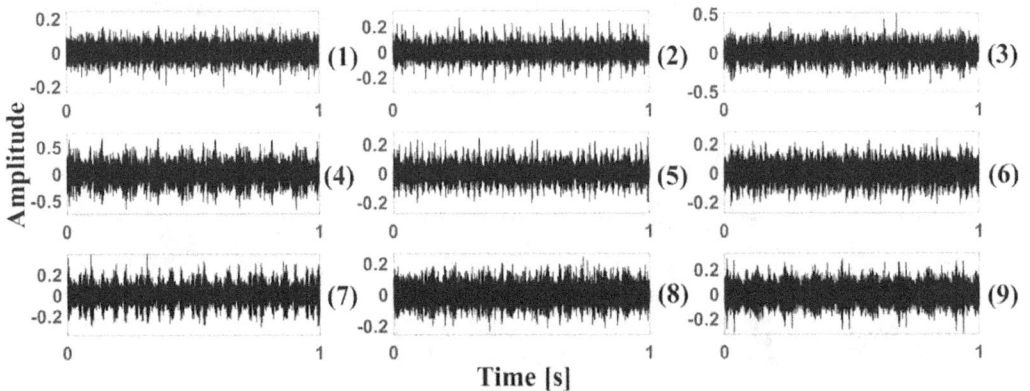

FIGURE 4.21   Time-domain waveforms of the nine working states of sun gears.

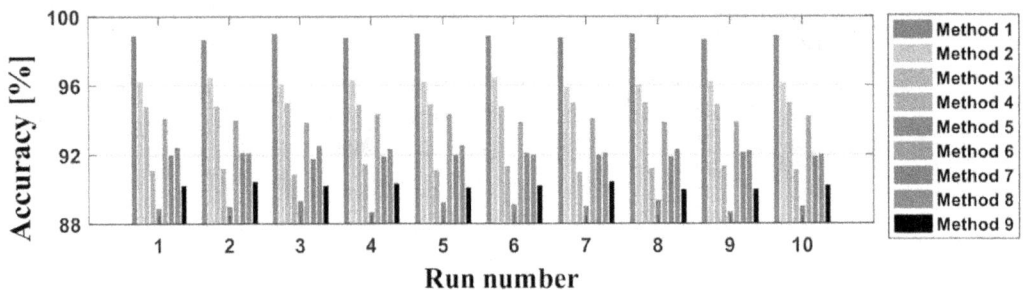

FIGURE 4.22   Repeated diagnosis results based on the nine methods in Case 1 (f_b=0.621 and f_c=3.114).

number of the testing samples whose true labels are known. As listed in Table 4.8, the average value of the overall testing accuracy of the proposed method (f_b=0.621 and f_c=3.114) reaches 98.86% (8,897/9000, where 8,897 is the total number of correctly classified samples during the 10 runs and 9,000=10×9×100 is the total number of testing samples during 10 runs), and it is higher than that for the eight contrastive methods, which are 96.20%, 94.90%, 91.18%, 89.02%, 94.07%, 91.98%, 92.27%, and 90.22%. For the first run, the testing accuracy of the proposed method is 98.89% (890/900), and F-measure is used to evaluate its diagnosis performance for each working state, as shown in Figure 4.23.

TABLE 4.8   Statistical Diagnosis Results of the Nine Methods in Case 1

Diagnosis Methods	Average Testing Accuracies
Method 1 (Proposed method)	**98.86% (8897/9000)**
Method 2 (SAE: Morlet with C^T)	96.20% (8658/9000)
Method 3 (CNN: LeNet-5)	94.90% (8541/9000)
Method 4 (Gaussian DBN)	91.18% (8206/9000)
Method 5 (Basic DBN)	89.02% (8012/9000)
Method 6 (SAE: ReLU with C^E)	94.07% (8466/9000)
Method 7 (SAE: ReLU with C^T)	91.98% (8278/9000)
Method 8 (SAE: Sigm with C^E)	92.27% (8304/9000)
Method 9 (SAE: Sigm with C^T)	90.22% (8120/9000)

"Bold" indicates that this method has the highest Average Testing Accuracies in comparison.

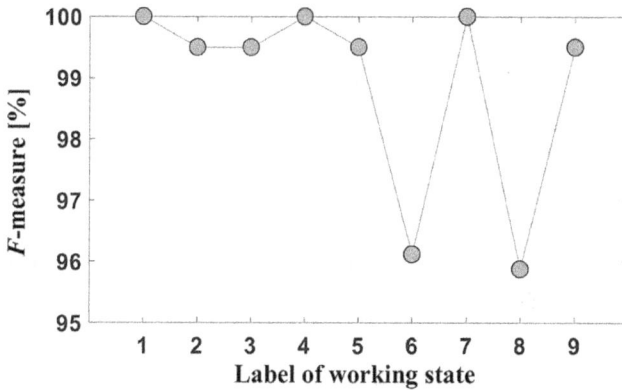

FIGURE 4.23   F-measures of the proposed method for different gear sun states.

Based on the contrastive results, it can be concluded that the proposed method exhibits better diagnosis performance than other existing deep learning methods for fault diagnosis of a sun gear. Specifically, the superiority of the nonnegative constraint is demonstrated by the comparison results provided by Method 1 and Method 2. Also, the comparisons between the enhanced cost function and the traditional cost function of the first MAE are given in Figure 4.24, which also shows the effectiveness of the former.

Many related studies have shown that SAEs with three hidden layers are often deep enough to perform high fault diagnosis accuracies. The model structure of the proposed MSAE is constructed as "1024-450-250-100-9" by setting the number of neurons in the hidden layers in descending order and about half of the neuron number of the previous layer. In this case study, the other hyperparameters are given in Table 4.9, and most of them are determined by experimental experience.

The importance of adaptive Morlet wavelet design is shown by the validation accuracies based on different combinations of parameters f_b and f_c, as shown in Figure 4.25. From Figure 4.25, it is seen that the validation accuracies are seriously affected by the parameters f_b and f_c. To show the superiority of FOA, two other algorithms are used for comparisons, which are genetic algorithm (GA) and particle swarm optimization (PSO).

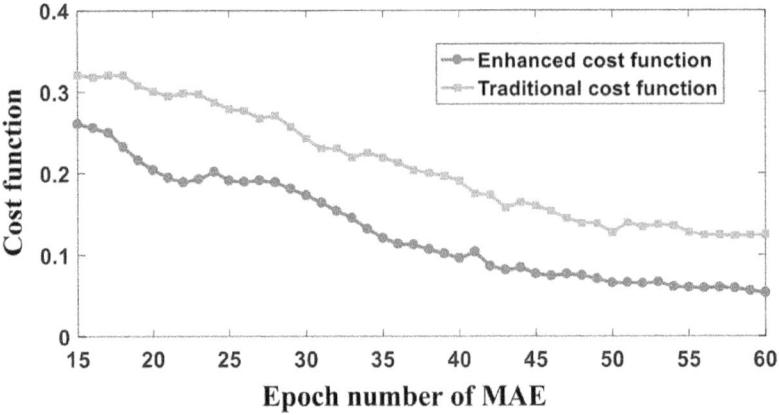

FIGURE 4.24   Comparisons between the enhanced cost function and the traditional cost function in Case 1.

TABLE 4.9   Parameters of the Proposed Method in Case 1

Description	Value
Maximum epoch number of the MSAE	60
Sparsity constant/sparse penalty coefficient	0.08/5
Weight decay factor/learning rate	0.003/0.01
The optimal parameter f_b of Morlet wavelet	**0.621(Given by FOA)**
The optimal parameter f_c of Morlet wavelet	**3.114(Given by FOA)**
The maximum generation number of FOA	20
The population size of FOA	25

"Bold" emphasizes that the setting method of this parameter is the innovation point. It is given by FOA according to the method mentioned above.

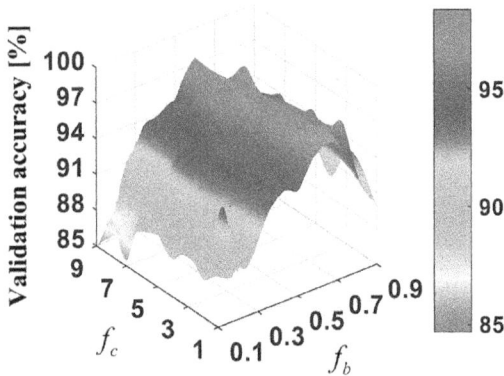

FIGURE 4.25   Effect of parameters f_b and f_c on the validation accuracies in Case 1.

Figure 4.26 shows the misclassification rates on the validation samples using different algorithms. The misclassification rates using FOA converge after about only 20 epoch times, while GA and PSO need more while starting from the same initial conditions. Besides, the minimum misclassification rate using FOA is smaller than the other three methods.

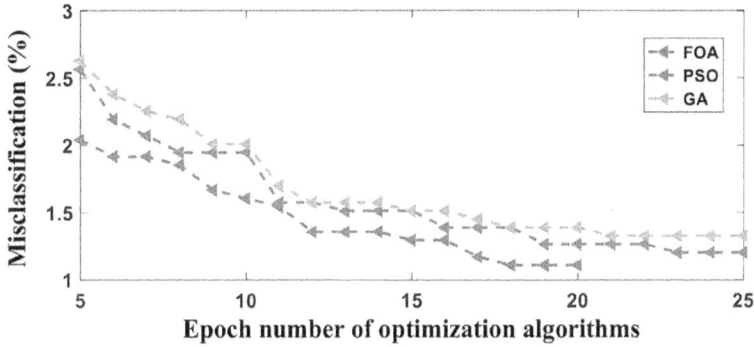

FIGURE 4.26   Misclassification rate of the validation samples using three types of optimization algorithms in Case 1.

TABLE 4.10   Comparisons between Other Wavelets and State-of-the-Art Methods under Different White Noise Levels in Case 1

Diagnosis Methods	Signal-Noise Ratio (SNR)			
	24 dB	20 dB	16 dB	12 dB
Method 1 (Proposed method)	**96.30%**	**94.48%**	**92.01%**	**88.16%**
Method 10 (SAE: Gaussian)	94.52%	93.64%	89.34%	85.86%
Method 11 (SAE: Haar)	81.88%	75.92%	71.50%	64.09%
Method 12 (SAE: Mexican Hat)	95.06%	93.97%	91.00%	86.42%
Method 13 (SAE: Shannon)	88.62%	85.07%	79.98%	72.06%
Method 14 proposed in [38]	88.29%	85.83%	81.38%	75.93%
Method 15 proposed in [39]	94.00%	91.94%	88.26%	84.03%
Method 16 proposed in [40]	93.38%	90.86%	87.90%	84.86%
Method 17 proposed in [41]	88.94%	86.50%	83.00%	78.34%
Method 18 proposed in [42]	87.14%	84.94%	81.06%	76.86%

"Bold" indicates that this method has the highest Average Testing Accuracies under a certain white noise level.

In addition to those methods listed in Table 4.8, the proposed method is also compared with state-of-the-art deep learning methods published in recent years. The influence of noise and comparisons between the other four types of popular wavelets holding explicit expressions are discussed here at the same time. The average testing accuracies of the 10 runs given under different white noise levels are shown in Table 4.10. Here, all the SAEs are designed with the enhanced cost functions, and in order to make fair comparisons, the input of each method is raw time-domain vibration data with no signal preprocessing or feature extraction. Some conclusions can be drawn from Table 4.10 as follows:

1. The proposed method based on Morlet wavelet shows higher diagnosis accuracies than Gaussian wavelet and Mexican Hat wavelet, while the results of Haar wavelet and Shannon wavelet are much worse.

2. The proposed method is more effective compared with the five state-of-the-art methods in analyzing the raw nonstationary vibration data under the influence of noise.

3. With the increase of noise, although the accuracies of all the methods decrease (from 96.30% to 88.16%), the proposed method shows the best antinoise capability.

This chapter mainly focuses on diagnosing faults based on the raw vibration data; however, it is believed that using specialized signal processing techniques for denoising can improve accuracy.

4.3.1.3.2 Case 2: Fault Diagnosis of a Roller Bearing Unit   In Case 2, the proposed method is used to analyze the vibration data collected from a bearing fault diagnosis test rig, Anhui University of Technology, China, as shown in Figure 4.27. The rotating speed is set at 900 rpm and the load is 2 kN.m. Vibration signals during the stable run are collected with a sampling frequency of 10 kHz using ICP INV9822 accelerometer. The specifications of the accelerometer including frequency range, measure range, and sensitivity are 0.5 Hz–8 kHz, ±50 g, and 100 mV/g, respectively. The model number of the tested bearing is 6205-2RS.

Nine working states of the roller bearing are collected and their samples are listed in Table 4.11, including different fault types and different damage degrees. Each sample

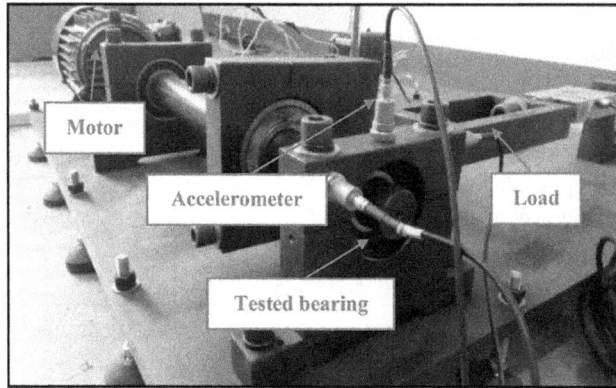

FIGURE 4.27   Fault diagnosis test rig of roller bearings.

TABLE 4.11   Nine Working Conditions of Roller Bearings

Working States of Roller Bearings	Number of the Training/Validation/ Testing Samples	Labels of the States
Normal	150/50/125	1
Ball (0.2 mm)	150/50/125	2
Ball (0.3 mm)	150/50/125	3
Inner race (0.2 mm)	150/50/125	4
Inner race (0.3 mm)	150/50/125	5
Inner race (0.4 mm)	150/50/125	6
Inner race (0.5 mm)	150/50/125	7
Outer race (0.2 mm)	150/50/125	8
Outer race (0.4 mm)	150/50/125	9

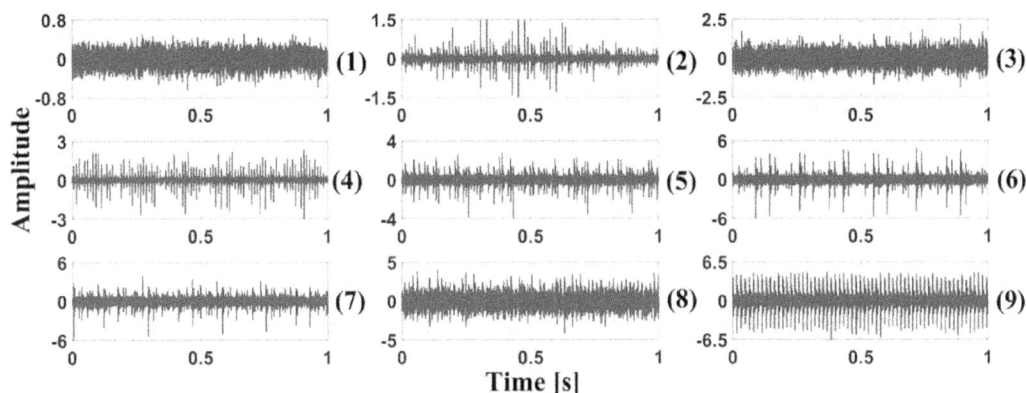

FIGURE 4.28   Time-domain waveforms of the nine working states of roller bearings.

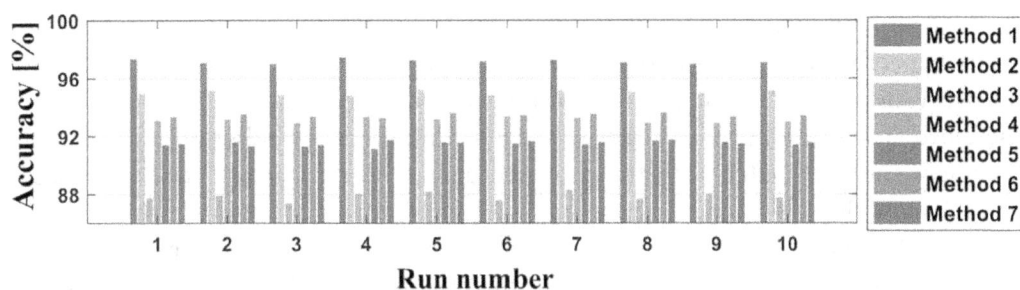

FIGURE 4.29   Repeated diagnosis results based on the nine methods in Case 2 (f_b=0.835 and f_c=2.421).

consists of 600 sampling data points, and the time-domain waveforms of the nine working states are plotted in Figure 4.28.

In this case study, LSTM, another two types of SAEs based on Leaky ReLU (LReLU) and exponential linear unit (ELU), and the five state-of-the-art deep learning methods listed in Table 4.4 are used for comparison.

As before, each method runs 10 trials. The detailed testing diagnosis accuracies of the proposed method, LSTM, and two SAEs (LReLU and ELU) are given in Figure 4.29, and their statistical results are listed in Table 4.12. The average testing accuracy provided by the

TABLE 4.12   Statistical Diagnosis Results of the Seven Methods in Case 2

Diagnosis Methods	Average Testing Accuracies
Method 1 (Proposed method)	**97.16% (10930/11250)**
Method 2 (SAE: Morlet with CT)	94.99% (10686/11250)
Method 3 (Deep LSTM)	87.84% (9882/11250)
Method 4 (SAE: LReLU with C^E)	93.09% (10473/11250)
Method 5 (SAE: LReLU with C^T)	91.43% (10286/11250)
Method 6 (SAE: ELU with C^E)	93.43% (10511/11250)
Method 7 (SAE: ELU with C^T)	91.54% (10298/11250)

"Bold" indicates that this method has the highest Average Testing Accuracies in comparison.

proposed method ($f_b=0.835$ and $f_c=2.421$) is 97.16% (10 930/11 250, 11 250=9*125*10), and it is higher than those of the six contrastive methods, which are 94.99%, 87.84%, 93.09%, 91.43%, 93.43%, and 91.54%, respectively. For the first run, the specific testing accuracy provided by the proposed method is 97.33% (1095/1125).

The confusion matrix is shown in Figure 4.30 (two decimal places) in which the horizontal and vertical axes are the predicted and true state labels, respectively. These values located in the diagonal represent the correct classification rate of each state, and the others represent the misclassification rates. The diagnosis accuracies for most of the classes are higher than 0.97. The result on the fifth condition is slightly lower where some testing samples are misclassified to conditions 2, 4, and 6. Table 4.13 shows the comparison results of the proposed method with the five state-of-the-art deep learning methods under different levels of white noise, which further confirms the superiority of the proposed method in the direct analysis of raw nonstationary vibration data under the influence of noise.

The hyperparameters of the MSAE are given in Table 4.14. According to Table 4.9 in Case 1 and Table 4.14 in Case 2, the optimal parameters f_b and f_c have changed. Therefore, it is meaningful to adaptively determine these two adjustable parameters to achieve the best match with the analyzed data. The comparison results in Cases 1 and 2 show that the proposed method is more effective than other state-of-the-art deep learning methods for fault diagnosis of rotating machinery key parts. Adaptive Morlet wavelet activation function enables the MSAE to establish an accurate nonlinear mapping between various working states and the raw nonstationary vibration data. Besides, the nonnegative constraint of connection weights helps to achieve the high-quality reconstruction of the MSAE.

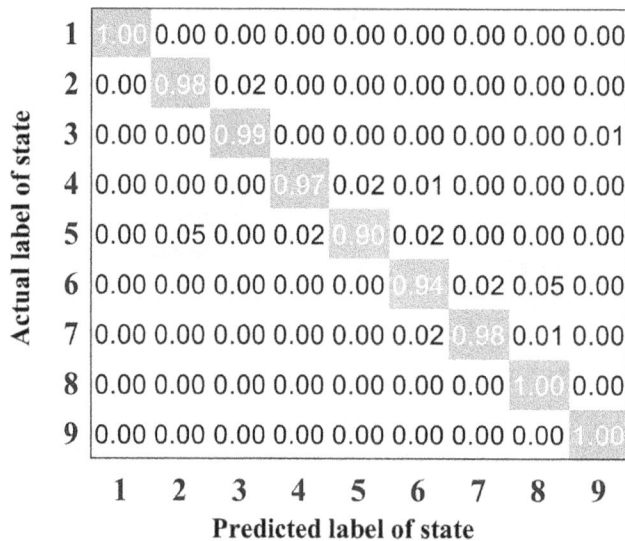

Actual label of state	Predicted label of state								
1	1.00	0.00	0.00	0.00	0.00	0.00	0.00	0.00	0.00
2	0.00	0.98	0.02	0.00	0.00	0.00	0.00	0.00	0.00
3	0.00	0.00	0.99	0.00	0.00	0.00	0.00	0.00	0.01
4	0.00	0.00	0.00	0.97	0.02	0.01	0.00	0.00	0.00
5	0.00	0.05	0.00	0.02	0.90	0.02	0.00	0.00	0.00
6	0.00	0.00	0.00	0.00	0.00	0.94	0.02	0.05	0.00
7	0.00	0.00	0.00	0.00	0.00	0.02	0.98	0.01	0.00
8	0.00	0.00	0.00	0.00	0.00	0.00	0.00	1.00	0.00
9	0.00	0.00	0.00	0.00	0.00	0.00	0.00	0.00	1.00

FIGURE 4.30 Confusion matrix of the proposed method for the first run in Case 2.

TABLE 4.13 Comparisons between Other Wavelets and State-of-the-Art Methods under Different White Noise in Case 2

Diagnosis Methods	Signal-Noise Ratio (SNR)			
	24 dB	20 dB	16 dB	12 dB
Method 1 (Proposed method)	**95.04%**	**93.85%**	**90.37%**	**85.25%**
Method 8 (SAE: Gaussian)	93.69%	91.28%	87.20%	82.08%
Method 9 (SAE: Haar)	80.29%	74.40%	68.67%	61.81%
Method 10 (SAE: Mexican Hat)	93.96%	91.52%	88.15%	83.25%
Method 11 (SAE: Shannon)	86.95%	84.75%	78.30%	70.38%
Method 12 proposed in [43]	88.01%	84.90%	80.08%	75.16%
Method 13 proposed in [44]	93.85%	91.00%	87.44%	82.68%
Method 14 proposed in [45]	92.79%	90.10%	86.95%	82.04%
Method 15 proposed in [46]	88.56%	85.48%	81.99%	76.95%
Method 16 proposed in [47]	86.48%	84.36%	80.20%	76.08%

"Bold" indicates that this method has the highest Average Testing Accuracies under a certain white noise level.

TABLE 4.14 Parameters of the Proposed Method in Case 2

Description	Value
Maximum epoch number of the MSAE	70
Sparsity constant/sparse penalty coefficient	0.10/ 5
Weight decay factor/learning rate	0.003/0.01
The optimal parameter f_b of Morlet wavelet	0.835 (Given by FOA)
The optimal parameter f_c of Morlet wavelet	2.421 (Given by FOA)
The maximum generation number of FOA	20
The population size of FOA	25

4.3.2 Semi-Supervised Fault Diagnosis Method Based on Label Propagation Strategy and Dynamic Graph Attention Network

Semi-supervised learning aims to mine the information contained in unlabeled samples by using the limited labeled samples [43], which has been researched in intelligent fault diagnosis of machinery over recent years. Existing research can cope with the challenge of insufficient labeled samples in fault diagnosis tasks to some extent [44–46]. However, they all analyze the multiple samples in isolation from the perspective of Euclidean space, resulting in the implicit related information between samples not being efficiently exploited [47,48]. Graph-based methods have been widely studied in intelligent manufacturing and fault diagnosis in recent years [49,50]. As one of these methods, GNNs can establish structures between nodes from non-Euclidean space, that is to say, the representation of nodes can be updated by learning the features of its neighbor nodes along the relationship structure [51–53].

In the past 2 years, a few scholars have gradually explored the application of GNNs to semi-supervised fault diagnosis of machinery [54–56]. Although the recent semi-supervised GNNs mentioned above have shown relatively high accuracies in intelligent fault diagnosis of machinery, there still exist the following limitations which need to be considered:

1. The joint dependencies between the small labeled samples and large unlabeled samples around the distribution are ignored in the above studies, resulting in underutilization of the limited label information. Theoretically, the samples with edge connection relationship probably have the same label category due to their relatively similar distribution characteristics [57]. Therefore, the existing scarce label information has great potential to be fully mined to assist in predicting the unlabeled samples.

2. The GNNs involved in recent researches have limited ability to well extract features of the different neighbor nodes [58]. Therefore, more powerful GNNs need to be constructed to extract features of different neighbor nodes in a more dynamic and expressive manner.

3. The above researches are limited to ideal diagnosis scenarios under steady speed, i.e., they do not involve speed fluctuation scenarios with complex mapping among signal characteristics and fault modes. However, in practical engineering, machinery equipment often goes through start–stop process or needs to change speeds due to specific task requirements [59]. Therefore, researches on semi-supervised fault diagnosis under speed fluctuation scenarios are more meaningful.

4. The labeled rates in all the GNNs-based semi-supervised fault diagnosis tasks mentioned above are no less than 5%. However, in engineering practice, due to the increasingly expensive cost of labeling data and the long-term accumulation of massive monitoring data, the available labeled samples will be more and more limited [60]. Therefore, GNNs-based semi-supervised fault diagnosis research under extremely low labeled rates is worth to further explore.

In this section, to deal with challenging scenarios with speed fluctuation and extremely low labeled rates, a new semi-supervised fault diagnosis method called LPS-DGAT is proposed [61].

4.3.2.1 Main Algorithms
4.3.2.1.1 Related Theories
4.3.2.1.1.1 Basic GNN Unlike CNN, GNN is the learning model based on graph structure data and can define the graph convolution based on connected relation of nodes in non-Euclidean space. The convolution difference between CNN and GNN is represented in Figure 4.31. The graph structure used in GNNs can be simplified as follows:

$$G=(H, A) \tag{4.25}$$

where $H=\{h_1, h_2, \cdots h_n\}\in R^{n\times d}$ denotes the node feature matrix; n is the number of nodes; d is the feature dimension; $A\in R^{n\times n}$ is the adjacency matrix representing the edge structure; and $A_{ij}\in A$ represents an edge connecting node h_i and nodeh h_j.

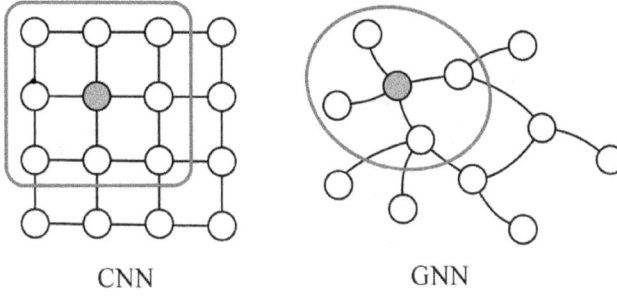

CNN   GNN

FIGURE 4.31   The convolution of CNN and GNN.

The GNN's transformation layer updates the node representation by learning its neighbor nodes' features. The layer accepts the graph structure G as the input and exports a new node feature matrix $H' = \{h_1', h_2', \cdots h_n'\}$, defined as follows:

$$h_i' = \Gamma\left(h_i, \mathrm{T}\left(\{h_j \mid j \in \mathrm{N}_i\}\right)\right) \tag{4.26}$$

where N_i is the neighbor node number of h_i; $\Gamma(\cdot)$ represents a nonlinear layer; and $\mathrm{T}(\cdot)$ represents a node aggregation mode.

4.3.2.1.1.2 Basic GAT   GAT is an advanced GNN model, and the following describes the detailed structure of the GAT layer. First of all, an operation is required to transform input features into higher-level features with better expression. A shared linear transformation parametrized by the weight matrix W is applied to all nodes. Then, a shared attention mechanism F is executed on nodes. After LReLU layer, the attention score is obtained. All attention scores are normalized by using softmax function to make them easily comparable among different nodes:

$$e_{ij} = \mathrm{LeakyReLU}\left(F\left[Wh_i \| Wh_j\right]\right) \tag{4.27}$$

$$\alpha_{ij} = \mathrm{softmax}_j\left(e_{ij}\right) = \frac{\exp\left(e_{ij}\right)}{\sum_{j' \in N_i \cup \{i\}} \exp\left(e_{ij'}\right)} \tag{4.28}$$

where $\|$ indicates a connected operation; the attention score e_{ij} denotes features' importance of node h_j to node h_i; α_{ij} is the normalized attention coefficient between node h_i and node h_j; and W and F are updated during iterations of the GAT layer. The calculation flow of α_{ij} is shown in Figure 4.32.

4.3.2.1.2 The Proposed Method

4.3.2.1.2.1 Design of the LPS   In the existing GNNs-based semi-supervised fault diagnosis frameworks, the joint dependencies between the small labeled samples and the large unlabeled samples around the distribution are usually ignored, i.e., the limited label information is not fully mined. Considering that these samples connected with edges probably share

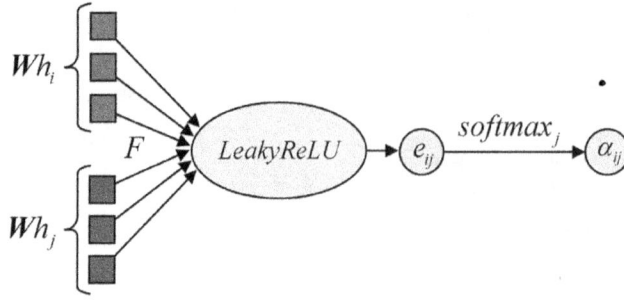

FIGURE 4.32 Calculation flow of α_{ij} in the basic GAT layer.

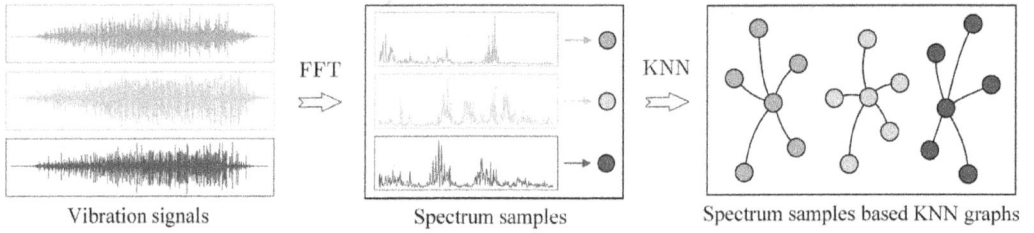

| Vibration signals | Spectrum samples | Spectrum samples based KNN graphs |

FIGURE 4.33 The process of constructing KNN graphs based on spectrum samples.

the same label due to their similar distribution characteristics, in this paper, the LPS is designed based on KNN and smoothing assumption (SA), which can propagate label information to neighbor samples along the edges and realize the full utilization of limited label information.

On one hand, the Euclidean distance is used to judge the adjacency relation between all samples to construct the spectrum samples based KNN graphs, which can well reflect local geometric properties between samples. The construction process is shown in Figure 4.33, and the distance measure formula of KNN algorithm can be expressed as follows:

$$L_{ij} = \left(\sum_{l=1}^{d} \left| h_i^{(l)} - h_j^{(l)} \right|^2 \right)^{\frac{1}{2}} \left(i \neq j \right) \tag{4.29}$$

where L_{ij} is the distance between node h_i and nodeh h_j; $h_i^{(l)}$ is the lth dimension feature of the nodeh h_i; here, a node refers to a spectrum sample after fast Fourier transformation (FFT). The edge structure is encoded into the adjacency matrix A, which can be expressed as follows:

$$A_{ij} = KNN\left(k, L_{ij}, \Omega_i \right) \tag{4.30}$$

where $\Omega_i = \{ L_{i1}, L_{i2}, \cdots L_{in} \}$ is the set of distances between h_i and all other nodes; k is the hyperparameter of KNN graphs; if L_{ij} is one of the k minimal values in Ω_i, $KNN(\cdot)=1$; otherwise, $KNN(\cdot)=0$.

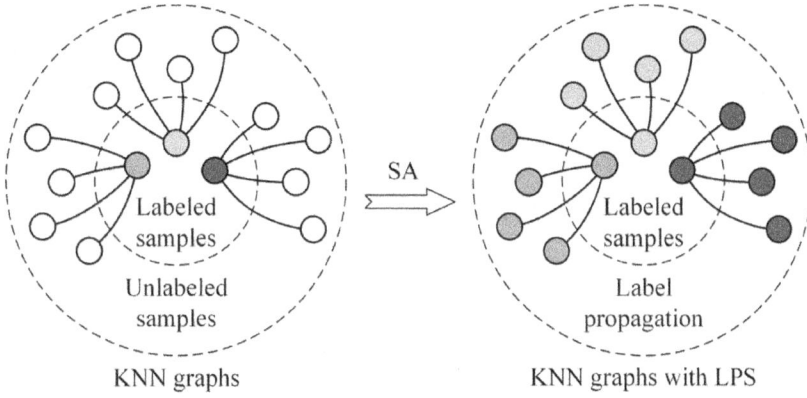

FIGURE 4.34 The designed LPS based on KNN and SA.

On the other hand, SA in semi-supervised learning holds that similar samples tend to have the same classification label. Therefore, we propagate label information to unlabeled samples along the edge, which can be expressed as follows:

$$y_j = SA\left(A_{ij}, \left(h_i, y_i\right), h_j\right) = \begin{cases} y_i & A_{ij} = 1 \\ \text{None} & A_{ij} = 0 \end{cases} \tag{4.31}$$

where y_i (one-hot encoding) is the true label of h_i; $y_j \in Y$ is the label of h_j judged by SA; and Y is the label set including the true labels and the labels judged by SA.

Finally, we design the LPS based on KNN and SA, which can spread the label information along the connected edges to unlabeled samples.

After that, the limited label information can be more fully exploited to improve the model learning ability. The process of the design LPS is shown in Figure 4.34.

4.3.2.1.2.2 Construction of the DGAT Despite that the basic GAT can iteratively update the node weights to distinguish the importance of information between neighbor nodes, different connected relations between central nodes will be ignored by using static attention, which may limit the expression ability. Therefore, the DGAT based on dynamic attention is constructed in this paper, which can improve the aggregation mode of central nodes to enhance the discriminatory ability of neighbor nodes.

The main problem of the basic GAT is that W and F in Eq. (4.27) are parameters shared globally for all the nodes and applied continuously, so they can be collapsed into a single linear layer, resulting in it being static attention, which can be expressed as follows:

$$e_{ij} = \text{LeakyReLU}\left(F_1 W h_i + F_2 W h_j\right) \tag{4.32}$$

$$F = \left[F_1 \| F_2\right] \tag{4.33}$$

where F_1, F_2 are the shared attention mechanism.

To well address this problem, the constructed DGAT applies W after concatenation and uses F after LReLU layer, thus for dynamic attention. The simple modification makes a significant difference in the expressiveness of the attention function, which can be expressed as follows:

$$e_{ij} = F \cdot \text{LeakyReLU}\left(W \cdot \left[h_i \| h_j\right]\right) \tag{4.34}$$

The specific difference between static attention given in Eq. (4.32) and dynamic attention given in Eq. (4.34) is described as follows. In basic GAT, the attention score set $E_i = \{e_{i1}, \cdots, e_{ik}\}(i=1,2,\cdots, n)$ calculates static scores of the central node set $C = \{c_1, \cdots, c_n\}$ and the neighbor node set $B = \{b_1, \cdots, b_k\}$. For every set E_i, there exists a neighbor node $b_{j\max} \in B$, which always has $e_{ij\max} \geqslant e_{ij}$ for all central nodes $c_i \in C$ and all neighbor nodes $b_i \in B$. In DGAT, every attention score set E_i calculates dynamic scores of the central node set C and the neighbor node set B. It can be explained that there is a mapping $\varphi : C \rightarrow B$, for every E_i, which has $e_{i\varphi(i)} \geqslant e_{ij}$ for all central nodes $c_i \in C$ and all neighbor nodes $b_i \in B$. $\varphi(i)$ is determined by the central node c_i. The above difference can be visually demonstrated as shown in Figure 4.35. As shown, the highest attention score of all central nodes in the basic GAT appears in neighbor node b4, while the highest attention score in the constructed DGAT appears in different neighbor nodes based on the feature information of central nodes.

Batch normalization (BN) is introduced in each DGAT layer to cope with the frequency shift and amplitude variation characteristics of spectrum samples under speed fluctuation to accelerate the network convergence, which is then followed by a nonlinear ReLU layer. The DGAT layer's calculation flow is shown in Figure 4.36, and the node update's formula is expressed as follows:

$$h_i' = \text{ReLU}\left(BN\left(\sum_{j \in N_i \cup \{i\}} \alpha_{ij} h_j \right) \right) \tag{4.35}$$

After obtaining $H' = \{h_1', \cdots, h_n'\}$, input it to the full connection (FC) layers to obtain the predicted label set $Z = \{z_1, \cdots, z_n\}$, which can be expressed as follows:

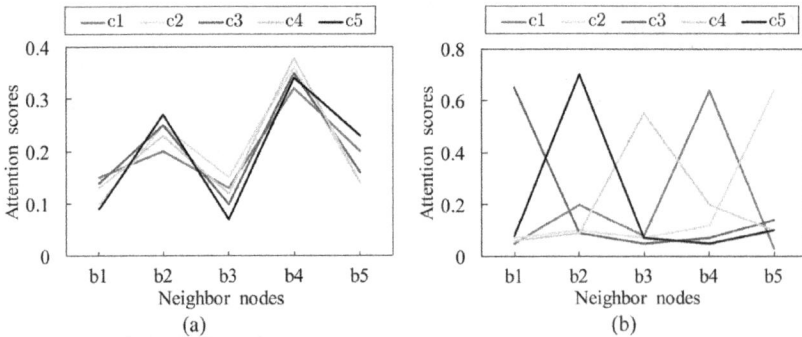

FIGURE 4.35　The attention scores in GAT and DGAT ($n=k=5$): (a) Basic GAT and (b) DGAT.

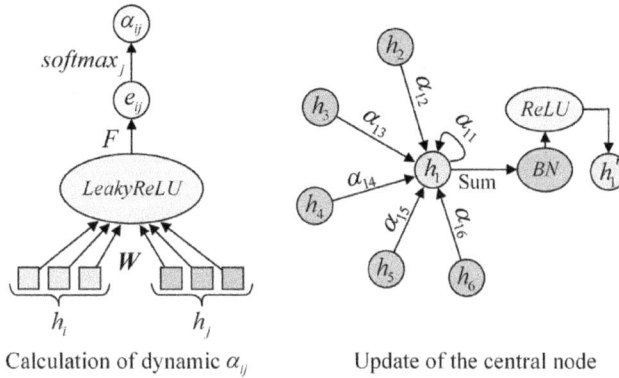

Calculation of dynamic α_{ij} Update of the central node

FIGURE 4.36 Calculation flow and update in the DGAT layer.

$$Z = FC(H')\tag{4.36}$$

Finally, calculate the loss function by the predicted label set Z and the label set Y as follows:

$$\text{Loss} = -\frac{1}{I}\sum_{i}^{I}\sum_{t}^{T}y_i^{(t)}\ln\left(z_i^{(t)}\right)\tag{4.37}$$

where I is the label number including the true labels and the labels judged by the designed LPS; T is the number of states; $y_i^{(t)}$ is the t dimension value of the label y_i (one-hot encoding); and $z_i^{(t)}$ is the t-dimension value of the predicted label z_i (one-hot encoding).

4.3.2.2 Overall Framework of the Proposed Method

The overall framework of the proposed method is shown in Figure 4.37, and the details are explained below:

Step 1: Collect the vibration signals of machinery key components in various health states under speed fluctuation, consisting of extremely low labeled samples and a large number of unlabeled samples. Then, the extremely low labeled samples combined with some unlabeled samples are used as the training set, while the rest of the unlabeled samples are used as the test set.

Step 2: Design the LPS based on KNN and SA. The vibration signals are passed through equations (4.29)–(4.31) to obtain two KNN graphs. One graph is the training set judged by the LPS, while the other is the unlabeled test set.

Step 3: Construct the DGAT model based on dynamic attention. Input the training set into two DGAT layers and two FC layers to obtain the predicted label set by equations (4.35) and (4.36). Then, calculate loss by equation (4.37) and update the model through the Adam algorithm.

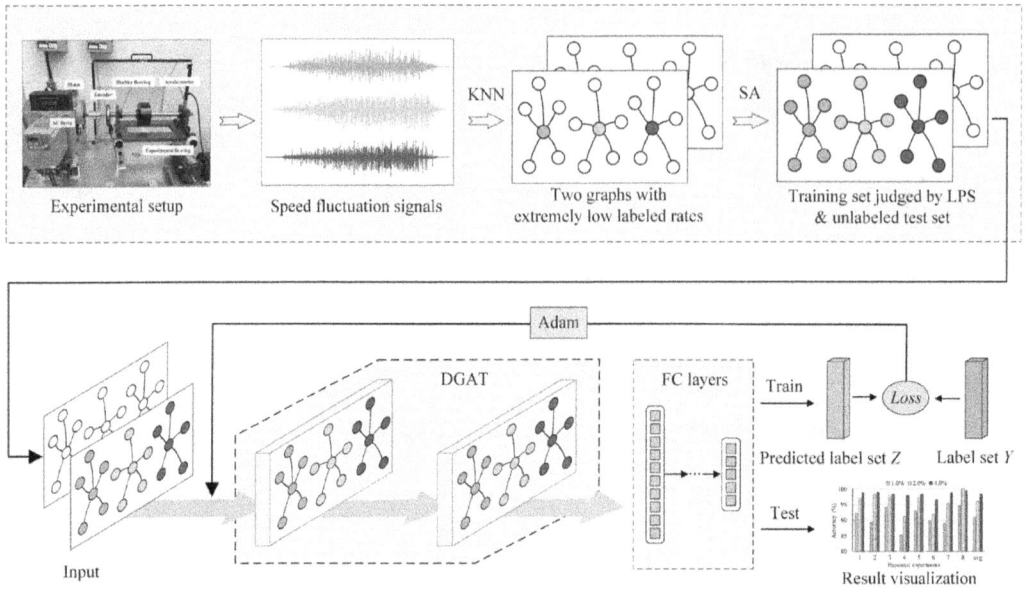

FIGURE 4.37 The framework of the proposed method.

Step 4: Feed the unlabeled test set into the trained model to obtain diagnostic results and compare with other semi-supervised fault diagnosis based on the commonly used GNNs.

4.3.2.3 Experimental Validation

4.3.2.3.1 Introduction to Datasets   Bearing and gear are two kinds of key components widely used in machinery. The superiority and effectiveness of the proposed method are verified with two semi-supervised fault diagnosis tasks of bearing and gear under speed fluctuation and extremely low labeled rates.

Case 1:

The dataset of gear is derived from the experimental bench of Xi'an Jiaotong University, and four root cracks with different degrees are prefabricated on the spur gears, as shown in Figure 4.38. Together with the normal state, a total of five health states of spur gears are considered and their vibration signals under speed fluctuation (0–1,200 rpm-0) are collected for analysis with sampling frequency of 10 kHz.

Case 2:

The dataset of bearing is derived from the SpectraQuest mechanical failure bench from University of Ottawa, which is shown in Figure 4.39.

Together with the normal state, a total of five health states of bearing including inner race defect, outer race defect, ball defect, combined defects on inner race, outer race, and ball are considered. Their vibration signals including two processes of first deceleration and then acceleration under speed fluctuation are collected for analysis with sampling frequency of 200 kHz.

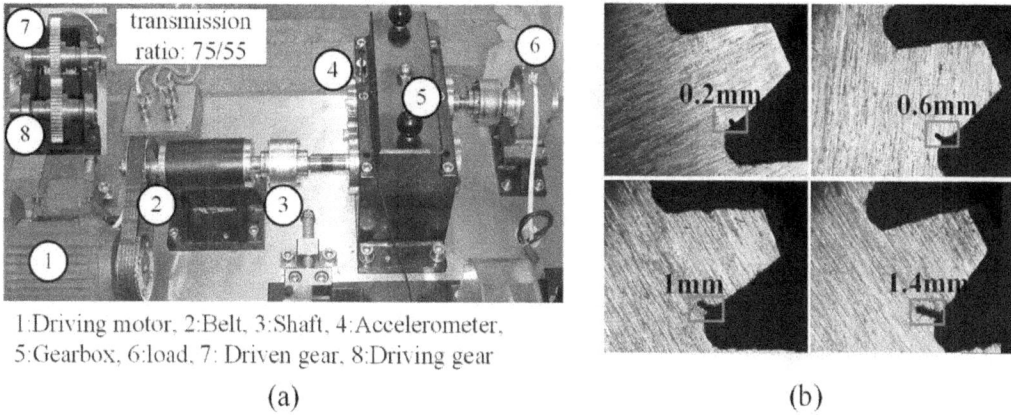

1:Driving motor, 2:Belt, 3:Shaft, 4:Accelerometer,
5:Gearbox, 6:load, 7: Driven gear, 8:Driving gear

(a)　　　　　　　　　　　　　　　　　(b)

FIGURE 4.38   Experimental setup of Case 1: (a) the experimental bench and (b) root cracks.

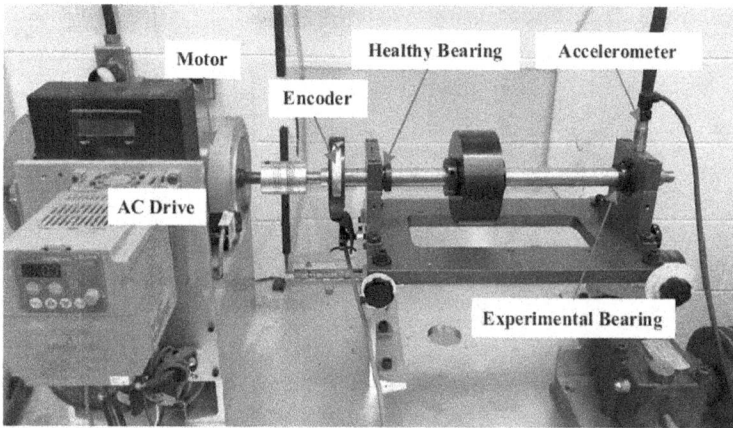

FIGURE 4.39   Experimental bench of Case 2.

In this paper, the introduction to the datasets of two cases is listed in Table 4.15. The total sample numbers for each health state are 200 in both cases. In Case 1, the length of each sample is 800 sampling points, and training samples for each health state are 40, of which there are only two labeled samples, with an extremely low labeled rate of 1%. As for Case 2, the length of each sample is 10,000 sampling points, and training samples for each health state are ten, of which there is only one labeled sample, with an extremely low labeled rate of 0.5%. The sample lengths are determined empirically based on the difference in sampling frequency. Vibration signals under speed fluctuation of two cases are shown in Figure 4.40, from which it can be seen that the vibration signals are complicated.

4.3.2.3.2 Comparisons with Other GNNs-Based Semi-Supervised Fault Diagnosis   In this part, to avoid experimental coincidence, eight replicate experiments are executed for all methods. The network structure and hyperparameters are set to the same for all GNNs.

TABLE 4.15   Introduction to Datasets of Two Cases

Cases	Speed Fluctuation (rpm)	Health States	Total Samples	Labeled Samples and Labeled Rates	Training Samples
Case 1	0-1200-0	Normal	1,000	10 (Labeled rate: 1%)	200
		Crack 0.2 mm			
		Crack 0.6 mm			
		Crack 1.0 mm			
		Crack 1.4 mm			
Case 2	1574-912-1262	Ball defect	1,000	5 (Labeled rate: 0.5%)	50
	1498-882-1345	Combined defects			
	1518-906-1118	Inner race defect			
	1476-840-1116	Normal			
	1512-894-1170	Outer race defect			

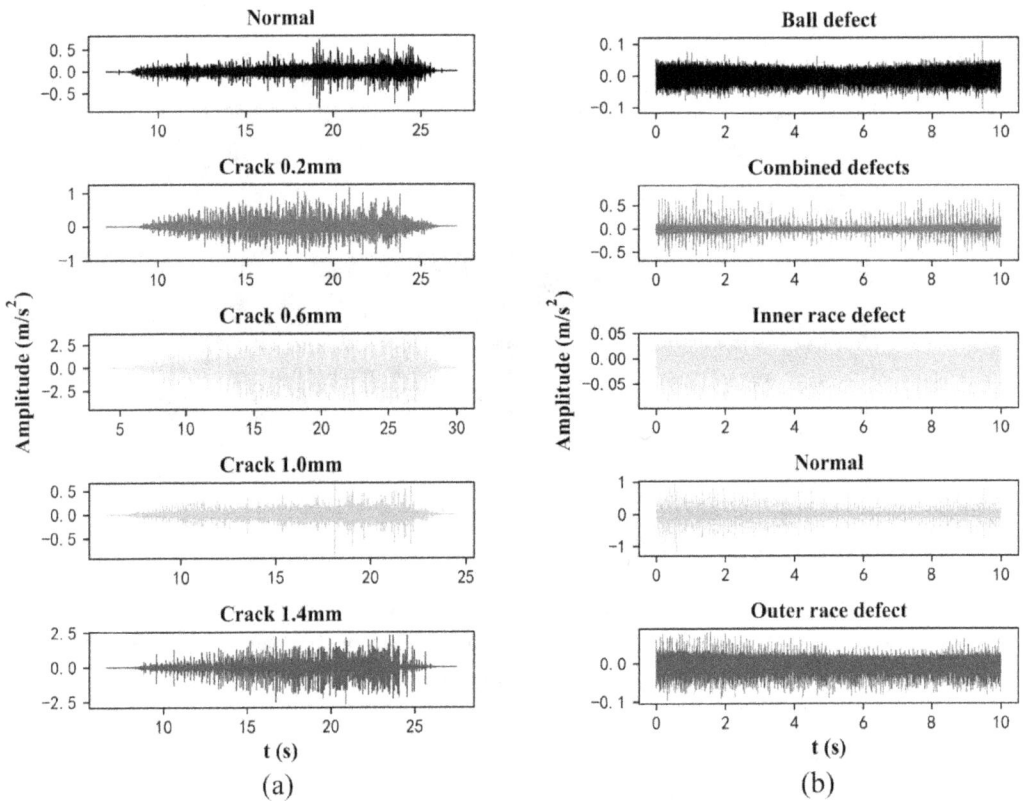

FIGURE 4.40   The collected vibration signals under speed fluctuation: (a) Case 1 and (b) Case 2.

The operating environments are described as follows: operation system: Windows 10; CPU: i5–10400F; GPU: GTX1650; and running software version: Pytorch1.7.1.

Based on the idea of greedy algorithm, we control the variables in order to find a relatively better combination of hyperparameters. The main hyperparameters of the model are set as follows: the number of iterations is set to 200; the learning rate is set to 0.01 and decays to 0.1 in 100 and 150 iterations, respectively.

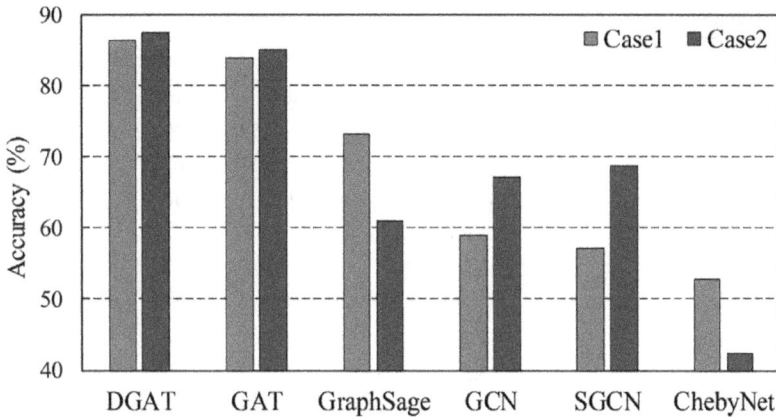

FIGURE 4.41   Diagnostic accuracies of the 6 GNNs in two cases.

First, the superiority of constructed DGAT is validated through comparisons with 5 GNNs that are widely studied, including basic GAT, GraphSage, GCN, simplifying graph convolutional networks (SGCN), and ChebyNet. The diagnostic results of 6 GNNs in two cases are shown in Figure 4.41. The accuracies of 6 GNNs in Case 1 are 86.33%, 83.83%, 73.24%, 58.97%, 57.17%, and 52.72%, respectively. In Case 2, the accuracies are 87.46%, 85.00%, 60.88%, 67.13%, 68.67%, and 42.32%, respectively. It can be found that the DGAT achieves the highest accuracies in both cases and improves over GAT by 2.50% and 2.46%, respectively. It demonstrates that the DGAT is better to extract features from nodes by dynamically assigning weights.

Then, the effectiveness of the designed LPS is validated by comparing the above 6 GNNs with or without LPS. The diagnostic results of all experiments are shown in Table 4.16. The proposed LPS-DGAT achieves the highest average accuracies in both cases. Especially, the accuracies are 90.99% and 93.96%, respectively, which improve over DGAT by 4.66% and 6.50%; the variances are 3.13% and 2.54%, which are 3.22% and 0.29% lower than DGAT. The reason is that the designed LPS takes full advantage of the label co-dependency between samples and increases the label information that can be learned by the model. In both cases, LPS improves the diagnostic accuracies of GAT by 3.75% and 5.80%, GraphSage by 11.01% and 9.38%, GCN by 20.04% and 11.90%, SGCN by 21.18% and 10.74%, and ChebyNet by 30.61% and 24.04%, respectively.

These comparison results indicate that the designed LPS is universal and robust to improve the semi-supervised diagnosis performance of different GNNs. The running times of all methods are shown in Table 4.16.

The training time and the test time of all methods are in the same order of magnitude, respectively, and the running times are fast.

To further demonstrate the effects of all the methods for different health states, we visualize the confusion matrixes for the eighth experiment of Case 1, which are shown in Figure 4.42. The number of test samples for each health state is 160, the horizontal coordinates are predicted labels and the vertical coordinates are true labels, where 0–4 refer to the five health states listed in Table 4.1 (from top to bottom), respectively. The results show that the proposed method gives the best diagnostic performance, and confirm that the

TABLE 4.16   Diagnostic Results and Running Times of All the Methods

Semi-Supervised Fault Diagnosis Methods	Diagnostic Results (%)		Training Time (s)		Test Time (s)	
	Case 1	Case 2	Case 1	Case 2	Case 1	Case 2
LPS-DGAT	**90.99 ± 3.13**	**93.96 ± 2.54**	3.90	0.02	5.02	0.03
DGAT	86.33±6.35	87.46±2.83	3.83	0.02	4.98	0.03
LPS-GAT	87.58±5.00	90.80±4.42	3.40	0.01	4.01	0.02
GAT	83.83±5.76	85.00±2.42	3.43	0.01	3.97	0.02
LPS-GraphSage	84.25±3.38	70.26±3.48	2.64	0.01	3.94	0.03
GraphSage	73.24±1.66	60.88±6.42	2.62	0.01	3.94	0.03
LPS-GCN	79.01±5.79	79.03±4.56	2.74	0.01	3.26	0.02
GCN	58.97±8.04	67.13±8.30	2.75	0.01	3.23	0.02
LPS-SGCN	78.35±4.88	79.41±5.59	2.63	0.01	3.22	0.04
SGCN	57.17±7.92	68.67±4.51	2.61	0.01	3.19	0.03
LPS-ChebyNet	83.33±4.99	66.36±8.24	3.06	0.02	4.41	0.05
ChebyNet	52.72±7.48	42.32±3.81	3.04	0.01	4.35	0.05

"Bold" indicates that this method has the highest Average Testing Accuracies in comparison.

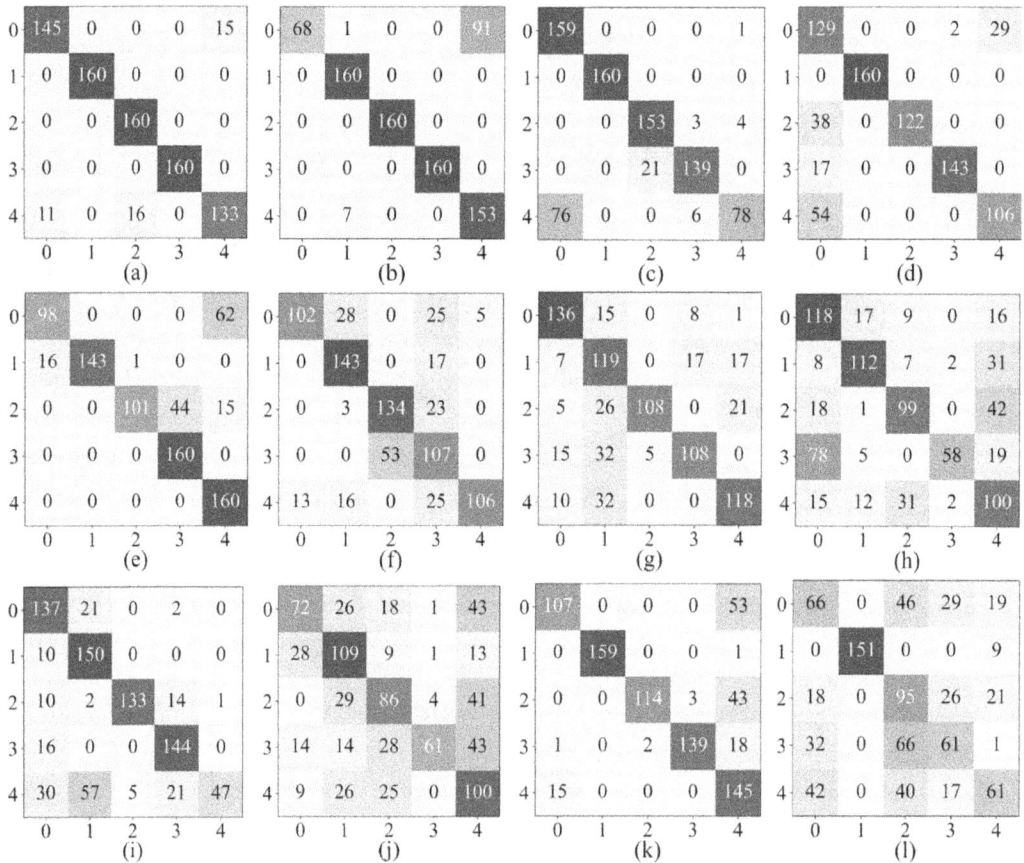

FIGURE 4.42   Confusion matrixes for all the methods on the eighth experiment of Case 1: (a) LPS-DGAT; (b) DGAT; (c) LPS-GAT; (d) GAT; (e) LPS-GraphSage; (f) GraphSage; (g) LPS-GCN; (h) GCN; (i) LPS-SGCN; (j) SGCN; (k) LPS-ChebyNet; and (l) ChebyNet.

designed LPS facilitates different GNNs for semi-supervised fault diagnosis of machinery under extremely low labeled rate.

To better demonstrate the feature extraction performance of all the methods, we reduce the output feature vectors to two dimensions (2D) by the t-SNE for the eighth experiment of Case 1, which are shown in Figure 4.43. As shown, the features extracted by DGAT are more representative than GAT, and the proposed method is more discriminative for all health states than other methods, which can prove that the DGAT combined with LPS is still effective under speed fluctuation where signal signs are complex.

As for Case 2, we use ROC curves to demonstrate the specific diagnostic effects of all the methods, which are shown in Figure 4.44. The horizontal axis and the vertical axis of the ROC curves are false positive rate and true positive rate, respectively. The area under the curve (AUC) is used to measure the quality of classification for each health state, which ranges from 0.5 to 1. Higher AUC values mean better diagnosis performance, which are listed in parentheses. The results show that all five curves of the proposed method are the closest to the upper left corner and have the highest AUC values compared with other methods, further confirming

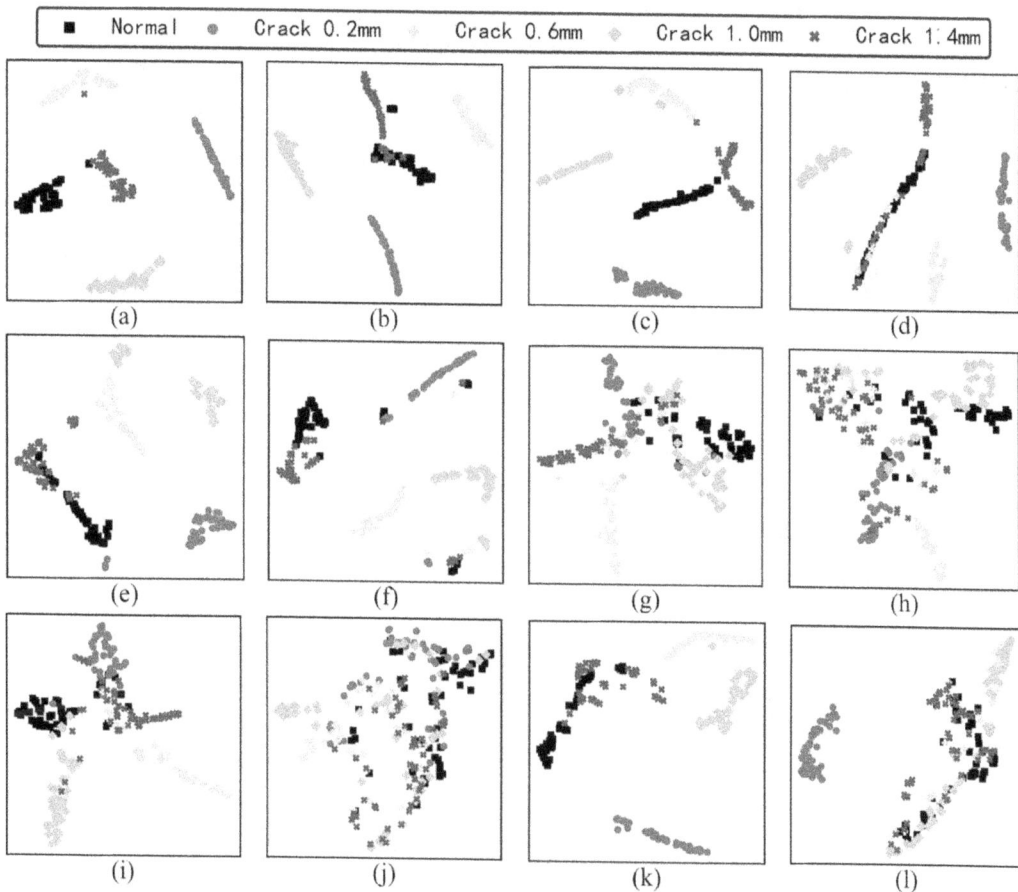

FIGURE 4.43   2D visualization of the output features for all the methods on the eighth experiment of Case 1: (a) LPS-DGAT; (b) DGAT; (c) LPS-GAT; (d) GAT; (e) LPS- GraphSage; (f) GraphSage; (g) LPS-GCN; (h) GCN; (i) LPS-SGCN; (j) SGCN; (k) LPS-ChebyNet; and (l) ChebyNet.

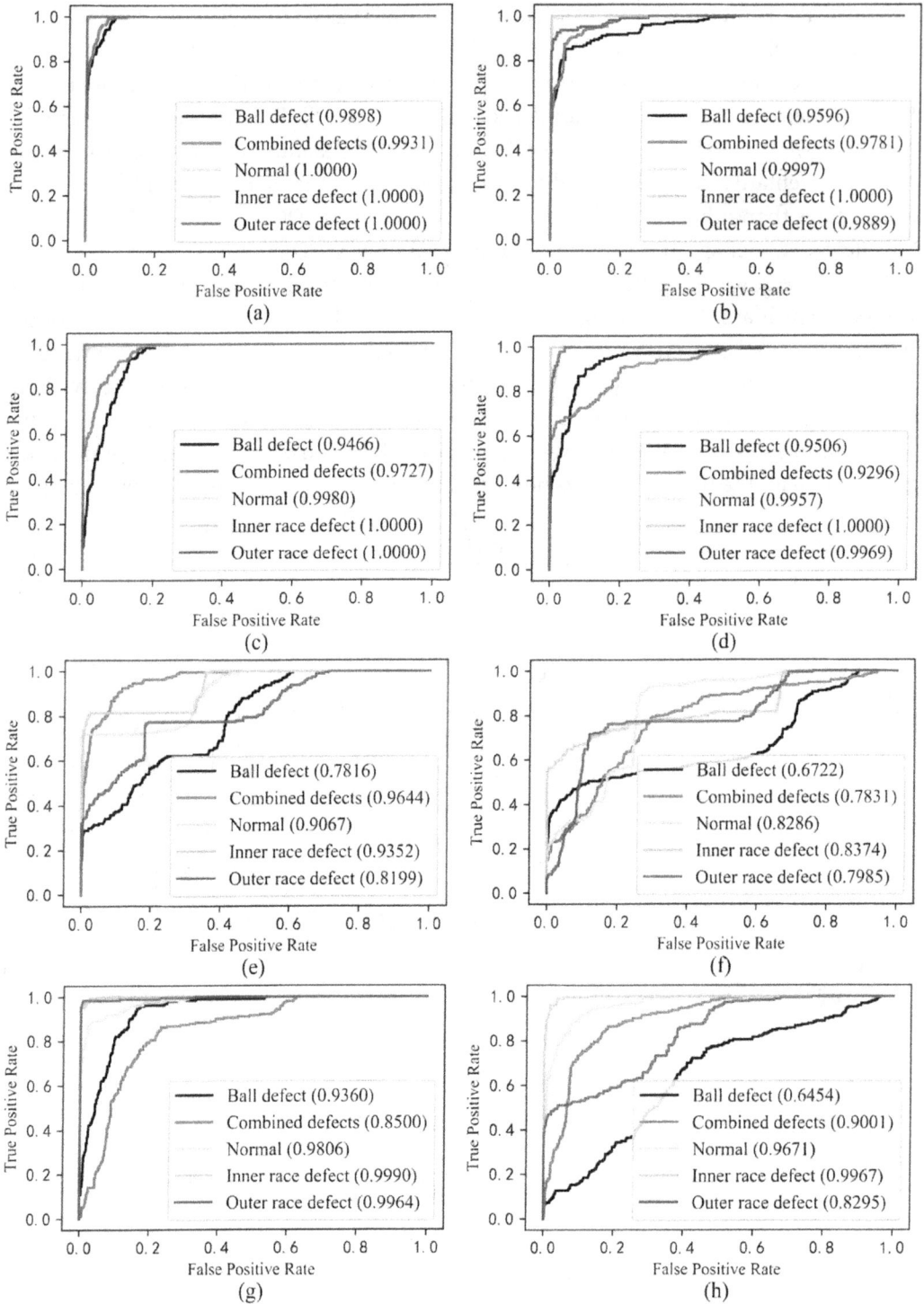

FIGURE 4.44 The ROC curves for all the methods on the eighth experiment of Case 2: (a) LPS-DGAT; (b) DGAT; (c) LPS-GAT; (d) GAT; (e) LPS-GraphSage; (f) GraphSage; (g) LPS-GCN; (h) GCN; (i) LPS-SGCN; (j) SGCN; (k) LPS-ChebyNet; and (l) ChebyNet.

(Continued)

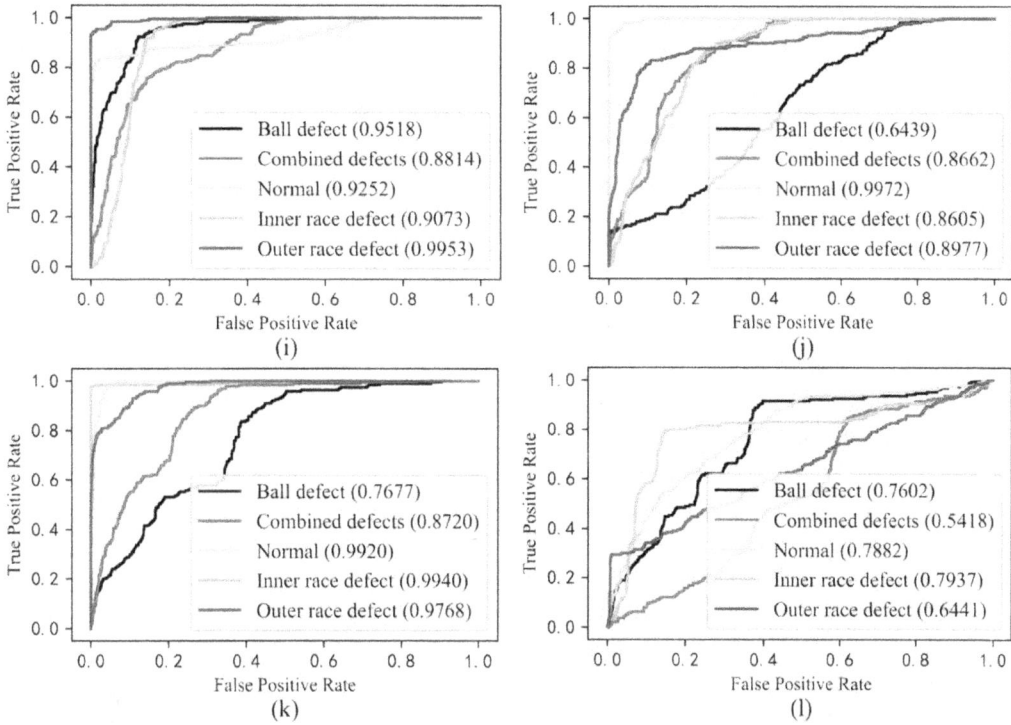

FIGURE 4.44 (*CONTINUED*) The ROC curves for all the methods on the eighth experiment of Case 2: (a) LPS-DGAT; (b) DGAT; (c) LPS-GAT; (d) GAT; (e) LPS-GraphSage; (f) GraphSage; (g) LPS-GCN; (h) GCN; (i) LPS-SGCN; (j) SGCN; (k) LPS-ChebyNet; and (l) ChebyNet.

the superiority and effectiveness of the proposed method. By adding the designed LPS, AUC values of all 6 GNNs are improved, further demonstrating the universality of the LPS.

4.3.2.3.3 Discussions on k Values and Low Labeled Rates   Next, we discuss the effects of k values used for the construction of spectrum samples based KNN graphs. Figure 4.45 shows diagnostic accuracies and label accuracies judged by LPS of the proposed method at different k values. As shown, different k values have certain influences on the diagnostic effects. For Case 1, $k = 5$ has the highest accuracy of 90.99%; for Case 2, $k = 3$ has the highest accuracy 93.96%. The label accuracies judged by LPS show a fluctuating downward trend with increasing k values, mainly because increasing k values will make the label number judged by LPS increase (the label number judged by LPS = k value ∗ the true label number). In both cases, when the k values are taken as 5 and 3, the label accuracies judged by LPS are 99.25% and 95%, respectively, with high confidence.

Finally, we explore the effects of low labeled rates to the proposed method. The labeled rates of Case 1 are set to 1%, 2%, and 4%, respectively, and the numbers of labeled samples for each health state are 2, 4, and 8, respectively. In Case 2, the labeled rates are 0.5%, 2%, and 4% respectively, and the numbers of labeled samples for each health state are 1, 4, and 8, respectively. The diagnostic accuracies of all eight experiments are shown in Figure 4.46. With increasing labeled rates, the diagnostic accuracies show an increasing trend. Case 1 contains gear vibration signals under speed fluctuation with the entire start–stop process. At labeled rates of 2% and 4%, the average accuracies are 96.08% and 98.39%, respectively.

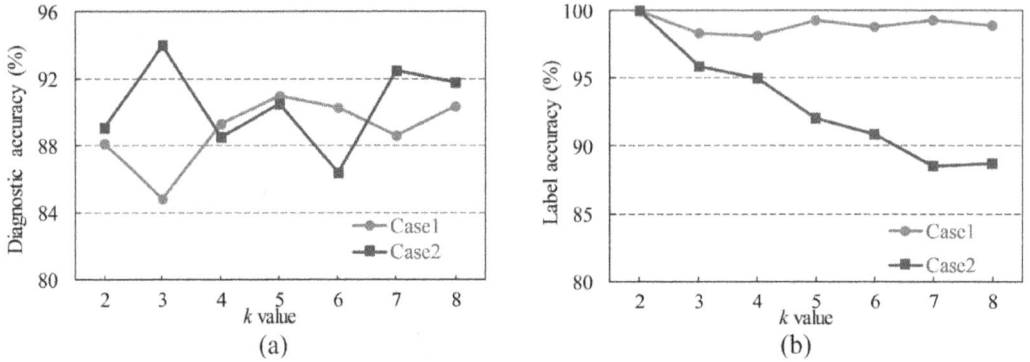

FIGURE 4.45 Accuracies of the proposed method at different k values: (a) diagnostic accuracies and (b) label accuracies judged by LPS.

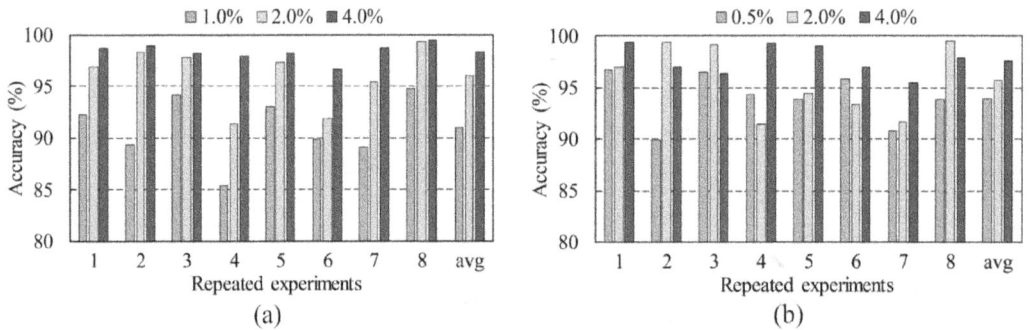

FIGURE 4.46 Accuracies of the proposed method with different low labeled rates: (a) Case 1 and (b) Case 2.

Case 2 contains bearing vibration signals under speed fluctuation with first deceleration and then acceleration.

With the labeled rate of only 0.5% and training samples of 10 for each health state, the proposed method has accuracies over 89.89% in all 8 experiments. It verifies that the proposed method can achieve fault diagnosis at lower label rates.

4.3.3 Unsupervised Fault Diagnosis Method with Deep Learning

Rotating machinery plays an important role in the field of manufacturing industries [37,62,63]. Therefore, real-time anomaly detection of rotating machinery is important for economical and reliable production. However, the excellent diagnostic performance of many deep learning models is inseparable from sufficient, labeled, and identically distributed samples, which limits their application in industrial scenarios.

Therefore, unsupervised cross-domain fault diagnosis provides more flexibility and extends the capability of the above-mentioned deep learning methods.

4.3.3.1 Unsupervised Fault Diagnosis Method Based on Novel Joint Transfer Network

Bearing is a common and key component of rotating machinery equipment, whose faults may cause economic loss and even human casualties. Therefore, bearing fault diagnostics is of great importance to the operational safety of mechanical equipment [7,15,64–66].

In practical engineering, it is difficult or even impossible to obtain many or even a few samples containing rich fault label information for reasons such as cost and safety. In addition, mechanical equipment often operates under variable conditions of load and rotational speed. As a result, the measured samples usually obey different distributions, so the deep learning models trained on samples under one distribution (source domain) are probably difficult to directly generalize to samples under another distribution (target domain) [60,67].

Unsupervised domain adaptation is an important transfer learning strategy that can be used to solve the above-mentioned problems, and it has received extensive attention from scholars worldwide [68–71]. These recent methods have shown good transfer learning ability in various unsupervised cross-domain fault diagnosis tasks of bearing. However, some challenging problems still need to be solved to further the diagnosis performance.

1. Most of the source-domain fault data of bearing in the current research come from the laboratory test rigs, which are more appropriate for investigating the general law of failure phenomena. However, the construction of laboratory test rigs with a certain accuracy requires a large amount of long-term resource investment, which makes it difficult to flexibly meet the requirements of fault data under different working conditions. Fortunately, with the help of numerical simulation technology, a large number of simulation data with rich fault label information can be easily obtained, which can also reduce the resource dependence on the laboratory test rigs [72,73].

2. The current studies usually only consider the marginal distribution alignment between the source and target domains while ignoring the conditional distribution alignment, which may cause fuzzy samples close to the class boundary or far from the class center in the target domain to be incorrectly classified [74,75]. Since there are no labeled samples in the target domain of unsupervised cross-domain fault diagnosis research, modeling its conditional distribution is extremely challenging. Therefore, it is urgent to design new loss functions that can achieve simultaneous alignments of conditional distribution and marginal distribution in unsupervised scenarios.

3. In the current research, each source-domain sample is usually assigned equal weight when performing domain adaptation. Even those source-domain samples that are quite different from the target domain are still given the same importance, which may cause negative transfer [76]. Consequently, it is necessary to develop a weight allocation mechanism for samples in the source domain, which is expected to adaptively assign different weights for each source-domain sample to suppress negative transfer and improve the diagnostic accuracy to a certain extent.

In this section, a new method called novel joint transfer network (NJTN) is proposed to diagnose various faults of bearing in unsupervised cross-domain scenarios automatically [77]. Compared with various methods in two cases from the simulation domain to the experimental domain, the results demonstrate that the proposed method can achieve the best diagnosis performance.

The innovations and contributions of the proposed method are summarized as follows:

4. Different from the existing transfer diagnosis research studies between experimental data, this chapter constructs the source domain by using simulation data with rich fault label information, reducing the dependence on the laboratory test rigs and exploring data–physical coupling driven fault diagnosis way.

5. An improved loss function embedded with joint MMD (JMMD) is designed to realize the simultaneous alignments of marginal distribution and conditional distribution in the unsupervised scenario.

6. A weight allocation mechanism is developed based on domain similarity, which assigns different weights to each source-domain sample to suppress negative transfer.

4.3.3.1.1 Main Algorithms

4.3.3.1.1.1 Unsupervised DANN DANN is mainly used to solve the problem of unsupervised cross-domain data distribution mismatch, and it reduces the distribution discrepancy between the source domain and the target domain by adapting their marginal distributions. The structure of DANN is shown in Figure 4.47, and it is mainly composed of feature extractor G_f, domain discriminator G_d, and classifier G_y in which G_f is used to learn domain-invariant and discriminative feature representations; G_d tries to discriminate whether the feature representations transformed by G_f are from the source domain or the target domain; and G_y is responsible for classification. In particular, G_f and G_d are connected by the gradient reversal layer.

Let $\mathcal{D}_s = \left\{ x_i^s, y_i^s \right\}_{i=1}^{n_s}$ be the source domain with n_s labeled samples, and $\mathcal{D}_t = \left\{ x_i^t \right\}_{i=1}^{n_t}$ be the target domain with n_t unlabeled samples. In this chapter, it is assumed that the source domain and the target domain share the same label space, which means that they have the same health state categories. The common loss function of DANN consists of categorical

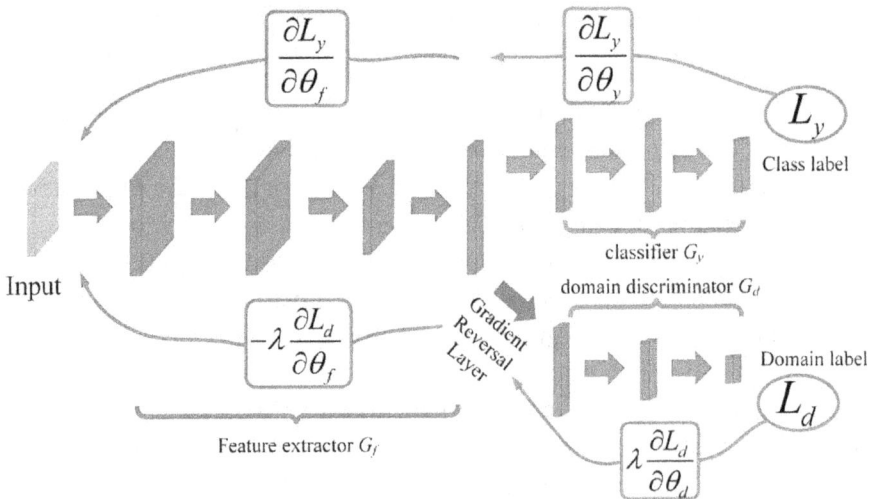

FIGURE 4.47   Classic DANN model diagram.

cross-entropy loss L_y and domain discriminative cross-entropy loss L_d, and it can be defined as follows:

$$L\left(\theta_f, \theta_y, \theta_d\right) = L_y\left(\theta_f, \theta_y\right) - \lambda L_d\left(\theta_f, \theta_d\right) \tag{4.38}$$

$$L_y\left(\theta_f, \theta_y\right) = -\frac{1}{n_s}\sum_{i=1}^{n_s}\sum_{k=0}^{K_s-1}I\left[y_i^s = k\right]\log\left(G_y^k\left(G_f\left(x_i^s\right)\right)\right) \tag{4.39}$$

$$L_d\left(\theta_f, \theta_d\right) = -\frac{1}{n_s}\sum_{i=1}^{n_s}\log\left(G_d\left(G_f\left(x_i^s\right)\right)\right)$$

$$\tag{4.40}$$

$$-\frac{1}{n_t}\sum_{j=1}^{n_t}\log\left(1 - G_d\left(G_f\left(x_j^t\right)\right)\right)$$

where θ_f, θ_d, and θ_y are the parameters of G_f, G_d, and G_y, respectively; λ is a regularization coefficient; $G_y^k(\cdot)$ represents the kth element of the output; $I[\cdot]$ is an indicator function; and K_s is the number of health states. When the parameters are updated, for $L_y : G_f$ and G_y expect it to be minimized to improve the classification ability of the network; for $L_d : G_d$ expects it to be minimized to enhance the domain discrimination ability and G_f expects it to be maximized to reduce the distribution discrepancy between domains. Therefore, the parameter updates are as follows:

$$\left(\hat{\theta}_f, \hat{\theta}_y\right) = \arg\min_{\theta_f, \theta_y} L\left(\theta_f, \theta_y, \hat{\theta}_d\right) \tag{4.41}$$

$$\left(\hat{\theta}_d\right) = \arg\max_{\theta_d} L\left(\hat{\theta}_f, \hat{\theta}_y, \theta_d\right) \tag{4.42}$$

where $\hat{\theta}_f$, $\hat{\theta}_y$, and $\hat{\theta}_d$ are the optimal parameters. The above-mentioned updates can be completed by using the stochastic gradient descent (SGD) optimization.

4.3.3.1.2 Overall Framework of the Proposed Method   This section is divided into four parts to introduce our proposed NJTN: construction of the source domain based on simulation data, design of an improved loss function embedded with JMMD, development of a weight assignment mechanism for source-domain samples, and model architecture and optimization algorithm.

4.3.3.1.2.1 Construction of Source Domain Based on Simulation Data   In this chapter, in order to obtain a large number of fault samples with rich label fault information and reduce the dependence on laboratory test rigs, a rotor-bearing system simulation model is built using the lumped-mass method to generate vibration responses of bearings under different states for constructing the source-domain dataset. The dynamic simulation model is shown in Figure 4.48. It is mainly composed of a disc, a shaft, a bearing, a base, and other

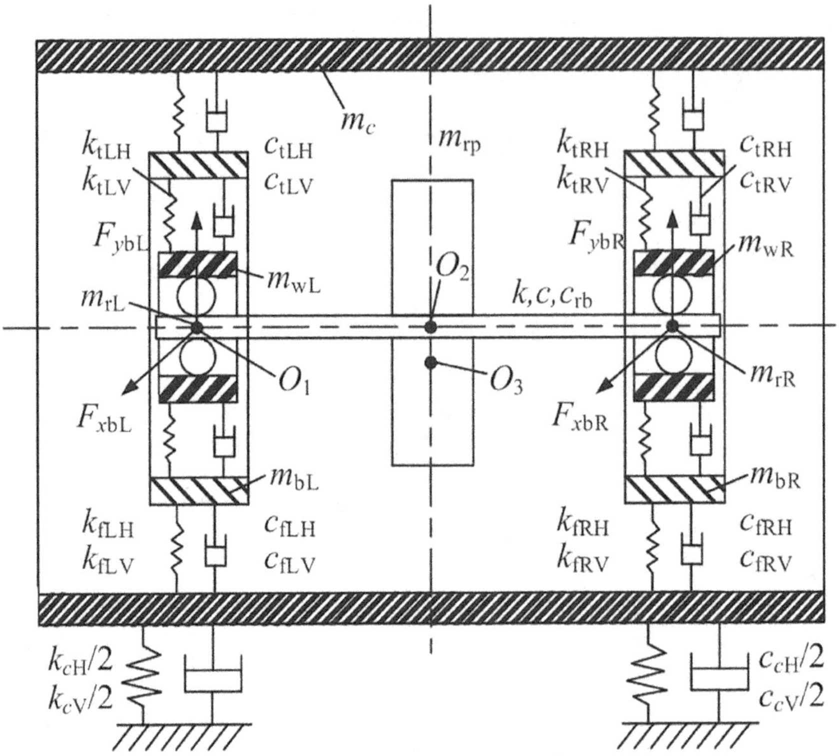

FIGURE 4.48 Schematic diagram of the simulated rotor-bearing system.

components. The outer ring of the bearing is fixed on the bearing seat, while the inner ring rotates with the shaft.

Assuming that the stiffness and damping of the main components are fixed values and the local defects occur at the left support bearing, six differential equations of motion related to the left support bearing can be listed:

$$m_{bL}\ddot{x}_{bL}+k_{fLH}\left(x_{bL}-x_c\right)+c_{fLH}\left(\dot{x}_{bL}-\dot{x}_c\right)$$
$$+k_{tLH}\left(x_{bL}-x_{wL}\right)+c_{tLH}\left(\dot{x}_{bL}-\dot{x}_{wL}\right)=0 \tag{4.43}$$

$$m_{bL}\ddot{y}_{bL}+k_{fLV}\left(y_{bL}-y_c\right)+c_{fLV}\left(\dot{y}_{bL}-\dot{y}_c\right)$$
$$+k_{tLV}\left(y_{bL}-y_{wL}\right)+c_{tLV}\left(\dot{y}_{bL}-\dot{y}_{wL}\right)=-m_{bL}g \tag{4.44}$$

$$m_{rL}\ddot{x}_{rL}+k\left(x_{rL}-x_{rp}\right)+c_{rb}\dot{x}_{rL}-F_{xbL}=0 \tag{4.45}$$

$$m_{rL}\ddot{y}_{rL}+k\left(y_{rL}-y_{rp}\right)+c_{rb}\dot{y}_{rL}-F_{ybL}=-m_{rL}g \tag{4.46}$$

$$m_{wL}\ddot{x}_{wL}+k_{tLH}\left(x_{wL}-x_{bL}\right)+c_{tLH}\left(\dot{x}_{wL}-\dot{x}_{bL}\right)+F_{xbL}=0 \tag{4.47}$$

$$m_{wL}\ddot{y}_{wL}+k_{tLH}\left(y_{wL}-y_{bL}\right)+c_{tLH}\left(\dot{y}_{wL}-\dot{y}_{bL}\right)+F_{ybL}=-m_{wL}g \tag{4.48}$$

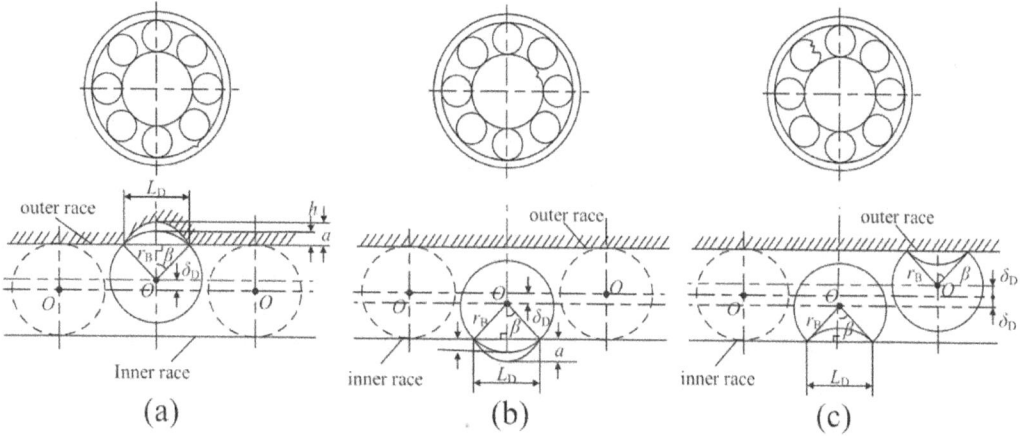

FIGURE 4.49 Schematic of the local defect area: (a) outer race fault, (b) inner race fault, and (c) Ball fault.

where m_{bL} is the mass of bearing support; k_{fLH} and K_{fLV} are the lateral and longitudinal support stiffnesses between casing and bearing support, respectively; c_{fLH} and c_{fLV} are the lateral and longitudinal support dampings between casing and bearing support, respectively; k_{tLH} and k_{tLV} are the lateral and longitudinal support stiffnesses between the bearing outer ring and the bearing support, respectively; c_{tLH} and c_{tLV} are the lateral and longitudinal squeeze film dampings between the bearing outer ring and the bearing support; m_{rL} is the equivalent mass of the rotor; k and c_{rb} are the shaft stiffness and the damping of the rotor at bearing, respectively; F_{xbL} and F_{ybL} are the support reaction forces of bearing; and m_{wL} is the mass of bearing outer ring. Specifically, the initial parameters of the rotor-bearing system are consistent with those in the literature.

Figure 4.49 shows a schematic diagram of bearings with local defect area on the outer ring, inner ring, and ball components. The cyclically varying contact forces F_{xbL} and F_{ybL} are the sum of the contact forces of the balls at different angles:

$$F_{xbL} = \sum_{j=1}^{Z} f_j \cos\theta_j, \ F_{ybL} = \sum_{j=1}^{Z} f_j \sin\theta_j \qquad (4.49)$$

where θ_j is the angular position of the j th ball, and f_j is the contact force between the j th ball and the raceway calculated according to the following formula:

$$f_j = C_b \left[\delta_j \right]^n = C_b \left(x\cos\theta_j + y\sin\theta_j - \delta_0 - \delta_D \right)^n$$
$$\times H \left(x\cos\theta_j + y\sin\theta_j - \delta_0 - \delta_D \right) \qquad (4.50)$$

where C_b is the Hertzian contact stiffness, $H(\cdot)$ is the Heaviside function, δ_j is the clearance between the ball and the raceway, δ_0 is the initial clearance, δ_D is the clearance caused by the local defect area, and n is the load-deformation coefficient, which is 2/3 for the ball bearing.

TABLE 4.17   Main Parameters of Bearing JIS6306

Descriptions of Parameters	Values of Parameters
Radius of outer race/mm	63.9
Radius of inner race/mm	40.1
Pitch diameter/mm	52
Diameter of rolling element/mm	11.9
Number of balls	8
Contact angle/°	0
Hertzian contact stiffness/N·m$^{-3/2}$	13.34×10^9

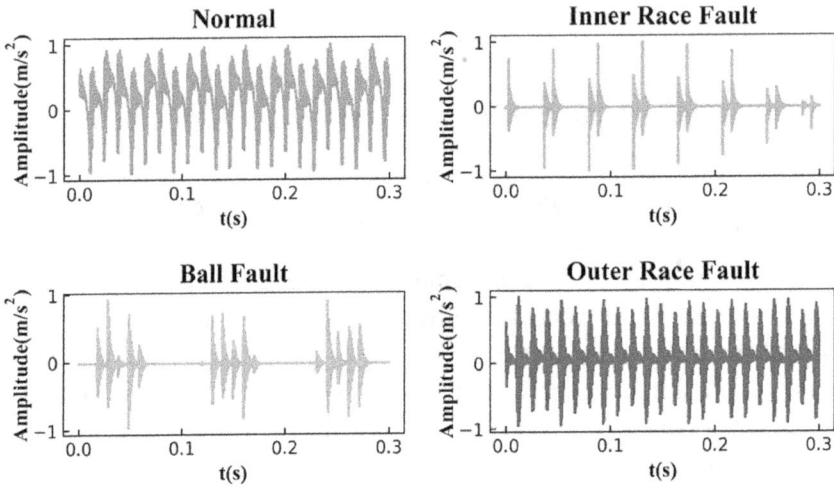

FIGURE 4.50   Time-domain waveforms of simulation data for four states of rolling bearing.

The simulation model adopts a JIS 6306 bearing, and its main parameters are listed in Table 4.17. The damage diameter L_D is set to 0.5334 mm, the damage depth a is set to 2.794 mm, and the sampling frequency is set to 10 kHz. Figure 4.50 shows a time-domain waveform diagram of the vibration acceleration signal collected by the left support bearing under four different health states when the angular velocity of the shaft is 150 rad/s. The angular velocity of the rotating shaft changes from 150 to 270 rad/s at 5 rad/s intervals, and the horizontal and vertical acceleration signals of the left support bearing are collected to construct the source-domain dataset.

4.3.3.1.2.2 Design of Improved Loss Function Embedded with JMMD   Most existing unsupervised domain adaptation methods may misclassify fuzzy samples near the class boundary of the target domain. They ignored the conditional distribution alignment; however, conditional distribution is extremely difficult to characterize as the target domain does not have any labels. Thus, the traditional methods are limited to the marginal distribution alignment of the source and target domains. To address this problem, this chapter adopts the JMMD approach to improve the common loss function of the DANN.

The design process of the improved loss function embedded with JMMD is mainly as follows. First, use the output of G_f and G_y in DANN to form the joint distribution of source domain $P_s\big(G_f(X^s),G_y(G_f(X^s))\big)$ (denoted as P_s) and the joint distribution of target domain $P_t\big(G_f(X^t),G_y(G_f(X^t))\big)$ (denoted as P_t), where $X^s=\{x_i^s\}_{i=1}^{n_s}$ and $X^t=\{x_i^t\}_{i=1}^{n_t}$. Second, calculate the discrepancy between P_s and P_t by reproducing kernel Hilbert space embeddings, denoted as $D(P_s,P_t)$. Finally, use the squared distance between kernel mean embeddings to compute the empirical estimate of $D(P_s,P_t)$, denoted as $\hat{D}(\theta_f,\theta_y)$. Let $\hat{f}_i^s=G_f(x_i^s),\hat{f}_i^t=G_f(x_i^t),\hat{y}_i^s=G_y(G_f(x_i^s))$, and $\hat{y}_i^t=G_y(G_f(x_i^t))$, then the following JMMD formula can be obtained:

$$\hat{D}(\theta_f,\theta_y)=\frac{1}{n_s^2}\sum_{i=1}^{n_s}\sum_{j=1}^{n_s}k\left(\hat{f}_i^s,\hat{f}_j^s\right)\cdot k\left(\hat{y}_i^s,\hat{y}_j^s\right)$$

$$+\frac{1}{n_t^2}\sum_{i=1}^{n_t}\sum_{j=1}^{n_t}k\left(\hat{f}_i^t,\hat{f}_j^t\right)\cdot k\left(\hat{y}_i^t,\hat{y}_j^t\right) \tag{4.51}$$

$$-\frac{1}{n_s n_t}\sum_{i=1}^{n_s}\sum_{j=1}^{n_t}k\left(\hat{f}_i^s,\hat{f}_j^t\right)\cdot k\left(\hat{y}_i^s,\hat{y}_j^t\right)$$

where $k(\cdot)$ is the kernel function, and considering the generality of Gaussian kernel, we choose Gaussian kernel function with bandwidth set to median pairwise squared distances on the training data.

Despite the fact that (4.51) captures the interaction between different variables in joint distributions P_s and P_t and explicitly calculates the joint distribution discrepancy between the source domain and the target domains, it is usually still not enough to accurately characterize the dynamically changing data distribution. In contrast, the discrepancy learned by DANN, which is an implicit distance, can be learned dynamically in the data. Therefore, by embedding (4.51) into (4.38) (i.e., the common loss function of the DANN) for improvement, on one hand, the above-mentioned problem can be alleviated to a certain extent; on the other hand, using JMMD to align the joint distributions of source and target domains and reducing their marginal distribution discrepancy through DANN can simultaneously help adapt their conditional distributions in the unsupervised scenario, which can overcome the limitations on separate marginal distribution alignment.

After that, the proposed improved loss function embedded with JMMD is

$$L(\theta_f,\theta_y,\theta_d)=L_y(\theta_f,\theta_y)-\lambda L_d(\theta_f,\theta_d)+\mu\hat{D}(\theta_f,\theta_y) \tag{4.52}$$

with

$$\lambda,\mu\sim\frac{2m}{\left(1+e^{-Kt}\right)}-m \tag{4.53}$$

where μ is also a regularization coefficient, and the growth trajectory of regularization coefficients λ and μ is given in equation (4.53) in which t is the current iterative training time, m is the upper limit of regularization coefficient, and K controls the rate of change.

4.3.3.1.2.3 Development of Weight Assignment Mechanism for Source-Domain Samples The classic unsupervised domain adversarial adaptation methods usually assign equal weight to each source-domain sample to adapt to the target domain. However, there may be samples in the source domain that are excessively different from the target domain. At this time, if the equal weight allocation strategy is still used, it is likely to cause a negative transfer of the diagnostic model.

In this chapter, in order to characterize the similarity of samples in different domains to suppress negative transfer, a weight assignment mechanism for source-domain samples is developed. Considering that the samples with higher similarity in different domains are more difficult to separate whereas samples with larger discrepancies are easier to distinguish for the domain discriminator, the specific weights assigned to each source-domain sample can be calculated according to the domain prediction error of the domain discriminator, which is defined as follows:

$$\omega_i^s = -\log\left(G_d\left(G_f\left(x_i^s\right)\right)\right) \tag{4.54}$$

where ω_i^s denotes the specific weight of the ith sample in the source domain, and after using min-max normalization, the normalized weight of ω_i^s is given as follows:

$$\tilde{\omega}_i^s = \frac{\omega_i^s - \omega_{i,m}^s}{\omega_{i,m}^s - \omega_{i,m}^s} \tag{4.55}$$

where $\omega_{i,\max}^s = \max\left(\omega_i^s\right)$ and $\omega_{i,\min}^s = \min\left(\omega_i^s\right)$. The normalized weights are introduced into (4.40), and then the new domain discriminative cross-entropy loss can be expressed by

$$\tilde{L}_d\left(\theta_f,\theta_d\right) = -\frac{1}{n_s}\sum_i^{n_s}\tilde{\omega}_i^s\log\left(G_d\left(G_f\left(x_i^s\right)\right)\right)$$

$$-\frac{1}{n_t}\sum_j^{n_t}\log\left(1-G_d\left(G_f\left(x_j^t\right)\right)\right). \tag{4.56}$$

The final improved loss function can be obtained by combining (4.56) and (4.52), which is defined as follows:

$$L\left(\theta_f,\theta_y,\theta_d\right) = L_y\left(\theta_f,\theta_y\right) - \lambda\tilde{L}_d\left(\theta_f,\theta_d\right) + \mu\hat{D}\left(\theta_f,\theta_y\right) \tag{4.57}$$

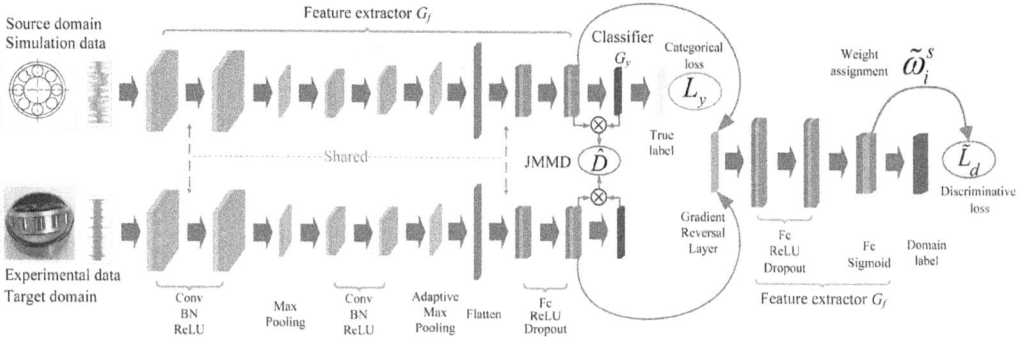

FIGURE 4.51 Model architecture of the proposed method.

4.3.3.1.2.4 Model Architecture and Optimization Algorithm Figure 4.51 shows the model architecture of the proposed method, which is mainly composed of three modules: feature extractor G_f, classifier G_y, and domain discriminator G_d. Specifically, G_f consists of four one-dimensional (1D) convolution (Conv) layers, two Max pooling layers, and two fully connected (Fc) layers. Each Conv layer comes with a 1D batch normalization layer and a ReLU activation function, and each Fc layer comes with a ReLU activation function and a dropout layer; G_y consists of an Fc layer; G_d consists of three Fc layers, two of which come with ReLU activation functions and dropout layers, and the remaining one comes with a Sigmoid activation function.

The proposed model is trained using the SGD algorithm, and the parameters θ_f, θ_d, and θ_y will be updated as follows:

$$\theta_f \leftarrow \theta_f - \varepsilon \left(\frac{\partial L_y}{\partial \theta_f} - \frac{\lambda \partial \tilde{L}_d}{\partial \theta_f} + \frac{\mu \partial \hat{D}}{\partial \theta_f} \right)$$

$$\theta_y \leftarrow \theta_y - \varepsilon \left(\frac{\partial L_y}{\partial \theta_y} + \frac{\mu \partial \hat{D}}{\partial \theta_y} \right) \tag{4.58}$$

$$\theta_d \leftarrow \theta_d - \frac{\varepsilon \lambda \partial \tilde{L}_d}{\partial \theta_d}$$

where ε is the learning rate.

In each epoch, after the training phase of the model, the target-domain test set will enter the model and they will only go through G_f and G_y; then G_y will give the classification results. The specific training and test procedures of the proposed model are shown in Algorithm 4.1 (Figure 4.52).

4.3.3.1.3 Experimental Validation

4.3.3.1.3.1 Introduction of Target-Domain Datasets and Comparison Methods The effectiveness of the proposed method is verified in two cases of unsupervised cross-domain bearing fault diagnosis in which Case 1 uses dataset from the QPZZ-II bearing fault test rig to construct Target domain 1 and Case 2 uses dataset from the electric locomotive bearing fault test rig to construct Target domain 2. Figures 4.53 and 4.54, respectively, show the experimental

---

Algorithm 1: Training and Test Procedures of the Proposed Model.

Input: The preprocessed source-domain labeled dataset $\mathcal{D}_s = \{x_i^s, y_i^s\}_{i=1}^{n_s}$ and the preprocessed target-domain unlabeled dataset $\mathcal{D}_t = \{x_i^t\}_{i=1}^{n_t}$

1:   Set model hyperparameters such as convolutional layer, pooling layer, activation function, learning rate, iteration number, and batch size

2:   Randomly initialize the weights and biases of the model

3:   Divide the training and test sets for the source and target domains respectively

4:   **For** each training epoch **do**

5:     **If** phase = train **then**

4:       **For** each data batch **do**

5:         Calculate the output of G_f and G_y

6:          Solve $\hat{D}(\theta_f, \theta_y)$ based on Eq. (14)

7:          Solve $L_y(\theta_f, \theta_y)$ based on Eq. (1)

8:         Calculate the output of G_f and G_d

9:          Solve $\tilde{\omega}_i^s$ based on Eq. (17) and Eq. (18)

10:         Solve $\tilde{L}_d(\theta_f, \theta_d)$ based on Eq. (19)

11:         Update model parameters $\theta_f, \theta_d, \theta_y$ based on Eq. (21)

12:      **End For**

14:     **If** phase = test **then**

16:       Calculate the output of G_y

13:   **End For**

Output: Optimized model parameters $\theta_f, \theta_d, \theta_y$ and classification results on the test set

---

FIGURE 4.52   Algorithm 4.1: Training and test procedures of the proposed model.

FIGURE 4.53   Bearings installed in different experimental devices: (a) QPZZ-II bearing fault test rig and (b) electric locomotive bearing fault test rig.

devices and three kinds of fault bearings in two cases. The sampling frequencies of the two cases are 25.6 and 12.8 kHz, respectively. In the two cases, the sample dimension is 1,000, and the sample setting of the source domain is consistent with the selected target domain. The time-domain waveforms of target-domain bearing under four health states in two cases are shown in Figures 4.55 and 4.56, respectively. The details of the two target-domain

FIGURE 4.54 Three kinds of fault bearings in different cases: (a) Case 1 and (b) Case 2.

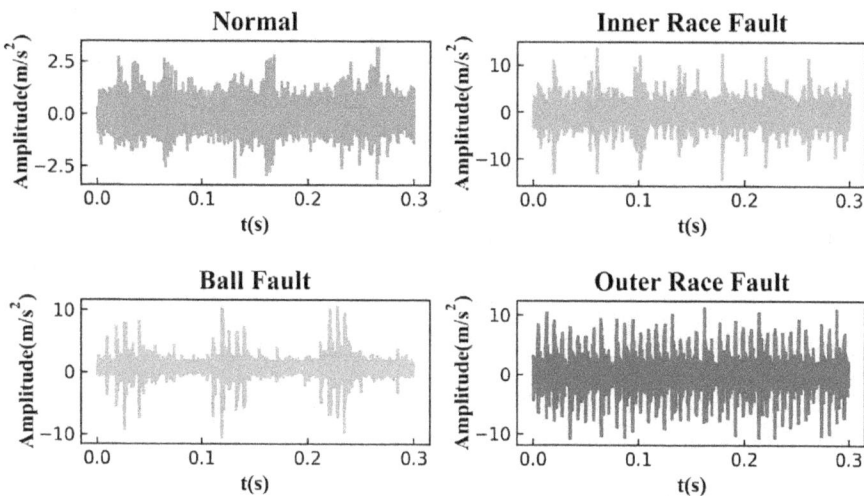

FIGURE 4.55 Time-domain waveforms of rolling bearing under four states in Case 1.

datasets are listed in Table 4.18. We perform de-mean, "0–1" normalized data process-ing for all data points under each fault class in the source and target domains before the samples are fed into the model, respectively.

In order to verify the superiority of the proposed method, it is compared with the other seven methods:

1. First, the proposed method (denoted as Method 1) is compared with the standard convolutional neural network (denoted as Method 2) to verify the necessity of a transfer diagnosis strategy.

2. Second, for reflecting the advantages of the proposed method, it is compared with some popular unsupervised cross-domain fault diagnosis methods, including

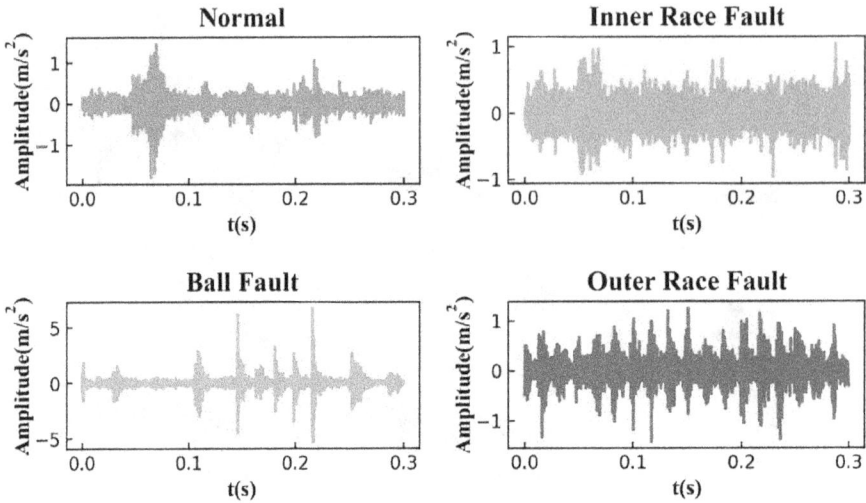

FIGURE 4.56   Time-domain waveforms of rolling bearing under four states in Case 2.

TABLE 4.18   Details of Target-Domain Datasets

Datasets	Bearing Types	Health States	Number of Training/Test Samples	Motor Speeds
Target domain 1 (Case 1)	NU205EM	Normal	320/80	1,500 rpm
		Inner race fault		
		Ball fault		
		Outer race fault		
Target domain 2 (Case 2)	552732	Normal	480/120	490 rpm
		Inner race fault		498 rpm
		Ball fault		531 rpm
		Outer race fault		490 rpm

DANN, deep adaptation network, joint adaptation network, and CORAL, which are successively denoted as Method 3 to Method 6.

3. Finally, to respectively illustrate the effectiveness of the designed improved loss function embedded with JMMD and the developed weight allocation mechanism for source-domain samples, Domain Adversarial Neural Network (DANN) combined with only the improved loss function and only the weight allocation mechanism are used for comparisons, which are denoted as Method 7 and Method 8.

Ten trials for each method are run in each case to reduce the random influence. In addition, based on the idea of a greedy algorithm, we experimentally choose the value of a variable under the premise that other hyperparameter variables are determined in order to find a relatively better combination of hyperparameters. In the two cases, the main parameters of the proposed method are set as follows: The learning rate is 0.0001, the batch size is 32, the iteration number N is 300, the upper limit of regularization coefficient m is 1, the rate of change K is 1/30, the value of dropout probability is 0.5 (default value), the optimizer is Adam (the values of first-order moment decay coefficient and second-order moment decay coefficient are 0.9 and 0.999, respectively), and the weight decay is 1×10^{-5}.

4.3.3.1.3.2 Analysis of the Comparison Result   In this section, the average accuracy, average F1-score, and confusion matrix are used to fully compare the diagnostic performance of each method on the test set of the target domain. The results are shown in Table 4.19 and Figures 4.57–4.61, respectively. In the confusion matrix, the labels of normal, inner race fault, ball fault, and outer race fault states are set as 0, 1, 2, and 3, respectively; the abscissa is the predicted label, and the ordinate is the real label.

TABLE 4.19   Comparisons of Unsupervised Cross-Domain Diagnosis Results of the Eight Methods in Two Cases

Methods	Case 1(%)	Case 2 (%)
Method 1	**85.532 ± 1.359**	**81.355 ± 1.880**
Method 2	25.812±0.648	24.250±2.284
Method 3	79.312±3.103	61.189±6.846
Method 4	81.281±2.688	73.334+14.556
Method 5	79.937±4.219	67.771±12.536
Method 6	47.094±10.120	39.270±11.399
Method 7	82.688±2.283	77.145±2.641
Method 8	82.156±2.138	75.645±2.092

"Bold" indicates that this method has the highest Average Testing Accuracies in Two Cases.

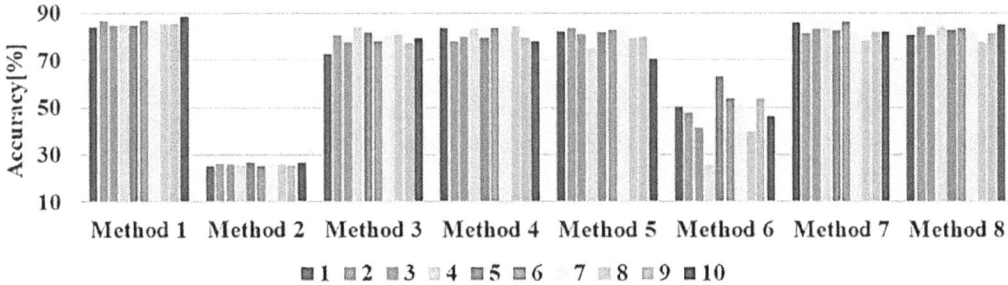

FIGURE 4.57   Unsupervised cross-domain diagnosis accuracies of the eight methods during the ten repetitions in Case 1.

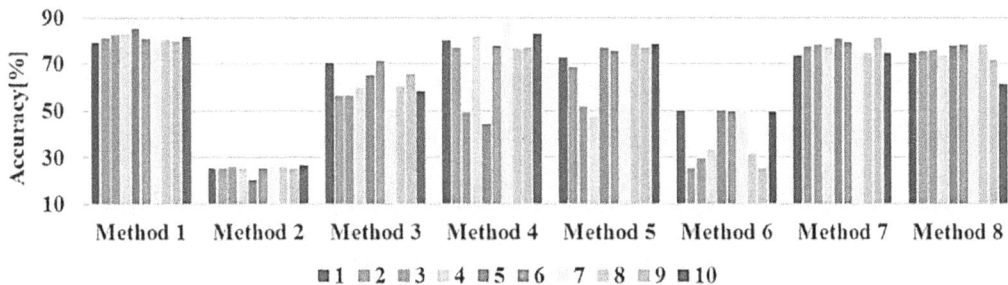

FIGURE 4.58   Unsupervised cross-domain diagnosis accuracies of the eight methods during the ten repetitions in Case 2.

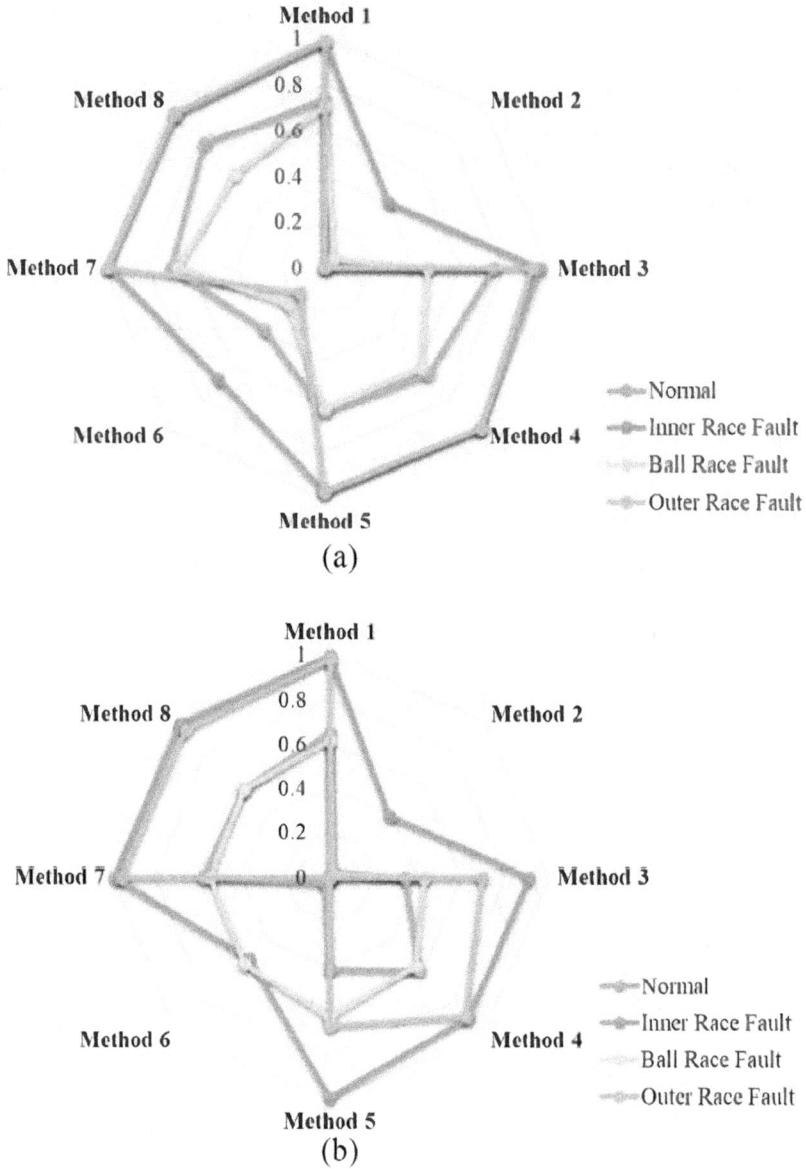

FIGURE 4.59   Average F1-scores of four states of each method in different cases: (a) Case 1 and (b) Case 2.

The following main conclusions can be drawn:

1. Through the comparisons between Methods 1 and 2–6, it can be seen that Method 1 has achieved the highest fault diagnosis accuracies. When faced with the unsupervised scenario, the features learned by the proposed method are more representative and easier to distinguish. Specifically, in the two cases, the average diagnostic accuracies of Method 1 are 85.532% and 81.355%, respectively; however, the highest accuracies of other methods are only 81.281% and 73.334%, respectively.

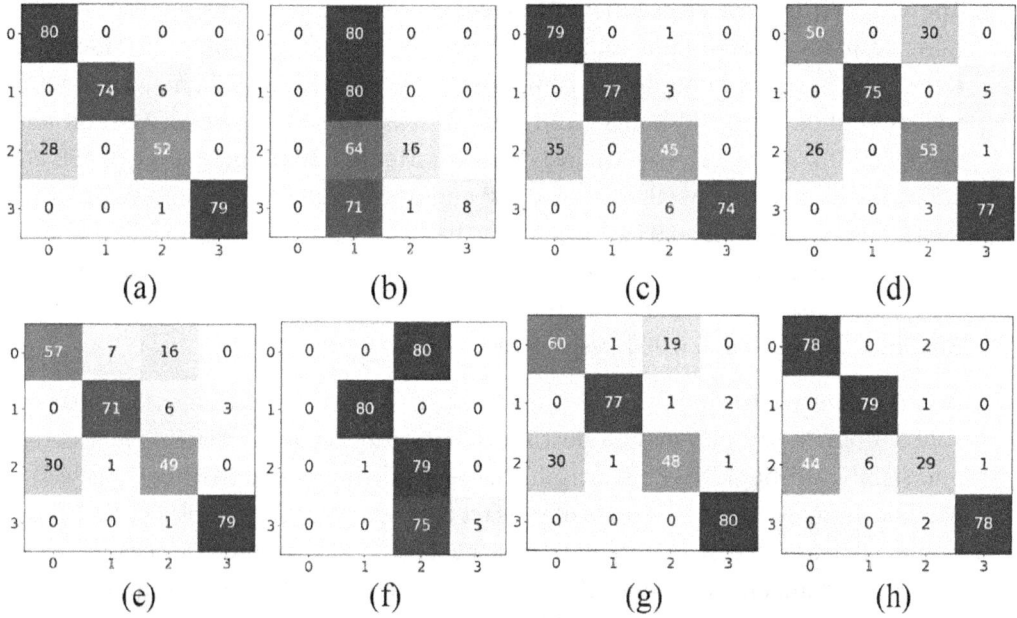

FIGURE 4.60 Confusion matrixes for the third trial of each method in Case 1: (a) Method 1, (b) Method 2, (c) Method 3, (d) Method 4, (e) Method 5, (f) Method 6, (g) Method 7, and (h) Method 8.

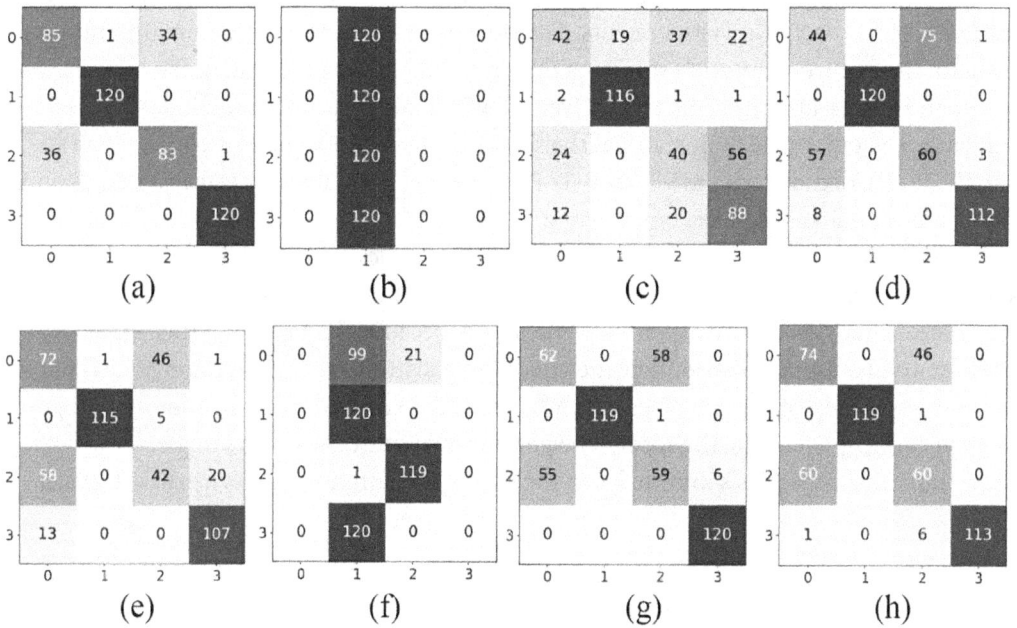

FIGURE 4.61 Confusion matrixes for the third trial of each method in Case 2: (a) Method 1, (b) Method 2, (c) Method 3, (d) Method 4, (e) Method 5, (f) Method 6, (g) Method 7, and (h) Method 8.

2. The designed improved loss function embedded with JMMD helps enhance the diagnosis accuracy of each fault category, especially the ball fault and outer race fault states, which can be found by the comparisons between Method 7 and Method 3, as well as Method 1 and Method 8. Concretely, in Case 1, the F1-scores of Method 7 are 0.664 and 0.987, respectively, whereas those of Method 3 are successively 0.458 and 0.928; the F1-scores of Method 1 are 0.708 and 0.989, respectively; however, Method 8 has those of 0.574 and 0.953, accordingly. In Case 2, the F1-scores of Method 7 are 0.545 and 0.962, respectively, whereas those of Method 3 are 0.434 and 0.697, respectively; the F1-scores of Method 1 are 0.616 and 0.980, respectively, whereas those of Method 8 are 0.530 and 0.904, respectively.

3. It is not difficult to find out that the developed weight allocation mechanism for source-domain samples plays a positive role in improving the diagnostic accuracy and stability through the comparisons between Method 8 and Method 3, as well as Method 1 and Method 7. In spite of the fact that adapting the samples in the source domain that are excessively different from the target domain can cause negative transfer, the mechanism can adaptively assign different weights for each source-domain sample by measuring their similarities with the target-domain samples to suppress negative transfer to a great degree.

In order to show the performance of feature adaptation of the proposed method, t-distributed stochastic neighbor embedding (t-SNE) is used to reduce the dimension of the output feature vector to achieve visual results. For the test sets of the source domain and target domain, the feature adaptation results of the third trial of each method in the two cases are shown in Figures 4.62 and 4.63, respectively, in which the labels of normal, inner race fault, ball fault, and outer race fault states in the source domain are set as S_0, S_1, S_2, and S_3, and the target-domain ones are T_0, T_1, T_2, and T_3. The abscissa is feature component 1, and the ordinate is feature component 2.

According to the comparison results of feature adaptation, the deep features of the source and target domains extracted by the proposed method are more similar than others, i.e., they have better domain-invariant properties. It is because the proposed method takes the adaptations of marginal distribution and conditional distribution into account, which significantly reduces the distribution discrepancy between domains. At the same time, the proposed method can adaptively give the specific weights for each source-domain sample, so as to selectively transfer the diagnosis knowledge in the source domain to suppress negative transfer.

4.3.3.2 Unsupervised Fault Diagnosis Method Based on Hybrid Robust Convolutional Autoencoder

Machine tools are important and fundamental constitutes for manufacturing industry, which are widely used in various processes such as turning, milling, and drilling [78,79]. In the era of Industry 4.0, machine tools are developing toward intelligence, automation, high precision, multi-function and high reliability, etc. [80–82]. However, machine tools may exhibit unpredictable anomalies which can cause production downtime, serious losses, or

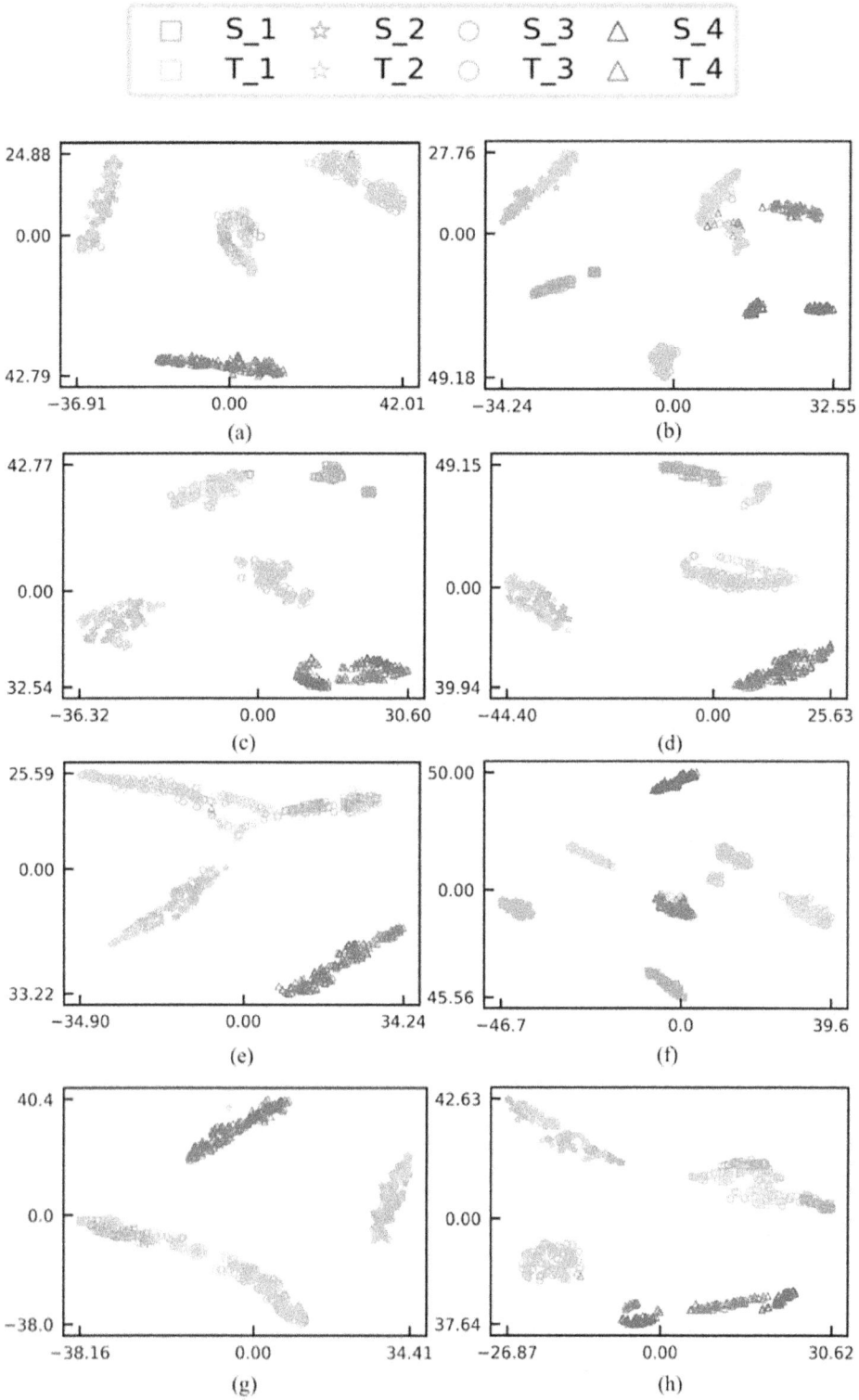

FIGURE 4.62   Visual feature adaptation results for each method using t-SNE in Case 1: (a) Method 1, (b) Method 2, (c) Method 3, (d) Method 4, (e) Method 5, (f) Method 6, (g) Method 7, and (h) Method 8.

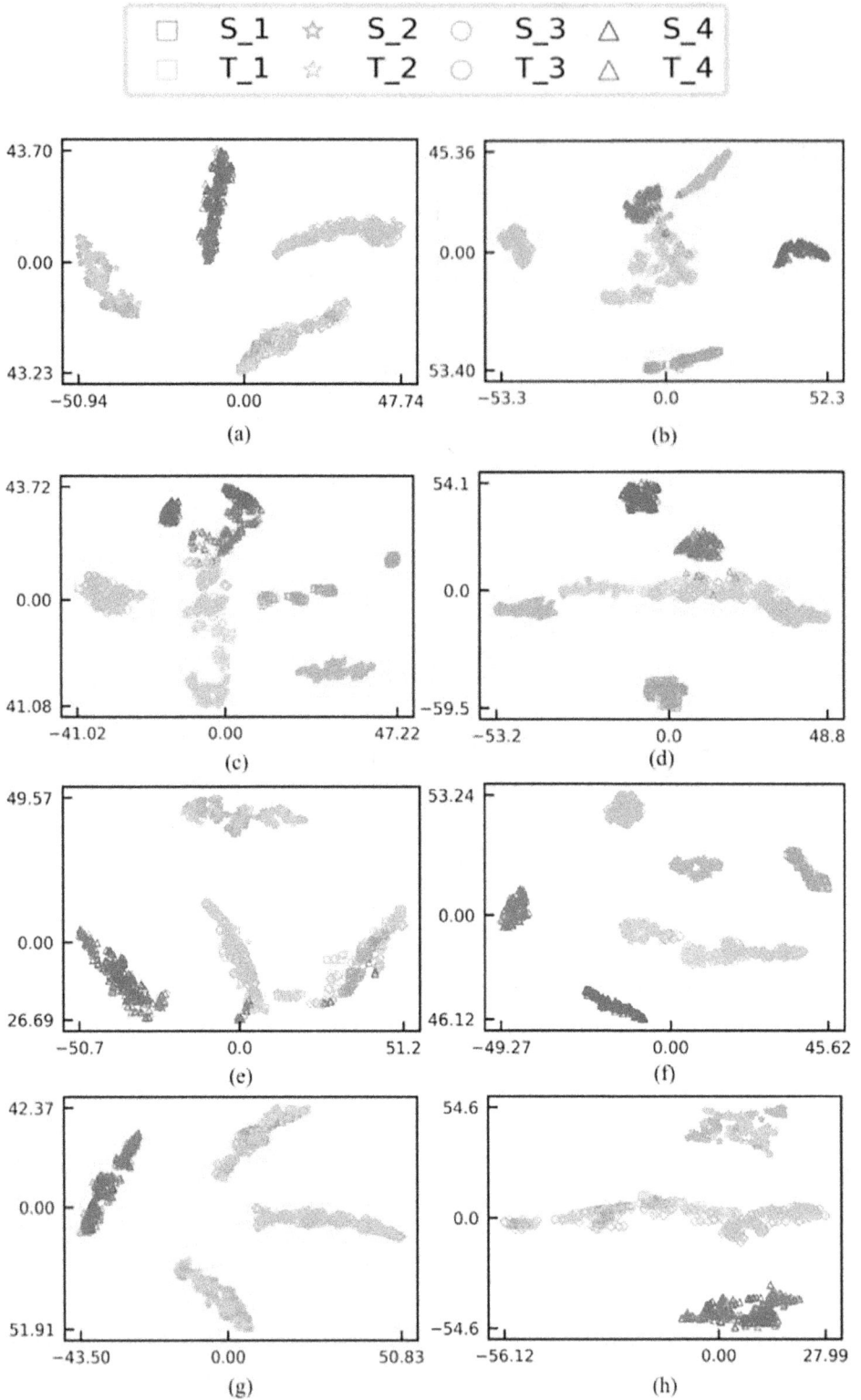

FIGURE 4.63 Visual feature adaptation results for each method using t-SNE in Case 2: (a) Method 1, (b) Method 2, (c) Method 3, (d) Method 4, (e) Method 5, (f) Method 6, (g) Method 7, and (h) Method 8.

even casualties if not detected in time. Therefore, real-time anomaly detection of machine tools is important for economical and reliable production.

Data-driven methods have been proven to be effective for anomaly detection of machine tools, which can be divided into three procedures: signal collection, feature extraction, and state detection. The collected signals can be vibration signals [83], current signals [84], acoustic signals [85], and temperature signals [86]. Deep learning-based intelligent methods have been widely studied in recent years for anomaly detection and condition monitoring of machine tools due to their adaptive feature mining capability and end-to-end convenience.

The recent deep learning models have shown decent results for dealing with different challenges in anomaly detection and condition monitoring of machine tools [87–90]. However, they all rely on sufficient labeled data, which limits their application in industrial scenarios. In the laboratory, it is easier to obtain data with accurate labels, but this is often difficult in the engineering practice [91]. Most of the machine tools are in normal status during actual operation, and labeled data is often very scarce or even unavailable. Therefore, unsupervised anomaly detection solely based on normal operating data provides more flexibility and extends the capability of the above-mentioned deep learning methods.

Autoencoder is a classical data reduction and reconstruction method, which has been explored and improved by many scholars in recent years to excel its performance in unsupervised anomaly detection of different fields [92–94]. In this section, a new improved autoencoder method called HRCAE is developed for unsupervised anomaly detection of machine tools under noises.

4.3.3.2.1 Basic CAE   CAE, as a classical unsupervised learning method, has been widely used in image denoising, data reconstruction, and other relevant fields. It can combine the feature extraction function of CNN with the unsupervised feature reconstruction function of autoencoder. In general, the CAE consists of multiple convolution encoders and deconvolution decoders, which can be sketched in Figure 4.64.

A convolution encoder usually consists of a convolution layer and a pooling layer to extract data feature, which can be expressed as follows:

$$H = \text{pool}\left(\sigma\left(\sum \left(X \odot w^i + b^i\right)\right)\right) \tag{4.59}$$

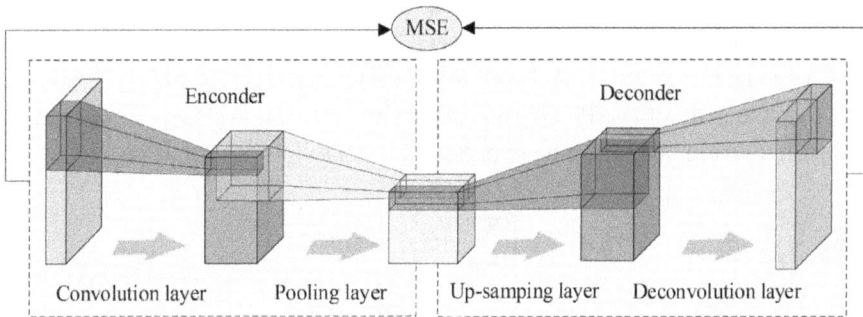

FIGURE 4.64   The structure of CAE.

where X is the input data; H is the feature obtained after encoding; w^i is the i th convolution kernel in the convolution layer, and b^i is the ith bias; $\odot$ indicates the convolution operation; σ is the activation function; and pool ($\cdot$) indicates the pooling operation.

A convolution decoder generally includes a deconvolution layer and an up-sampling layer, which can be expressed as follows:

$$\hat{X} = \text{ups}\left(\sigma\left(\sum \left(H \otimes \hat{w}^i + \hat{b}^i \right) \right) \right) \tag{4.60}$$

where $\hat{X}$ is the data returned after decoding; $\hat{w}^i$ is the ith deconvolution kernel in the deconvolution layer, and $\hat{b}^i$ is the ith bias; $\otimes$ indicates the deconvolution operation; and ups ($\cdot$) indicates the up-sampling operation.

The reconstruction error, which is used to optimize the network's parameters, is denoted as the Euclidean distance between X and $\hat{X}$ as follows:

$$\text{MSE} = \frac{1}{n} \sum_{i=1}^{n} \left(\hat{X}_i - X_i \right)^2 \tag{4.61}$$

where n is the data number; $\hat{X}_i$ is the ith reconstructed data; X_i is the ith input data; and MSE is the reconstruction error, and the network aims to minimize it.

4.3.3.2.2 Overall Framework of the Proposed Method   We develop the HRCAE driven by multi-sensor information for unsupervised anomaly detection of machine tools under noises. The proposed method consists of two steps: constructing a PCDF module and designing a FDD loss function.

4.3.3.2.2.1 Construction of the PCDF Module   The model training in the existing studies to anomaly detection of machine tools relies on labeled data, yet this is often very scarce or even unavailable in engineering practice. We construct the PCDF module that can cope with the above shortcomings, as described below.

The vibration data of machine tools with noises collected by multiple sensors is acquired and inputs to the convolutional network in multiple channels for the fusion to obtain richer health information. Inspired by the Inception network with parallel convolutional blocks, we form a PCDF module by paralleling two convolutional networks to learn richer characteristics. The network width is enlarged by parallel training to improve network robustness, which can better fit the distribution of normal data for unsupervised anomaly detection. The structure of the PCDF module is shown in Figure 4.65, where the convolution and deconvolution layers can be expressed by the following equations:

$$H = \sigma\left(\sum \left(X \odot w^i + b^i \right) \right) \tag{4.62}$$

$$\hat{X} = \sigma\left(\sum \left(H \otimes \hat{w}^i + \hat{b}^i \right) \right) \tag{4.63}$$

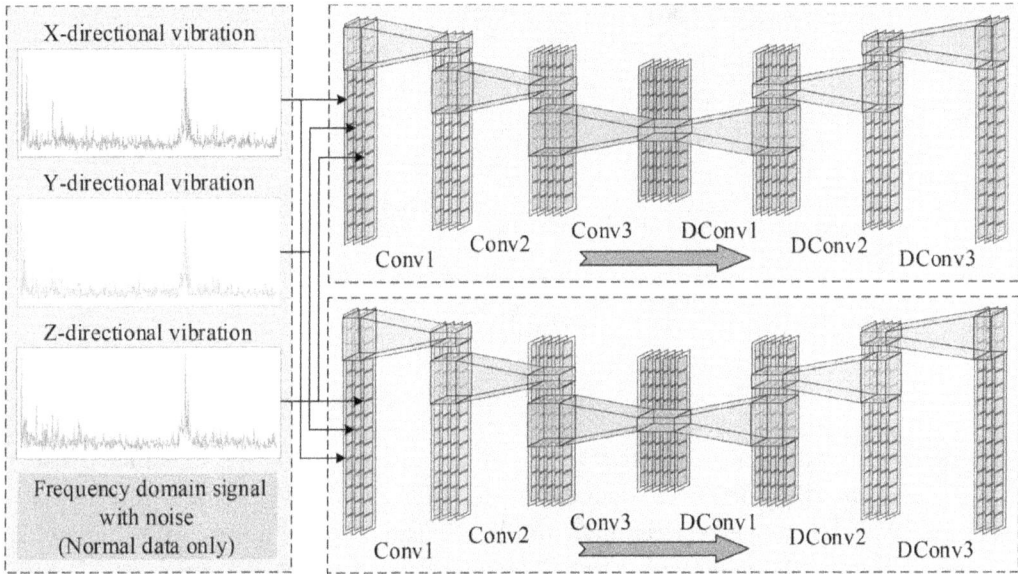

FIGURE 4.65   The structure of PCDF module.

TABLE 4.20   The Structural Parameters of PCDF Module

No.	Layer	Channel Size	Kernel Size/Stride	Batch Normalization	Activation Function
1	Convl	3×16	3/1	Yes	LeakyReLU
2	Conv2	16×32	3/1	Yes	LeakyReLU
3	Conv3	32×64	3/1	Yes	LeakyReLU
4	Dconvl	64×32	3/1	Yes	LeakyReLU
5	Dconv2	32×16	3/1	Yes	LeakyReLU
6	Dconv3	16×3	3/1	No	No

where equation (4.62) is the convolution layer; w^i and b^i are the ith one-dimensional convolution kernel and the corresponding bias here. Equation (4.63) is the deconvolution layer; $\hat{w}^i$ and $\hat{b}^i$ are the ith one-dimensional deconvolution kernel and the corresponding bias here. Compared to equations (4.59) and (4.60), we discard the pooling layer and the up-sampling layer because they are accompanied by a certain degree of data feature loss, which is not conducive to the distribution fitting of the normal data. The specific parameters of the PCDF module are shown in Table 4.20.

4.3.3.2.2.2 Design of the FDD Loss Function   The current prevalent CAE mainly chooses equation (4.61) as the reconstruction error. However, Euclidean distance is difficult to faithfully describe the similarity of complex feature spaces. We design a new FDD loss function, which can measure the distribution similarity between data from the perspectives of angle and distance to effectively suppress the effect of noises, thus further improving the model robustness for anomaly detection of machine tools.

Several studies have demonstrated that the effective combination of Euclidean distance and Cosine similarity can achieve good results in the fields of image recognition, language

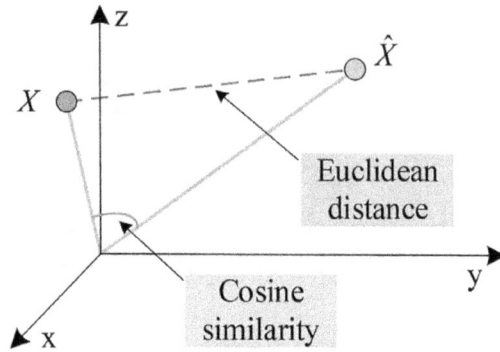

FIGURE 4.66 The difference between Euclidean distance and Cosine similarity.

conversion, music classification, etc. The Euclidean distance is a spatial distance between data, focusing on the difference in values for each dimension. The Cosine similarity measures the geometric angle between data and is insensitive to the numerical difference in each dimension, enabling better capture of feature difference in direction. A visual representation of the difference between Cosine similarity and Euclidean distance is shown in Figure 4.66. The formula for calculating the Cosine similarity is expressed as follows:

$$CS = \frac{\sum_{i=1}^{n} \left(\hat{X}_i \times X_i \right)}{\sqrt{\sum_{i=1}^{n} \left(\hat{X}_i \right)^2} \times \sqrt{\sum_{i=1}^{n} \left(X_i \right)^2}} \tag{4.64}$$

where $CS \in [-1,1]$ is the Cosine similarity between data, and the smaller value indicates the larger directional difference. CS can complement MSE well, and the distance and angle differences between data are fully considered. Thus, the designed FDD loss function can be obtained as follows:

$$FDD = MSE + \lambda \frac{(1 - CS)}{2} \tag{4.65}$$

where λ is an adjustable hyperparameter, which is used to balance the weight of distance and angle difference between data. The loss value is fed into Adam optimizer to automatically optimize the network's learning parameters, which can be expressed as follows:

$$m_t = \beta_1 m_{t-1} + (1 - \beta_1) \nabla_\theta J(\theta_{t-1}) \tag{4.66}$$

$$v_t = \beta_2 v_{t-1} + (1 - \beta_2) \left(\nabla_\theta J(\theta_{t-1}) \right)^2 \tag{4.67}$$

$$\theta_t = \theta_{t-1} - \alpha \frac{\dfrac{m_t}{\left(1 - \beta_1^t\right)}}{\left(\sqrt{\dfrac{v_t}{\left(1 - \beta_2^t\right)}} + \varepsilon \right)} \tag{4.68}$$

where θ_t is the network's learning parameters at the tth iteration; α is the learning rate; $\nabla_\theta J(\theta_{t-1})$ is the computed gradient of the tth iteration; m_t is the exponential moving average of the calculated gradient for the tth iteration; and v_t is the exponential moving average of the squared calculated gradient for the tth iteration; in general, $\beta_1 = 0.9, \beta_2 = 0.999, \varepsilon = 10^{-8}$.

4.3.3.2.2.3 General Steps of the HRCAE The general steps of the HRCAE are shown in Figure 4.67 and explained in detail as below.

Data preparation: Acquire CNC machine tool vibration data collected by multiple sensors, adding Gaussian random noises, performing normalization operation and FFT transform. The data with noises is then divided into training set and validation set containing only normal data, and a test set containing both normal and anomaly data.

Model initialization: Set the model hyperparameters such as weight coefficient λ, learning rate α, iteration number T, and PCDF module structure. The learning parameters θ of the network are randomly initialized.

Model offline training: The training set with only normal data is fed into the constructed PCDF module in multiple channels, and two convolutional distribution fitting networks are trained in parallel. The reconstructed data $\hat{X}$ is obtained based on equations (4.62) and (4.63). The FDD loss values of the two networks are calculated and added based on equations (4.61), (4.64), and (4.65), and the learning parameters θ of the network are iteratively updated based on equations (4.66)–4.68).

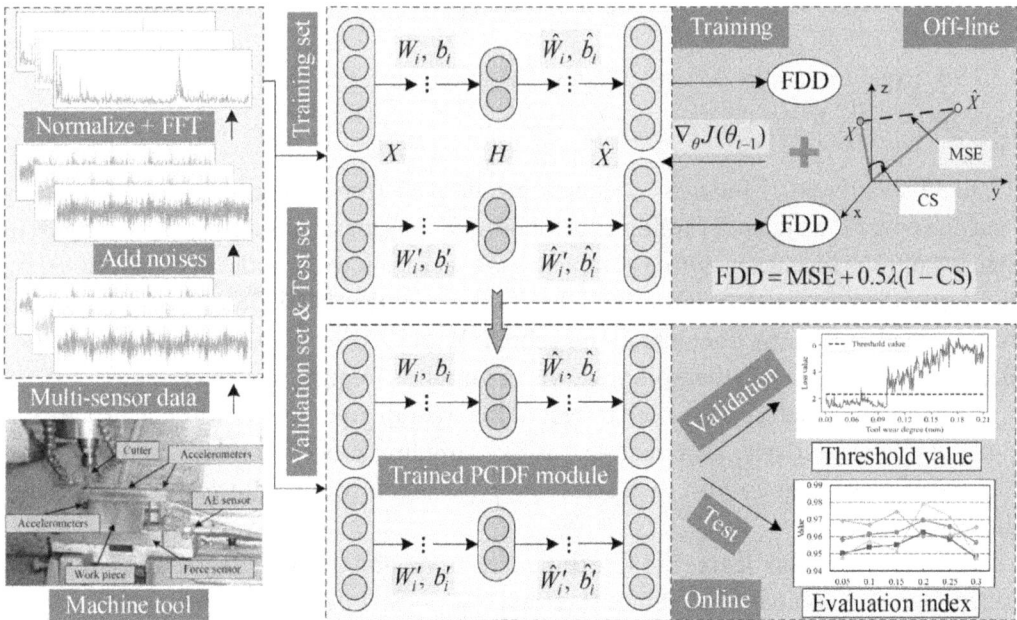

FIGURE 4.67 General steps of the HRCAE.

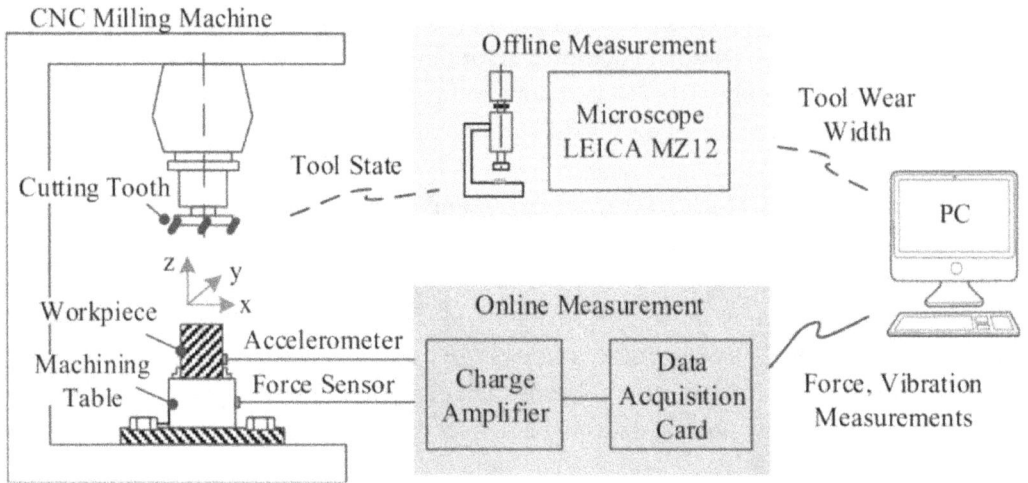

FIGURE 4.68   Data collection system

Model online validation: The validation set with only normal data is fed into the trained PCDF module, and after obtaining the reconstructed data, the sum of the FDD loss value for each sample in the two networks is calculated. Because there may be very few unfitted noise samples and considering the model's robustness, the loss value containing 95% of normal data is chosen as the threshold value for anomaly data.

Model online testing: The test set containing normal and anomaly data is fed into the trained PCDF module, and the sum of FDD loss value is calculated for each test sample to determine the normal or anomaly based on the threshold value. Finally, evaluation indexes such as accuracy, recall, precision, and *F*-score are displayed.

4.3.3.3.3 Experimental Validation   We test the effectiveness of the proposed method on a high-speed CNC machine.

4.3.3.3.3.1 Description of Datasets   Datasets are derived from a high-speed CNC milling machine operating under dry milling operation, and the data acquisition system is shown in Figure 4.68. The specific operating parameters are set as follows: spindle speed is 10,400 rpm; *X*-direction feed rate is 1,555 mm/min; *Y*-direction cutting depth is 0.125 mm; and *Z*-direction cutting depth is 0.2 mm. Tool vibration in the *X*, *Y*, and *Z* directions is measured by piezo-electric accelerometers on the workpiece with a sampling frequency of 50 kHz.

A total of three datasets are used to record the wear degree of three tools, which are C1, C4, and C6, respectively. Each dataset has 315 sampling processes that record the variation of tool wear degree, with each sampling process recording tool wear degree in three directions.

We average the wear degree in these three directions, and the wear degree above 0.10 mm is considered as anomaly data. Since there are few normal data in C6 dataset, we conduct unsupervised anomaly detection on the C1 and C4 datasets, and the dataset information is shown in Table 4.21. The C1 dataset contains data for tool wear degree from 0.04 to

TABLE 4.21   Introduction to Datasets

Datasets	Training Set (Only Normal Data)	Validation Set (Only Normal Data)	Test Set (Normal and Anomaly Data)
C1	600	600	1,300
	(6×100)	(6×100)	(13×100)
	0.04–0.09 mm	0.04–0.09 mm	0.04–0.16 mm
C4	700	700	1,800
	(7×100)	(7×100)	(18×100)
	0.03–0.09 mm	0.03–0.09 mm	0.03–0.20 mm

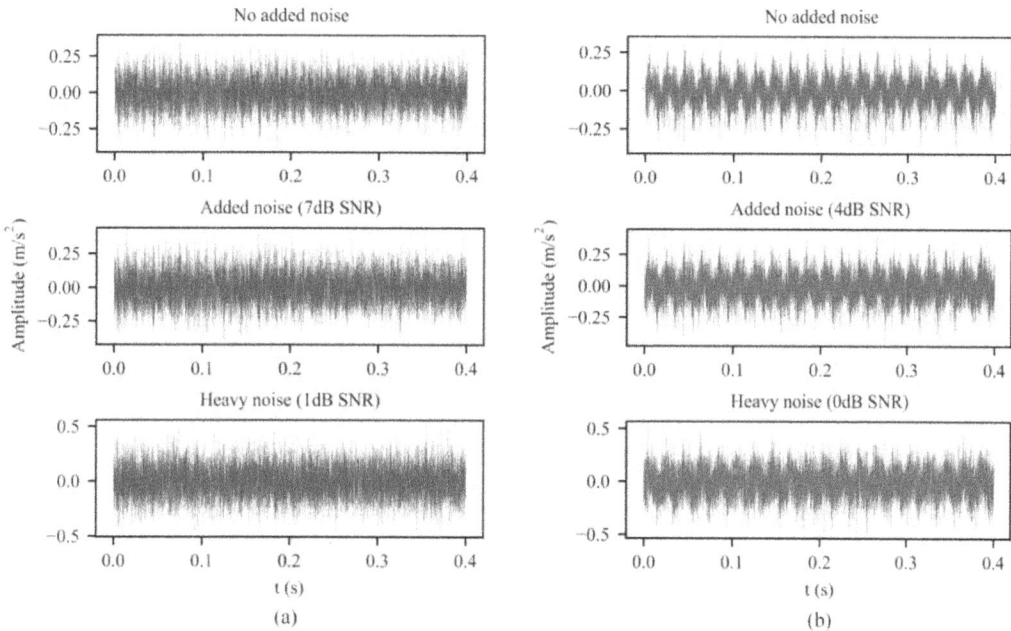

FIGURE 4.69   Vibration signals of datasets under different noises: (a) C1 dataset and (b) C4 dataset.

0.16 mm. In order to simulate the full degradation process, for the training and validation sets, tool wear degree varies from 0.04 to 0.09 mm with normal data only, 100 samples randomly selected at 0.01 mm intervals, 600 samples in total; the test set of tool wear degree contains the full range of 0.04 to 0.16 mm, with 100 samples randomly selected at 0.01 mm intervals, for a total of 1,300 samples. The length of each sample is 1,024 sampling points; the training set, validation set, and test set are from different sampling processes. The C4 dataset contains data with tool wear degree from 0.03 to 0.20 mm.

The training set, validation set, and test set are divided by the same rules as the C1 dataset in Table 4.21. It is important to emphasize that we only divide the dataset by tool wear degree and do not need to use specific wear degree labels in the training of the model, so the model training is unsupervised.

Figure 4.69 shows the vibration signals of the two datasets under different noises. As noises increase, the destructive effect on the vibration signals becomes more apparent. Under the heavy noise, the amplitudes of the vibration signals in both datasets are almost doubled compared to the signals without the addition of noise, which puts high demands on the robustness of the model. The value of signal-to-noise ratio (SNR) is calculated as below:

$$SNR = 10 \log_{10} \left(\frac{P_s}{P_n} \right) \tag{4.69}$$

where P_s is the data power; P_n is the noise power; and SNR is the signal-to-noise ratio in dB and its smaller value means more heavy noise.

4.3.3.3.3.2 The Validity of the HRCAE   In this section, to reduce the random effect for experiments, all the methods are repeated five times. The algorithms are run in Pytorch 1.7.1 of Windows 10, with GPU GTX1650 4G. Based on the greedy algorithm idea, we search for a relatively better combination of hyperparameters by controlling the variables. The main hyperparameters are set as follows: the number of iterations is set to 300; the learning rate is set to 0.01, which decays 0.1 at 20 and 200 iterations, respectively; the hyperparameter λ in FDD loss function is set to 0.2. In the future, we will explore simpler and more general hyperparameter setting methods based on the existing study.

First, we examine the unsupervised anomaly detection capability of the developed HRCAE under heavy noise and compare it with the commonly used unsupervised methods based on autoencoder, including CAE, MAE, DAGMM, sparse autoencoder (SAE), and denoising autoencoder (DAE) in which the superiority of the PCDF module and the FDD loss function are also verified. The evaluation indexes of anomaly detection on the C1 and C4 datasets for each method are calculated as follows:

$$\text{Accuracy} = \frac{TP + TN}{TP + TN + FP + FN} \tag{4.70}$$

$$\text{Precision} = \frac{TP}{TP + FP} \tag{4.71}$$

$$\text{Recall} = \frac{TP}{TP + FN} \tag{4.72}$$

$$F\text{-score} = 2 \times \frac{\text{Precision} \times \text{Recall}}{\text{Precision} + \text{Recall}} \tag{4.73}$$

where TP indicates correctly detected as anomaly data; FP indicates error detection as anomaly data; TN means correctly detected as normal data; FN means error detection as normal data. Accuracy can reflect the quality of global classification; and F-score combines two indexes of precision and recall, which is an important index of how well anomaly data is detected.

Table 4.22 shows the evaluation indexes of each method for anomaly detection on the C1 dataset. The accuracy and F-score of the proposed method reach 95.95% and 96.21%

TABLE 4.22   Evaluation Indexes and Times of Each Method on the C1 Dataset with SNR = 1 dB

Unsupervised Methods	Evaluation Indexes (%)				Times (s)	
	Accuracy	Precision	Recall	F-score	Offline	Online
HRCAE	95.95±0.75	96.85±0.25	95.60±1.50	**96.21 ±0.74**	36	2.5
PCDF-MSE	94.85±1.61	96.15±0.98	94.20±2.75	95.15±1.58	36	2.5
CAE-FDD	94.68±1.65	96.05±0.75	94.00±3.59	94.89±1.64	19	1.9
CAE-MSE	88.90±4.46	92.91±5.75	86.31±6.56	89.29±4.29	19	2
MAE-FDD	94.00±1.75	96.02±0.84	92.71±3.63	94.30±1.78	78	2.4
MAE-MSE	85.79±3.96	90.74±4.67	82.37±9.15	86.03±4.37	78	2.5
DAGMM	84.51±1.56	78.35±1.17	**98.43 ± 2.51**	87.24±1.38	355	3.1
SAE-FDD	90.62±6.23	96.76±1.16	85.37±11.35	90.41±6.84	20	2.0
SAE-MSE	83.72±10.05	90.97±7.97	77.74±15.67	83.19±11.13	20	2.0
DAE-FDD	88.11±8.62	**97.81 ± 0.61**	79.69±16.25	87.03±10.98	19	1.9
DAE-MSE	82.03±11.67	95.02±2.20	70.11±22.15	79.23±14.96	19	2.0

"Bold" indicates that this method has the highest Evaluation Index (Accuracy, Precision, Recall or F-score) in comparison.

under heavy noise with SNR = 1 dB $(P_n/P_s \approx 80\%)$, which are the optimal and the most stable among all methods. Although the recall of DAGMM reaches 98.43% and the precision of DAE-FDD reaches 97.81%, the other three evaluation indexes of them are unsatisfactory. The PCDF-MSE shows an improvement of 5.95%, 3.24%, 7.89%, and 5.86% in each evaluation index compared to the CAE-MSE; the HRCAE shows an improvement of 1.27%, 0.80%, 1.60%, and 1.32% in each evaluation index compared to the CAE-FDD.

These improvements indicate that the constructed PCDF module has stronger robustness than the single convolutional distribution fitting network. After replacing MSE with FDD loss function, four evaluation indexes of MAE are improved by 8.21%, 5.28%, 10.34%, and 8.27%, respectively; SAE's evaluation indexes are improved by 6.90%, 5.79%, 7.63%, and 7.22%, respectively; DAE's evaluation indexes are improved by 6.08%, 2.79%, 9.58%, and 7.80%, respectively; CAE's evaluation indexes are improved by 5.78%, 3.14%, 7.69%, and 5.60%, respectively; PCDF module's evaluation indexes are improved by 1.10%, 0.70%, 1.40%, and 1.06%, respectively, and the stability of all methods is improved. The stability of anomaly detection results for the above methods is also improved. It shows that the designed FDD loss function has universal applicability to various unsupervised improved autoencoder methods and can effectively measure the reconstruction errors of data from distance and angle directions to suppress the impact of heavy noise. The running times of all methods are shown in Table 4.22. The offline training time of DAGMM is the longest, and the online validation test times of all methods are within 3 seconds except DAGMM.

Figure 4.70 shows the visual test results of methods on the C1 dataset with SNR = 1 dB. The horizontal coordinates indicate the tool wear degree of the test set, which ranges from 0.04 to 0.16 mm, while the vertical coordinates are the loss function value of each sample. Tool wear degrees less than 0.10 mm are normal samples; those above 0.10 mm are anomaly samples, and the dotted line is the threshold value for detecting anomaly samples. To express the aesthetics, the average loss value of five adjacent samples is taken as a point for plotting the curve. Because of different model structures and loss functions of each

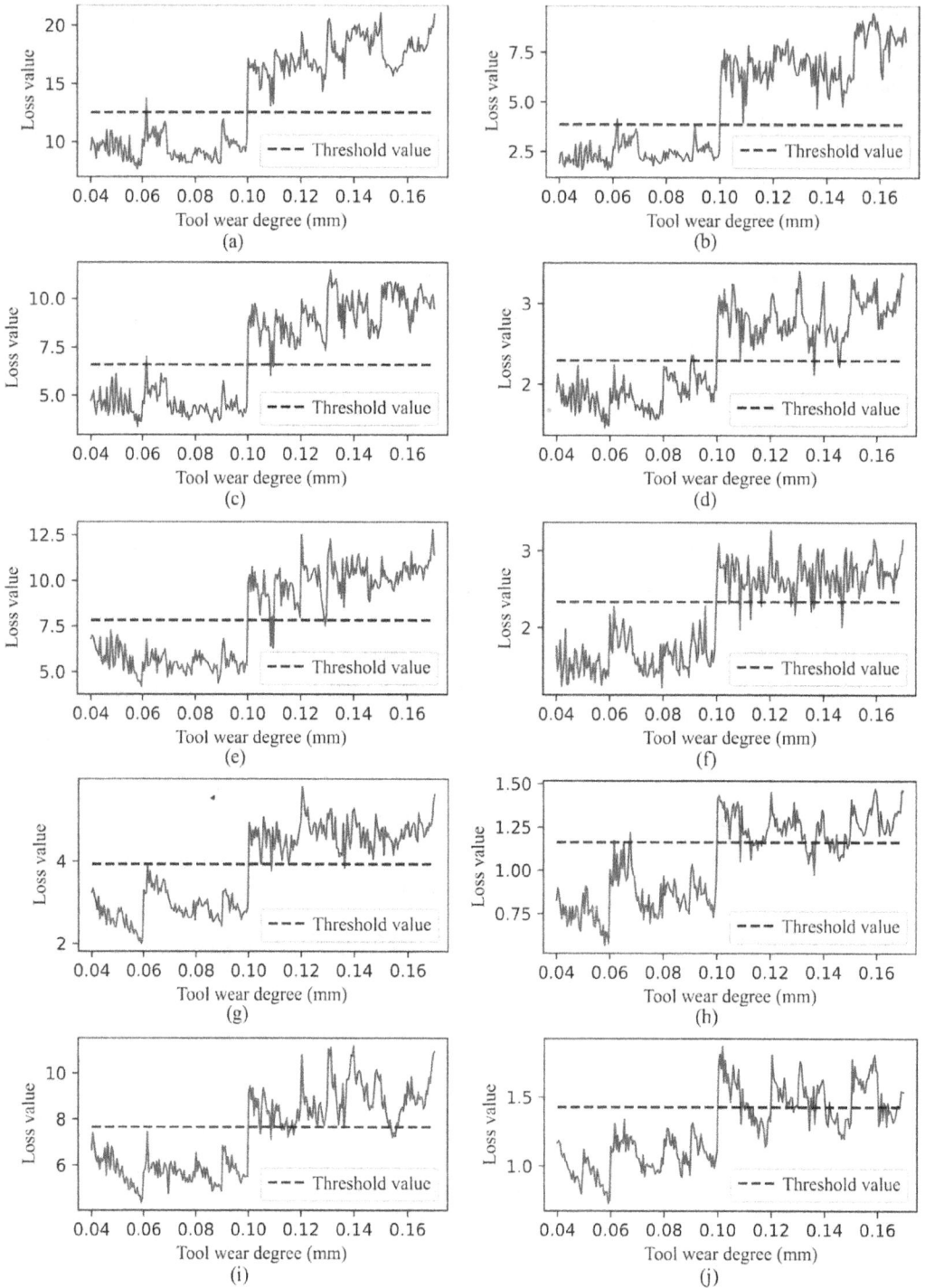

FIGURE 4.70 The visual test results of each method on the C1 dataset with SNR = 1 dB: (a) HRCAE, (b) PCDF-MSE, (c) CAE-FDD, (d) CAE-MSE, (e) MAE-FDD, (f) MAE-MSE, (g) SAE-FDD, (h) SAE-MSE, (i) DAE-FDD, and (j) DAE-MSE.

method, as well as different optimal spaces of each stochastic training, there are differences in the loss values of each method. It can be observed that the visual test results of each method can correspond to the evaluation results in Table 4.3. The proposed method HRCAE and the PCDF-MSE differentiations between normal and anomaly samples are relatively more obvious compared to other methods. This indicates that the constructed PCDF module can better fit the distribution of normal samples to distinguish the anomaly samples. After replacing MSE with FDD loss function, the discriminative degree of each method for normal and anomaly samples is improved to some extent, which again demonstrates the robustness of FDD loss function under heavy noise.

On the C4 dataset, four evaluation indexes of all methods for anomaly detection are shown in Table 4.23. Under the heavy noise with $\text{SNR} = 0\,\text{dB}\left(P_n/P_s = 100\%\right)$, the proposed method HRCAE achieves 96.25%, 96.01%, 97.95%, and 96.95% for the four evaluation indexes, respectively, showing superior and stable anomaly detection results over other methods. Although the DAGMM has the highest precision, its other three evaluation indexes are very bad, and the volatility is very high compared to the C1 dataset. After replacing the MSE with the FDD loss function, all methods show the improvement in four evaluation indexes and have better stability, which are consistent with the performance on the C1 dataset. On the C4 dataset, the PCDF module also shows advantages compared to other methods. The running time regularity of all methods on the C4 dataset is consistent with that on the C1 dataset.

Figure 4.71 shows the corresponding visual test results of six methods on the C4 dataset. Tool wear degree for the test set ranges from 0.03 to 0.20 mm, and wear degrees above 0.10 mm are anomaly samples. Other plotting principles are the same as Figure 4.70. On the C4 dataset, the PCDF module can better discriminate between normal and anomaly samples than other models, and the FDD loss function can further enhance the model's discrimination ability, which is consistent with Figure 4.70.

TABLE 4.23　Evaluation Indexes and Times of Each Method on the C4 Dataset with SNR = 0 dB

Unsupervised Methods	Evaluation Indexes (%)				Time (s)	
	Accuracy	Precision	Recall	F-score	Offline	Online
HRCAE	96.25±1.17	96.01±1.09	**97.95 ± 2.32**	**96.95 ± 0.99**	43	2.9
PCDF-MSE	93.91±2.42	96.72±0.48	93.20±4.15	94.89±2.18	42	2.9
CAE-FDD	93.00±2.79	95.71±2.52	92.89±6.85	94.10±2.58	23	2.3
CAE-MSE	88.31±5.48	96.11±2.95	84.26±7.84	89.67±5.03	23	2.3
MAE-FDD	94.43±1.83	96.34±0.89	94.49±3.31	95.38±1.59	94	2.9
MAE-MSE	91.14±4.32	95.78±1.65	89.56±8.49	92.35±4.08	94	2.9
DAGMM	53.22±2.63	**97.81 ± 0.32**	23.98±4.32	38.37±5.39	424	3.9
SAE-FDD	92.21±1.65	96.25±1.08	90.80±2.55	93.43±1.47	23	2.3
SAE-MSE	78.95±8.05	95.69±1.19	68.53±12.91	79.37±9.20	23	2.3
DAE-FDD	88.66±6.68	95.21±3.63	85.60±8.54	89.90±6.12	23	2.3
DAE-MSE	77.57±8.47	94.66±2.98	66.89±13.39	77.87±10.09	23	2.3

"Bold" indicates that this method has the highest Evaluation Index (Accuracy, Precision, Recall or F-score) in comparison.

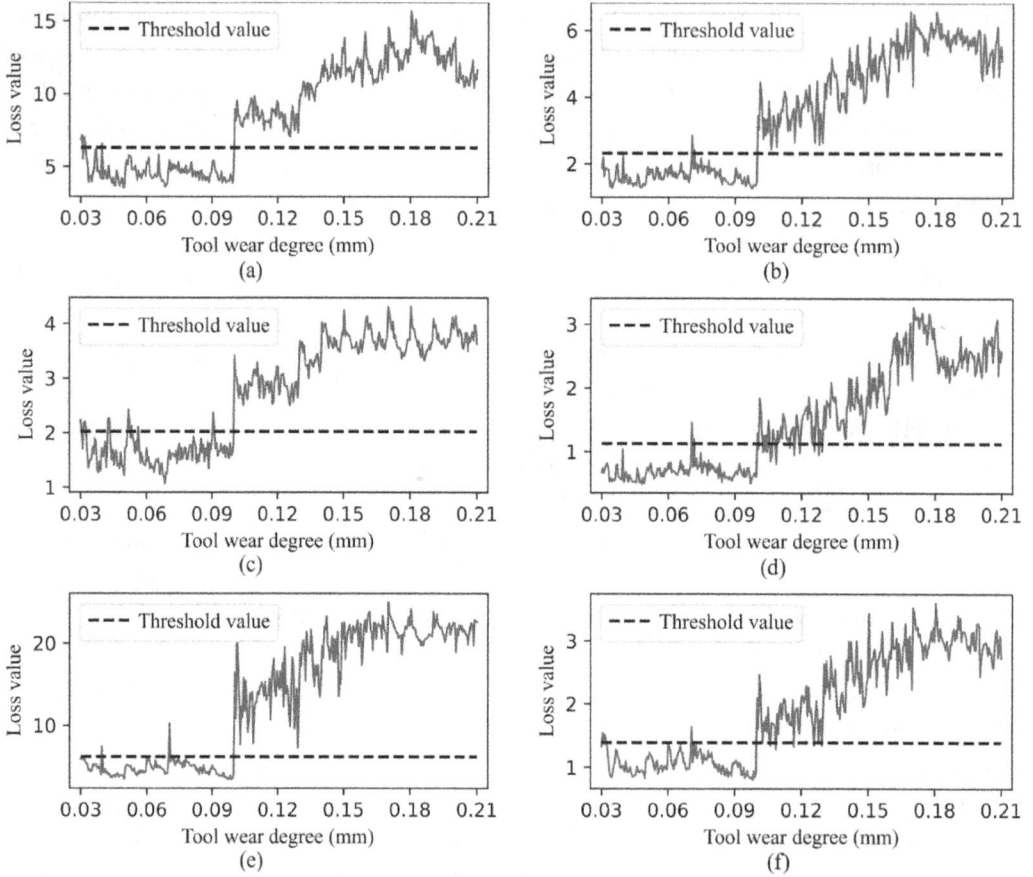

FIGURE 4.71 The visual test results of each method on the C4 dataset with SNR = 0 dB: (a) HRCAE, (b) PCDF-MSE, (c) CAE-FDD, (d) CAE-MSE, (e) MAE-FDD, and (f) MAE-MSE.

Subsequently, we plot ROC curves and confusion matrixes to show the anomaly detection results of methods under different noise effects. Figure 4.72 shows the ROC curves for different methods on the C4 dataset with SNR of 0, 1, 2, and 4 dB (P_n/P_s is approximately 100%, 80%, 60%, and 40%). The AUC is used to measure the anomaly detection performance of each method and ranges from 0.5 to 1 [61]. The higher AUC value represents the better performance of the method, and the AUC values for methods are listed in parentheses of right corner. At SNR values of 0, 1, 2, and 4 dB, the curves of the proposed method are closest to the upper left corner, and their AUC values are 99.38%, 99.19%, 99.54%, and 99.91%, respectively, which are stable and better than other methods. As for the PCDF module, the AUC values of the HRCAE compared to the CAE-FDD are improved by 1.08%, 1.17%, 0.57%, and 0.87% with SNR of 0, 1, 2, and 4 dB; the AUC values of the PCDF-MSE compared to the CAE-MSE are improved by 2.56%, 1.05%, 2.38%, and 1.99%, respectively. After replacing MSE with FDD loss function, the AUC values of DAE are improved by 5.84%, 4.25%, 4.14%, and 2.60%, respectively; SAE are improved by 4.62%, 4.18%, 3.82%, and 2.77%, respectively; CAE are improved by 2.37%, 1.28%, 2.92%, and 2.28%, respectively; PCDF module are improved by 0.89%, 1.40%, 1.11%, and 1.16%, respectively.

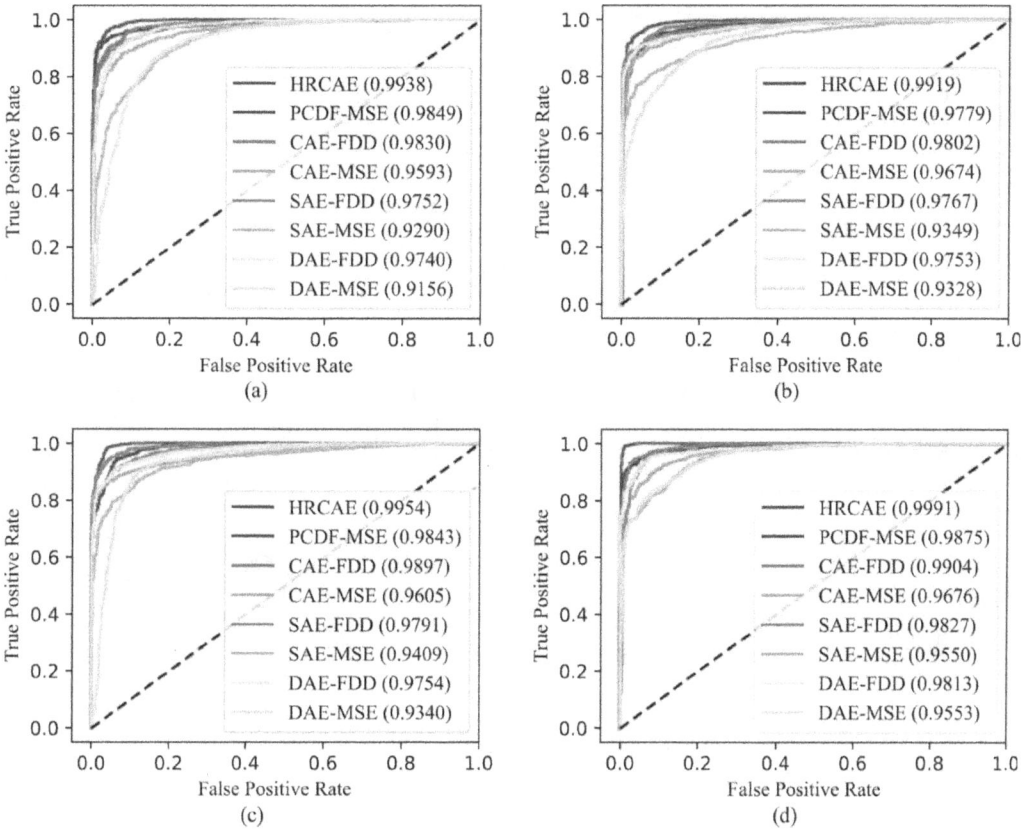

FIGURE 4.72   The ROC curves of methods under different noises on the C4 dataset: (a) SNR=0 dB, (b) SNR=1 dB, (c) SNR=2 dB, and (d) SNR=4 dB.

Figure 4.73 shows the confusion matrixes for methods on the C1 dataset with SNR of 1, 2, 4, and 7 dB (P_n/P_s is approximately 80%, 60%, 40%, and 20%). Coordinate 0 represents normal samples, with the total number of 600, while coordinate 1 represents anomaly samples, with the total number of 700. The horizontal coordinate is the predicted value and the vertical coordinate is the true value. We can intuitively see that the proposed method HRCAE has good classification results under different noises. At the SNR values of 1, 2, 4, and 7 dB, the proposed method achieves F-score of 96.50%, 98.30%, 97.68%, and 98.24%, with small volatility under different noises. Both the constructed PCDF module and the designed FDD loss function show enhancement effects, further demonstrating the robustness of the proposed method for anomaly detection under different noises.

4.3.3.3.3.3 Discussions on Multi-Sensor and λ Values   Next, we verify the advantage of fusing multi-sensor information under different noises. Figure 4.74 shows the accuracy and F-score of the proposed method under different sensor data. The figure legend *XYZ* indicates the multi-sensor data in three vibration directions, while *X*, *Y*, and *Z* indicate the single-sensor data in the vibration direction along that axis.

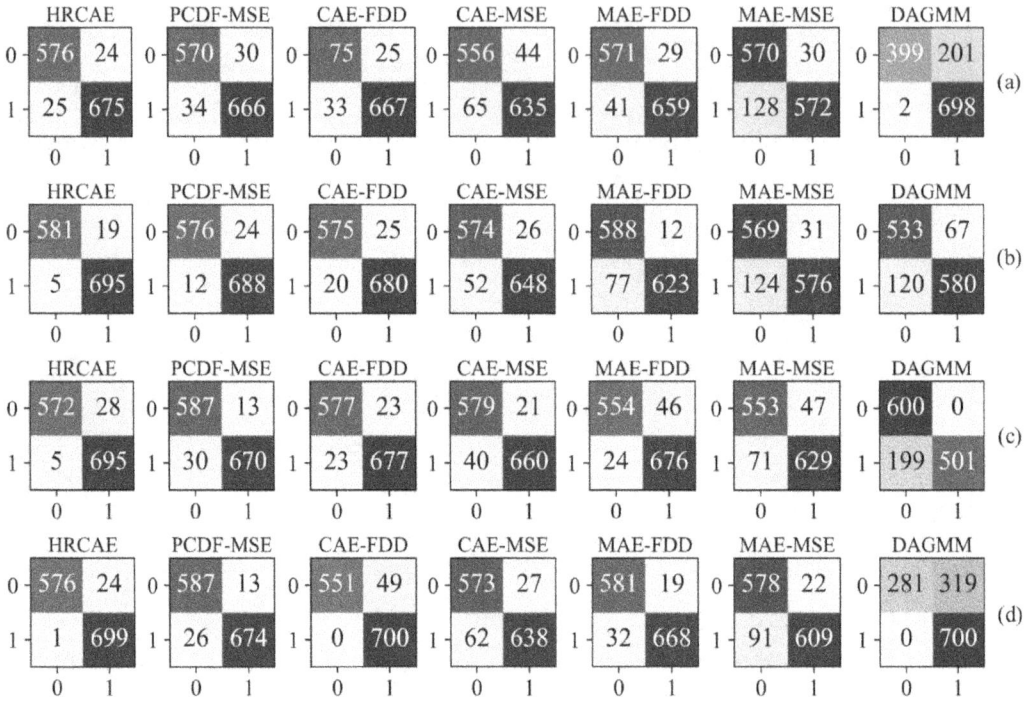

FIGURE 4.73 Confusion matrixes of methods under different noises on the C1 dataset: (a) SNR=1 dB, (b) SNR=2 dB, (c) SNR=4 dB, and (d) SNR=7 dB.

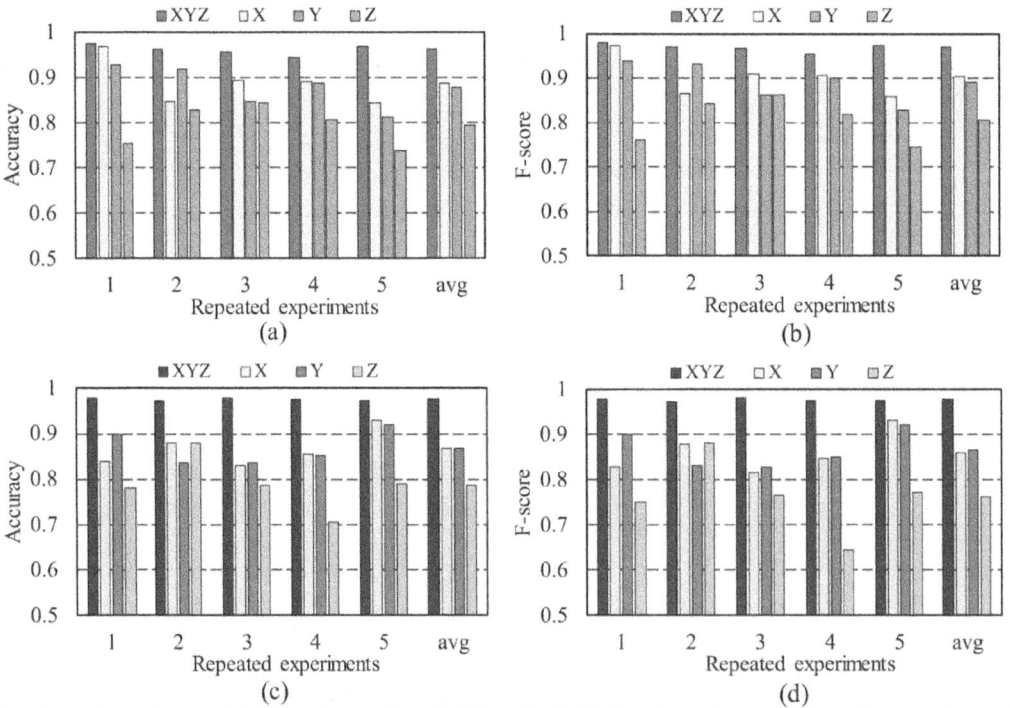

FIGURE 4.74 The evaluation indexes of the proposed method under different sensor data: (a) Accuracy on the C4 dataset with SNR=0 dB; (b) F-score on the C4 dataset with SNR=0 dB; (c) accuracy on the C1 dataset with SNR=7 dB; and (d) F-score on the C1 dataset with SNR=7 dB.

Under the heavy noise with SNR=0 dB on the C4 dataset, the average accuracies are 96.25%, 88.99%, 87.98%, and 79.49% for multi-sensor data and single-sensor data in X, Y, and Z directions, respectively; the F-values are 96.95%, 90.36%, 89.27%, and 80.64%, respectively. Under the influence of noise with SNR=7 dB on the C1 dataset, the average accuracies are 97.60%, 86.66%, 86.85%, and 78.77% for multi-sensor data and single-sensor data in X, Y, and Z directions, respectively; the F-values are 97.80%, 86.08%, 86.73%, and 76.31%, respectively. Under different noises in both datasets, the proposed method can effectively fuse multi-sensor information, and the anomaly detection results are significantly better than those of single-sensor data, reducing the requirements for sensor location arrangement and selection.

Finally, we investigate the values of the hyperparameter λ in the FDD loss function. Figure 4.75 shows the evaluation indexes of the proposed method at different λ values under the heavy noise on the C1 and C4 datasets. To ensure that both distance and angle differences are useful, the MSE and CS loss values need to be of the same order of magnitude.

According to data preparation and Eqs. (4.61), (4.64), and (4.67), we choose λ values in the range of 0.05–0.3. With the increase of λ values, each evaluation index has a certain fluctuation. The fluctuation of recall is more intense compared with the other three evaluation indexes, which indicates that anomaly samples are relatively more sensitive to λ values; the accuracy and F-score first increase and then decrease. When λ=0.2, the proposed method is optimal in terms of accuracy, recall, and F-score on both datasets.

4.3.4 Intelligent Fault Diagnosis under Varying Working Conditions with Transfer Learning

Due to automatic detecting capability, more and more intelligent fault diagnosis techniques have been proposed and applied to health monitoring of rotor-bearing system and other industrial equipment [95,96]. Through the literature review, it can be found that most of the existing methods focus on vibration analysis of the rotor bearing [97–100]. However, vibration analysis has the problem of affecting the equipment structures and the difficulty

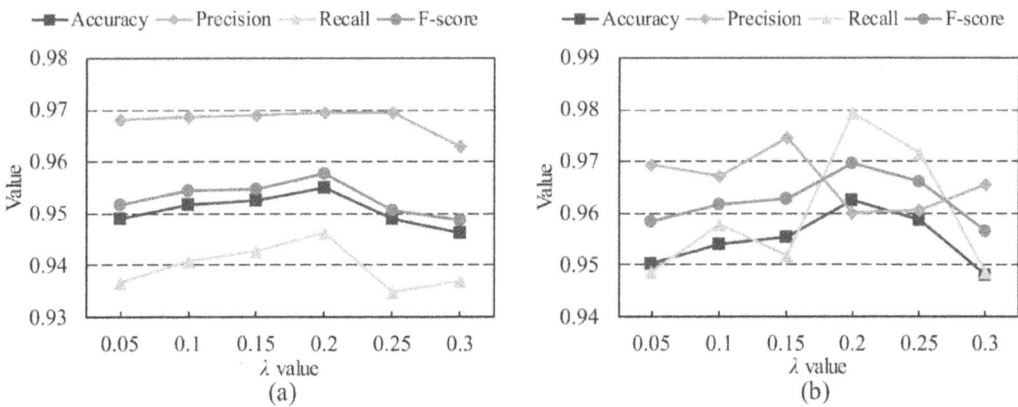

FIGURE 4.75 The evaluation indexes of the proposed method at different λ values: (a) SNR=1 dB on the C1 dataset and (b) SNR=0 dB on the C4 dataset.

in installing sensors. Besides, the processing of vibration signals is very complicated due to the long transfer path of the signal, changeable working conditions, and strong noise in real applications [101,102].

Infrared thermography (IRT) has provided an advanced tool for equipment condition monitoring in recent years [103–105]. Compared with vibration monitoring, IRT holds unique superiorities such as noncontact, simple installation, high precision, and high sensitivity [102]. Considering the special abilities of processing two-dimensional (2-D) images, CNNs have been gradually used for fault diagnosis of industrial equipment using the infrared thermal images in the past few years [106–108].

However, the current CNN-based diagnosis approaches with thermal images can only deal with the same working condition that is rarely the case in real applications. Also, these methods are all based on the availability of a large number of training samples which are difficult and expensive to acquire. In engineering practice, the operating conditions of equipment frequently change, which will result in different distributions of the collected samples [109]. In addition, it is challenging to train an excellent CNN from scratch only using a few samples [70]. Thus, how to enable the CNNs trained with limited thermal images to achieve satisfactory fault diagnosis accuracy of rotor-bearing system under varying working conditions has become an urgent task.

Transfer learning is considered to have great potential to complete different but similar tasks from the source domain to the target domain [14,89]. Parameter transfer, the most widely applied transfer learning technique, aims to provide valuable parameter knowledge for the target model from a well-pretrained model (source model) [70,110]. With well-located initial parameters and a small number of target samples, the target model can be quickly fine-tuned to solve the target task. Since 2017, the transfer diagnosis performance of CNNs integrated with parameter transfer has been demonstrated by a few case studies [70,89]. Thus, CNN and parameter transfer can be investigated for fault diagnosis of rotor-bearing system under different working conditions with limited samples.

In this section, a modified transfer CNN driven by thermal images is proposed to diagnose faults of rotor-bearing system under varying working conditions [111].

4.3.4.1 Main Algorithms

4.3.4.1.1 The Classical CNN Theory   Among different types of CNN models, the LeNet-5 is the most classical with less trainable parameters, more mature theory, and higher computational efficiency. Besides, LeNet-5 is specifically designed to process 2-D grayscale images, which mainly consists of an input layer, two convolutional layers, two pooling layers, and a fully connected layer, as shown in Figure 4.76. For the convolutional layer in the lth layer, its output is

$$a_{i,j}^{k,l} = f\left(\sum_{k=1}^{K}\sum_{i,j=1}^{N_W}\sum_{r,s=1}^{N_W} W_{r,s}^{k,l} * x_{i+r-1,j+s-1}^{k,l-1} + b^{k,l}\right) \quad (4.74)$$

$$f(x) = \frac{m(0,x) \text{ or } 1}{\left(1+e^{-x}\right)} \quad (4.75)$$

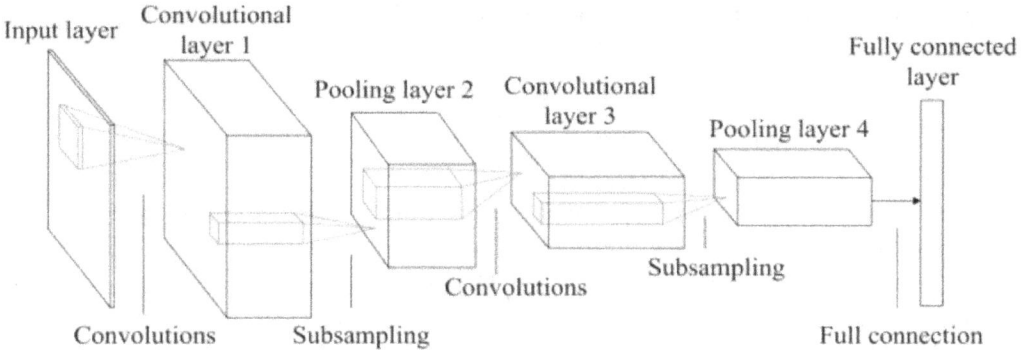

FIGURE 4.76   Model architecture of the classical CNN.

where $a_{i,j}^{k,l}$ means the element in the i th row and the j th column of the k th output feature map in the lth layer, $x_{i+r-1,j+s-1}^{k,l-1}$ and $W_{r,s}^{k,l}$ are the element of the k th input feature map and convolution kernel located in the corresponding position, respectively, $N_W \times N_W$ is the size of each $W_{r,s}^{k,l}$, $b^{k,l}$ is the bias, $*$ is a convolution operation, and $f(\cdot)$ is the activation function.

The subsampling operation is used to obtain compressed feature representation and decrease calculation cost, which is usually carried out by max pooling or average pooling as

$$f_{i,j}^{k,l} = \underset{(i,j)\in R_P}{\mathrm{maxpool}}\left(a_{i,j}^{k,l}\right) \text{ or } \underset{(i,j)\in R_P}{\mathrm{avepool}}\left(a_{i,j}^{k,l}\right) \tag{4.76}$$

where maxpool(.) is the max pooling, avepool(.) is the average pooling, $f_{i,j}^{k,l}$ is the element of the k th output feature map after pooling, and R_P is the pooling region with each size of $N_P \times N_P$. Transformed by the convolution and pooling layers, deep feature maps are finally fed into a fully connected layer and a softmax layer for classification.

4.3.4.1.2 The Proposed Method
4.3.4.1.2.1 Modified CNN Design   In this section, to overcome the problems of the classical CNN, modifications of the basic CNN are carried out in two aspects: Pooling strategy and activation function.

Max and average operations are the two popular pooling strategies in the basic CNN. Some potential information is ignored using max pooling since only the strongest feature elements are selected. Although average pooling considers all the elements, they are always treated equally, reducing the contributions of those important elements. Besides, the certainties caused by these two pooling strategies are more likely to result in overfitting.

As a new subsampling strategy, stochastic pooling can address the problems of max and average operations by fusing all the feature elements according to their contributions, and the sampled probabilities are

$$p_{i,j} = \frac{a_{i,j}}{\left(\sum_{(i,j)\in R_S} a_{i,j}\right)} \tag{4.77}$$

where $p_{i,j}$ is the probability of element and R_s is the pooling region. In the backpropagation phase, the fused element is weighted by the probabilistic form of averaging as

$$f_{i,j} = \sum_{(i,j)\in R_s} \left(p_{i,j} \cdot a_{i,j} \right) \tag{4.78}$$

In addition to the pooling strategy, the activation function also has an impact on CNN performance. The gradient vanishing and the computational complexity are the main limitations of sigmoid (Sigm). The neurons of ReLU always stop learning when the inputs are negative. LReLU is an enhanced variant of ReLU, which can address these problems by giving a small positive value α when the inputs are negative, helping neurons continue working.

$$f(x) = \begin{cases} x, & \text{if } x > 0 \\ \alpha x, & \text{if } x \leq 0 \end{cases}. \tag{4.79}$$

By now, the modified CNN has been designed, as shown in Figure 4.77. In order to improve the training performance, stochastic gradient descent algorithm with learning rate decay and momentum is applied to update weights and biases:

$$\theta_{t+1} = \theta_t - \eta_t \nabla E(\theta_t) + \mu_t (\theta_t - \theta_{t-1}) \tag{4.80}$$

$$\eta_{t+1} = \frac{\eta_t}{\varepsilon} \tag{4.81}$$

$$\mu_t = \begin{cases} 0.5, & \text{if } t < \text{mom\_epoch} \\ 0.95, & \text{if } t \geq \text{mom\_epoch} \end{cases} \tag{4.82}$$

where t is the epoch number, θ_{t+1} is the trained parameters of the modified CNN, $\nabla E(\theta_t)$ is the derivative of cross-entropy error between the true and predicted labels, η_t is the learning rate, ε is the decay factor, μ_t is the momentum, and mom_epoch is the boundary epoch number.

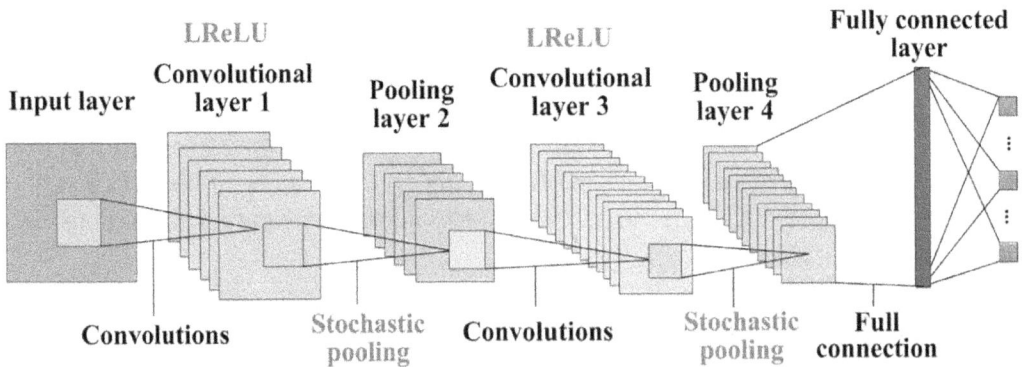

FIGURE 4.77 Modified CNN using stochastic pooling and LReLU.

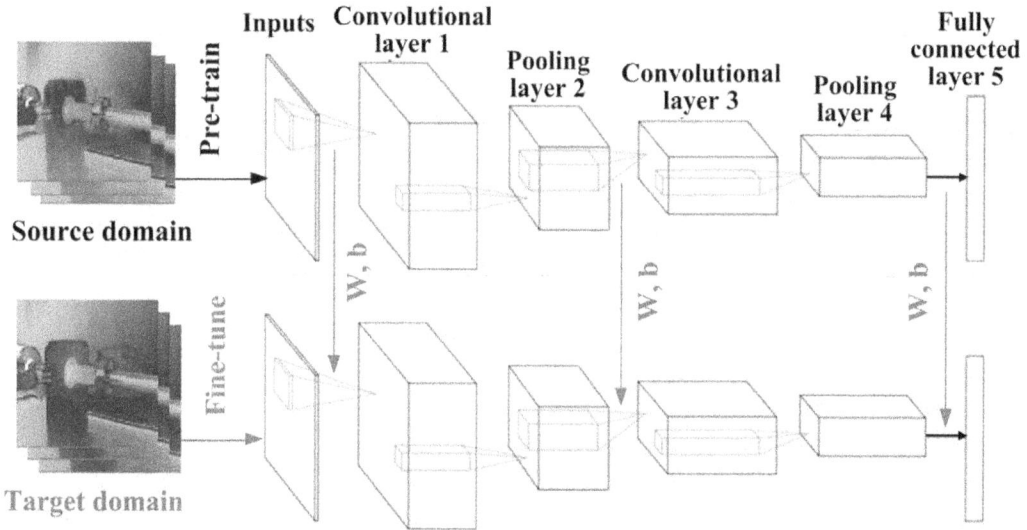

FIGURE 4.78   Construction of modified transfer CNN.

4.3.4.1.2.2 Construction of Modified Transfer CNN   As shown in Figure 4.78, by introducing parameter transfer into the modified CNN, a modified transfer CNN can be constructed as follows: (1) Enough source-domain samples are used to pretrain a source modified CNN by minimizing cross-entropy error between the true and predicted labels according to equations (4.80)–(4.82). (2) Prepare a target modified CNN holding the same structure and hyperparameters as the source model. (3) Transfer all the well-learned weights and biases from the source modified CNN to the target. (4) A few target-domain samples are used to fine-tune the target modified CNN. During the fine-tuning stage, all the weights are adjusted.

4.3.4.2 Overall Framework of the Proposed Method
As shown in Figure 4.79, this chapter presents modified transfer CNN and thermal images for fault diagnosis of rotor-bearing system cross-working conditions and the following are the main procedures:

Step 1: Collect thermal images of rotor-bearing system under different working conditions, which are first translated into grayscale images and then divided as the source domain and target domain. The source domain contains enough labeled samples, while the target domain has a few labeled samples.

Step 2: Stochastic pooling and LReLU are combined to design modified CNN.

Step 3: A source modified CNN model is first well-pretrained using enough samples from the source domain.

Step 4: Transfer all the trained weights and biases provided by the source modified CNN to initialize a target modified CNN with the same structure and hyperparameters.

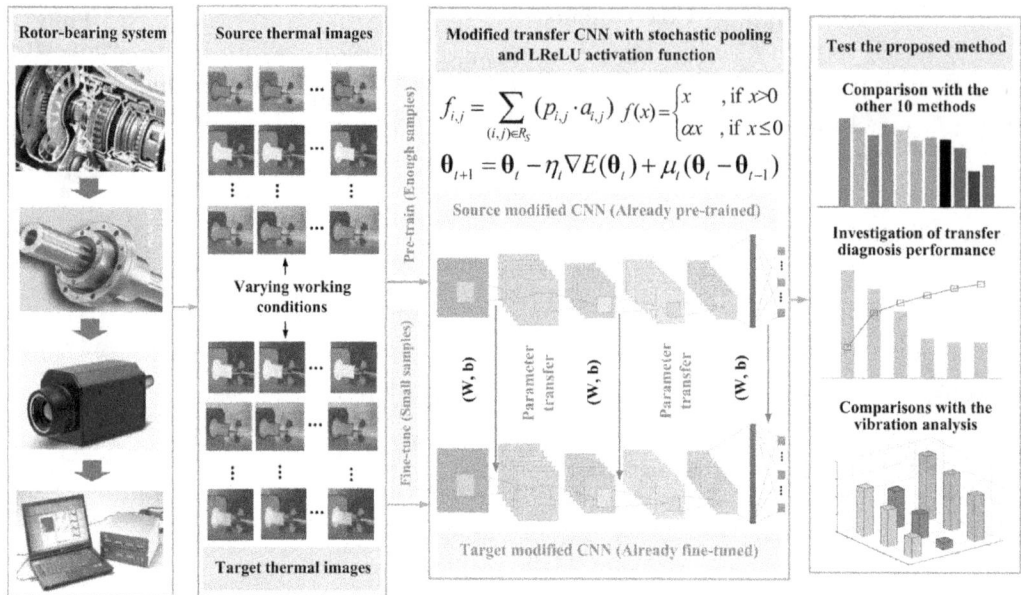

FIGURE 4.79   Overall framework of the proposed method.

Step 5: Fine-tune the target modified CNN using small samples from the target domain to further adjust the weights and biases.

Step 6: Test the diagnosis performance of the fine-tuned target modified CNN using the rest samples from the target domain.

4.3.4.3 Experimental Validation

4.3.4.3.1 Thermal Images of Rotor-Bearing System   In this chapter, the experimental data is from the GUNT PT500 rotating machinery shown in Figure 4.80, which mainly consists of driven motor, rotor shaft, tested bearing, transmission belt, and thermal camera (FLIR A5 with a resolution of 320×156 pixels). The thermal camera is installed 40 cm away from the fixed rotor shaft.

In the experiments, eight health conditions of rotor-bearing system are created, as shown in Table 4.24. Three mass blocks (5g) are placed on a side of a disk fixed on the rotating rotor to simulate unbalanced cases, as shown in Figure 4.81a.

As listed in Table 4.25, the thermal images collected under 2,000 and 3,000 rpm are employed as the source and target domains, respectively. We acquire a total of 150 source samples and 104 target samples. Each target condition only has two training (fine-tuning) samples. A raw thermal image of condition 3 with its region of interest is shown in Figure 4.81b. According to the restricted region, the collected thermal image samples of four health conditions (conditions 3, 4, 7, and 8) from the source and target domains are plotted in Figure 4.82. To meet the 2-D input forms of CNN, all of these thermal images are first translated into the grayscale images with sizes of 104×104.

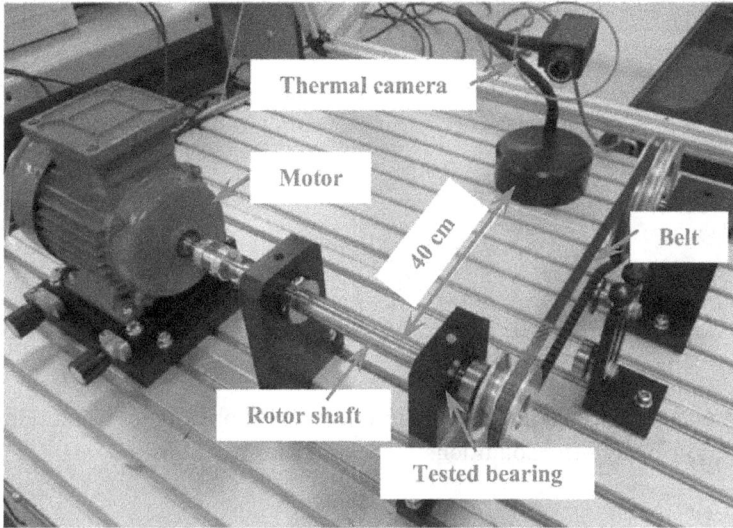

FIGURE 4.80   Diagnosis system of GUNT PT500 rotating machinery.

TABLE 4.24   Eight Health Conditions of Rotor-Bearing System

Health Conditions of Rotor-Bearing System	Labels of Conditions
Normal bearing and normal rotor	Condition 1 (C1)
Outer race fault of bearing and normal rotor	Condition 2 (C2)
Inner race fault of bearing and normal rotor	Condition 3 (C3)
Ball fault of bearing and normal rotor	Condition 4 (C4)
Normal bearing and unbalanced rotor	Condition 5 (C5)
Outer race fault of bearing and unbalanced rotor	Condition 6 (C6)
Inner race fault of bearing and unbalanced rotor	Condition 7 (C7)
Ball fault of bearing and unbalanced rotor	Condition 8 (C8)

FIGURE 4.81   (a) Disc fixed on the rotating rotor with three mass blocks. (b) Raw thermal image of condition 3 under steady operation of 2000 rpm.

4.3.4.3.2 Comparisons with Other Methods   In order to prove the feasibility of the modifications in the proposed method, several currently popular models are utilized for comparisons, including another eight types of CNNs built with different pooling strategies and activation functions, as well as two unsupervised deep learning models, i.e., DBN and SAE.

TABLE 4.25 Detailed Information of the Source and Target Datasets

Datasets of the Thermal Images	Rotating Speeds	Health Conditions	Sizes of the Training/Testing Samples
Source domain (S)	2,000 rpm	Conditions 1–8	100*8/50*8
Target domain (T)	3,000 rpm	Conditions 1–8	2*8/102*8

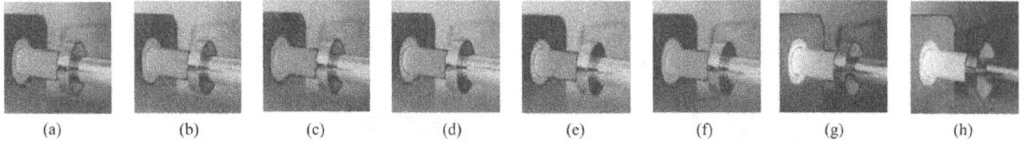

(a) (b) (c) (d) (e) (f) (g) (h)

FIGURE 4.82 Collected thermal images of four health conditions. (a)–(d) Conditions 3, 4, 7, and 8 from the source domain. (e)–(h) Conditions 3, 4, 7, and 8 from the target domain.

TABLE 4.26 Transfer Fault Diagnosis Results of Different Methods

Diagnosis Methods	Diagnosis Strategies	Average Diagnosis Result
Method 1 (CNN: LReLu & SP, Proposed method)	Supervised learning and parameter transfer	95.55% (7797/8160)
Method 2 (CNN: LReLu &MP)	Supervised learning and parameter transfer	92.52% (7550/8160)
Method 3 (CNN: LReLu & AP)	Supervised learning and parameter transfer	90.17% (7358/8160)
Method 4 (CNN: ReLU & SP)	Supervised learning and parameter transfer	9.3.66% (7643/8160j
Method 5 (CNN: ReLU & MP)	Supervised learning and parameter transfer	91.48% (7465/8160)
Method 6 (CNN: ReLu & AP)	Supervised learning and parameter transfer	88.93% (7257/8160)
Method 7 (CNN: Sigm & SP)	Supervised learning and parameter transfer	90.12% (7354/8160)
Method 8 (CNN: Sigm &MP)	Supervised learning and parameter transfer	88.95% (7258/8160)
Method 9 (CNN: Sigm & AP)	Supervised learning and parameter transfer	86.83% (7085/8160)
Method 10 (Basic DBN)	Unsupervised learning and parameter transfer	80.36% (6557/8160)
Method 11 (Basic SAE)	Unsupervised learning and parameter transfer	81.68% (6665/8160

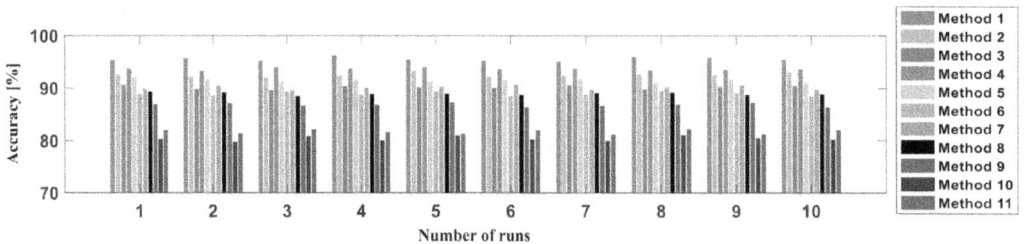

FIGURE 4.83 Detailed transfer diagnosis results of different methods.

The diagnosis goals of all the methods are to classify the testing samples in the target domain using the deep models pretrained by the source-domain samples.

Each method is performed for ten repeated runs to avoid contingency and particularity. The transfer fault diagnosis results of various methods are recorded in Table 4.26, and the detailed information is available in Figure 4.83. Through the statistical analysis, the max and mean values of the diagnosis results given by the proposed method are 96.20% (785/816) and 95.55% (7,797/8,160, 8,160=102×8×10), respectively. The average accuracies based on the ten contrastive methods are 92.52%, 90.17%, 93.66%, 91.48%, 88.93%, 90.12%, 88.95%, 86.83%, 80.36%, and 81.68%, respectively, and they are all lower than the proposed method.

From Figure 4.83, the best diagnosis result of the proposed method occurs at the fourth run with corresponding confusion matrix shown in Figure 4.84. In Figure 4.84, the horizontal x-axis and the vertical y-axis refer to the predicted label and the true label, respectively, and the diagonal elements are the accuracies of each individual state. The precision rate, recall rate, and F-score of the proposed method for the fourth run are calculated in Figure 4.85 according to the following formulas:

$$Precision = \frac{TP}{TP+FP} \tag{4.83}$$

$$Recall = \frac{TP}{TP+FN} \tag{4.84}$$

$$F\text{-score} = \frac{2\times Precision \times Recall}{Precision + Recall} \tag{4.85}$$

in which TP, FP, and FN denote the sizes of the true positive, false positive, and false negative samples, respectively. Every individual condition has a corresponding F-score, and a higher F-score represents better classification performance. It can be found from Figures 4.84 and 4.85 that except conditions 2 and 3, the F-score values provided by the proposed method for other conditions are very high.

According to our experience, two convolutional layers and two pooling layers are deep enough in this case study. To strike a good balance between testing accuracy and

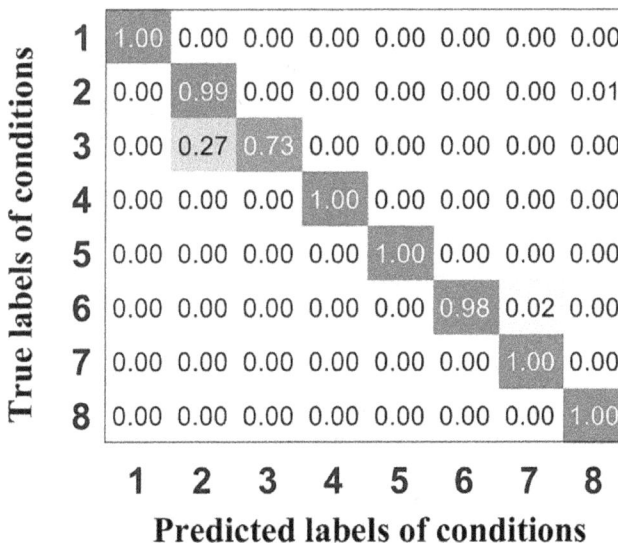

FIGURE 4.84 Confusion matrix of the proposed method for the fourth run.

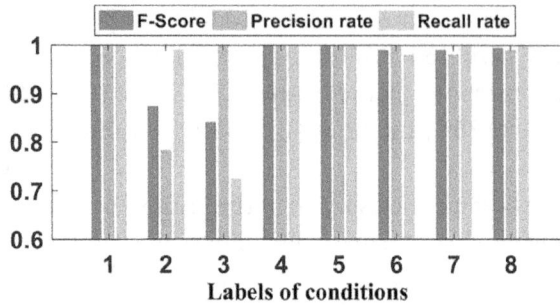

FIGURE 4.85   Precision rate, recall rate, and *F*-score of the proposed method for the fourth run.

TABLE 4.27   Structure Setup of the Proposed CNN Model

Parameters	Descriptions
Input feature map	104×104
Convolution kernels in layer 1	$(5 \times 5) \times 6$
Activation function in layer 1	LReLU
Stochastic pooling in layer 2	$(2 \times 2) \times 6$
Convolution kernels in layer 3	$(5 \times 5) \times 12$
Activation function in layer 3	LReLU
Stochastic pooling in layer 4	$(2 \times 2) \times 12$
Softmax layer	8

computing time, the model architectures of the constructed CNN are given in Table 4.27. The other hyperparameters are selected as follows:

The initial learning rate is 0.008, the initial momentum is 0.5, the final momentum is 0.95, the epoch number is 500, the boundary epoch is 8, the decay factor is 5, and the small positive value of LReLU is 0.004, which are mainly determined according to experiences and experimentations. The number of fine-tuning samples from the target domain is set as 2. The relationships between average diagnosis accuracies, standard deviations, and the numbers of fine-tuning samples are investigated in Figure 4.86. It can be observed that the diagnosis accuracies and standard deviations show, respectively, steady upward and downward trends as the number of target training samples becomes larger.

4.3.4.3.3 Superiority of Infrared Thermal Images Analysis   The superiority of thermal images over vibration analysis is investigated in this section. The piezo-electric accelerometer is placed on the side of the support with a sampling frequency of 32,768 Hz. Similar to Part C-(1), working conditions of 2,000 and 3,000 rpm are used as the source and target domains. Data samples of each health condition constructed by the raw vibration signals and time-frequency signals are shown in Figures 4.87 and 4.88, respectively. Each raw data sample refers to a sequence containing 1,024 (32×32) data points without overlap.

The numbers of the source-domain training samples for each health condition are set as 50 (50S), 80 (80S), and 100 (100S), respectively, while the samples sizes in the target

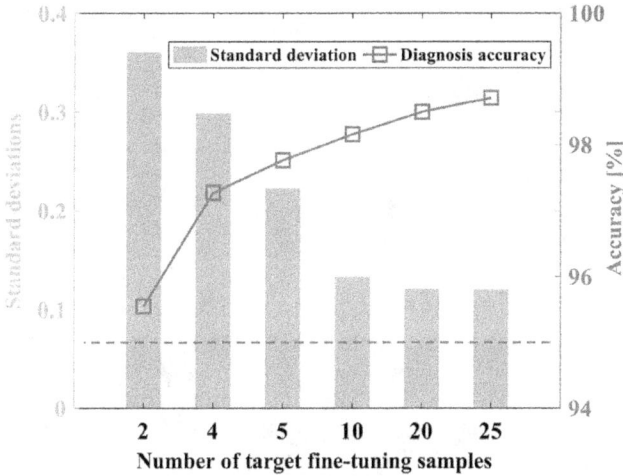

FIGURE 4.86 Transfer diagnosis performance under different fine-tuning samples from the target domain.

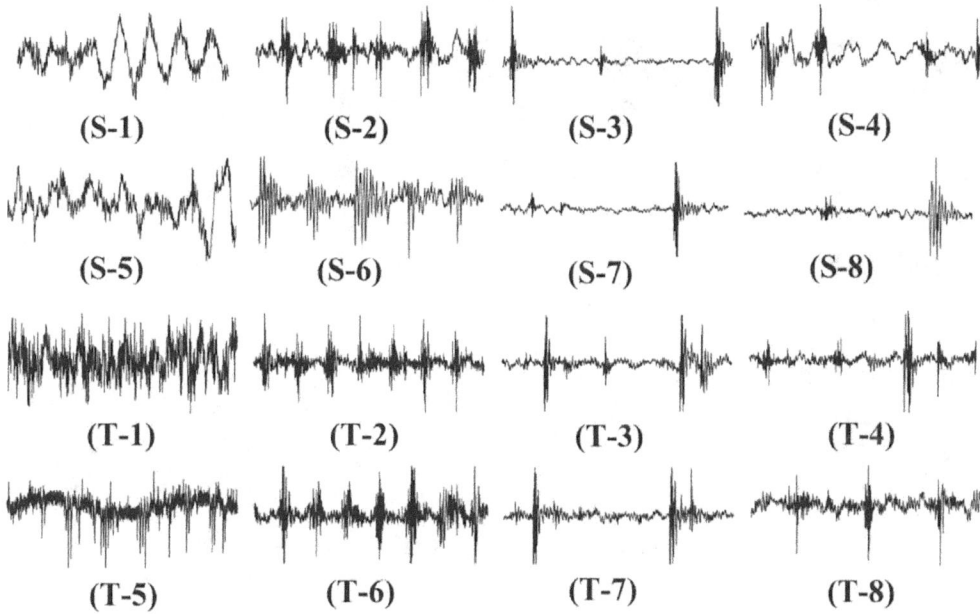

FIGURE 4.87 Data samples of each health condition constructed by raw vibration signals (S, Source domain; T, Target domain).

domain are the same as in Part C-(1). Based on the transfer diagnosis during the ten repeated times, the average diagnosis accuracies of the proposed method for three kinds of inputs are listed in Table 4.28.

Figure 4.89 shows the best accuracies using different inputs during the ten repeated times. From Table 4.28, two main conclusions are drawn as follows: (1) With the increase

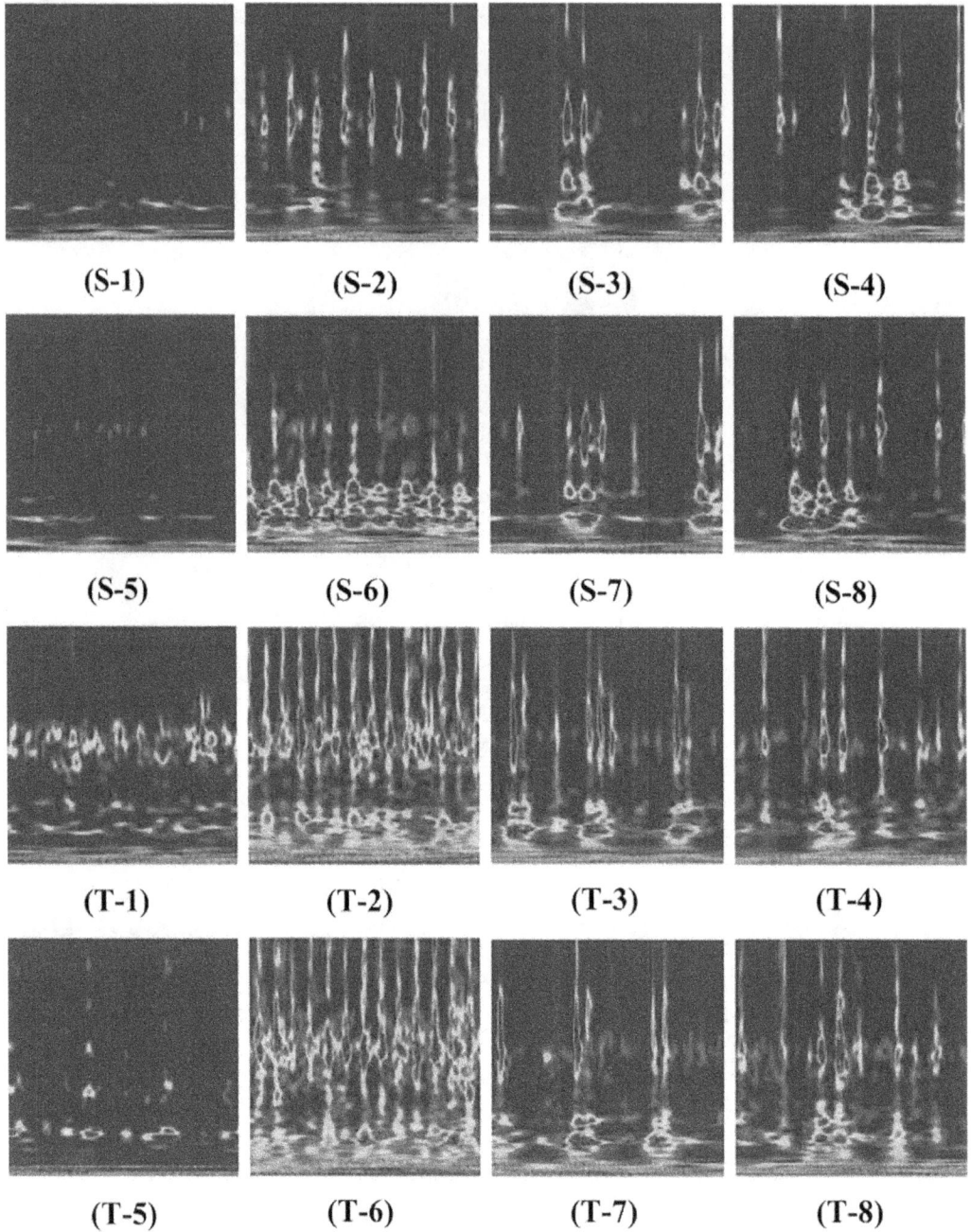

FIGURE 4.88 Data samples of each health condition constructed by time-frequency signals (S, Source domain; T, Target domain).

of the numbers of the source-domain training samples, all of the accuracies based on different inputs become higher. (2) The accuracies using infrared thermal images (Input 1) are always higher than the other two inputs (Input 2 and Input 3).

TABLE 4.28   Average Transfer Diagnosis Accuracies of the Proposed
Method for Three Kinds of Inputs

Different Inputs of the Proposed	Sizes of the Source Training Samples		
CNN Model	5o (50S)	75(75S)	100 (100S)
Infrared thermal images (Input l)	85.18%	91.53%	95.55%
Raw vibration signals (Input 2)	62.11%	72.93%	79.04%
Time-frequency images (Input 3)	70.96%	80.26%	86.18%

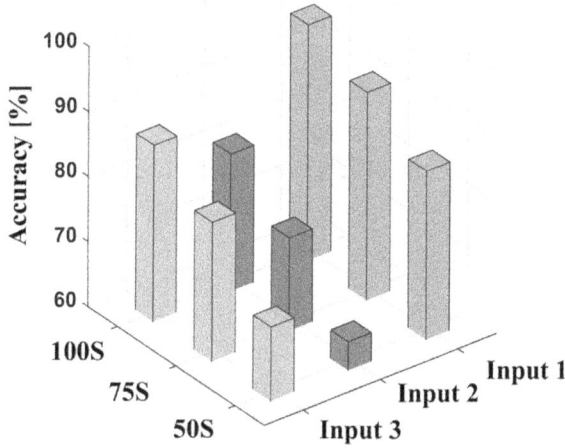

FIGURE 4.89   Best accuracies using different inputs during the ten repeated times.

4.3.4.3.4 Limitations of the Proposed Method   Despite the proposed method showing better performance for transfer fault diagnosis of rotor-bearing system than the basic CNN and vibration analysis, we have to admit some limitations during the experiments.

1. **Fluctuation problem**: From the detailed results shown in Figure 4.83, it can be found that the proposed method provides different results for each run. Though the average and highest accuracies of the proposed method are both larger than others, the standard deviation is not. The randomness of the stochastic pooling and particularity of the two fine-tuned samples randomly selected from the target domain are the main reasons for fluctuation problems.

   With the increase of the fine-tuned samples from the target domain, the standard deviations will become small to a great degree, as shown in Figure 4.86.

2. **Hyperparameter selection**: Although the basic CNN is modified by using some skill, more hyperparameters have to be predetermined simultaneously, including the initial learning rate, initial momentum, final momentum, boundary epoch, and decay factor. In this chapter, these hyperparameters are all selected by experimentations.

   Hyperparameter selection is an inherent problem in designing different deep learning models, which will more or less be solved with the accumulation of experience and knowledge.

3. **Region of interest for thermal images**: From Figure 4.81b, it can be observed that the raw thermal images always contain heavy background noise, i.e., the unrelated information, which will largely affect diagnosis performance. Advanced image segmentation techniques can select high-quality regions of interest to focus on fault characteristics of rotor-bearing system. However, it is carried out manually in this chapter.

4.4 FAULT PREDICTION AND INTELLIGENT MAINTENANCE UNDER INDUSTRY 4.0

An intelligent production line of industry 4.0 needs to meet the requirements of multi-variety, small batch, and personalized customization production modes. Meanwhile, it puts forward stringent requirements for the safety, reliability, and maintainability of the manufacturing equipment. However, the use of intelligent equipment will inevitably lead to its performance degradation. The predictive maintenance, as an important part in manufacturing maintenance, plays a vital role in scheduling, maintenance management, and quality improvement.

In this section, we introduce the methods of fault prediction and intelligent maintenance in combination with big data, digital twin, and cloud technology under industry 4.0 [62].

4.4.1 Equipment Electrocardiogram Mechanism for Condition Monitoring and Remaining Life Prediction

With the help of a mature data acquisition system, as shown in Figure 4.90, the visualization of the production process contributes to making the production process more transparent.

Inspired by the human electrocardiogram (ECG), we propose using an equipment ECG, called EECG, to determine performance degradation rules for manufacturing equipment.

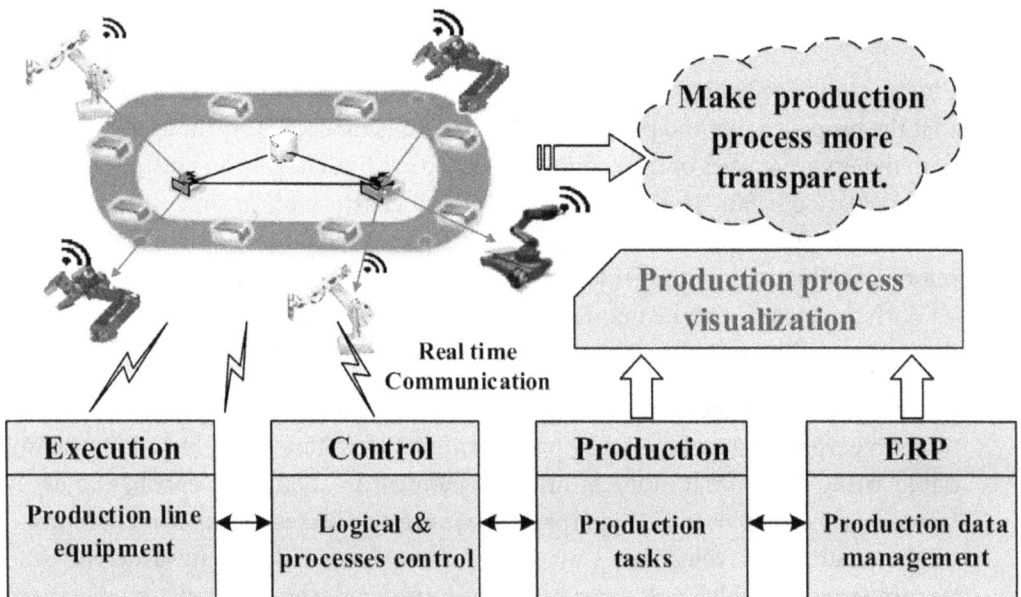

FIGURE 4.90 Process visualization system model.

As the equipment operates over fine-grained time slots, the goal of using the EECG is to monitor the performance degradation and optimize the cycle time of the production line. The EECG can identify potential issues, assist the intelligent production line in achieving preventive maintenance before an equipment fault actually occurs, and eventually form a closed loop of the "monitoring–diagnosis–maintenance" strategy [112].

4.4.1.1 System Architecture

The real-time acquisition of dynamic data information enables the EECG to monitor the operation status of an equipment and provide the time-series prediction of the equipment status. Figure 4.91 shows the system architecture of the EECG for a mixed flow manufacturing production line in a discrete workshop. As shown in the figure, the architecture has three layers. First, the workshop automation network is utilized for data collection from the underlying equipment, including high-speed data collection by using programmable logic controller (PLC) and OPC UA-based data unified transmission.

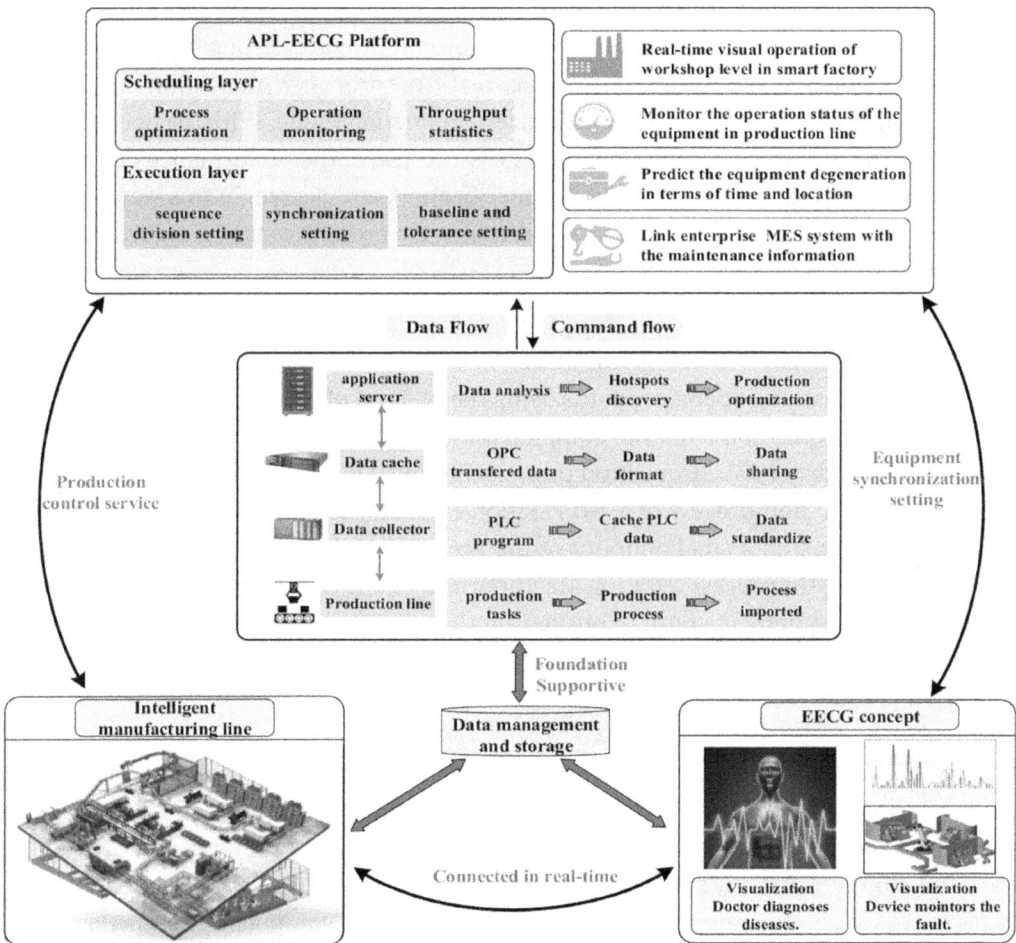

FIGURE 4.91   System architecture of EECG in smart factory.

Second, the data server is used for format management, data caching, and data sharing. Last, the application server is deployed to provide the continuum analysis of the data, explore the hotspots, and further improve the production cycle. The EECG represents an intuitive visual display of data, which contributes to providing an in-depth perception of the equipment status. To build the APL-EECG, it is crucial to develop novel techniques based on the following aspects:

1. scientifically divide the granularity of the manufacturing process;

2. choose the online matching algorithm for time-series process sequences;

3. set the baseline for a single equipment to complete a process or a task;

4. determine the tolerance degree for the fluctuation of the operation time.

The platform of the APL-EECG includes scheduling and execution layers, which are used for setting the parameters and optimizing the production, respectively. The EECG can clearly show the duration during the operation. Monitoring the operation status reflects the degradation trend of the equipment performance. It provides strong support for a new mode fault prediction. The APL-EECG realizes real-time process visualization in workshop and can be applied to a personalized customized production mode. With the operation status monitoring of the production line, the EECG can reveal the performance degradation and predict the evolution rule in terms of time, location, and degree. In this way, when the EECG is connected to the PLM or MES, it can provide preventive maintenance guideline for the workshop equipment.

4.4.1.2 Equipment Electrocardiogram Mechanism

The EECG monitors the operation status using the method of fine-grained processes visualization. Data acquisition from the equipment forms the basis of the EECG. Figure 4.92 shows the flowchart of data acquisition processes in the APL-EECG. In discrete manufacturing, the production process of a workstation requires the equipment to complete the production task just in a small amount of the production time. The granularity division and fluctuation of the process duration are important EECG characteristics that can reflect a change in the equipment status. Considering these factors, this section mainly focuses on the EECG mechanism including the multigranularity division method, the matching strategy, and other work characteristics.

4.4.1.2.1 Granular Division of Process Duration   The time length of the operation duration forms the basis for building the EECG. The product manufacturing process is often subdivided into multiple processing tasks. There may be many processing tasks between the start and stop of a station, and each processing task can be further subdivided into one or more sequences. The raw data collected by PLC or by the equipment itself is the operation duration with time-series feature. Rough set theory has the formal definition for the information system (IS). When it comes to the data collection based on multigranularity,

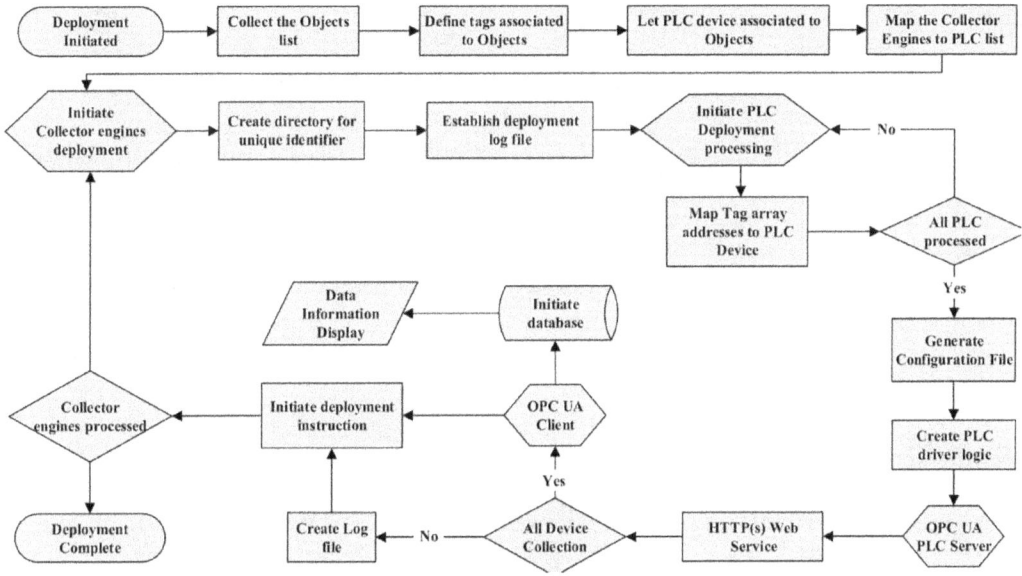

FIGURE 4.92   Flowchart of data acquisition in APL-EECG.

we were inspired by the rough set theory to define the EECG information system. In this section, we introduce the multigranularity division method based on the rough set theory.

Definition 4.1

EECG IS for a single equipment is represented by a four-tuple set $\text{GRS} = (U, P, V, f)$ and the IS of the APL-EECG can be described as

$$\text{MGRS} = \left\{ \text{GRS}_i \mid \text{GRS}_i = \left(U, P_i, \left\{ (V_p) \, p \in ATi, f_i \right\} \right) \right\}.$$

In this system, $U = \{e_1, e_2, e_3, \ldots, e_n\}$ denotes the nonempty finite object set of the equipment function, and it is also known as universe of the rough set. The process set denoted by $P = \{p_1, p_2, p_3, \ldots, p_m\}$ can be completed by the equipment function, while V_p denotes the cycle length of the process with $V = \cup_{p \in P} V_p$. The function $f : U \times P \to V$ is a mapping information function, and $f(e, p) \in V_p$.

Definition 4.2

$$\exists p_i(t) \in P, \text{ and } \forall e \in U \text{ constitutes } U = \left\{ e_i \mid e_i(t) = \sum_{0 \leq i \leq m} \left(\sim \hat{p}_i(t) \right) \right\}$$

The cycle length of the operation duration for the ith process is denoted by $p_i(t)$. Based on the cycle length, the entire process is segmented, and the function of the equipment is divided into sequences of operations. The completion time of the equipment function depends on the total number of operation sequences.

Definition 4.3

Define that $f = \{f_i \mid f_i : U \times P_i \to V_{p_i}, \forall i \in \text{div}(T)\}$ is a mapping function and the process set P belongs to the function set U. The set $e = \left(p_1, f(e, p_1)\right), \left(p_2, f(e, p_2)\right), \ldots, \left(p_n, f(e, p_n)\right)$ shows that the total time consumed by a function is composed of times consumed by different operations.

The operation and nonoperation durations are divided according to the functional characteristics of the equipment. For noncontinuous operation equipment, such as the empty stroke of the equipment, the change of time length can be effectively captured, and the signal will not be lost when the network transmission channel is highly utilized.

Definition 4.4

Assume that the total time consumed by an equipment during a manufacturing process is $X_{\text{total}} = \sum_{i=1}^{N} \sum_{j=1}^{M} f(x_{ij})$ and the total operation duration is $E_p = \sum_{i=1}^{N} f(P_i)$, and then the efficiency of the granularity division of the equipment process can be represented by $\mu_p = \ln\left(\dfrac{E_p}{X_{\text{total}} - E_p}\right)$. The channel utilization of the transmission network can be represented by $\mu_i = \dfrac{1}{MN} \sum_{i=1}^{N} \sum_{j=1}^{M} x_{ij}$.

In the preceding expressions, x_{ij} is the node of the operation division, and μ_p and μ_i are used to measure the granularity efficiency. The effective operation duration should be as long as possible according to the functional characteristics, so that $\mu_p > 0$. To avoid increase in the network load, the action interval of the equipment completing the process is ignored after the process is divided, such as the reset action of the manipulator during assembly and the empty stroke of the welding manipulator during welding.

The division of process duration is a complicated step. Definitions 4.1–4.3 give a general method to represent the structure of the EECG information, and Definition 4.4 gives a method to divide the process based on the functional characteristics of the equipment. The process division provides basic support for the determination of the subsequent cycle length.

4.4.1.2.2 Matching Strategy for Process Sequences   The EECG system accurately reflects the process progress in real time and meets the time critical requirements for anomaly detection. For an intelligent production line with variable mass customization, the product variety is changing in the running of the production line. The process sequences for different products will change in real time. For the certain production categories, the process sequences for the reference templates are collected in advance, and the reference templates are randomly selected from the collected process sequences. When the reference templates are known beforehand, the desired process will be matched with the reference templates online.

To identify templates of different processes, dynamic matching of the process template and real-time sequence of the time-series signal needs to be carried out in the EECG system. Inspired by the distortion measure in audio matching, we know that DTW is applied to measure the similarity of two voice sequences in which Euclidean distance is used to calculate the distance between elements of sequences. For the intelligent production line, the process flow is more complex. The elements of the matching sequence include station sequence and its process sequence. We use the matrix to represent the 2-D matching sequence. It is difficult to realize 2-D sequence matching by using Euclidean distance in the classical DTW method. Considering these factors, this section introduces a vector improved DTW algorithm for developing the matching strategy. It exploits the time-series characteristics of the manufacturing process.

The DTW is used to compare the similarity of two sequences. The main idea behind this algorithm is to provide the optimal path for nonlinear alignment by minimizing the accumulated distance between the two similar sequences. Assume that Q is the test template and C is the reference template, as shown in the following:

$$\begin{cases} Q = q_1, q_2, \ldots, q_i, \ldots, q_n \\ C = c_1, c_2, \ldots, c_j, \ldots, c_m \end{cases} \tag{4.86}$$

DTW deals with the time-series sequences correspondence problem between Q and C by using the time normalization function, as shown in the following:

$$d_{i,j} = \left(x(i) - y(j) \right)^2 \tag{4.87}$$

$$D = \min_{C} \frac{\sum_{n=1}^{K} \left[d\left(x_{i(n)}, y_{j(n)} \right) \cdot w_n \right]}{\sum_{n=1}^{K} w_n} \tag{4.88}$$

In the above-mentioned equation, $W = \left(w_1, w_2, \ldots, w_k, \ldots, w_K \right)$ denotes the warping path of the DTW. The minimum distance between the test template and the reference template is given as D. It refers to the similarity of the two sequences.

The time-series sequences of the EECG system meet the preconditions of sequence matching by DTW, but these sequences have their particular features in production processes. Assume that a work piece in the whole production line is completed by S workstations and each workstation is composed of one or at most N processes. The mapping function $f : U \times P \to V$ denotes the operation duration, and when the number of the processes is less than N, the margin will be replaced by a zero value. The whole process is represented by a matrix of size $S \times N$, as shown in the following:

$$FP = \begin{bmatrix} f\left(e_1, p_{(1,1)}\right) & f\left(e_1, p_{(1,2)}\right) & \cdots & f\left(e_1, p_{(1,n)}\right) \\ f\left(e_2, p_{(2,1)}\right) & f\left(e_2, p_{(2,2)}\right) & \cdots & f\left(e_2, p_{(2,n)}\right) \\ \vdots & \vdots & \ddots & \vdots \\ f\left(e_s, p_{(s,1)}\right) & f\left(e_s, p_{(s,2)}\right) & \cdots & f\left(e_s, p_{(s,n)}\right) \end{bmatrix} \tag{4.89}$$

Let $\gamma_i = \left(f\left(e_i, p_{(i,1)}\right), f\left(e_i, p_{(i,2)}\right), \ldots, f\left(e_i, p_{(i,n)}\right) \right)$ be the ith row of the above-mentioned matrix, and then (4) can be rewritten as $FP = [\gamma_1, \gamma_2, \ldots, \gamma_i, \ldots, \gamma_s]^T$. The production processes of the test template FP_{test} and the reference template FP_{refer} can be expressed as follows:

$$
\begin{aligned}
FP_{test} &= [\alpha_1, \alpha_2, \ldots, \alpha_i, \ldots, \alpha_n]^T \\
FP_{refer} &= [\beta_1, \beta_2, \ldots, \beta_j, \ldots, \beta_m]^T
\end{aligned}
\tag{4.90}
$$

From the above-mentioned equation, the estimated distance between two components of the test template and reference template is $d_{i,j} = \alpha_i - \beta_j^2$. It is defined as a component in the following:

$$
D_{i,j} = d_{i,j} + m \begin{cases} D_{i,j-1} \\ D_{i-1,j} \\ D_{i-1,j-1} \end{cases} \left(2 \le i \le M, 2 \le j \le N \right)
\tag{4.91}
$$

The warping path $\phi_{(\alpha,\beta)}$ for the distance accumulation is determined by α_i and β_j, $\phi_{(\alpha,\beta)}(1)=(1,1)$ and $\phi_{(\alpha,\beta)}(K)=(M,N)$. The step-length $\psi(k)$ is subject to the constraint $0 \le \psi(k) - \psi(k-1) \le 1$

The matching strategy for the process sequences can be summarized as follows. When the production sequences of the equipment are collected by the APL-EECG, the matching strategy is adopted for template identification. The distance between the test template and each reference template presented in the template library is calculated. The shortest distance indicates the highest similarity of the production sequence. The most similar template is chosen for real-time alignment of the time-series sequences.

Figure 4.93 (Algorithm 4.1) shows the vector improved DTW algorithm.

4.4.1.2.3 Important Operating Characteristics The matching strategy for the process sequences ensures the alignment of sequence in the acquired time-series data. The length rulers of the production sequences are needed to evaluate the acceptability of the operation duration. The baseline represents the expected or designed duration of each operation, and the tolerance represents the floating value of the baseline after the baseline is determined. When the baseline and tolerance are set up, the operation duration is graded into either of the four levels: good, warning, warning, and fault. The determination of the baseline and tolerance is closely related to the time length of the operation duration. This section presents the theoretical basis of these operating characteristics for the APL-EECG.

The baseline is used to demarcate the expected cycle length of the process, and it is also a base value to measure the normality of the equipment beat. We use the sampling statistics to calculate the base value of an operation duration. The root mean square (RMS) value of the sampling data is adapted as the base value. Equation (4.92) reflects the cycle length of

---

Algorithm 1: Vector Improved DTW Algorithm.

Input: $FP_{net} = [\alpha_1, \alpha_2, \ldots \alpha_i, \ldots \alpha_n]^T$

Input: reference template set: $\left(FP_{refer}^1, FP_{refer}^2, FP_{refer}^3, \ldots, FP_{refer}^l\right)$

Output: target reference template: FP_{refer}'

1   **Begin**
2   Initialization   *//initialized test template;*
3   for $i \leftarrow 1$ to l *//match every reference template;*
4      $[1; K] \leftarrow [1; M] \times [1; N]$
5      $\phi_{(\alpha, \beta)}(1) = (1,1); \phi_{(\alpha, \beta)}(K) = (M, N)$ *//alignment of sequence*
6      $ands; \phi_{(\alpha, \beta)}(k) = DWT(\alpha_x(k), \beta_y(k))$ *//calculate the weight;*
7   for $n \leftarrow 1$ to K
8        $d(\alpha_{x(n)}, \beta_{y(n)}) = \|\alpha_{x(n)} - \beta_{y(n)}\|^2$
9        $Sum(D_n) = Sum(D_{n-1}) + d(\alpha_{x(n)}, \beta_{y(n)}) \cdot w_n$ *// w_n is the weighting funication*
10      $Sum(W_n) = Sum(W_{n-1}) + w_n$
11   end for
12     $Dist(FP_{test}, RP_{refer}^i) = {Sum(D_n)}\big/{Sum(W_n)}$ *//calculate the similarity;*
13   if $Dist(FP_{test}, RP_{refer}^{i-1}) \geq Dist(FP_{test}, RP_{refer}^i)$
14   target reference template: $FP_{refer}' = FP^i$
15   return FP_{refer}'
16   end if
17   **End**

FIGURE 4.93   Algorithm 4.1: Vector improved DTW Algorithm.

the sampling process of a random Nth times. The cycle length of the ith sampling in the process j is given by $p_j^{(i)}$, and U_{BL} represents the base value of a process in EECG as

$$U_{BL} = \sqrt{\frac{\sum_{i=1}^{N}\left[f\left(e, p_j^{(i)}\right)\right]^2}{N}} \tag{4.92}$$

Tolerance is closely related to the operation duration. It is the benchmark for judging whether the equipment beat around the baseline is within a normal range. The modified sample variance is used to express the tolerance of the operation duration offset baseline, as shown in the following:

$$S_n^* = \sqrt{\frac{1}{n-1}\sum_{i=1}^{n}\left(f\left(e, p_j^{(i)}\right) - U_{BL}\right)^2} \tag{4.93}$$

In the above-mentioned equation, S_n^* represents the tolerance value for fluctuation of the operation duration. The operation duration is graded according to the tolerance value. A serious deviation of a normally operating equipment from the baseline is a small-probability event. According to the long-term sampling observations, the change of the operation duration follows a normal distribution. In this part, the 3σ principle is used to evaluate the fluctuation of operation duration and $\sigma \approx S_n^*$.

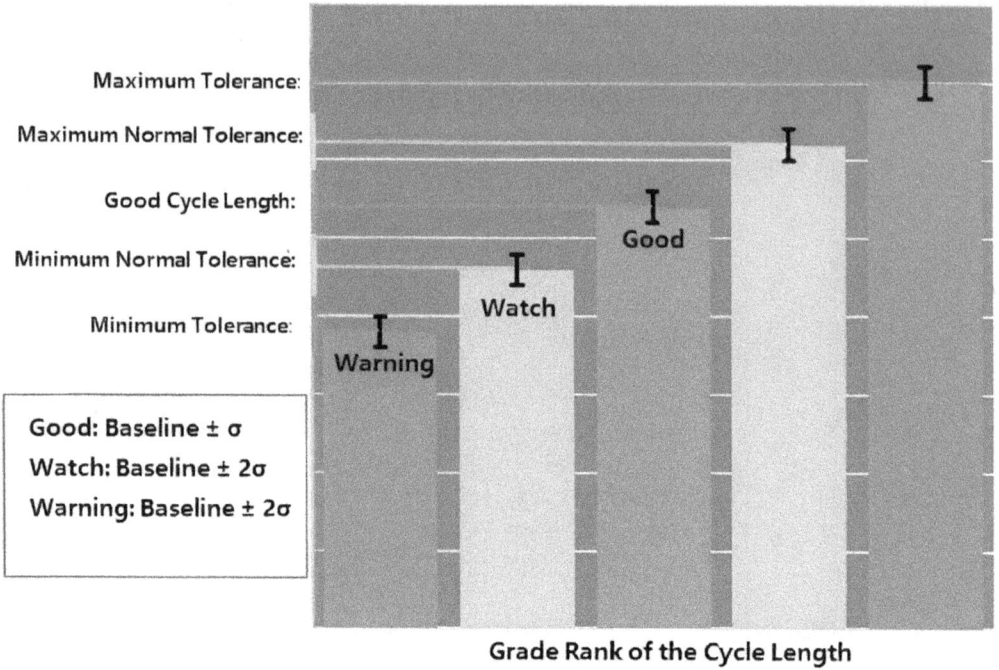

FIGURE 4.94   Grading of the cycle length based on tolerance.

As shown in Figure 4.94, the X-axis reflects the graded for adjusting the values of the cycle length. The Y-axis reflects the tolerance value based on the 3σ principle. The grade division of the X-axis is accompanied with the changing tolerance value of the Y-axis. When the operation duration is within the baseline tolerance, it will be marked as "good." The time length higher than the baseline tolerance will be marked as "watch," while that higher than the "nTol" value will be marked as "warning." When the operation duration is higher than the "Tol" value, it means that the equipment would malfunction.

In the APL-EECG system, the envelope is an important feature that reflects the floating range of the operation duration. It directly represents the extent to which the operation duration exceeds the baseline, and the value is calculated according to (9). In the following equation, U_{Evp} denotes the limiting value of the envelope:

$$
\begin{cases}
U_{\text{Evp}} = U_{\text{BL}} \pm m \sqrt{\sum_{m=1}^{k} \left(\dfrac{f\left(e, p_j^{(m)}\right) - f\left(P_j\right)}{f\left(P_j^{m}\right) - f\left(P_j^{(m)}\right)} \right)^2} \\
j \neq 1, j \in P_j
\end{cases}
\tag{4.94}
$$

Under random sampling carried out k times, $P_j = \left(p_1, p_2, p_3, \ldots, p_J \right)$ and $f\left(P_J\right)$ denotes the sampled value. In this work, the higher limit of the envelope is adopted.

4.4.1.3 Performance Monitoring and Production Optimization Using EECG

4.4.1.3.1 Automatic Production Line—EECG Optimization of Cycle Time   The cycle time mainly determines the production rate of the intelligent production line. The operation performance of the equipment and logical settings of the operation instructions significantly impact the cycle time. In the welding production line of a car plant, the factors that can delay the equipment beat include inefficient waiting area, unreasonable interference area, loosening of clamping equipment, fluctuation of welding manipulator, and aging of equipment sensors. The APL-EECG can effectively reflect the change in operation duration and then quickly identify the action steps to improve the operation cycle time.

Using the APL-EECG, the method of mathematical programming is adapted to optimize the cycle time. When the equipment executes an external task, the collection of the equipment beat can be regarded as the sequence flow of the process. The entire cycle time is composed of the work piece delivery time and the processing time. The operation duration will be reflected in the APL-EECG in a fine-grained fashion.

The mathematical optimization model of the cycle time in the APL-EECG is given in the following:

$$\min Z = \sum_{i=1}^{m}\sum_{j=1}^{m} x_{ij}\left(c_{ij}+o_{ij}\right)$$

$$\text{s.t.}\begin{cases} \sum_{i=1}^{m} x_{ij}=1, i=1,2,\ldots,m, \\ \sum_{j=1}^{m} x_{ij}=1, j=1,2,\ldots,m, \\ \sum_{i=1}^{s}\sum_{j=1}^{s} x_{ij}\leq s-1, 2\leq s\leq m-1, \\ s\subseteq\{1,2,\ldots,m\} \\ x_{ij}\in\{0,1\}, i,j=1,2,\ldots,m, i\neq j \end{cases} \tag{4.95}$$

The optimization target is to minimize the operation duration and choose the appropriate conveying path of the work piece. It also makes the task execution sequence more compact when building the APL-EECG. In equation (4.95), the production task set is $T=\{t_1,t_2,\ldots,t_m\}$, and the operation instructions depend on the type of this task set. The transmission path of the work piece between stations i and j is denoted by c_{ij}, which is limited by the conveying path of the actual production line. The time consumed on the station is given by o_{ij}, and x_{ij} denotes that task t_i will be executed after the execution of task t_j and also represents the direction in which the tasks continue. A global optimization algorithm used for solving (4.95) is given in Figure 4.95 (Algorithm 4.2).

4.4.1.3.2 Online Monitoring of Equipment Performance   Decay Healthy operation of intelligent equipment is a guarantee for the highly efficient production line. Traditional FIS has

---

Algorithm 2: Global Optimization Algorithm for Solving (10).

Input: key sequence value $x_{ij}^o \Rightarrow c_i \rightarrow c_j$;

Task Set $T \subseteq$ Command Set C

Output: the optimal solution $x_{ij}^* \Rightarrow c_i \rightarrow c_j$

1	**Begin**
2	Initialize the unknown sequence x_{ij}^o
3	for $k \leftarrow 1$ to K *//K denotes the alternative operation sequence*
3	select the feasible value $x_{ij}^{(k)} \in x_{ij}^o$
4	for $i \leftarrow 1$ to m *//upgoing command value*
5	for $j \leftarrow 1$ to m *//downgoing command value*
6	$x_{ij}^{(k)} \Rightarrow c_i \rightarrow c_j \parallel T \subseteq C^{(k)}$
7	$Z^{(k)} = Z^{(k)} + x_{ij}^{(k)}(c_{ij} + o_{ij})$ *//calculate the Z value*
8	end for
9	end for
10	If $(Z^{(k)} \leq Z^{(k-1)})$ then *//loop through to get the minimum*
11	$Z^{min} = Z^{(k)}$
12	end if
13	end for
14	return $x_{ij}^* \rightarrow Z^{min}$ *//return the optimized sequence*
15	**End**

FIGURE 4.95   Algorithm 4.2: Global optimization algorithm for solving (4.95).

a single early warning mode, which only records early warning events corresponding to a finite number of time nodes. Thus, it is highly likely to lose a number of important event alarms. The APL-EECG system can continuously monitor the equipment performance decay and help predict the potential failure trend of the equipment visually. Changes in the continuous state are judged by the boundary condition. Deviation from the normal excessively or repeatedly will be marked as hotspots.

In terms of future performance degradation of the intelligent equipment, the APL-EECG adopts a grading scheme based on the emergency level. Figure 4.96 reflects the online monitoring of equipment performance decay for several processes in the production line. The upper side of the horizontal axis shows the cycle length of a process, and the upper envelope gives the range of the cycle length that can be floated upward. The lower side of the horizontal axis shows the grading level used for the cycle length evaluation. Those processes that are continuously graded as "warning" will be also labeled as hotspots.

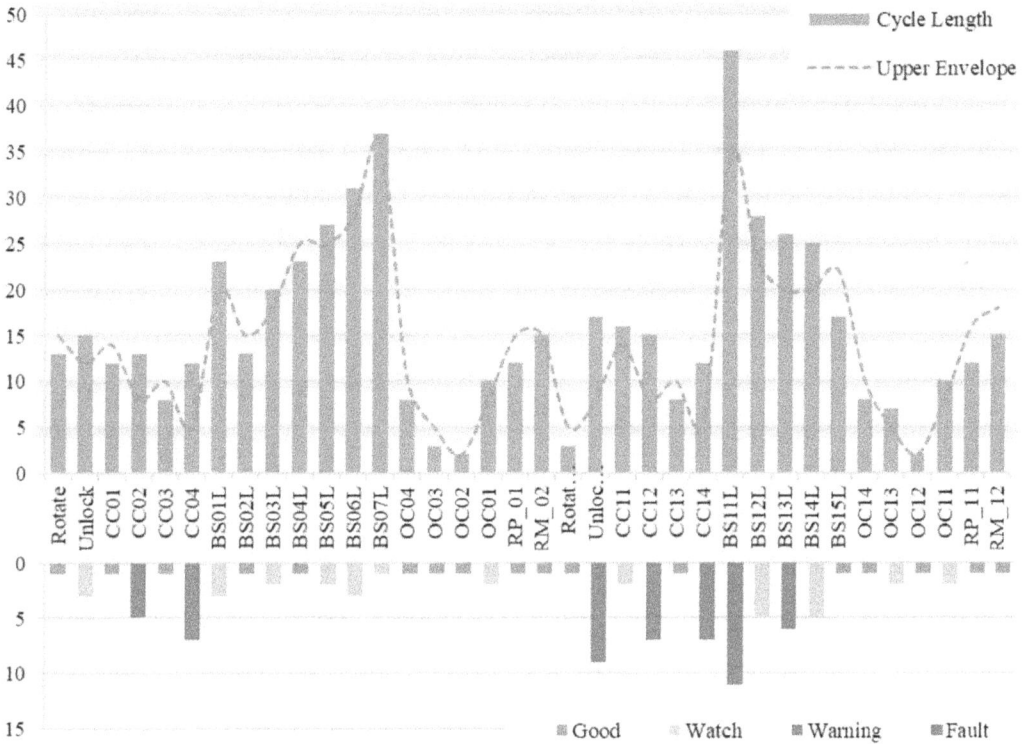

FIGURE 4.96 Online monitoring of equipment performance decay.

Hotspots are important references for developing effective preventive maintenance strategies for the equipment. During operation of the production line, the performance of the equipment decays continuously. The online alarm mode of the APL-EECG can directly reflect the increase of the operation duration and accurately locate the source of an upcoming future performance degradation.

4.4.1.4 Implementation of EECG
4.4.1.4.1 Implementation Scenarios   An intelligent production line is established in our intelligent manufacturing laboratory. It involves workpiece carving and assembly, and patterns laser marking. The production line is a typical small batch discrete production line. The layout of the production line in the laboratory is shown in Figure 4.97. The production stations are distributed discretely between the main line and the branch line, and RFID tags are used to store the operation information. The workpiece is transferred by the conveyor belts or AGV continuously.

In this section, the implementation platform of the APL-EECG is described, which was deployed based on our application scenario. As shown in Figure 4.95, every execution unit is deployed as the workstation of EECG, which can be divided into 13 sequences (SE) including nine manipulators, two CNCs, a marking machine, and an assembly machine. The inherent beat of each equipment is tested in advance because the operation times of different types of products vary greatly. For the testing purposes, the type of products

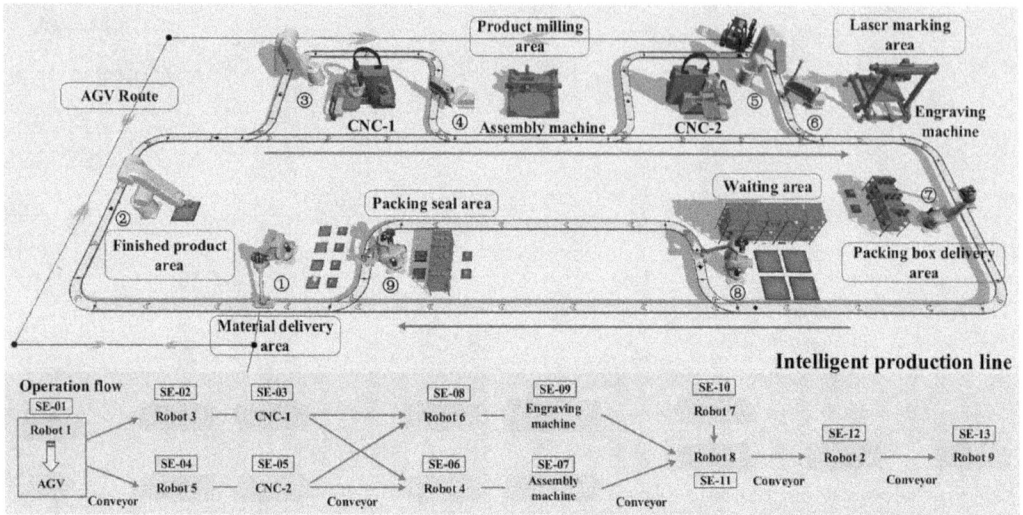

FIGURE 4.97   Implementation platform of the APL-EECG in the manufacturing laboratory

TABLE 4.29   Beat of the Main Workstations

Workstation	Minimum Beat	Average Beat	Beat Std
Robot 1	2.3	3.5	1.7
Robot 3	1.2	2.3	0.9
Robot 5	2.4	3.7	1.3
Robot 8	3.1	4.6	1.3
Robot 9	3.2	5.1	1.6
CNC-1	35.2	52.7	14.5
CNC-2	32.7	50.6	15.3
Assembly Machine	28.6	42.7	20.3

processed by CNC or engraving machine is determined. The beat of the main working station is listed in Table 4.29. The main validation schemes are listed as follows:

1. deployment of the APL-EECG;

2. improvement of the equipment beat;

3. performance degradation monitoring of the equipment; and

4. maintenance guidance for the production line.

4.4.1.4.2 Implementation Results   This section shows the performance of the APL-EECG implementation based on the existing prototype platform. Tests were carried out over the duration lasting more than 90 days.

Adjacent operation cycles showing large variation of the cycle length are selected to reflect the production process of two different products, as shown in Figure 4.98. The figure shows that the cycle length of the SE-07 increases continuously and exceeds

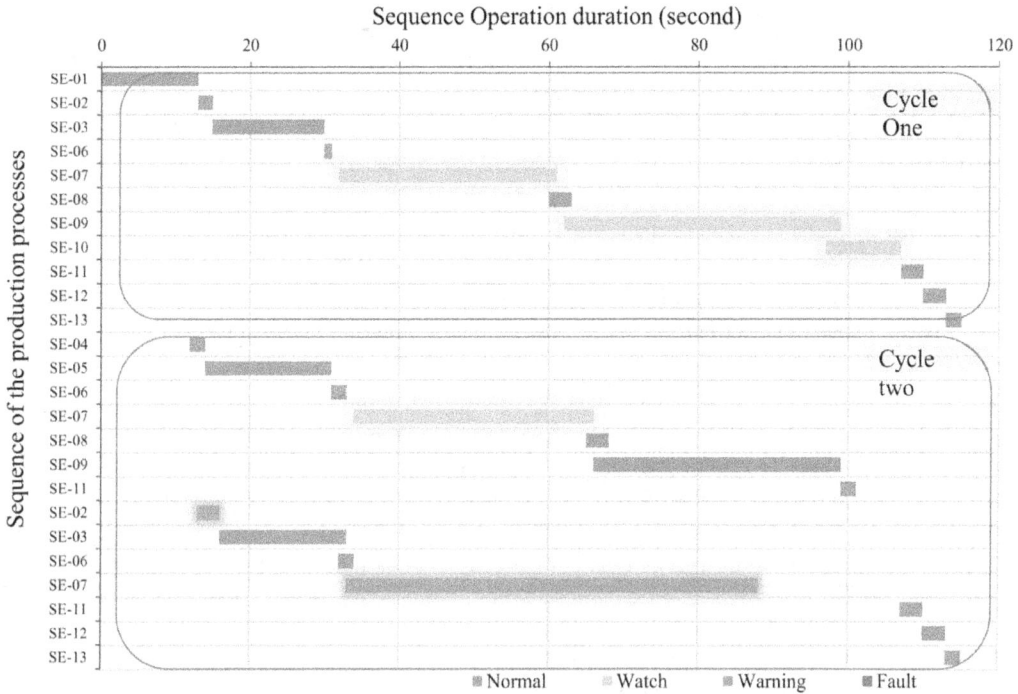

FIGURE 4.98   Experimental results obtained from the APL-EECG platform.

the minimum tolerance. Its grading information changes from "watch" to "warning," and consequently, its dynamic state will be displayed in the maintenance interface as a hotspot. The cycle length of SE-02 shows large fluctuations, which reflect the instability in its operation.

Table 4.30 lists parts of the alarm information of the APL-EECG. The table reflects the increase caused by aging equipment components in the operation duration. Based on the maintenance information of SE-03, its process changes from the "over cycle" and "blocked internally" to the final "blocked down" state. This demonstrates that the EECG platform can provide continuous monitoring information. Eventually, it was discovered that the "blocked down" state was caused by the damage to manipulator caliper. As the traditional FIS is based on the timed feedback or manual operation, it is difficult to guarantee all-round relevance of the equipment status. The FIS does not reflect the decay process of the equipment but only its final status. On the contrary, the APL-EECG can provide continuous alarm information according to the fluctuation of the cycle length. Therefore, the APL-EECG can help in developing a reasonable maintenance strategy to avoid sudden "blocked down" state of the equipment.

Based on Section 4.4.1.3.1, it can be seen that the deployment of the APL-EECG can provide the fine-grained operation status of the equipment and optimize the cycle time by using the mathematical programming methods. From the aspect of workstation, the delay can be reduced after the dwell time is reduced. In the early stage, the unreasonable

TABLE 4.30   Alarm Information of the APL-EECG

Time	Sequence	State	Duration	Status Alert	Hotspot	Assist
Tues. 15:07:13	SE-03	Over Cycle	46 seconds	Watch	No	No
Tues. 16:08:43	SE-02	Over Cycle	5 seconds	Watch	No	No
Wed. 09:12:55	SE-08	Over Cycle	7seconds	Watch	No	No
Wed. 11:23:15	SE-03	Internal Blocked	52 seconds	Warning	Yes	No
Wed. 15:09:20	SE-04	Over Cycle	5 seconds	Watch	No	No
Wed. 15:30:32	SE-01	Out of Auto	53 seconds	Warning	Yes	Charging
Wed. 17:27:08	SE-09	Over Cycle	72 seconds	Warning	Yes	Tool Change
Thur. 09:21:07	SE-11	Over Cycle	18seconds	Watch	No	No
Thur. 10:14:32	SE-02	Over Cycle	5 seconds	Watch	No	No
Thur. 15:07:23	SE-03	Internal Blocked	50 seconds	Warning	Yes	No
Thur. 16:20:17	SE-13	Over Cycle	8 seconds	Warning	Yes	No
Fri. 10:23:31	SE-01	Over Cycle	63 seconds	Watch	No	No
Fri. 13:35:12	SE-04	Internal Block	12 seconds	Warning	Yes	Replace valve
Fri. 15:42:55	SE-03	Blocked Down	127 seconds	Fault	Yes	Caliper replacement
Mon. 11:17:23	SE-13	Internal Blocked	17 seconds	Warning	Yes	Fixture Change

program setting can be detected intuitively on the EECG panel. Aging components of the equipment may increase dwell time during the operation.

For example, if loosening and clamping of a part are checked simultaneously, tightening of the part can be accelerated, and the time between tightening and loosening can be reduced.

From the aspect of mathematical planning, the information of workstation or manipulator that the action synchronization, program action, and interference area can be optimized will feedback to the operation and maintenance system. Figure 4.99 shows the optimization effect of the APL-EECG on the cycle time in our prototype platform. Figure 4.99a demonstrates that the average operation duration of the four processes is gradually decreasing. From Figure 4.97b, it is evident that the standard deviation of the four processes is gradually decreasing, which indicates increasing stability of the equipment. Figure 4.100 shows the comparison of the performance of the FIS and APL-EECG over the complete test duration.

During the 90 days test, the number of times "over cycle" and "blocked down" occur in the APL-EECG system is counted. Under a typical FIS, the monitoring of the production line would be limited to see when a cycle has started and stopped through the cycle time. APL-EECG monitors the time from the start to stop and monitors the actual underlying processes during the cycle. Compared with the FIS, Figure 4.100 shows that the APL-EECG can effectively reduce both "over cycle" and "blocked down" counts, and the results show that the APL-EECG can improve the continuity of the intelligent production line.

In the experimental results, we selected a wide range change of the cycle time to demonstrate the effect of implementing the APL-EECG. Practically, the cycle time of the production line is usually measured in seconds for automobile or packaging. Even a slight improvement of the equipment beat has the potential to increase the production line

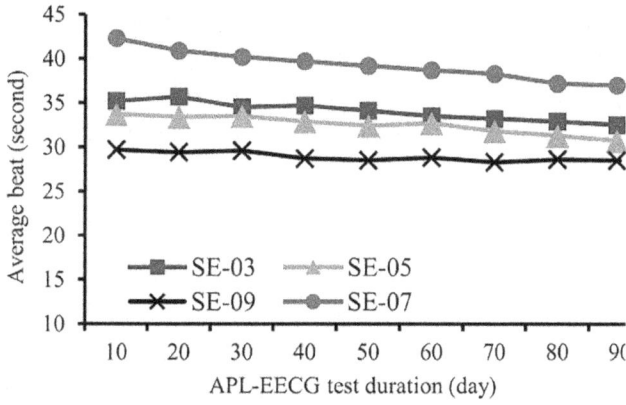

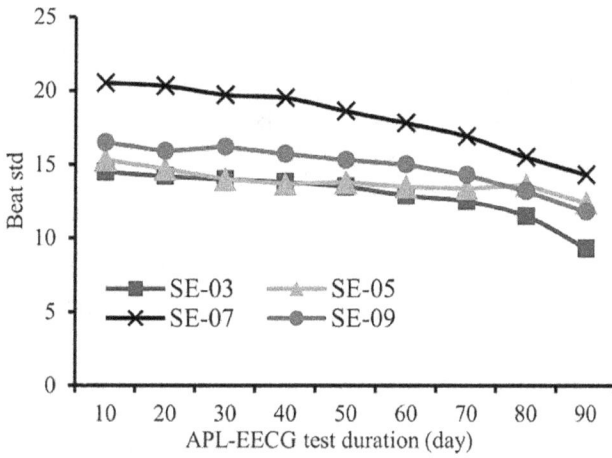

FIGURE 4.99   Optimization effect of the APL-EECG on cycle time: (a) average beat and (b) beat std.

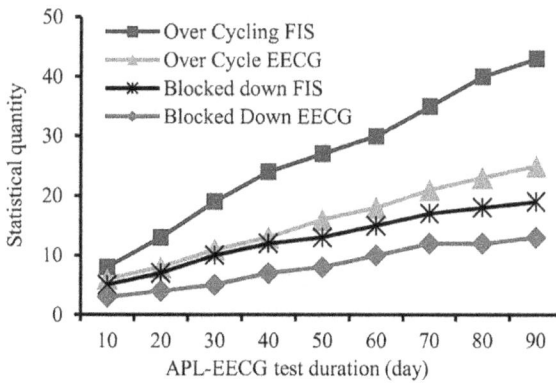

FIGURE 4.100   Comparison of the performance of the FIS and APL-EECG over the complete test duration.

efficiency and bring huge economic benefits for enterprises. The equipment beat can be optimized by reducing the dwell time or modifying its unreasonable procedure settings. Considering the long-life cycle of the equipment, the test on the equipment itself is not significant when most of the test data are normal. Because most intelligent equipment ages gradually, the APL-EECG can effectively monitor the operation status of the equipment in the long run and identify potential faults in the equipment.

4.4.2 Fault Prediction Method with Digital Twins and Machine Learning

Complex equipment, such as an aerospace vehicle, complex electromechanical equipment, and railway transportation equipment, is the important basic equipment in the manufacturing industry chain. Product life cycle management and service of complex equipment are important for the transformation and upgrading of the global manufacturing industry. The main objective is to establish a digital twin for complex equipment and manage the full life cycle of complex equipment by using a calculation model driven by both models and data. In this way, the current status of complex equipment could be better understood, monitored, tracked, and mastered. Therefore, it is necessary to provide required services based on information interaction feedback [113].

4.4.2.1 Proposed Digital Twin Technology Architecture

The establishment and application of the digital twin depend on the development and innovation of many technologies, such as perception control technology, data acquisition and transmission technology, massive data management and processing technology, and high-performance computing technology. The machine learning methods can be embedded into a digital twin technology system to establish two-way information connection and integration, where a two-way connection indicates a two-way relationship between the physical and virtual worlds and the digital twin of sub-models in the virtual space. This is helpful for the landing use and innovative applications of a digital twin. This paper presents a machine learning-based digital twin technology system for complex equipment. The proposed system is shown in Figure 4.101, where it can be seen that it consists of three layers, namely, data layer, edge-cloud computing layer, and service layer. It should be noted that there is a progressive relationship between the three layers. Starting from the edge-cloud computing layer, the construction of each layer is based on the previous layers, providing further enrichment and expansion of the previous layers' functions. The functions of the three layers are explained in detail in the following.

- **Data layer**

  The data layer represents the basis and underlying support of the entire digital twin technology system. It includes the data acquisition module and the data transmission module. The data needed for the operation of the whole digital twin system are provided by the data acquisition module, and these data include real-time state sensing data, historical operating data, and user interaction data. Data collection equipment (e.g., RFID tags and sensors) collect data from the physical site. High-performance sensors and distributed sensing technology ensure the accurate

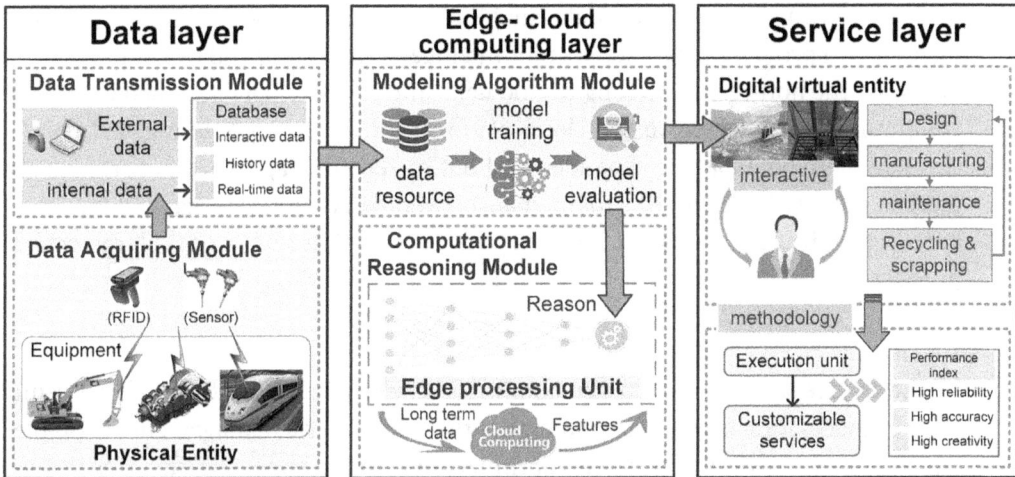

FIGURE 4.101   The ML-based digital twin technology system of complex equipment.

acquisition of real-time physical characteristics of the equipment. The historical operation data of a system include historical information on the design, manufacturing, service, and recycling of the same type systems. All historical information is filed and stored for subsequent usage. The user demand interaction data are obtained through users' interaction with physical products via active behaviors characterized by certain setting preference parameters and feedback on usage. These data reflect user experience and suggestions. All these data are transmitted to the upper system via the data transmission module to support the operation of the upper system.

The application of 5G and other advanced technologies shortens data transmission time and improves the communication service quality and real-time follow-up performance of a digital twin system. For example, the 5G communication technology has the advantages of high bandwidth, low delay, strong connection, and high reliability.

- **Edge-cloud computing layer**

    The realization of the assistant reality function of a digital twin depends on a powerful computing function. The computing layer is constructed based on edge-cloud collaboration, and it is composed of the modeling algorithm module and the computing reasoning module. After the modeling algorithm module obtains data provided by the data layer, it uses a data-driven method to establish a structured knowledge network data model with full links, multiple disciplines, and cross-domains. This layer combines intelligent learning and optimization algorithms such as machine learning-based algorithms to establish data mining and knowledge discovery models. The computing reasoning module uses a super-realistic data representation model to perform knowledge learning and reasoning and provides the multi-angle decision-making results and analysis conclusions needed for the realization of the functional layer of the digital twin.

The real-time, short-period data analysis task is submerged in complex equipment, which provides better support for intelligent processing and execution of local business. The non-real-time, long-period data, and effective data obtained from edge computing are uploaded to the cloud platform for further processing and feature extraction. The database is expanded and optimized. This approach ensures advantages in areas such as periodic maintenance, overall decision-making, and feedback-based optimization of virtual "small cycles" production and actual "large cycles" production.

- **Service layer**

    The service layer represents a customizable service window that can provide functional services for variable requirements of the full life cycle management of complex equipment. Considering different optimization perspectives, the functional layer provides services directly to reality based on the reasoning results of the computing layer. The service meets requirements for multiple performance indicators, including high reliability, high accuracy, high real-time, and high-value creativity. This is the embodiment of the core value of the digital twin system.

    Complex equipment has the characteristics of personalization and customization, and its digital twin serves manufacturers and users. The service layer is based on the massive, multi-dimensional data of the data layer and the powerful information interface of the computing layer. It can customize value-added business functions according to the actual needs and can also improve the intelligence level and user experience of manufacturing enterprises.

4.4.2.2 Digital Twin Architecture for Complex Equipment

One of the primary objectives of a digital twin is to integrate cyberspace and physical space, which is to apply the digital twin to the life cycle management of complex equipment and understand, analyze, monitor, and optimize the state or behavior of physical equipment. This can be realized through virtual interactivity and feedback, data mining and analysis, and multi-source data association and fusion, and this approach can satisfy the requirements of complex modern equipment manufacturing. A product digital twin of complex equipment is constructed to perform different functions, including geometric perception, behavior simulation, and process recovery of physical equipment and its architecture, as shown in Figure 4.102. The proposed architecture establishes a full life cycle data center for complex equipment and uses a single data source to realize the bidirectional connection between physical and information spaces and data communication between all stages of the full life cycle.

The digital twin sub-models of design, manufacturing, maintenance, recycling, and scrapping are integrated into the digital twin of complex equipment. These models provide necessary model support for the life cycle management of complex equipment. These four sub-models complement each other and evolve together; they use the analytic hierarchy process (AHP), Bayesian estimation, expert system, D-S reasoning method, artificial intelligence, and other related methods to analyze and utilize data sources. This provides the

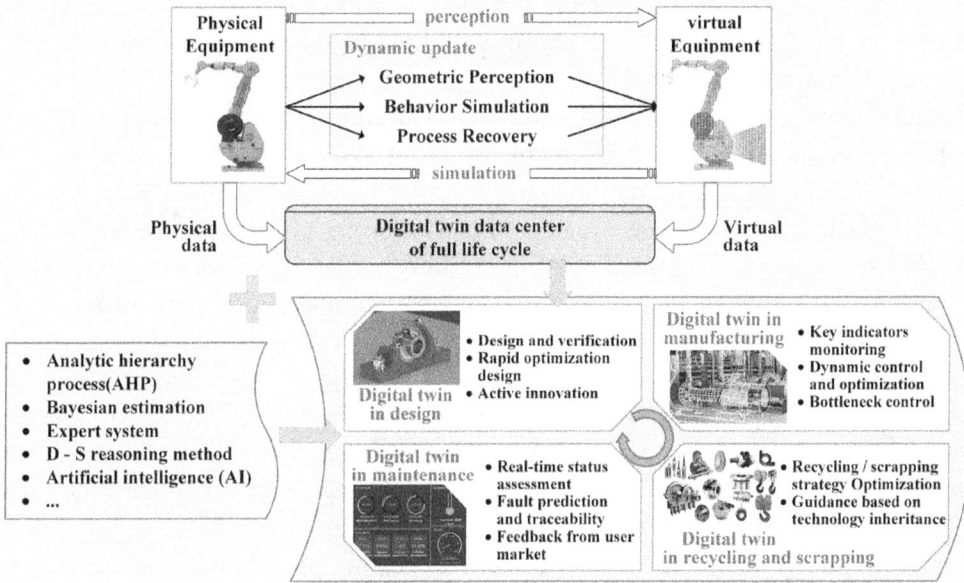

FIGURE 4.102 Digital twin architecture of complex equipment.

high-potential added value introduced by the virtual data and services behind physical products and realizes dynamic fusion for updating and monitoring. Thus, it is essential to perform the whole life cycle process in real time through intelligent development.

4.4.2.2.1 Digital Twin in Design   Product design is based on the target function of the design object and user requirements. After discussion, analysis, and design processing, the requirements are transformed into a specific text or graphic expression to provide reasonable planning for the production stage. Driven by the product's digital twin data, the Digital Twin in Design (DTD) provides a two-way interactive connection between physical and virtual products for data sharing. It also strengthens the synergy between these two and continuously explore innovative, unique, and valuable product optimization design schemes. Thus, it can achieve dynamic innovation that meets the customized requirements ahead of the time.

The digital twin is not just a digital product display model; it can also improve the accuracy of the design and assign the behavior of the actual product to the corresponding virtual product. In addition, it can verify the performance of a product in a real environment through simulation. The constructed design originates from practice and serves practice. The DTD emphasizes the virtual–physical integration of the complete life cycle. It establishes links between products, production tools, environments, and processes and integrates information of different stages of the product life cycle into the design phase while providing a coherent overview of the development cycle. Therefore, the DTD can promote decision-making and improve design quality and efficiency in many aspects through its various functions, thus improving and optimizing products and their manufacturing systems.

4.4.2.2.2 Digital Twin in Manufacturing   It is necessary to construct the Digital Twin in Manufacturing (DTMF) before performing actual production tasks. It is also necessary to extract manufacturing-related process information from the DTD along with associated manufacturing resources information, such as information on processing equipment, process parameters, tooling, and molds. To realize the DTMF, it is important to simulate the production process of complex equipment under agreed conditions through virtual production in advance. This process can predict bottlenecks of resources, workpieces, and stages. Thus, in this way, bottleneck control can be achieved, and the speed and accuracy of a new product can be improved during the production process through the redistribution of bottleneck resources and optimization of component combination and operation sequence performed before production.

The customized and sophisticated manufacturing process of complex equipment requires effective process planning and key indicator monitoring. In the manufacturing stage, the connection is established between the manufacturing entity and the virtual product by the DTMF. The measured data related to the product production, processing, and quality control processes, such as geometric measurement, vibration measurement, force measurement, and progress data, are mapped to the virtual product and displayed in real time. This process ensures a comprehensive collection of production factors and manufacturing information and achieves online process control as well as monitoring based on the actual production data.

The manufacturing process is characterized by dynamics and uncertainties, such as tool wear, aging of processing equipment, and changes in material properties, which are caused by environmental conditions. Despite careful planning, these situations can cause a gap between the actual production results and expected results. The DTMF helps to integrate process planning data with the measured production data while monitoring key manufacturing parameters.

Therefore, when an abnormal situation occurs, such as a violation of plans or discovering a better production plan through learning, the intelligent decision-making module provides the corresponding treatment and scheme adjustment timely. The new adjustment scheme is fed back to the manufacturing entity to improve the manufacturing quality and to realize dynamic control and self-optimization, thus achieving the purpose of the virtual control.

4.4.2.2.3 Digital Twin in Maintenance   Complex equipment typically has the characteristic of high cost, which is related to costly maintenance. However, both users and enterprises expect to achieve the maximum value of the equipment at the lowest possible cost. In the use and service phase of a product, the Digital Twin in Maintenance (DTMT) is used to track and monitor the product status in real time. The DTMT acquires real-time data that characterize the quality and functional status of the monitored equipment during operation, such as engine rotor speed, vibration value, lubricating oil temperature, and exhaust pressure, by using smart sensors. It also helps in constructing a remote monitoring system with quantitative indicators and establishes a hierarchical health management system for complex equipment parts, subsystems, systems based on equipment historical use and maintenance data. It analyzes and predicts equipment operating conditions, remaining life, and potential failures and assets failure time in advance. It also performs beforehand maintenance to improve the efficiency of spare parts management and to reduce or avoid customer losses and problems caused by unexpected equipment shutdown. A virtual product can adjust itself to adapt constantly to changes in the physical space conditions. It reflects and evaluates the health status of physical service products through virtual products. In this way, the risk of manual evaluation using a one-way physical information flow is reduced. The insights obtained by the application of the DTMT can also be used to improve product design and manufacturing processes.

For physical equipment with a fault or a quality problem, it is better to use a virtual product to quickly locate the fault or the problem. This approach can also be combined with the historical operation data before the failure and can both provide the simulation traceability of a virtual product and analyze the causes of quality problems from their root. It is also evident to develop discovery methods and solutions to the problem of the same type and to verify the feasibility of these solutions. It also feeds the final results back to the physical space to achieve the early warning and troubleshooting.

The application of complex equipment has a strong driving force for both enterprise production and social progress. The rationality of complex equipment design parameters and operating parameter settings, as well as the adaptability to different working conditions, determines the functional level, quality advantages, and customer satisfaction of equipment.

Equipment providers collect real-time data during product service by the DTMT. Using these data, they construct an empirical model for parameter optimization for different application scenarios and provide customers with guidance on its configuration. This can improve user satisfaction, enhance user experience, and maximize the functionality of the equipment. At the same time, customers can connect with the product through the DTMT and thus provide positive feedback to manufacturers. By considering the customer preferences, manufacturers can obtain insight into potential, future, real-time demands of customers for products and thus can meet them in advance. Therefore, it can avoid making the research and development decisions from offsetting market demand, shorten the design time and introduction cycle of new products, and enhance market competitiveness.

4.4.2.2.4 Digital Twin in Recycling and Scrapping   The application of the Digital Twin in Recycling and Scrapping (DTRS) meets the requirements of green ecology and sustainable development. The out-of-service equipment record data related to scrapping and recycling, such as reasons for scrapping and recycling, expected product life, and actual life, can be obtained through digital twin. The DTRS provides a reference for product scrapping and recycling operations by relative data analysis. The related data include material data of parts (e.g., scarce resources, recyclable materials, or toxic materials) in the DTD and information related to the working environment of components and equipment (e.g., working conditions such as exposure to uranium and other radioactive elements) in the DTMT. When the physical product is scrapped or recycled, the corresponding virtual product is kept and archived in the digital form. Following the concept of technological inheritance, the production and application data of the previous generation are passed to the next generation. Historical information from different stages of the product life cycle is integrated into the design stage, forming the closed-loop management of the product life cycle data. Based on specific product requirements, it guides the design improvement and functional innovation of the next generation and increases the efficiency of new product development and production. The historical normal-operation and failure data of scrapping and recycling equipment can provide a helpful basis for quality analysis and health management of the same or similar product groups.

4.4.2.3 Case Analysis

This part presents the application example of the proposed complex equipment management method based on machine learning and DTMT to verify the effectiveness of the proposed method.

4.4.2.3.1 DTMT Application to Operation and Maintenance of Health Management   Complex equipment represents important equipment support for modern economic industries. However, unexpected shutdowns caused by failures can affect both the economy and society. Therefore, the health management of complex equipment during its service is particularly important. Using a combination of the DTMT and deep learning technology, this study performs real-time monitoring and predictive evaluation of equipment status to improve stability and reliability of the equipment operation. As shown in Figure 4.103, mapping of the physical operating status data of the equipment entity to the DTMT is performed by the synchronous simulation operation in the virtual space. This enables real-time monitoring of equipment status and simulation of future operation and avoids interference of redundant monitoring data on equipment operation status. Data denote the basic and core driving force for the DTMT application to health monitoring and fault prediction of real equipment. A multilayer nonlinear deep learning-based network structure has advantages in discovering useful information and essential characteristics of data and has high accuracy and completion degree in tasks like fault identification and prediction. The hierarchical abstraction capabilities of data features of deep learning algorithms

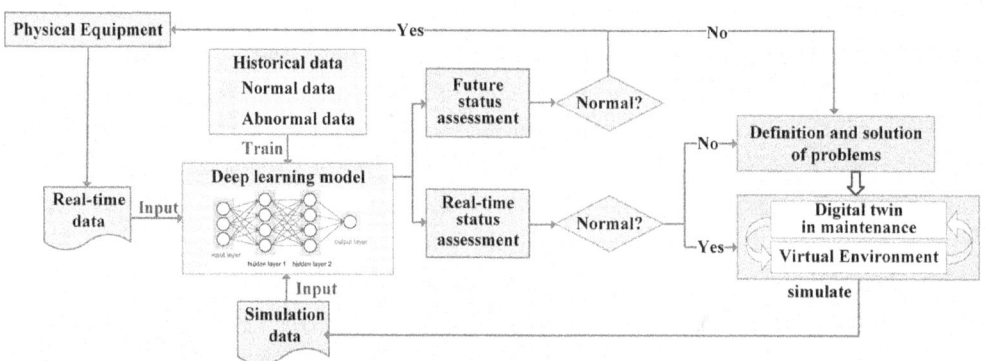

FIGURE 4.103   Application of the DTMT to the operation and maintenance health management.

can be useful to extract differences in high-dimensional association features between the monitoring data under normal and abnormal conditions using the historical data. It establishes association rule models between data and failures to monitor and predict the behavior and status of equipment. When an abnormality in equipment operation is detected by the real-time data analysis, i.e., when a fault has occurred, it automatically identifies and classifies the fault based on the correlation model. It analyzes the type and cause of the detected fault, formulates a maintenance plan fast, and restores or guarantees the normal operation of the monitored equipment. In case of simulation data anomalies at the time of a potential fault in the DTMT, it inputs the corresponding solution to the DTMT for simulation validation. If according to the simulation results, the monitoring data return a normal value, that indicates that the potential fault has been eliminated. Then, the corresponding decision-making plan is sent to the physical entity to guide the maintenance process, and prior repair is realized. This can reduce the equipment maintenance cost and uncertainty in the operation process and can achieve self-perception, self-prediction, and self-decision of equipment.

4.4.2.3.2 Health Management Mechanism of DTMT   Monitoring and changing trend prediction of characteristic parameters of main components help to optimize the complex equipment control. In this way, the operation state of complex equipment can be predicted better. Bearing, as a key component of transmission, exists in the gearbox, axle box, motor, and other key parts of complex equipment; it plays an irreplaceable role in maintaining motion accuracy and improving mechanical efficiency. Therefore, fault diagnosis and prediction of bearings could significantly reduce the maintenance cost and time. Meanwhile, bearing temperature is an effective index that can directly reflect the bearing condition, and the temperature characteristics have the advantages of thermal inertia and strong anti-interference ability. Therefore, this study selects the temperature characteristic of bearing to judge whether a fault has occurred or not and then performs the health assessment.

A locomotive is the main power provider in the railway transportation system, and it is an important piece of equipment that affects traffic safety. In this study, the running gear of the CDD5B1 diesel locomotive is taken as a research object, and the DTMT is used for predictive maintenance of diesel locomotive bearings. The specific implementation process is presented in Figure 4.104, where it can be seen that the process consists of three main stages: real-time data collection, establishment of a calculation model based on the DTMT, and bearing health assessment.

- **Data acquisition and database**

    The accuracy of fault prediction and health management depends on the relevance, adequacy, and quality of the data used. Therefore, it is necessary to establish a predictive maintenance database for the DTMT using different data, such as historical data from the same source, historical maintenance data, and real-time operating data. The characteristics of the train operation data are periodic during the operation cycle.

    After being denoised and normalized, the train operation data of the same type are uploaded to the cloud and stored as sample data for future use. Homologous

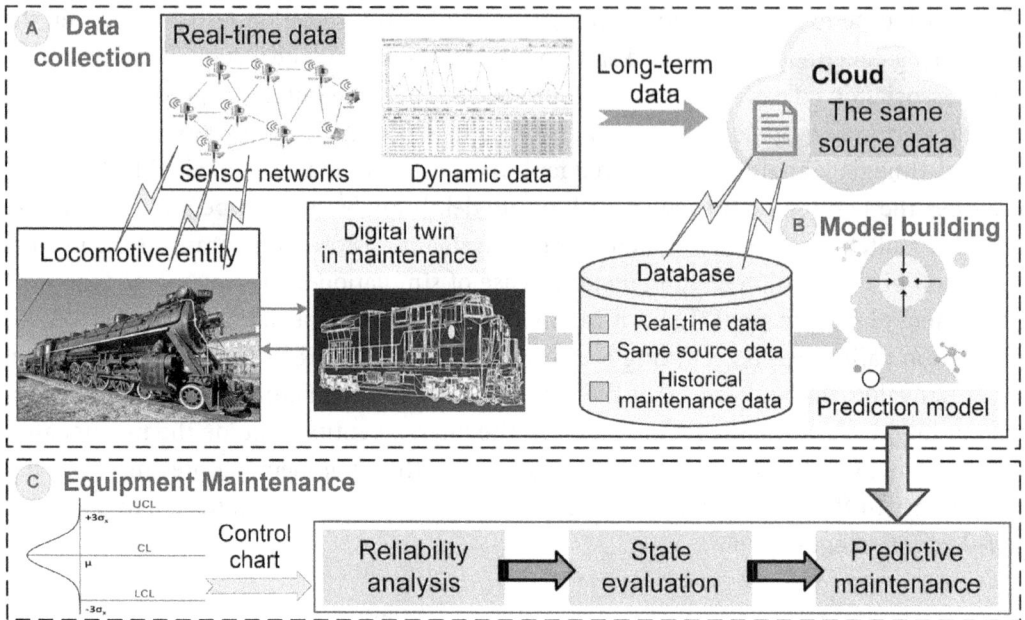

FIGURE 4.104   Health management implementation process based on the DTMT.

historical data are backed up and updated on the cloud in time to ensure data consistency. Sink data are sent to the edge of the digital twin and used when needed. Historical maintenance data include detailed information on the maintenance activities performed by equipment, replaced components, and failures. According to the requirements for dynamic health assessment of the locomotive running department, this study establishes a sensor network system to percept and capture real-time data.

- **Prediction model based on DTMT**

    The overall framework of the prediction model based on the DTMT is presented in Figure 4.105, where it can be seen that it consists of three parts: the data preprocessing part, the single model prediction part, and the model integration part.

    Part I: Preprocessing of the original data. Identify and delete duplicate values and outliers in the original data; use the Lagrangian interpolation to fill the missing values; use the Savitzky–Golay filter to denoise and normalize the data.

    The original data are divided into training and test sets. The training set is used to train the proposed model, and the test set is used to test the performance of the trained model.

    Part II: Single model prediction results. Use three machine learning methods, namely, the LASOO, SVR, and XGBoost, to design an axle temperature prediction model in the DTMT and obtain the prediction results of the sub-series.

    Part III: The combination prediction results. Obtain the final prediction result by integrating the prediction results of the three machine learning networks. In this

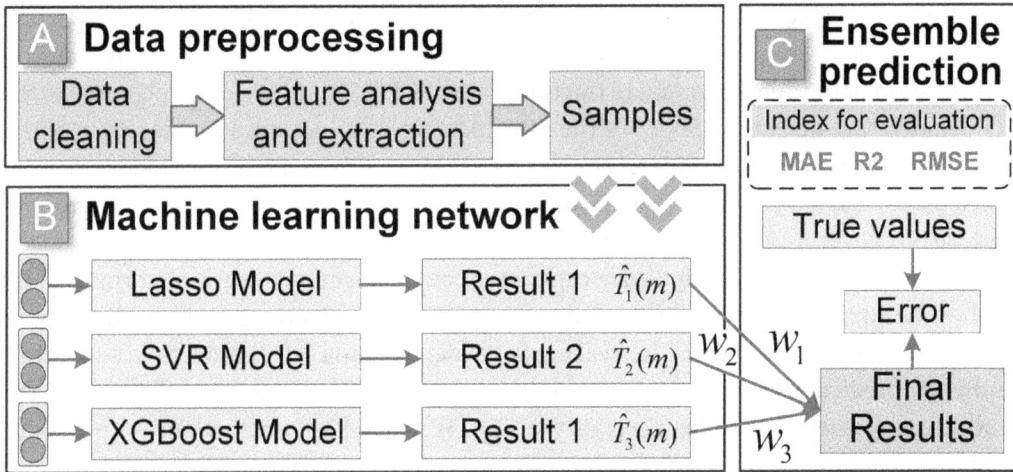

FIGURE 4.105 The flowchart of the DTMT-based prediction model.

study, the model set is realized by setting the weight coefficients of the prediction results of the learning networks. The corresponding calculation formulas are as follows:

$$\hat{T}(m)=w_1\hat{T}_1(m)+w_2\hat{T}_2(m)+w_3\hat{T}_3(m) \tag{4.96}$$

where $w_i(i=1,2,3)$ denotes the weight coefficient of a learning network $i,\hat{T}_i(m)(i=1,2,3)$ denotes the prediction results of a learning network, and $\hat{T}(m)$ represents the sensor data output by the DTMT.

- **Assessment of bearing health status**

  The normal axis temperature data conform to the normal distribution. The nature of the normal distribution shows that the probability of data appearing outside the three standard deviations of the mean is only 0.2%, which is very small. According to the principle of probability theory "taking small-probability events as practically impossible," it can be known that the occurrence of such an event is due to abnormal causes that make the overall distribution deviate from the normal position, indicating that a failure has occurred. Therefore, the $X-S$ control char can be used to control the normal temperature range of the axle with the upper control limit of $\mu+3\sigma_x$ and the lower control limit of $\mu-3\sigma_x$.

4.4.2.3.2 Operation and Maintenance Health Management Application   The on-board safety protection system loaded on the CDD5B1 diesel locomotive collects parameters that can reflect the operating status of the monitored target in real time and provides long-distance transmission and anti-interference protection. The monitoring status information includes but is not limited to the vehicle number, axle number, and temperature information, alarm prompt information, and vehicle speed. In addition, there are six axles in the locomotive, each of which has six temperature measurement points. When the temperature of a certain axis reaches or exceeds 90°C, the system will issue an alarm. On March 26, 2018, during the operation of the dual-locomotive traction train of CDD5B1–5109 +5111 locomotive, the axle temperature detection device of the 5109 locomotive announced 51 bearing-temperature-rise alarms, and the temperature was 89°C. After the on-site inspection by maintenance personnel, it was found that the outer bearing cylinder of the axle box had felt into the axle box cover, and the cage was severely deformed.

The axle temperature alarm device can announce an alarm only when abnormal values of physical quantity have been detected to remind drivers to stop and check the vehicle, which causes a postponement. The bearing faults denote cumulative processes, and in this study, the predictive maintenance method based on the DTMT is used to perform the health management of the bearings of the CDD5B1 diesel locomotive and to pre-judge faults in advance.

In this section, the data used for case implementation is the original data in Excel format provided by CRRC Qishuyan Locomotive Co., LTD. (Changzhou, Jiangsu, China). The original data includes the normal (no alarm of axle temperature detection device) data of CDD5B1–5111 locomotive in March 2018 and the fault (alarm of axle temperature detection device) monitoring data of CDD5B1–5109 locomotive before March 26, 2018. The axle temperature monitoring and alarm system of a diesel locomotive has 67 data monitoring items.

Data is collected about every 30 seconds, but repeated collection may occur at the same time point. According to the daily running time of locomotive, 1,000–2,000 data items may be collected, with a total of nearly 10,000 data points.

The CDD5B1–5111 locomotive runs synchronously with the CDD5B1–5109 locomotive, and their two states are similar. Therefore, the historical operation data of the CDD5B1 5111 locomotive under normal conditions can be used as homologous historical data. These data are added to the CDD5B1–5109 locomotive DTMT database to build the axle temperature prediction model.

When a single axle is taken as the monitoring object, the feature items that can be selected are listed in Table 4.31.

When a single measurement point is taken as the monitoring object, the feature items that can be selected are listed in Table 4.32.

For a single axle as a monitoring object, the changes in the normal data of the No. 5 axle of the CDD5B1–5111 locomotive and the abnormal data of the 5,109 locomotive over time are presented in Figures 4.106 and 4.107, respectively. For a single measuring point as a monitoring object, the changes in the normal data of the No. 5 axle of the

TABLE 4.31   Characteristic Terms of a Single Axle

Term Number	Term Name
1	The environment temperature/°C
2	Main generator temperature/°C
3	GPS speed/km/h
4–9	Temperature of measuring points 1–6 of axis X/°C

TABLE 4.32   Characteristic Terms of a Single Measuring Point

Term Number	Term Name
1	The environment temperature/°C
2	Main generator temperature/°C
3	GPS speed/km/h
4	Temperature of measuring point M on axis X/°C

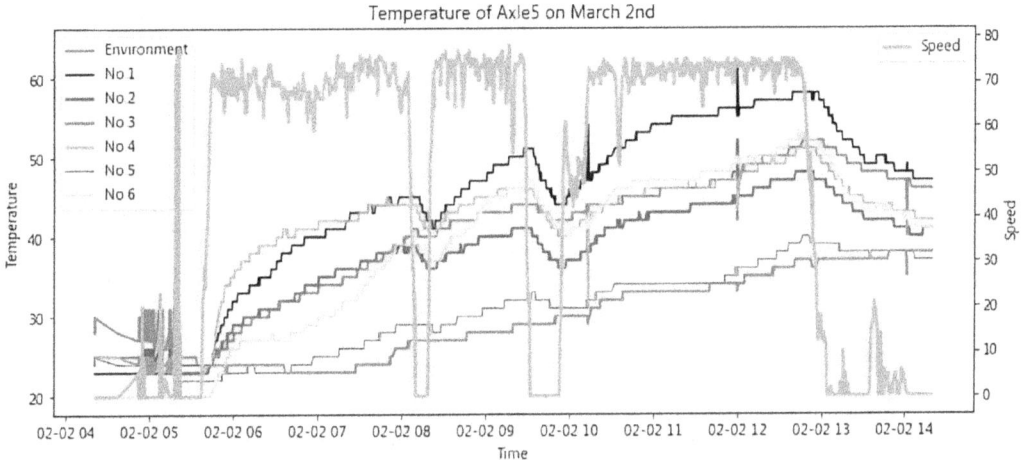

FIGURE 4.106   Normal data of the No. 5 axle over time.

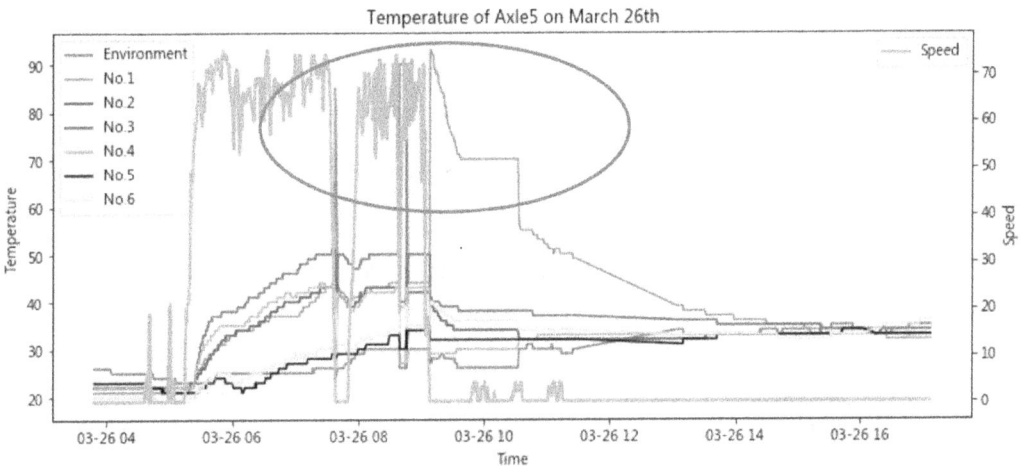

FIGURE 4.107   Abnormal data of the No. 5 axle over time.

CDD5B1–5111 locomotive and the abnormal data of the 5,109 locomotive over time are presented in Figures. 4.108 and 4.109, respectively.

In Figures 4.106 and 4.108, it can be seen that under normal circumstances, the changing trends of the temperature of the same measuring points on the same axle and the same measuring point on different axles are roughly the same. Also, there is a certain correlation with the environmental temperature. When a fault occurs, the temperature

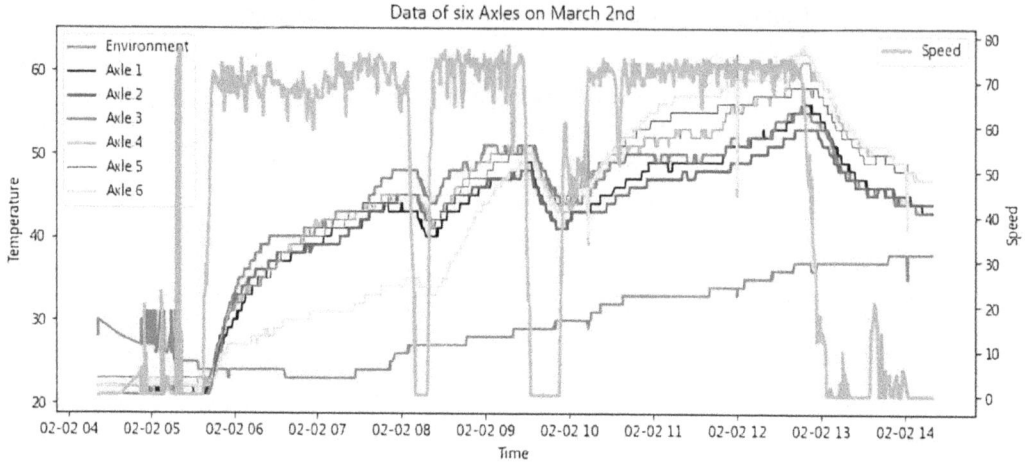

FIGURE 4.108   Normal data of the No.1 point of six axles over time.

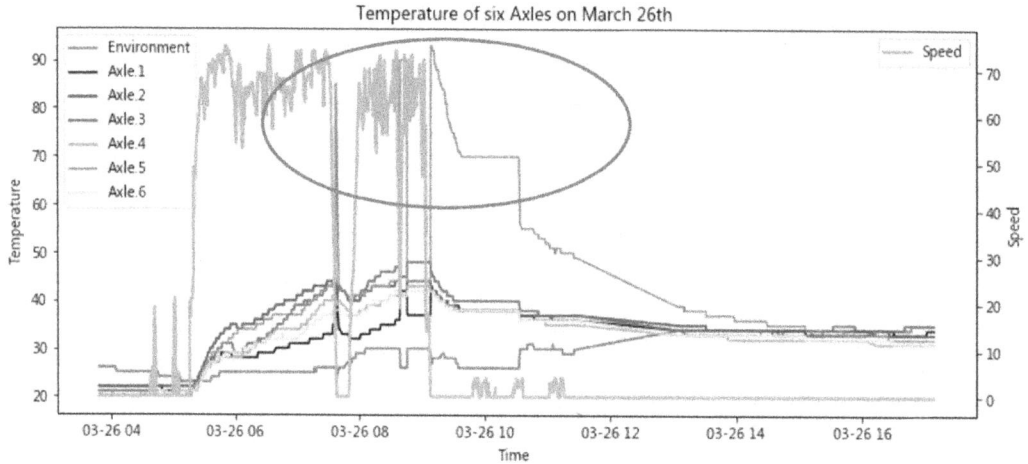

FIGURE 4.109   Abnormal data of the No.1 point of six axles over time.

data of a certain measuring point of the faulty axle deviates from the temperature change range of the remaining measuring points of the axle. Namely, it rises sharply, resulting in an "outlier" phenomenon. In this study, a temperature-speed model is established based on measured point 1 of axle 5, and the measured point temperature is used as an effective index to judge the working status of the No. 5 axle. In this way, the working conditions and operational safety of the entire diesel locomotive can be evaluated.

Figure 4.110 shows the temperature change of measuring point 1 on the six axles with the ambient temperature and locomotive speed in the week before the fault has occurred; it can be seen that although there was no alarm from the axle temperature monitoring device before March 26, the data of the No. 5 axle 1 measurement point had been in an "outlier" state for a long time. Thus, it can be inferred that the axle temperature at this time had deviated from the normal axle temperature data.

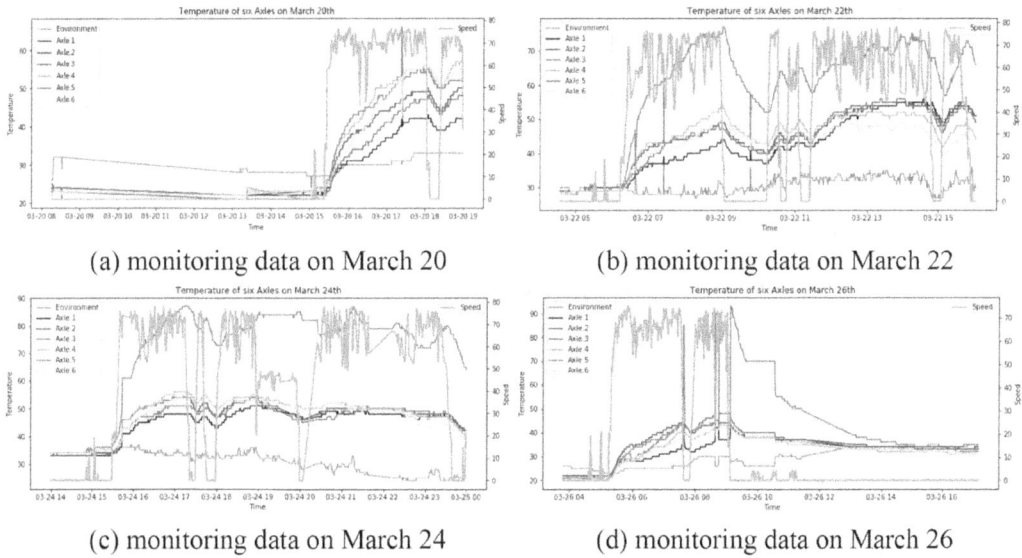

(a) monitoring data on March 20

(b) monitoring data on March 22

(c) monitoring data on March 24

(d) monitoring data on March 26

FIGURE 4.110 Monitoring data of the CDD5B1-5109 locomotive a week before the failure has occurred.

Although the temperature did not exceed the normal temperature range and predefined maximum temperature, there might be a potential failure; since at that time, the temperature could not correctly reflect the safe operation of the locomotive. These results provide feasible support for simulating the correct temperature and predicting potential failures using the DTMT.

The results show that there is a certain delay between the axle temperature change and the speed change. In this study, a time-series index is used to select non-parking phase data (i.e., GPS speed of zero) from the overall locomotive operation data. The environmental temperature and the main generator temperature denote the environmental data; the GPS speed is used as control data. The extracted data are used as input data of the axle temperature prediction model. Considering the delay between the axle temperature change and the speed change, the input variables are set to the current speed and the speed of the previous four-time steps. Aiming to reduce the impact of the delay, the DTMT considers the operating conditions of the locomotive, such as starting, accelerating, running smoothly, braking, or decelerating.

In this study, the monitoring data of the CDD5B1–5111 locomotive are randomly selected from the normal-operation data of 1 day, and the axle temperature-speed prediction models are designed using the LASOO, SVR, and XGBoost algorithms. The series method is employed to determine the model weight coefficients. The modeling process is shown in Figure 4.111. From the overall data, 10% of the samples are randomly selected as a test set, and the prediction results of different models on the test set are shown in Figures 4.112–4.115. All experiments are performed on the Python 3.8 platform installed on a personal computer.

The Root Mean Square Error (RMSE), Mean Absolute Error (MAE), and Coefficient of Deviation (R2) are used to evaluate the performance of the three single prediction models,

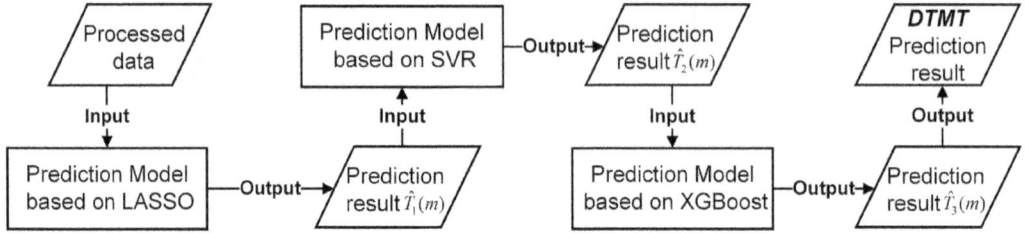

FIGURE 4.111   The flowchart of the series prediction model.

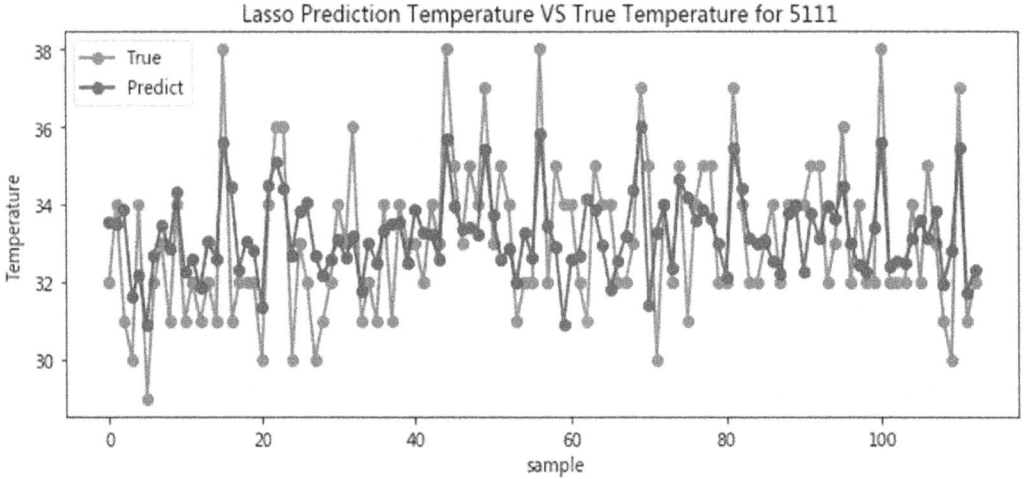

FIGURE 4.112   Prediction results of the LASOO model.

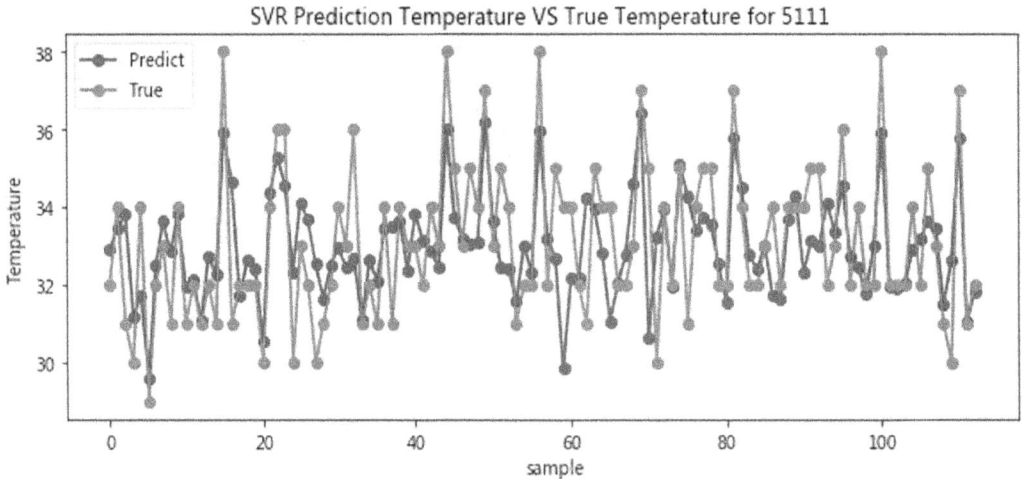

FIGURE 4.113   Prediction results of the SVR model.

the LASOO, SVR, and XGBoost models, and their combined models. The RMSE, MAE, and R2 are calculated by:

$$\text{RMSE} = \sqrt{\frac{1}{m} \sum_{i=1}^{n} \left(p_i - y_i \right)^2} \qquad (4.97)$$

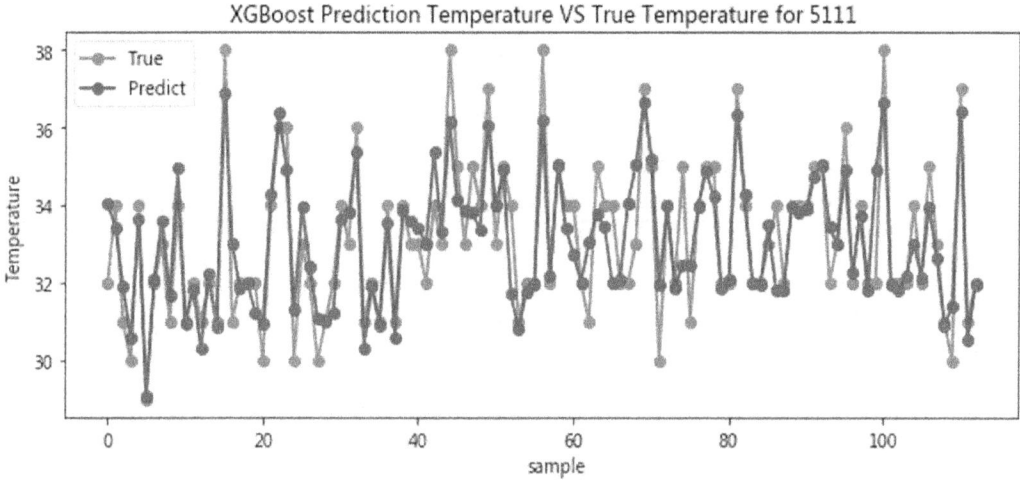

FIGURE 4.114   Prediction results of the XGBoost model.

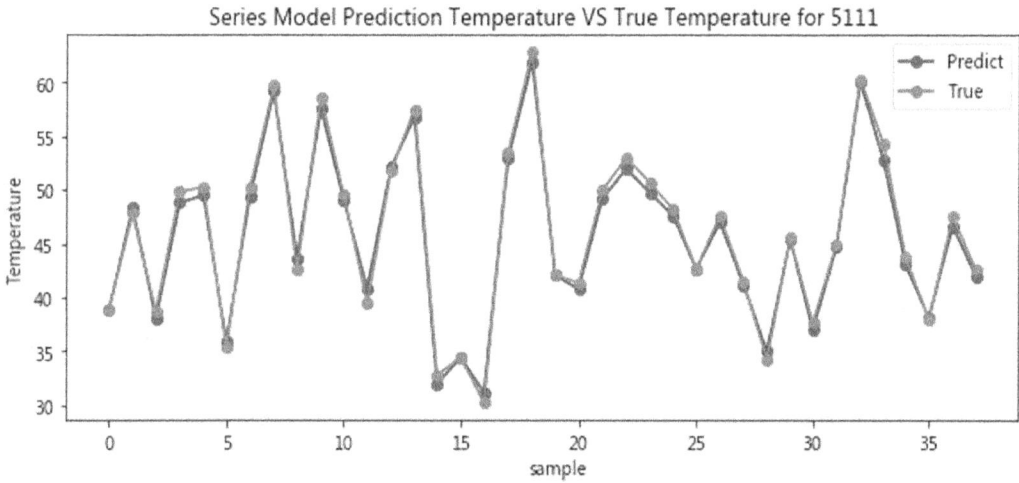

FIGURE 4.115   Prediction results of the proposed series model.

$$\text{MAE} = \frac{1}{m} \sum_{i=1}^{n} \left| \left(p_i - y_i \right) \right| \tag{4.98}$$

$$R2 = \frac{\sum_{i=1}^{n} \left(p_i - \overline{y} \right)^2}{\sum_{i=1}^{n} \left(y_i - \overline{y} \right)^2} \tag{4.99}$$

where m denotes the number of predicted samples, $\overline{y}$ is the mean of the true values of all the samples, and p_i and y_i are the predicted value and true value of the ith data sample.

The RMSE, MAE, and R2 values of the four prediction models are given in Table 4.33.

TABLE 4.33   The RMSE, MAE, R2 of the Four Prediction Models

	RMSE	MAE	R2
LASO0 prediction model	1.5035	1.0597	0.6926
SVR prediction model	1.5148	1.1361	0.6475
XGBoost prediction model	0.9364	0.6815	0.7283
Series Combination prediction model	0.7105	0.4333	0.8739

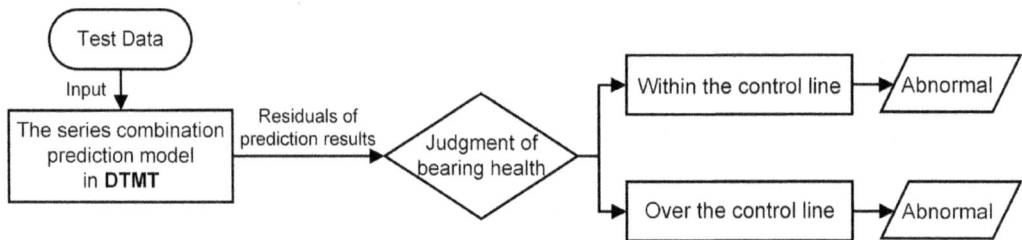

FIGURE 4.116   The block diagram of the verification process of the combined prediction model.

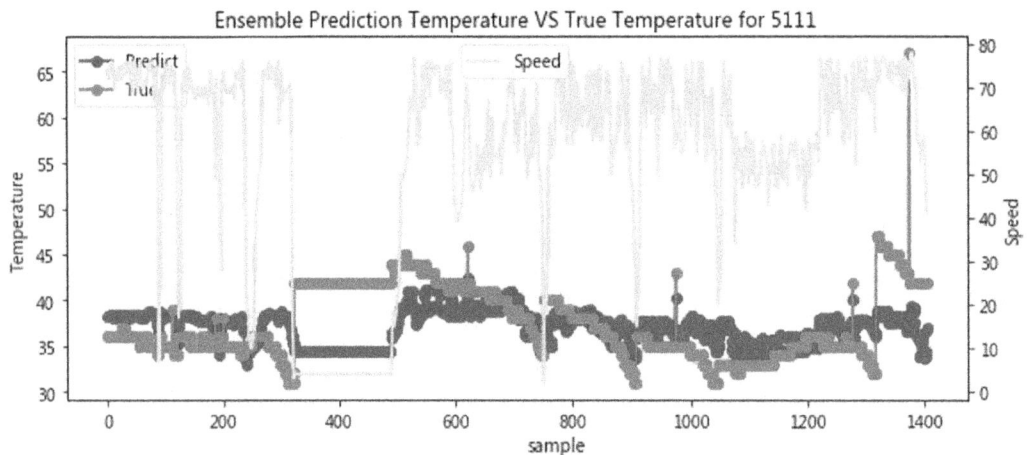

FIGURE 4.117   Normal data prediction results.

As shown in Table 4.33, the RMSE and MAE of the series combination forecasting model are the smallest, and its R2 value is closest to 1. These results show that the forecasting accuracy of the series combination forecasting model is the highest among all the models. Therefore, a series combination prediction model based on the DTMT is designed to predict a failure and warn about the axle temperature of a diesel locomotive.

The normal axle temperature data and fault data are used to verify the established axle temperature–vehicle speed series combination prediction model. The block diagram of the verification process is shown in Figure 4.116.

The prediction results and residual error distributions of the axle temperature during a normal operating day of the CDD5B1–5111 locomotive are shown in Figures 4.117 and 4.118, respectively. The prediction residuals of the normal data are all within the control

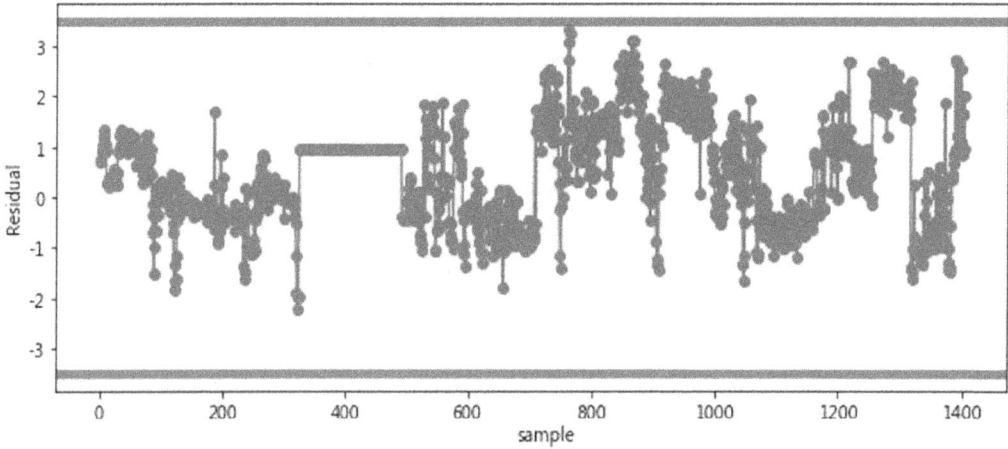

FIGURE 4.118 Normal data prediction result residuals. (The two lines above and below are control lines).

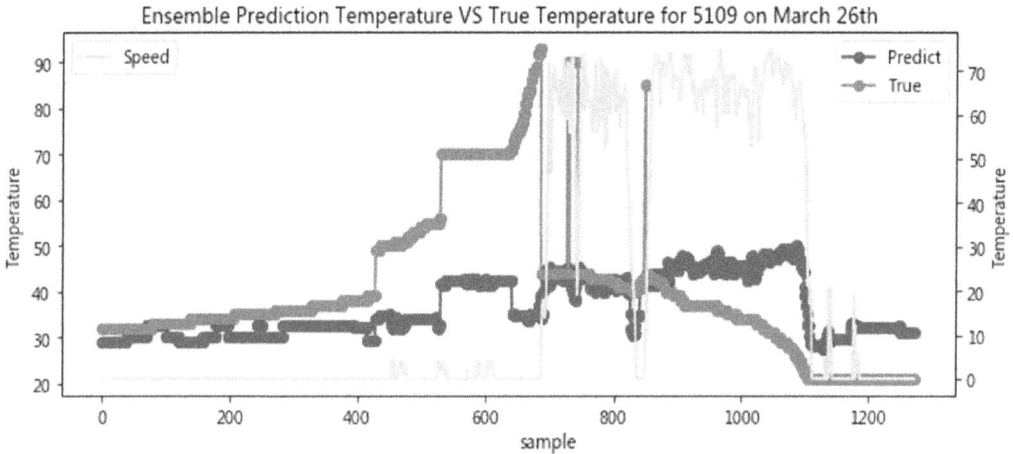

FIGURE 4.119 Abnormal data prediction result.

range. This result shows that the interior of the locomotive system is in line with the normal speed–axle temperature mechanism. There is no undetected problem of failure, and the locomotive is safe and reliable.

The predicted results of the failure data and the distribution of residual error when the CDD5B1–5109 locomotive failed on March 26 are shown in Figures 4.119 and 4.120, respectively.

The prediction residuals of the fault data have mostly exceeded the control line. This shows that at this time, the interior of the locomotive's system did not meet the normal speed–axle temperature mechanism, which resulted in a failure alert.

Through the prediction of the axle temperature under the normal and failure conditions of the locomotive, the correctness of the proposed model for the axle health status assessment has been verified. When the DTMT is used to predict the axle temperature of the CDD5B1–5109 locomotive on March 20, the prediction results and residual distribution are as shown in Figures 4.121 and 4.122, respectively.

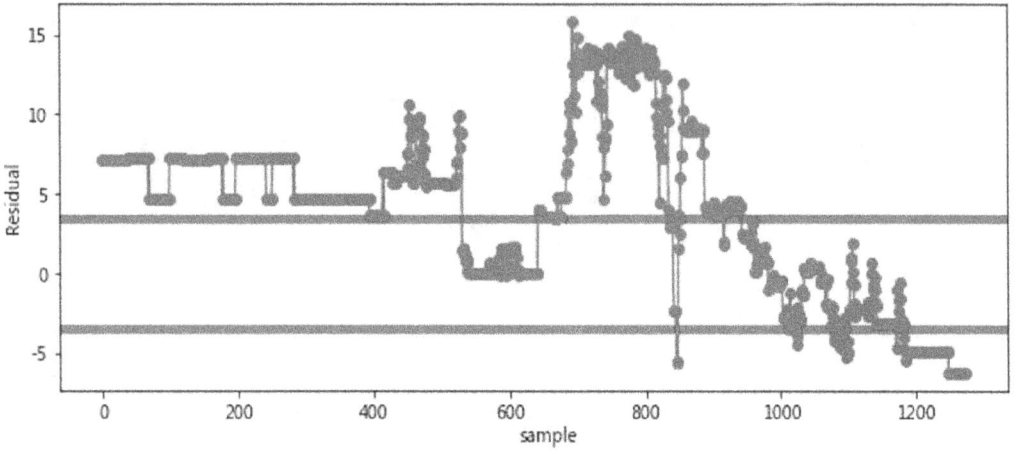

FIGURE 4.120 Abnormal data prediction result residual. (The two lines above and below are control lines).

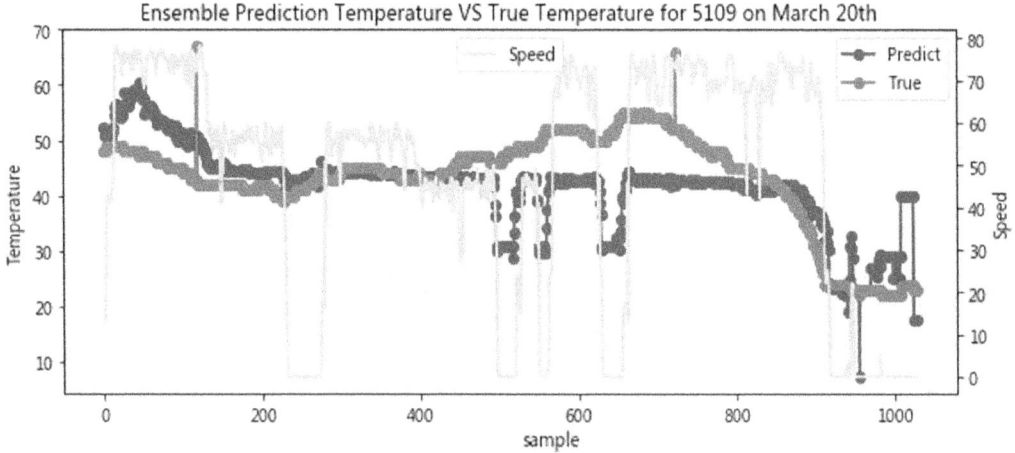

FIGURE 4.121 Axle temperature prediction results on March 20.

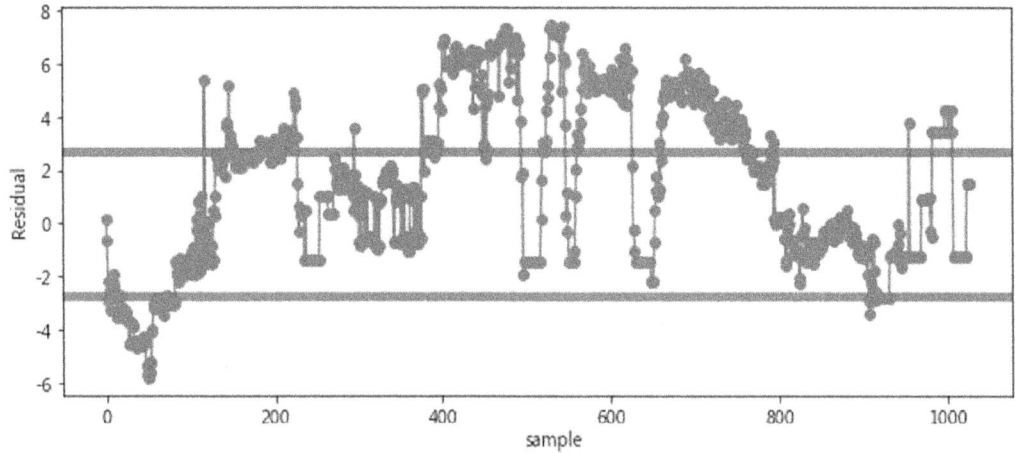

FIGURE 4.122 Axle temperature prediction results residual on March 20. (The two lines above and below are control lines).

As shown in Figures 4.121 and 4.122, although the axle temperature monitoring device did not alarm on March 20, the residual error of the axle temperature prediction result has partly exceeded the normal control range, which was abnormal.

The first warning occurred during the first acceleration of the locomotive, after which frequent abnormal warnings occurred. This shows that the interior of the locomotive at that time has failed or has a failure trend, so it needed to be stopped for maintenance purposes.

Therefore, by using the management method based on a combination of digital twin and machine learning, the abnormal axle temperature can be alarmed about 1 week in advance, hidden problems can be discovered, and maintenance can be performed according to the situation beforehand. In this way, the occurrence of function failures can be reduced, and the safety of the locomotive operation can be ensured.

In summary, under the background of a digital twin, this paper proposes a machine learning-based management method of complex equipment. The application example demonstrates the effectiveness and practical value of the proposed method in health management maintenance. The proposed method can be extended to the field of life cycle management.

4.4.3 Cloud-Assisted Active Preventive Maintenance Based on Manufacturing Big Data

Equipment maintenance plays an important role in intelligent manufacturing and directly affects the service life of equipment and its production efficiency. Current methods for equipment maintenance depend on system alarms, and an operator reports faults to equipment maintenance personnel. The fault then needs to be exactly located and resolved, leading to a shutdown in the production process. With the support of manufacturing big data, device data can be collected in real time, including device alarms, device logs, and device status, in order to evaluate the health condition of manufacturing equipment and preemptively detect breakdowns. Therefore, active preventive maintenance is proactive and can find problems earlier. Active preventive maintenance for intelligent manufacturing is now feasible through data collection and big data analysis. This section will focus on how to perform data processing in the cloud and improve the equipment maintenance process by introducing a new service mechanism [6].

4.4.3.1 System Architecture

4.4.3.1.1 Cloud-Assisted Architecture   With the support of industrial wireless networks, cloud computing, and big data analysis, the mode of maintenance used can be transferred from traditional maintenance (e.g., decentralized maintenance) to big data-based active preventive maintenance. Figure 4.123 shows the system architecture for manufacturing big data using active preventive maintenance. In the traditional mode, there are three layers to the maintenance: production line maintenance, workshop maintenance, and factory maintenance, and failures or problems are reported from the bottom layer to the top layer in a non-real-time fashion.

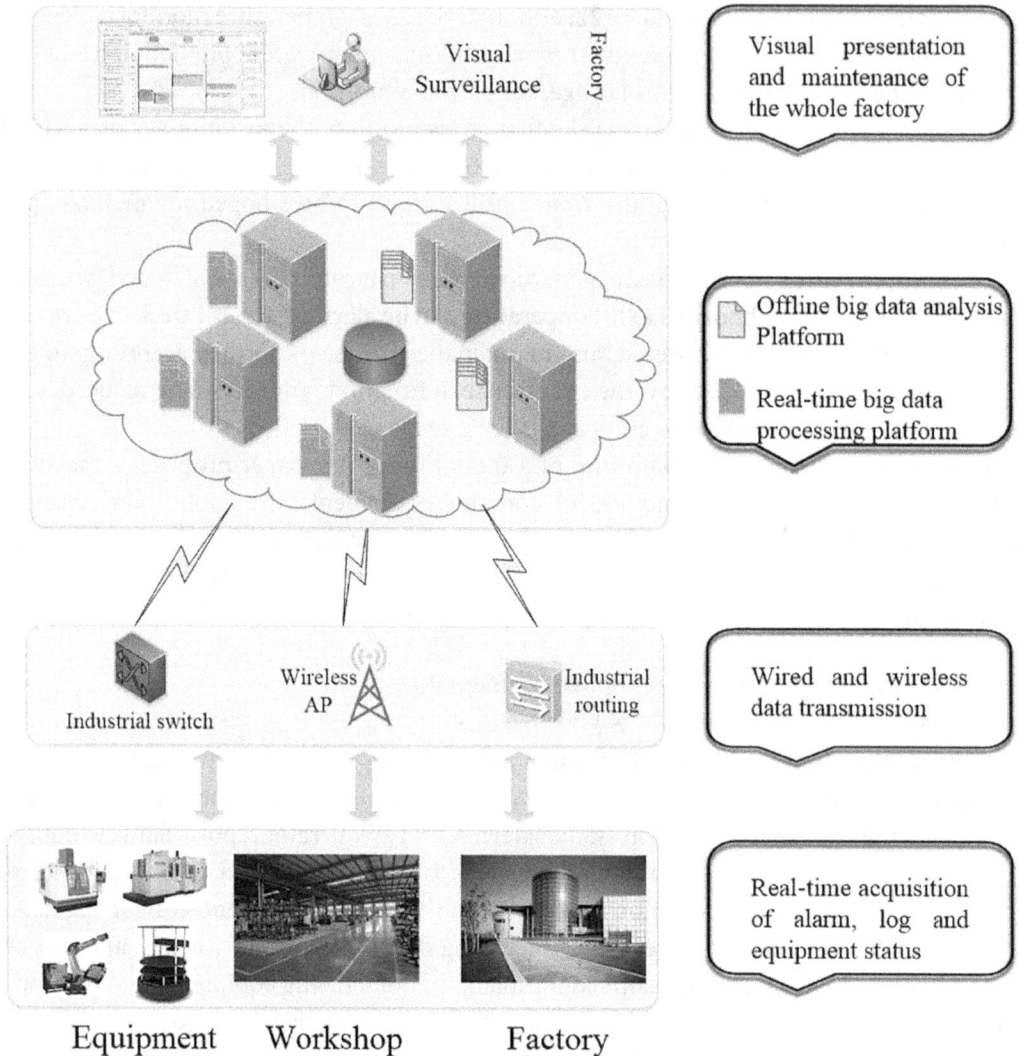

FIGURE 4.123 System architecture for manufacturing big data for active preventive maintenance.

Many more recent devices have the ability to send an alarm to a manager to resolve emerging problems as soon as possible after a failure or problem happens. However, this mode still lacks in predictability.

For big data-based active preventive maintenance, all related data, including product data, device status, facility logs, device alarms, and production process data, can be collected and forwarded to the cloud through an industrial wireless network. In the cloud, correlation analysis can be performed to evaluate internal relationships. For example, device status combined with a facility log may be used to further analyze the equipment health status. In addition, the analysis results may be visualized and displayed on a large screen or sent to the relevant manager through mobile services. In this architecture, the key research areas are the method for collection of manufacturing big data and the big data analysis. These two elements are the main differences between traditional maintenance and big data-based preventive maintenance.

4.4.3.1.2 Collection of Manufacturing Big Data   In this section, we mainly focus on data processing in the cloud. The collection of manufacturing big data is briefly introduced. For big data-based preventive maintenance, all related data needs to be collected, including alarms, device logs, and device status. These three types of data obviously have different features. Alarms are random events which should usually be processed in real time. Device status is periodically gathered, and device logs may be aperiodically forwarded to the cloud. Data formats and data preprocessing of all devices are crucial to ensure convenient transformation and further processing. To make it happen, OPC UA, the service-oriented architecture for industrial applications, has been regarded as one of the most promising solutions for data integration. Based on the wide use of the OPC Classic, OPC UA was designed to surpass the capabilities of the OPC Classic specifications and enhances aspects of security, expansibility, and platform independence. The attribute of platform independence makes compatible communication possible under different types of hardware and software platforms, which exhibits a good prospect for fine-grained data gathering and superior data analysis.

Another significant aspect of data gathering is network communication. Currently, Zigbee or WiFi are typically chosen to transmit collected data in practical applications. A Zigbee node requires a coordinator to access the cloud. Therefore, WiFi is more widely used because a coordinator node is not required and WiFi has the ability to directly access the cloud. If devices have poor interface information interaction ability, it is necessary to add an auxiliary device to improve data acquisition. All the data collection nodes must support the Restful protocol in order to actively send data to the cloud. In addition, data collection nodes also support adjustable sampling frequency and configurable application layer protocols.

4.4.3.2 Data Processing in the Cloud

4.4.3.2.1 Architectures of the Cloud Layer   As shown in Figure 4.124, there are two types of active maintenance based on industrial big data: the mechanism of real-time active maintenance (MRAM) and the mechanism of offline prediction and analysis (MOPA). The former method can be used to meet the high real-time requirements of operation and maintenance. Based on the characteristic data of production line equipment, workshops, and factory supervision, MRAM undertakes analysis of real-time states and provides real-time responses for alarming. Finally, MRAM schedules maintenance resources to achieve real-time active operation and maintenance. The latter method is suitable for mining potential hazards of maintenance projects and focuses on prediction of failures of equipment, workshops, or factories. Additionally, MOPA is also responsible for active implementation of the corresponding maintenance required.

The cloud layer is the key to realization of both MRAM and MOPA. For real-time processing of data and data mining, the cloud layer requires a high degree of parallelism, a high throughput of data, and a high robustness of communication. Therefore, in this

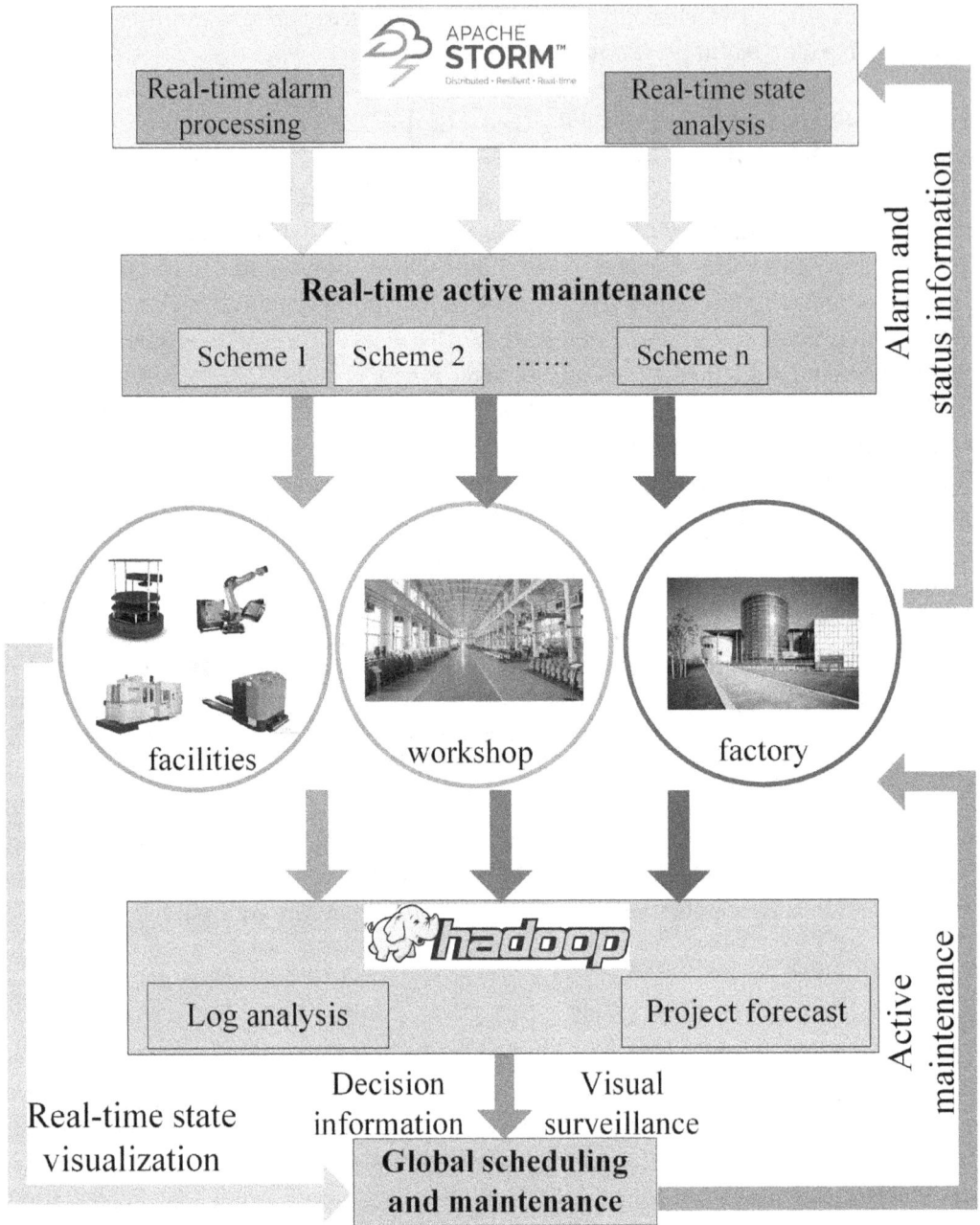

FIGURE 4.124   Mechanism of data processing for active preventive maintenance.

paper, the distributed real-time computing system STORM and the distributed parallel batch computing system Hadoop are deployed.

The lowest part of cloud layer is the distributed file system (HDFS) based on Hadoop, which is used to receive the upload log, the historical data, and other relevant data from the production equipment, workshop, and factory. In order to achieve efficient data access, a non-relational database based on Hbase is deployed based on the distributed file system.

The Sqoop components share data between the HDFS and a Mysql monitoring database in order to facilitate the maintenance of personnel monitoring.

The main nodes for STROM and Hadoop are deployed within the cloud computing cluster, which are nimbus and namenode, respectively. The corresponding computing nodes are supervisor and datanode. Deployment of STORM and Hadoop establishes the storage, monitoring, and calculation modules.

When the cloud layer is operational, the STORM cluster receives feature data from the devices, workshop, and factory, then processes alarms and performs real-time analysis of status information. Based on the processing logic that has been already defined by the maintenance personnel, MRAM schedules maintenance resources for any corresponding equipment requiring maintenance. In contrast, MOPA combines current working status and historical information stored in the cloud distributed file system, and failure prediction of maintenance projects is provided by the Hadoop cluster using the MapReduce batch calculation. Based on the results, visual analysis reports are delivered to a backstage supporter so that maintenance personnel can immediately dispatch related maintenance resources to implement active maintenance.

4.4.3.2.2 Real-Time Processing Algorithm   Active preventive maintenance contains two aspects: the mechanism of real-time active maintenance (MRAM) and the mechanism of offline analysis and prediction (MOAP).

For MRAM, the underlying devices, workshops, and factories are regarded as maintenance elements (items) on the same level. Based on the characteristics for maintenance of each element, the alarm signals and key state information are encapsulated as the items requiring maintenance. During a factory's operation, there is usually a higher real-time requirement to process alarm signals from underlying devices and the workshop. Therefore, these alarm signals have the highest priority in MRAM.

Compared with these alarms, real-time state data collected from underlying devices, the workshop, and the factory is used in a more intuitive way to reflect the real-time state of each maintenance element. By continuously monitoring key data of each maintenance element to achieve real-time assessment, MRAM realizes active maintenance by scheduling maintaining resources.

In this algorithm (Figure 4.125), N is used to represent the total number of items to be maintained. Since the underlying elements upload state information in real time, the array state[N] is used to represent the data values of the key state information of each item. Additionally, the array stateThreshold[N] is used to represent the threshold of each item's state value. For each item, the array alarm[N] is used to identify its alarm state (0 indicates normal, 1 indicates abnormal). When alarming or the value of an item is above a threshold value, MRAM will be triggered. MRAM will firstly call qualified available maintenance equipment to perform the maintenance task. After a period of time, MRAM will test whether the maintenance task has been completed, and if so, the corresponding item of alarm[N] is set to 0. If the maintenance equipment cannot solve the problem, qualified available maintenance staff will be called, and the corresponding item of alarm[N] will then also be set to 0 (Table 4.34).

Algorithm1: pseudocode of the real-time data processing

Initialization

While:

 for $i \leftarrow 1$ to N

   if (alarm(i)==1 || $state(i) >= stateThreshold(i)$)

   $alarm(i) \leftarrow 1$; //alarm or state is greater than threshold

   //call corresponding available maintenance equipment

   to item

   for $j \leftarrow 1$ to $numFaci[i]$

    //available maintenance equipment

    if ($maintainFaci[i, j]==0$)

    call $maintainFacility(i, j)$; // call

    $maintainFacility(i,j) \leftarrow 1$; //Set to work status

    //Wait for maintenance equipment to work

    wait();

    if ($state(i) < stateThreshold(i)$)

     $alarm(i) \leftarrow 0$; $maintainFacility(i, j)$

     $\leftarrow 0$;

    else // Equipment has not completed the task

     for $k \leftarrow 1$ to $numStaff[i]$

      //idle personnel

      if ($maintainStaff[i,k]==0$)

      call $maintainStaff(i,k)$; //call

      //Set working status

      $maintainStaff(i,k) \leftarrow 1$;

      $alarm(i) \leftarrow 0$;

      $maintainStaff(i,k) \leftarrow 0$;

      Break;

     End for

    End if

   End for

   if ($alarm\,[i]==0$)// Warning end

    Break; // exit current project maintenance

  End for

End While

FIGURE 4.125   Algorithm 4.1: Pseudocode of the real-time data processing.

4.4.3.2.3 Offline Prediction Algorithm   The offline prediction algorithm (Figure 4.126) is based on historical data stored in the distributed file system. The remaining effective working time of the current equipment is predicted by combining its working status and historical information. Therefore, the offline prediction algorithm can enhance

TABLE 4.34   All Variables during the Experiment

	Variable list
alarm$[N]$	Warning signal for each maintenance item
m	Number of processed procedures when η^k is first bigger than θ^k
N	Number of items which require maintenance
maintainFaci$[N,i]$	The sign of each item's maintenance equipment which represents whether it is ready to work
maintainStaff$[N,j]$	The sign of each item's maintenance staff which represents whether it is ready to work
numFaci$[N]$	Number of each item's maintenance equipment
numStaff$[N]$	Number of each item's maintenance staff
state$[N]$	Value of state for each maintenance item
stateThreshold$[N]$	The threshold value of state information for each maintenance item
t_i^k	Processing time of the Ith working procedure for the Kth maintenance item
T_i^k	Working life of the Ith working procedure for the Kth maintenance item
$\bar{T}^k$	Statistical average working life of the Kth maintenance item log stored in HDFS
T_{remain}^k	Predicted remaining working time of the Kth item
η_i^k	Relative work–life consumption ratio of the Ith working procedure for the Kth maintenance item
η^k	Cumulative work–life consumption ratio of the Kth maintenance item
θ^k	The threshold value of the cumulative work–life consumption ratio for the Kth maintenance item
f_i^k, v_i^k, ap_i^k	Input parameters for neural network
i, j, k, flag	Defined variable in for loop

the prediction accuracy for maintenance and smooth the production activities. Within a flexible manufacturing system (FMS), the maintenance elements usually need to be compatible with different processing materials and different processing parameters. The traditional estimation of equipment life is usually based on production experience to decide when maintenance is required, or else wait for a failure alarm of the key components. The former method takes security, operational fluency, and other factors into consideration, while the general experience-estimated value is usually conservative, resulting in a waste of spare parts. Although the latter method ensures that all parts get fully used, the strategy of "failure then replace" leads to a suspension in production, also causing production cost increases.

In production practice, the theoretical equipment life is different under different processing conditions. Therefore, large deviations are seen when using cumulative processing time under different processing conditions to determine whether to carry out maintenance. Additionally, this approach is likely to lead to conservative or excessive use.

To address this situation, the relative life loss rate of the equipment is proposed for different processing scenarios, which accumulates over time. When the threshold is reached, the remaining effective working time of the device is estimated by analyzing the historical life stored in the distributed file system. Finally, when failure occurs, the actual working life of the current device is uploaded to the distributed file system to provide feedback.

Algorithm2: pseudocode of off-line analysis and prediction

Initialization

for $k \leftarrow 1$ **to** N //for the Kthmaintenance project

   While

       $T_i^k \leftarrow$ **neuralNetwork**(f_i^k, v_i^k, ap_i^k);

       $\eta_i^k \leftarrow t_i^k / T_i^k$;

       $\eta^k \leftarrow \eta^k + \eta_i^k$;

       if ($\eta^k > \theta^k$)

       Break;

   End While

   // **sum()** function sums up T_i^k while i is between 1 and m

   $T_{remain}^k \leftarrow (1 - \eta^k) \cdot (\partial \cdot \overline{T}^k + (1 - \partial) \cdot$ **sum**(T_i^k)/m);

   wait(T_{remain}^k); // continue to use until maintained

   for $j \leftarrow 1$ **to** $numStaff[k]$ //call maintenance staff

       if ($maintainStaff[k,j]$==0)

          call $maintainStaff(k, j)$;

          $maintainStaff(k,j) \leftarrow 1$; //Set to work status

          wait();//Wait for staff to finish the maintenance

          $maintainStaff(k,j) \leftarrow 0$;

          Break;

       End if

   End for

   //update the actual working life to the distributed file system

   update(T^k);

End for

FIGURE 4.126   Algorithm 4.2: Pseudocode of offline analysis and prediction.

For the Kth maintenance item, the relative lifetime total consumption rate η^k of the equipment is the sum of the consumption rate under different processing conditions. t_i^k is the processing time under the Ith processing conditions and T_i^k is the processing life under these conditions. In the algorithm of MOAP, a neural network is established which is responsible for calculating life time under specific processing conditions. In general cutting processes, the classic three cutting elements (cutting speed v_i^k, feed f_i^k, and cutting depth ap_i^k) approximately reflect the characteristic under specific process condition which implies life time. Thus, this algorithm chooses the three cutting elements as the neural network's input. Considering different dimensions between three cutting elements, as a typical procedure in machine learning, the value of the three should firstly be normalized. Additionally, the algorithm chooses sigmoid function as activation function as logistic regression does. In the training phase, the output of the neural network is the recorded life time under specific processing condition. To quantify the recorded life time into discrete form so that the algorithm could train the neural network, the output layer has eight neural which represents 20, 40, 60, 80, 2, 4, 6, and 8 minutes, respectively. As the structure discussed above, error might be generated when quantifying the recorded life time into defined output form. By using backpropagation algorithm, we can train the neural network using batch gradient descent and the overall cost function as follows:

$$J(W,b) = \left[\frac{1}{m} \sum_{i=1}^{m} \left(\frac{1}{2} h_{W,b}(x) - y^2 \right) \right] + \frac{\lambda}{2} \sum_{l=1}^{n_--1} \sum_{i=1}^{s_s} \sum_{j=1}^{s_s+1} \left(W_{ji}^l \right)^2 \tag{4.100}$$

The first term in the definition of $J(W,b)$ is an average sum-of-squares error term. The second term is a regularization term that tends to decrease the magnitude of the weights and helps prevent overfitting.

After training the neural network, when transferring processing parameters into the neural network, the MOAP calculates the processing life under specific processing conditions by transferring the processing parameters into a neural network established by MOAP (the training samples are stored in the distributed file system). The total consumption rate is then updated by adding t_i^k / T_i^k. The value of η^k is obtained by the following formula:

$$\eta^k = \sum \eta_i^k = \sum \frac{t_i^k}{T_i^k} \tag{4.101}$$

As shown in the above formula, t_i^k and T_i^k are the processing time and life time of the Kth unit of equipment under the Ith production condition, respectively. In reality, continuous processes on a processing unit might have an impact on each other. While the consumption rate η^k in this paper is defined by sum of several η_i^k directly which is an assumption in order to simplify the analysis.

When the total consumption rate η^k reaches a threshold θ^k, MOAP begins to estimate the remaining effective working time T_{remain}^k:

$$T_{\text{remain}}^k = \left(1 - \eta^k \right) \cdot \left(\partial \cdot \overline{T}^k + (1 - \partial) \cdot \frac{1}{m} \sum_{i=1}^{m} T_i^k \right) \tag{4.102}$$

As formula (4.102) indicates, $\overline{T}^k$ is the average life expectancy of the equipment history based on statistics. $\partial(0 \leq \partial \leq 1)$ is the proportion factor. ∂ implies the impact of $\overline{T}^k$ on the predictive life time of the current equipment (The greater ∂ is, the greater the historical records impact). To some degree, the value of ∂ relies on the empirical which means that when the equipment is process sensitive, ∂ might be smaller. While, if the equipment is steady enough, we may tend to take the historical records into consideration, namely bigger ∂. The above algorithm is shown in Figure 4.126.

4.4.3.3 Analysis and Experiment

4.4.3.3.1 Prototyping Platform   The experimental prototype platform is composed of a physical resource layer, a network communication layer, and a cloud layer. The physical resource layer mainly includes different types of robots, mobile robots, and other processing equipment; the network communication layer mainly comprises routing devices, data packet forwarding nodes, and other network devices; and the cloud layer is based on the XenServer and Hadoop cloud ecosystem framework which has been already established as a private cloud platform and a big data processing platform. In summary, an industrial prototype platform of big data active operation and maintenance has been built, from the bottom level to the top level.

The processing equipment used for experimental verification is a machining center. Remote control and data transmission of the machining center was realized by connecting the embedded equipment and the machining center. In order to realize data acquisition, data preprocessing, data upload, and real-time processing, the embedded devices that were connected to the machining center were incorporated into a distributed data acquisition and transmission system based on Flume. Based on this, the data collected by Flume was added to a distributed message queue based on Kafka. The queued messages were then entered into a real-time computing system STORM according to a queue rule. The programming interface of STORM was implemented with real-time detection and active maintenance logic in order to achieve real-time active operation and maintenance by the experimental prototype platform. In addition, the introduction of STORM and Hadoop is significant to realize offline analysis and prediction. As indicated by Algorithm 2 (Figure 4.126), a neural network was established by STORM to ensure timely input parameters. In order to calculate the remain working time T_{remain}^k, Hadoop, the batch computing system, was responsible for implementing the logic of formula (4.102) using the historical information stored in the HDFS. It should be noted that the historical data had already been previously collected by Flume. The data was then uploaded into HDFS at particular time intervals. Based on this, the analysis report showed that a maintenance resource was about to be called using the MOAP.

4.4.3.3.2 *Analysis and Result*   Due to its active fault reporting and other features, real-time active maintenance has significantly improved benefits that are superior to the traditional maintenance mode features of decentralized maintenance, reporting after discovery, and

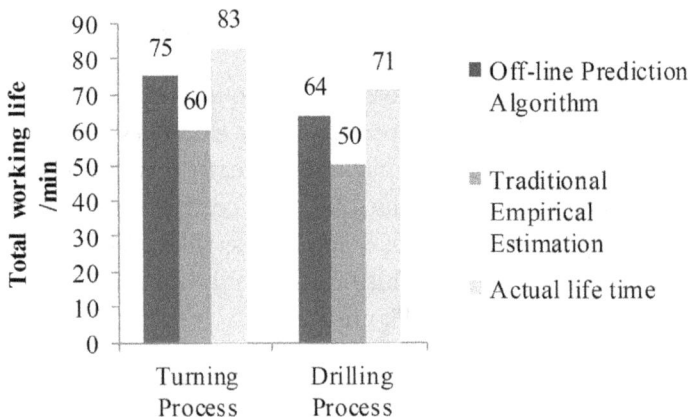

FIGURE 4.127   Comparison of tool working time based on MOAP, traditional experience estimation, and actual life time.

layer-level reporting. Traditional maintenance methods usually emphasize the logical relationship between equipment, workshop, and factory within the operation and maintenance process. In contrast, real-time active maintenance removes logical dependencies between the three entities. By removing these relationships, the individual elements are mapped only to corresponding maintenance resources, significantly simplifying the operational logic. Additionally, through a visualized unified management, real-time maintenance can achieve dynamic monitoring of the state of the whole factory. Compared with decentralized maintenance, reporting after discovery and other features of traditional maintenance mode, active operation and maintenance has obvious advantages in terms of response time. Finally, in contrast with layer-level fault reporting in traditional maintenance mode, the advantages of active reporting and the response of active maintenance to real-time active maintenance reflect more superior efficiency and flexibility. Real-time active maintenance mode and traditional maintenance mode are briefly compared in Table 4.35.

The offline prediction algorithm was verified using a machining center experiment. Effective use of the tool life was compared for the offline prediction algorithm and the traditional empirical estimation using two typical processes: turning and drilling. Both the tool and the material that was processed were represented by the Chinese National Standard (GB). A YT15 carbide cutting tool was used as the turning tool. Since the actual machining process is usually accompanied by different processing materials and processing parameters, stainless steel Y1Cr18Ni9, carbon structural steel Q235, and gray cast iron HB190 were designated as the processing materials according to the material cutting difficulty level. Similarly, for drilling, YG6 carbide drill was used as the cutting tool. Cemented carbide YG6A, medium carbon quenched and tempered steel 45th steel, and carbon tool steel T10A were designated as the processing materials. The specific parameters are shown in Table 4.36.

TABLE 4.35   Comparison of Real-Time Active Maintenance and Traditional Maintenance

	Real-Time Active Maintenance and Operation	Traditional Maintenance and Operation
Relationship between equipment, workshop, and factory	Flattened relationship	Hierarchical relationship
Management mode of operation maintenance	Centralized and unified	Decentralized operation and maintenance
Response time of operation and maintenance	Short	Long
Response mode	Active reporting	Reporting afterwards
Monitoring model	Visualized and unified presentation	Layer-level reporting

TABLE 4.36   Test Parameters for the Offline Prediction Experiment

	Turning Process	Drilling Process
Tool Material	YT15	YG6
Machined Material I	YlCrl8Ni9	YG6A
Machined Material II	Q235	45#
Machined Material III	HB190	T10A

The calculation of remaining working time of traditional method takes the "Concise Manual of Cutting Dosage" into consideration, which is authoritative and usually taken as reference in machining field. On the other hand, experimental performance is representative to reflect lifetime of specific tools. Considering the tools' performance in our experiment and content of "Concise Manual of Cutting Dosage," the empirical life expectancy of hard alloy tool and carbide drill in our experiment are predicted as 60 and 50 minutes, respectively.

For new method proposed in the chapter, when the total relative consumption rate η^k reaches the threshold θ^k, the MOAP will trigger prediction operation. By combining the historical data $\left(\bar{T}^k\right)$ stored in the distributed file system and the estimated life of the worked procedures $\left(T_i^k, i=1,2\ldots,m\right)$, the remaining effective life of the current tool T_{remain}^k is estimated. In this experiment, the threshold θ^k is set to 0.8 and the proportion factor ∂ set to 0.4. As the experiment indicates, by adding the consumed processing time and the predicted remaining lifetime, the total working time of turning process and drilling process are obtained as 75 and 64 minutes, respectively. As the object of reference, we added actual working life of the two experiments as 83 and 71 minutes. The experimental results are shown in Figure 4.127.

Using the tool and machining material specified in the experiment, when the relative consumption rate accumulated to a threshold value (0.8), the residual effective time prediction was triggered by MOAP. When the estimated remaining useful time had elapsed, the tool was now due for replacement. Comparison between traditional experienced estimation and offline prediction has been shown as in Figure 4.125. The actual life time in Figure 4.125 is regarded as a base line or the object of reference. As shown in Figure 4.125, compared with the traditional experienced processing time, the total amount of processing based on the offline prediction algorithm was significantly improved. The main reason for this is that traditional estimation methods are usually conservative to ensure the fluency and safety of the process. In addition to using historical data, the latter algorithm has a greater focus on the specific processes which have actually been performed by the current tool. Therefore, the total processing time can be increased while still guaranteeing processing stability.

4.4.3.3.3 Discussions   In this section, we concentrate on active preventive maintenance with industrial big data, which conforms to the trend of smart factory and digital manufacturing. The integration of Hadoop and Storm proposes a relatively comprehensive maintenance solution from aspects of both offline and online circumstances. In the later part of this paper, we take turning process and drilling process for example to illustrate the effect with introduction of data-driven maintenance, which reflects the contribution of data-driven method, while the experiment may not thoroughly embody the high-performance maintenance effect because of the difficulty to experiment with all layers of the proposed architecture.

Data collection, the first step and the basis of industrial big data, plays a vital role in industrial data-driven maintenance. With the integration of devices from different equipment suppliers, diversity of communication protocol, different degrees of device openness, and different intelligent levels between devices make it hard to establish fine-grained data flow for later statistical analysis and machine learning.

Fortunately, as mentioned in Part A, the appearance of OPC UA based on OPC Classic brings us a ray of light.

With the attribution of platform Independence, extensible model, and flexible Client/Server communication mode, OPC UA demonstrates a compatible and robust architecture, which may be the promising breakthrough for industrial big data and relative issues.

In this paper, we proposed to use neural network with historical data stored in HDFS to evaluate the total working time. With the increasing generated data stored in HDFS, the neural network cannot make fully use of those data because of its limited model complexity which is defined before training. With information technologies permeating all aspects of industry, more and more advanced algorithms are being taken into consideration in active preventive maintenance. However, in the industrial environment, requirement of robustness and real-time processing plays more important roles. For example, deep learning, which is proved successful in image recognition and natural language processing, could make use of stored data but may not be appropriate in active preventive maintenance. Algorithms which may not be that advanced but with expert knowledge might lead to better performance.

That is why formula (4.102) was applied to evaluate the remaining effective working time, and ∂, which is determined by expert knowledge, is the most influential factor to model accuracy.

4.5 SUMMARY

This book chapter significantly discusses the intelligent fault diagnosis and maintenance methods in smart manufacturing factory. The problems and challenges of existing fault diagnosis methods are summarized. It introduces how to effectively use advanced machine learning algorithms for supervised, semi-supervised, and unsupervised intelligent fault diagnosis to deal with the existing dilemma. Based on the reliability and maintainability requirements of smart production, it discusses how big data, digital twin, and cloud technology can help manufacturing equipment failure prediction and intelligent maintenance.

REFERENCES

[1] X. Dai and Z. Gao, "From model, signal to knowledge: A data-driven perspective of fault detection and diagnosis," *IEEE Transactions on Industrial Informatics*, vol. 9, no. 4, pp. 2226–2238, 2013.

[2] J. Wan, D. Zhang, S. Zhao, L. T. Yang and J. Lloret, "Context-aware vehicular cyber-physical systems with cloud support: Architecture, challenges, and solutions," *IEEE Communications Magazine*, vol. 52, no. 8, pp. 106–113, 2014.

[3] X. Li, D. Li, J. Wan, A. V. Asilakos, C. F. Lai and S. Wang, "A review of industrial wireless networks in the context of industry 4.0," *Wireless Networks*, vol. 23, no. 1, pp. 23–41, 2017.

[4] Z. Shu, J. Wan, D. Zhang and D. Li, "Cloud-integrated cyber-physical systems for complex industrial applications," *Mobile Networks and Applications*, vol. 21, no. 5, pp. 865–878, 2016.

[5] Y. Xu, Y. Sun, J. Wan, X. Liu and Z. Song, "Industrial big data for fault diagnosis: Taxonomy, review, and applications," *IEEE Access*, vol. 5, pp. 17368–17380, 2017.

[6] J. Wan, S. Tang, D. Li et al., "A manufacturing big data solution for active preventive maintenance," *IEEE Transactions on Industrial Informatics*, vol. 13, no. 4, pp. 2039–2047, 2017.

[7] E. T. Esfahani, S. Wang and V. Sundararajan, "Multisensor wireless system for eccentricity and bearing fault detection in induction motors," *IEEE/ASME Transactions on Mechatronics*, vol. 19, no. 3, pp. 818–826, 2014.

[8] Y. Hu, W. Bao, X. Tu, F. Li and K. Li, "An adaptive spectral kurtosis method and its application to fault detection of rolling element bearings," *IEEE Transactions on Instrumentation and Measurement*, vol. 69, no. 3, pp. 739–750, 2020.

[9] Z. Huo, M. Martínez-García, Y. Zhang, R. Yan and L. Shu, "Entropy measures in machine fault diagnosis: Insights and applications," *IEEE Transactions on Instrumentation and Measurement*, vol. 69, no. 6, pp. 2607–2620, 2020.

[10] L. Wang, G. Cai, J. Wang, X. Jiang and Z. Zhu, "Dual-enhanced sparse decomposition for wind turbine gearbox fault diagnosis," *IEEE Transactions on Instrumentation and Measurement*, vol. 68, no. 2, pp. 450–461, 2019.

[11] Y. Hao, L. Song, B. Ren, H. Wang and L. Cui, "Step-by-step compound faults diagnosis method for equipment based on majorizationminimization and constraint sca," *IEEE/ASME Transactions on Mechatronics*, vol. 24, no. 6, pp. 2477–2487, 2019.

[12] H. Shao, H. Jiang, H. Zhang and T. Liang, "Electric locomotive bearing fault diagnosis using a novel convolutional deep belief network," *IEEE Transactions on Industrial Electronics*, vol. 65, no. 3, pp. 2727–2736, 2018.

[13] F. Jia, Y. Lei, J. Lin, X. Zhou and N. Lu, "Deep neural networks: A promising tool for fault characteristic mining and intelligent diagnosis of rotating machinery with massive data," *Mechanical Systems and Signal Processing*, vol. 72–73, pp. 303–315, 2016.

[14] S. Shao, S. McAleer, R. Yan and P. Baldi, "Highly accurate machine fault diagnosis using deep transfer learning," *IEEE Transactions on Industrial Informatics*, vol. 15, no. 4, pp. 2446–2455, 2019.

[15] M. Xia, T. Li, T. Shu, J. Wan, C. W. de Silva and Z. Wang, "A two-stage approach for the remaining useful life prediction of bearings using deep neural networks," *IEEE Transactions on Industrial Informatics*, vol. 15, no. 6, pp. 3703–3711, 2019.

[16] M. Xia, T. Li, L. Xu, L. Liu and C. W. de Silva, "Fault diagnosis for rotating machinery using multiple sensors and convolutional neural networks," *IEEE/ASME Transactions on Mechatronics*, vol. 23, no. 1, pp. 101–110, 2018.

[17] S. Chen, Y. Meng, H. Tang, Y. Tian, N. He and C. Shao, "Robust deep learning-based diagnosis of mixed faults in rotating machinery," *IEEE/ASME Transactions on Mechatronics*, vol. 25, no. 5, pp. 2167–2176, 2020.

[18] G. Jiang, P. Xie, H. He and J. Yan, "Wind turbine fault detection using a denoising autoencoder with temporal information," *IEEE/ASME Transactions on Mechatronics*, vol. 23, no. 1, pp. 89–100, 2018.

[19] J. Wang, P. Fu, L. Zhang, R. Gao and R. Zhao, "Multilevel information fusion for induction motor fault diagnosis," *IEEE/ASME Transactions on Mechatronics*, vol. 24, no. 5, pp. 2139–2150, 2019.

[20] X. Pang, X. Xue, W. Jiang and K. Lu, "An investigation into fault diagnosis of planetary gearboxes using a bispectrum convolutional neural network," *IEEE/ASME Transactions on Mechatronics*, 2020, doi: 10.1109/TMECH.2020.3029058.

[21] X. Zhao, M. Jia, P. Ding, C. Yang, D. She and Z. Liu, "Intelligent fault diagnosis of multichannel motor-rotor system based on multi-manifold deep extreme learning machine," *IEEE/ASME Transactions on Mechatronics*, vol. 25, no. 5, pp. 2177–2187, 2020.

[22] J. Xiong, Q. Liang, J. Wan, Q. Zhang, X. Chen and R. Ma, "The order statistics correlation coefficient and PPMCC fuse non-dimension in fault diagnosis of rotating petrochemical unit," *IEEE Sensors J.*, vol. 18, no. 11, pp. 4704–4714, 2018.

[23] J. B. Xiong, Q. H. Zhang, G. X. Sun, X. T. Zhu, Z. L. Li and M. Liu, "An information fusion fault diagnosis method based on dimensionless indicators with static discounting factor and KNN," *IEEE Sensors J.*, vol. 16, no. 7, pp. 2060–2069, 2016.

[24] Q. H. Zhang, L. Shao, H. Li et al., "Research of concurrent faults diagnosis technology of rotating machinery based on the non-dimensional parameter," *Journal of Huazhong University of Science and Technology (Natural Science Edition)*, vol. 37, no. 1, pp. 156–159, 2009.

[25] H. Shao, H. Jiang, L. Ying and X. Li, "A novel method for intelligent fault diagnosis of rolling bearings using ensemble deep auto-encoders," *Mechanical Systems and Signal Processing*, vol. 102, pp. 278–297, 2018.

[26] Y. Li, K. Feng, X. Liang and M. J. Zuo, "A fault diagnosis method for planetary gearboxes under non-stationary working conditions using improved Vold-Kalman filter and multi-scale sample entropy," *Journal of Sound and Vibration*, vol. 439, pp. 271–286, 2019.

[27] H. Shao, H. Jiang, X. Li and S. Wu, "Intelligent fault diagnosis of rolling bearing using deep wavelet auto-encoder with extreme learning machine," *Knowledge-Based Systems*, vol. 140, pp. 1–14, 2018.

[28] X. Wen, Q. Miao, J. Wang and Z. Ju, "A multi-resolution wavelet neural network approach for fouling resistance forecasting of a plate heat exchanger," *Applied Soft Computing*, vol. 57, pp. 177–196, 2017.

[29] Y. Lei, Z. He and Y. Zi, "EEMD method and WNN for fault diagnosis of locomotive roller bearings," *Expert Systems with Applications*, vol. 38, no. 6, pp. 7334–7341, 2011.

[30] B. Tang, W. Liu and T. Song, "Wind turbine fault diagnosis based on Morlet wavelet transformation and Wigner-Ville distribution," *Renewable Energy*, vol. 35, no. 12, pp. 2862–2866, 2010.

[31] Y. Jiang, B. Tang, Y. Qin and W. Liu, "Feature extraction method of wind turbine based on adaptive Morlet wavelet and SVD," *Renewable Energy*, vol. 36, no. 8, pp. 2146–2153, 2011.

[32] S. Hassairi, R. Ejbali and M. Zaied, "A deep stacked wavelet autoencoders to supervised feature extraction to pattern classification," *Multimedia Tools and Applications*, vol. 4, pp. 1–17, 2017.

[33] W. T. Pan, "A new fruit fly optimization algorithm: Taking the financial distress model as an example," *Knowledge-Based Systems*, vol. 26, pp. 69–74, 2012.

[34] Y. Pan, R. Hong, J. Chen and W. Wu, "A hybrid DBN-SOM-PF-based prognostic approach of remaining useful life for wind turbine gearbox," *Renewable Energy*, vol. 152, pp. 138–154, 2020.

[35] E. Hosseini-Asl, J. M. Zurada and O. Nasraoui, "Deep learning of part-based representation of data using sparse autoencoders with nonnegativity constraints," *IEEE Transactions on Neural Networks and Learning*, vol. 27, no. 12, pp. 2486–2498, 2016.

[36] A. Afan and Y. Fan, "Automatic modulation classification using deep learning based on sparse autoencoders with nonnegativity constraints," *IEEE Signal Processing Letters*, vol. 24, no. 11, pp. 1626–1630, 2017.

[37] H. Shao, M. Xia, J. Wan and C. W. de Silva, "Modified stacked autoencoder using adaptive Morlet wavelet for intelligent fault diagnosis of rotating machinery," *IEEE/ASME Transactions on Mechatronics*, vol. 27, no. 1, pp. 24–33, 2022.

[38] W. Deng, H. Liu, J. Xu, H. Zhao and Y. Song, "An improved quantum-inspired differential evolution algorithm for deep belief network," *IEEE Transactions on Instrumentation and Measurement*, vol. 69, no. 10, pp. 7319–7327, 2020.

[39] M. Zhao, S. Zhong, X. Fu, B. Tang, S. Dong and M. Pecht, "Deep residual networks with adaptively parametric rectifier linear units for fault diagnosis," *IEEE Transactions on Industrial Electronics*, vol. 68, no. 3, pp. 2587–2597, 2021.

[40] F. Jia, Y. Lei, J. Lin, N. Lu and S. Xing, "Deep normalized convolutional neural network for imbalanced fault classification of machinery and its understanding via visualization," *Mechanical Systems and Signal Processing*, vol. 15, pp. 349–367, 2018.

[41] R. Zhao, D. Wang, R. Yan, K. Mao, F. Shen and J. Wang, "Machine health monitoring using local feature-based gated recurrent unit networks," *IEEE Transactions on Industrial Electronics*, vol. 65, no. 2, pp. 1539–1548, 2018.

[42] Z. Zhao, T. Li, J. Wu et al., "Deep learning algorithms for rotating machinery intelligent diagnosis: An open source benchmark study," *ISA Transactions*, vol. 107, pp. 224–255, 2020.

[43] J. E. van Engelen and H. H. Hoos, "A survey on semi-supervised learning," *Machine Learning*, vol. 109, no. 2, pp. 373–440, 2019.

[44] K. Yu, H. Ma, T. R. Lin and X. Li, "A consistency regularization based semi-supervised learning approach for Intelligent Fault diagnosis of rolling bearing," *Measurement*, vol. 165, p. 107987, 2020.

[45] P. Liang, C. Deng, J. Wu, G. Li, Z. Yang and Y. Wang, "Intelligent fault diagnosis via semisupervised generative adversarial nets and wavelet transform," *IEEE Transactions on Instrumentation and Measurement*, vol. 69, no. 7, pp. 4659–4671, 2020.

[46] X. Wu, Y. Zhang, C. Cheng and Z. Peng, "A hybrid classification autoencoder for semi-supervised fault diagnosis in Rotating Machinery," *Mechanical Systems and Signal Processing*, vol. 149, p. 107327, 2021.

[47] Y. Tang, X. Zhang, G. Qin et al., "Graph cardinality preserved attention network for fault diagnosis of induction motor under varying speed and load condition," *IEEE Transactions on Industrial Informatics*, vol. 18, no. 6, pp. 3702–3712, 2022.

[48] J. Liu, K. Zhou, C. Yang and G. Lu, "Imbalanced fault diagnosis of rotating machinery using autoencoder-based SuperGraph feature learning," *Frontiers of Mechanical Engineering*, vol. 16, no. 4, pp. 829–839, 2021.

[49] M. Lyu, X. Li and C.-H. Chen, "Achieving knowledge-as-a-service in IIOT-driven Smart Manufacturing: A crowdsourcing-based continuous enrichment method for industrial knowledge graph," *Advanced Engineering Informatics*, vol. 51, p. 101494, 2022.

[50] M. Liu, X. Li, J. Li, Y. Liu, B. Zhou and J. Bao, "A knowledge graph-based data representation approach for IIOT-enabled cognitive manufacturing," *Advanced Engineering Informatics*, vol. 51, p. 101515, 2022.

[51] T. Li, Z. Zhao, C. Sun, R. Yan and X. Chen, "Hierarchical attention graph convolutional network to fuse multi-sensor signals for remaining useful life prediction," *Reliability Engineering & System Safety*, vol. 215, p. 107878, 2021.

[52] C. Yang, J. Liu, K. Zhou, X. Jiang and X. Zeng, "An improved multi-channel graph convolutional network and its applications for rotating machinery diagnosis," *Measurement*, vol. 190, p. 110720, 2022.

[53] K. Zhang, J. Chen, S. He, F. Li, Y. Feng and Z. Zhou, "Triplet metric driven multi-head GNN augmented with decoupling adversarial learning for intelligent fault diagnosis of machines under varying working condition," *Journal of Manufacturing Systems*, vol. 62, pp. 1–16, 2022.

[54] X. Zhao, M. Jia and Z. Liu, "Semisupervised graph convolution deep belief network for fault diagnosis of Electormechanical system with limited labeled data," *IEEE Transactions on Industrial Informatics*, vol. 17, no. 8, pp. 5450–5460, 2021.

[55] Y. Gao, M. Chen and D. Yu, "Semi-supervised graph Convolutional Network and its application in intelligent fault diagnosis of rotating machinery," *Measurement*, vol. 186, p. 110084, 2021.

[56] Y. Tang, X. Zhang, Y. Zhai et al., "Rotating machine systems fault diagnosis using semisupervised conditional random field-based graph attention network," *IEEE Transactions on Instrumentation and Measurement*, vol. 70, pp. 1–10, 2021.

[57] Y. Chong, Y. Ding, Q. Yan and S. Pan, "Graph-based semi-supervised learning: A review," *Neurocomputing*, vol. 408, pp. 216–230, 2020.

[58] S. Brody, U. Alon and E. Yahav, "How attentive are graph attention networks," vol. arXiv.2105, pp. 14491, 2022.

[59] H. Cao, H. Shao, X. Zhong, Q. Deng, X. Yang and J. Xuan, "Unsupervised domain-share CNN for machine fault transfer diagnosis from steady speeds to time-varying speeds," *Journal of Manufacturing Systems*, vol. 62, pp. 186–198, 2022.

[60] Z. Zhao, Q. Zhang, X. Yu et al., "Applications of unsupervised deep transfer learning to intelligent fault diagnosis: A survey and comparative study," *IEEE Transactions on Instrumentation and Measurement*, vol. 70, pp. 1–28, 2021.

[61] S. Yan, H. Shao, Y. Xiao, J. Zhou, Y. Xu and J. Wan, " Semi-supervised fault diagnosis of machinery using LPS-DGAT under speed fluctuation and extremely low labeled rates," *Advanced Engineering Informatics*, vol.53, p. 101648, 2022.

[62] H. Shao, M. Xia, G. Han, Y. Zhang and J. Wan, "Intelligent fault diagnosis of rotor-bearing system under varying working conditions with modified transfer convolutional neural network and thermal images," *IEEE Transactions on Industrial Informatics*, vol. 17, no. 5, pp. 3488–3496, 2021.

[63] H. Shao, W. Li, M. Xia et al., "Fault diagnosis of a rotor-bearing system under variable rotating speeds using two-stage parameter transfer and infrared thermal images," *IEEE Transactions on Instrumentation and Measurement*, vol. 70, p. 3524711, 2021.

[64] L. Sun, H. Wang and P. Chen, "Vibration-based intelligent fault diagnosis for roller bearings in low-speed rotating machinery," *IEEE Transactions on Instrumentation and Measurement*, vol. 67, no. 8, pp. 1887–1899, 2018.

[65] W. Wang, Y. Lei, T. Yan, N. Li and A. Nandi, "Residual convolution long short-term memory network for machines remaining useful life prediction and uncertainty quantification," *Journal of Dynamics, Monitoring and Diagnostics*, vol. 1, no. 1, pp. 2–8, 2021.

[66] X. Li, H. Shao, H. Jiang and J. Xiang, "Modified Gaussian convolutional deep belief network and infrared thermal imaging for intelligent fault diagnosis of rotor-bearing system under time-varying speeds," *Structural Health Monitoring*, vol. 21, no. 2, pp. 339–353, 2022.

[67] J. Zhu, N. Chen and C. Shen, "A new multiple source domain adaptation fault diagnosis method between different rotating machines," *IEEE Transactions on Industrial Informatics*, vol. 17, no. 7, pp. 4788–4797, 2021.

[68] Y. Lei, F. Jia, J. Lin, S. Xing and S. X. Ding, "An intelligent fault diagnosis method using unsupervised feature learning towards mechanical big data," *IEEE Transactions on Industrial Electronics*, vol. 63, no. 5, pp. 3137–3147, 2016.

[69] J. Li, R. Huang, G. He, Y. Liao, Z. Wang and W. Li, "A two-stage transfer adversarial network for intelligent fault diagnosis of rotating machinery with multiple new faults," *IEEE/ASME Transactions on Mechatronics*, vol. 26, no. 3, pp. 1591–1601, 2021.

[70] B. Yang, Y. Lei, F. Jia and S. Xing, "An intelligent fault diagnosis approach based on transfer learning from laboratory bearings to locomotive bearings," *Mechanical Systems and Signal Processing*, vol. 122, pp. 692–706, 2019.

[71] Z. An, S. Li, Y. Xin, K. Xu and H. Ma, "An intelligent fault diagnosis framework dealing with arbitrary length inputs under different working conditions," *Measurement Science and Technology*, vol. 30, no. 12, p. 125107, 2019.

[72] K. Yu, Q. Fu, H. Ma, T. Lin and X. Li, "Simulation data driven weakly supervised adversarial domain adaptation approach for intelligent cross-machine fault diagnosis," *Structural Health Monitoring*, vol. 20, no. 4, pp. 2182–2198, 2020.

[73] Y. Gao, X. Liu and J. Xiang, "FEM simulation-based generative adversarial networks to detect bearing faults," *IEEE Transactions on Industrial Informatics*, vol. 16, no. 7, pp. 4961–4971, 2021.

[74] Y. Zhou, Y. Dong, H. Zhou and G. Tang, "Deep dynamic adaptive transfer network for rolling bearing fault diagnosis with considering cross-machine instance," *IEEE Transactions on Instrumentation and Measurement*, vol. 70, p. 3525211, 2021.

[75] T. Han, C. Liu, W. Yang and D. Jiang, "Deep transfer network with joint distribution adaptation: A new intelligent fault diagnosis framework for industry application," *ISA Transactions*, vol. 97, pp. 269–281, 2020.

[76] W. Li, Z. Chen and G. He, "A novel weighted adversarial transfer network for partial domain fault diagnosis of machinery," *IEEE Transactions on Industrial Informatics*, vol. 17, no. 3, pp. 1753–1762, 2021.

[77] X. Li, H. Jiang, K. Zhao and R. Wang, "A deep transfer nonnegativity-constraint sparse autoencoder for rolling bearing fault diagnosis with few labeled data," *IEEE Access*, vol. 7, pp. 91216–91224, 2019.

[78] C. Liu, H. Vengayil, R. Y. Zhong and X. Xu, "A systematic development method for cyber-physical Machine Tools," *Journal of Manufacturing Systems*, vol. 48, pp. 13–24, 2018.

[79] W. Liu, C. Kong, Q. Niu, J. Jiang and X. Zhou, "A method of NC Machine Tools Intelligent Monitoring System in smart factories," *Robotics and Computer-Integrated Manufacturing*, vol. 61, p. 101842, 2020.

[80] X. Xu, "Machine tool 4.0 for the new era of manufacturing," *The International Journal of Advanced Manufacturing Technology*, vol. 92, no. 5–8, pp. 1893–1900, 2017.

[81] L. Wang, "From intelligence science to intelligent manufacturing," *Engineering*, vol. 5, no. 4, pp. 615–618, 2019.

[82] C. Liu, P. Zheng and X. Xu, "Digitalisation and servitisation of machine tools in the era of industry 4.0: A review," *International Journal of Production Research*, pp. 1–33, 2021.

[83] J. Ratava, M. Lohtander and J. Varis, "Tool condition monitoring in interrupted cutting with acceleration sensors," *Robotics and Computer-Integrated Manufacturing*, vol. 47, pp. 70–75, 2017.

[84] G. Li, Y. Fu, D. Chen, L. Shi and J. Zhou, "Deep anomaly detection for CNC machine cutting tool using spindle current signals," *Sensors*, vol. 20, no. 17, p. 4896, 2020.

[85] C. Zhou, K. Guo and J. Sun, "Sound singularity analysis for milling tool condition monitoring towards sustainable manufacturing," *Mechanical Systems and Signal Processing*, vol. 157, p. 107738, 2021.

[86] T. Li, T. Shi, Z. Tang et al., "Real-time tool wear monitoring using thin-film thermocouple," *Journal of Materials Processing Technology*, vol. 288, p. 116901, 2021.

[87] Y. C. Liang, S. Wang, W. D. Li and X. Lu, "Data-driven anomaly diagnosis for machining processes," *Engineering*, vol. 5, no. 4, pp. 646–652, 2019.

[88] Z. Huang, J. Zhu, J. Lei, X. Li and F. Tian, "Tool wear predicting based on multi-domain feature fusion by deep convolutional neural network in milling operations," *Journal of Intelligent Manufacturing*, vol. 31, no. 4, pp. 953–966, 2019.

[89] C. Sun, M. Ma, Z. Zhao, S. Tian, R. Yan and X. Chen, "Deep transfer learning based on sparse Autoencoder for remaining useful life prediction of tool in manufacturing," *IEEE Transactions on Industrial Informatics*, vol. 15, no. 4, pp. 2416–2425, 2019.

[90] M. Cheng, L. Jiao, P. Yan et al., "Intelligent tool wear monitoring and multi-step prediction based on Deep Learning Model," *Journal of Manufacturing Systems*, vol. 62, pp. 286–300, 2022.

[91] X. Li, Z. Zhang, L. Gao and L. Wen, "A new semi-supervised fault diagnosis method via deep coral and transfer component analysis," *IEEE Transactions on Emerging Topics in Computational Intelligence*, vol. 6, no. 3, pp. 690–699, 2022.

[92] B. Zong, Q. Song, M. R. Min et al., "Deep autoencoding gaussian mixture model for unsupervised anomaly detection," *Proc. of ICLR*, pp. 1–6, 2018.

[93] D. Gong, L. Liu, V. Le et al., "Memorizing normality to detect anomaly: Memory-augmented deep autoencoder for unsupervised anomaly detection," *2019 IEEE/CVF International Conference on Computer Vision (ICCV)*, 2019.

[94] M. Thill, W. Konen, H. Wang and T. Bäck, "Temporal convolutional autoencoder for unsupervised anomaly detection in time series," *Applied Soft Computing*, vol. 112, p. 107751, 2021.

[95] D. Li, H. Guo, Z. Wang and Z. Zheng, "Unsupervised fake news detection based on autoencoder," *IEEE Access*, vol. 9, pp. 29356–29365, 2021.

[96] M. Zhao, J. Jia and J. Lin, "A data-driven monitoring scheme for rotating machinery via self-comparison approach," *IEEE Transactions on Industrial Informatics*, vol. 15, no. 4, pp. 2435–2445, 2019.

[97] L. Wan, G. Han, L. Shu, S. Chan and N. Feng, "PD source diagnosis and localization in industrial high-voltage insulation system via multimodal joint sparse representation," *IEEE Transactions on Industrial Electronics*, vol. 63, no. 4, pp. 2506–2516, 2016.

[98] A. Abid, M. Khan and M. Khan, "Multidomain features-based GA optimized artificial immune system for bearing fault detection," *IEEE Transactions on Systems, Man, and Cybernetics: Systems*, vol. 50, no. 1, pp. 348–359, 2020.

[99] M. Saufi, Z. Ahmad, M. Leong and M. Lim, "Gearbox fault diagnosis using a deep learning model with limited data sample," *IEEE Transactions on Industrial Informatics*, vol. 16, no. 10, pp. 6263–6271, 2020.

[100] J. Pan, Y. Zi, J. Chen, Z. Zhou and B. Wang, "LiftingNet: A novel deep learning network with layerwise feature learning from noisy mechanical data for fault classification," *IEEE Transactions on Industrial Electronics*, vol. 65, no. 6, pp. 4973–4982, 2018.

[101] J. Jiao, M. Zhao, J. Lin and C. Ding, "Deep coupled dense convolutional network with complementary data for intelligent fault diagnosis," *IEEE Transactions on Industrial Electronics*, vol. 66, no. 12, pp. 9858–9867, 2019.

[102] C. Ding, M. Zhao, J. Lin, J. Jiao and K. Liang, "Sparsity based algorithm for condition assessment of rotating machinery using internal encoder data," *IEEE Transactions on Industrial Electronics*, vol. 67, no. 9, pp. 7982–7993, 2020.

[103] Z. Jia, Z. Liu, C. Vong and M. Pecht, "A rotating machinery fault diagnosis method based on feature learning of thermal images," *IEEE Access*, vol. 7, pp. 12348–12359, 2019.

[104] M. Delgado-Prieto, J. Carino-Corrales, J. Saucedo-Dorantes, R. RomeroTroncoso and R. Osornio-Rios, "Thermography-based methodology for multifault diagnosis on kinematic chain," *IEEE Transactions on Industrial Informatics*, vol. 14, no. 12, pp. 5553–5562, 2018.

[105] V. Tran, B. Yang, F. Gu and A. Ball, "Thermal image enhancement using bi-dimensional empirical mode decomposition in combination with relevance vector machine for rotating machinery fault diagnosis," *Mechanical Systems and Signal Processing*, vol. 38, no. 2, pp. 601–614, 2013.

[106] M. Karakose and O. Yaman, "Complex fuzzy system based predictive maintenance approach in railways," *IEEE Transactions on Industrial Informatics*, vol. 16, no. 9, pp. 6023–6032, 2020.

[107] O. Janssens, R. Walle, M. Loccufier and S. Hoecke, "Deep learning for infrared thermal image based machine health monitoring," *IEEE/ASME Transactions on Mechatronics*, vol. 23, no. 1, pp. 151–159, 2018.

[108] O. Janssens, M. Loccufier and S. Hoecke, "Thermal imaging and vibration-based multisensor fault detection for rotating machinery," *IEEE Transactions on Industrial Informatics*, vol. 15, no. 1, pp. 434–444, 2019.

[109] A. Nasiri, A. Taheri-Garavandb, M. Omida and G. Carlomagno, "Intelligent fault diagnosis of cooling radiator based on deep learning analysis of infrared thermal images," *Applied Thermal Engineering*, vol. 163, p. 114410, 2019.

[110] Z. Chen, K. Gryllias and W. Li, "Intelligent fault diagnosis for rotary machinery using transferable convolutional neural network," *IEEE Transactions on Industrial Informatics*, vol. 16, no. 1, pp. 339–349, 2020.

[111] Y. Xiao, H. Shao, S. Y. Han, Z. Huo and J. Wan, "Novel joint transfer network for unsupervised bearing fault diagnosis from simulation domain to experimental domain," *IEEE/ASME Transactions on Mechatronics*, vol. 27, no. 6, pp. 5254–5263, 2022.

[112] B. Chen, J. Wan, M. Xia and Y. Zhang, "Exploring equipment electrocardiogram mechanism for performance degradation monitoring in smart manufacturing," *IEEE/ASME Transactions on Mechatronics*, vol. 25, no. 5, pp. 2276–2286, 2020.

[113] Z. Ren, J. Wan and P. Deng, "Machine-learning-driven digital twin for lifecycle management of complex equipment," *IEEE Transactions on Emerging Topics in Computing*, vol. 10, no. 1, pp. 9–22, 2022.

CHAPTER 5

Resource Dynamic Scheduling in Manufacturing

C URRENTLY, THE MANUFACTURING INDUSTRY faces a number of challenges, some of which are as follows: traditional mass production is not able to adapt to the rapid production of personalized products; resource limitations, environmental pollution, global warming, and an aging global population have become more prominent. The customer-to-manufacture paradigm reflects the characteristics of customized production where a manufacturing system directly interacts with a customer to meet his/her personalized needs. The goal is to realize the rapid customization of personalized products. The new generation of intelligent manufacturing technology offers improved flexibility, transparency, resource utilization, and efficiency of manufacturing processes. Compared with mass production, production organization of customer-to-manufacture is more complex, quality control is more difficult, and energy consumption needs attention. The resource dynamic scheduling has become one of the most prominent characteristics. The concept of a smart factory aims at the rapid manufacturing of a variety of products in small batches. Since the product types may change dynamically, system resources need to be dynamically reorganized. A multi-agent system (MAS) is introduced to negotiate resource scheduling and reconfiguration. What's more, other key technologies including Industrial Internet of Things (IIOT), Artificial Intelligence, and Digital Twin (DT) will also be deeply involved to support flexible customization manufacturing.

5.1 OVERVIEW

Intelligent manufacturing will be greatly beneficial to the integration of distributed competitive resources (e.g., manpower and diverse automated technologies) so that resource dynamic scheduling responding to market changes is possible. Therefore, in smart manufacturing, it is imperative to realize dynamic configurations of manufacturing resources [1]. First, dynamic scheduling mechanism based on edge-cloud collaboration is discussed. Edge computing in the context of IIOT is used to realize resource awareness. With the help

 DOI: 10.1201/9781003460992-5

of cloud platform, cloud robotics support a series of complex scenarios with high computing power. Edge-cloud collaboration helps to realize dynamic scheduling in custom manufacturing. Then, with the development of knowledge reasoning, knowledge graph, transfer learning, and other AI algorithms, dynamic resource reconfiguration can be achieved in different application scenarios. Ontology is introduced to share knowledge between manufacturing resources, which also improves cognitive ability of them. What's more, data-driven methods are integrated to provide decisions during the resource reconfiguration. The details are described as follows:

- **Edge awareness for load balance and scheduling:** In the context of intelligent manufacturing, the proliferation of terminal network devices has given rise to new challenges for operation and maintenance, scalability, and reliability of the data centers. The growth of edge computing has moved the computing from centralized data centers to the periphery of the network. Edge computing aims to address these challenges by creating an open platform with the capability of integrating core capabilities such as networking, computing, storage, and application. It has enabled intelligent services close to the manufacturing unit to meet the key requirements such as agile connection, data analytics via edge nodes, highly responsive cloud services, and privacy-policy strategy [2]. Edge computing can make full use of embedded computing capabilities of field devices to achieve equipment autonomy based on distributed information processing. Furthermore, edge computing supports the need of digital manufacturing enterprises for rapid configuration of the smart factory, which must adapt to personal demands of users and dynamic changes in production conditions.

- **Cloud robotics for service provisioning:** The basic component of CPS is the robots. A robot is an autonomous entity that can perceive the external environment, make intelligent decisions, and trigger physical actions. Although a robot is equipped with computing resources to conduct small-scale data processing, it cannot carry out large-scale computations on its own [3]. In case of widespread deployment, the maintenance of persistent communication among the robots becomes difficult due to their mobility. The lack of storage within the robots causes further disruptions in exchanging and preserving the large volume of data during robot-to-robot interactions. To overcome these limitations, the concept of cloud computing has been extended to multi-robot system, which is termed as cloud robotics [4]. In this paradigm, cloud offers the computing resources such as virtual machines or containers and engages resources from both local robots and remote data centers for scalable and extensive data processing [5].

- **Dynamic scheduling integrating industrial cloud and edge computing:** The development of Industry 4.0 has provided the possibility to meet frequent changes in product type and batches, a sharp decline in the delivery cycle, constraints of quality cost, and other relevant parameters of customized production mode. To ensure the mixed-flow production of diverse products at the same time, dynamic scheduling plays an important role in a flexible and scalable manufacturing system. Therefore, an intelligent manufacturing architecture integrating industrial cloud and edge computing is presented. The intelligent production edge is designed to provide the traditional devices the abilities of data access and self-decision making. Besides, the proposed architecture is modeled as a MAS with the edge intelligence support, describing the agent-based dynamic scheduling mechanism from the three aspects, namely, agent interaction, agent behavior, and negotiation mechanism.

- **Knowledge sharing for resource reconfiguration:** Due to the massive amount of data generated from the manufacturing devices, it is nearly impossible to consider all the manufacturing device resources. Thus, it is important to construct a new manufacturing description model to realize the knowledge sharing based reconfiguration of various manufacturing resources. In this model, the resources can be easily adjusted by running the model. Therefore, ontology modeling is conducted on a device and related attributes of an intelligent production line. An ontology represents an explicit specification of a conceptual model [6] by using a classical symbolic AI reasoning method (i.e., an expert system). Modeling an application domain knowledge through an expert system provides a conceptual hierarchy that supports system integration and interoperability via an interpretable way [7]. The manufacturing resources are mapped to different functions with various attributes. The knowledge sharing based resource reconfiguration framework is proposed to describe the intelligent manufacturing resources. The architecture consisted of four layers: the data layer, the rule layer, the knowledge layer, and the resource layer.

- **Data-driven cognition enhancement and decision-making:** The network technologies of the cognitive IIoT system help build the information relationship between the real world and the virtual space, such as the perceptual control technology, network communication technology, and information processing technology. The data-driven ML methods are embedded in IIoT to set up the innovative applications including intelligent production, networking collaboration, personalized customization, and service expansion. To enable intelligent applications of edge-enhanced IIoT, the cognitive IIoT provides rich datasets for ML to infer, predict, and make decisions when the external variables change. The cognitive IIoT acting on network transmission includes data marking, semantic, and feature abstraction. Moreover, the online learning or offline training is already promoting the typical IIoT applications (e.g., active operation and maintenance). By using the reasoning and prediction based on data, the intelligent decision-making is achieved according to the learning rules. Edge computing helps advanced data modeling and predictive analytics migrate to network edge for mainstream IIoT products. Data collection at the network edge also helps the application model to perceive the environment, promote the improvement of the application model, and help the model adapt to the dynamic environmental change.

5.2 KEY TECHNOLOGIES OF RESOURCE DYNAMIC SCHEDULING IN SMART MANUFACTURING

The cyber-physical production system (CPPS), which combines information communication technology, cyberspace virtual technology, and intelligent equipment technology, is accelerating the path of Industry 4.0 to transform manufacturing from traditional to intelligent. Intelligent manufacturing usually involves complex processing steps due to the long service cycle. This results in the implementation of the control process that is vulnerable to the interference of load, fault, environmental changes, and other uncertain factors. With the advent of the Internet of Things (IoT) era, effective dynamic scheduling of manufacturing resources is facing enormous challenges, especially in hybrid manufacturing. It is important to achieve the optimal resource allocation to cope with overloading and other serious scheduling issues.

In this section, the key technologies of MAS, IIoT, machine learning (ML), and DT will be introduced. These technologies have provided a new way for a wide range of manufacturing resources to optimize management and dynamic scheduling. The IIoT is industrial networks (e.g., wireless sensor networks) that can perceive physical resources in order to achieve reliable operation of the machine through real-time and effective monitoring in a smart factory. Due to the complex order and different functions of equipment, it is impossible to follow some predefined rules to complete an ordered task when the process of mixed orders flows into the smart factory. The multi-agent technology is introduced to solve the problems of order delay and task waiting in the hybrid manufacturing scenario. In order to make IIoT have the ability of perception and knowledge discovery, ML methods are proposed to improve the cognitive ability of IoT. The cognitive ability is explored to make IoT have the ability of understanding and decision-making at the network edge. Furthermore, with the help of real-time data acquisition, the DT can visualize and update the real-time status of manufacturing elements in the workshop, which is very useful for monitoring status of resources and optimizing the production scheduling. Therefore, dynamic scheduling and resources management based on the above technologies provides a solution for complex resource allocation problems in the current manufacturing scenarios, establishing a technical foundation for the implementation of intelligent manufacturing in Industry 4.0.

5.2.1 Multi-Agent Technology for Cooperation

A flexible manufacturing model of dynamic scheduling enables smart factory to equip with a variety of manufacturing function. However, hybrid manufacturing of multi-task and multi-object is the main problem of cooperative control. Due to the complex order and different functions of equipment, it is impossible to follow some predefined rules to complete an ordered task when the process of mixed orders flows into the smart factory. The master–slave control architecture between manufacturing system and equipment, which was used in the past, can make it difficult to ensure high operation efficiency.

Therefore, the multi-agent technology is introduced to solve the problems of order delay and task waiting in the hybrid manufacturing scenario [8]. In the job shop, the edge nodes are deployed on the device side to make the equipment form the MAS. Through the request for an intelligent agent of the task, a self-organized coordination between intelligent equipment is realized. The multi-agent technology for solving the job shop scheduling problem has the following advantages: (1) the edge nodes are used to make each agent independent and autonomous, and an appropriate strategy is chosen by reasoning and planning to determine the next step of the order task; (2) the MAS is a coordinated system, wherein each agent negotiates with each other to solve large-scale complex problems in parallel;

(3) the MAS is constructed of multi-level and diversified agents using the object-oriented method, which reduces complexity of the entire manufacturing system. Meanwhile, the problem complexity can be reduced for every agent [9].

An agent is equipped with a reasoning module and a knowledge base (KB), offering basic AI functionalities, such as inferencing and computing. Moreover, with the support of new communication technologies (e.g., 5G mobile networks and high-speed industrial wired networks), all agents and edge computing servers can be interconnected. Agents run on edge computing servers to guarantee low-latency services for data analytics. The agent edge servers are connected by high-speed IIoT to achieve low latency. What's more, edge computing servers support a variety of AI applications for MAS. The relevant application examples include personalized product identification based on DL, quantitative energy-aware model, process planning and dynamic reconfiguration through knowledge reasoning, etc.

Cooperation among multiple agents is necessary to dynamically construct collaborative groups for the completion of customized production tasks. Multiple agents provide a better option than a single device to build a collaborative operation to realize flexible manufacturing. Therefore, by combining the edge computing-assisted intelligent agents and different AI algorithms, a novel cooperative operation can be constructed. The strategy of cooperative operation by multiples agents can be divided into the order of submission, task decomposition, cooperative group, and subgroup assignment.

The working process of a flexible manufacturing production line can be described as follows. First, according to the customers' requirements, the customized product orders are issued to the manufacturing system through the recommender system. After receiving the product orders, the AI-assisted task decomposition algorithms take the product orders as the input, the device working procedure as the output, and the product manufacturing time as a constraint; these algorithms are mainly executed at the remote cloud server. A product order can be divided into multiple subtasks that are sent to all the agents via the industrial network. After the negotiation, agents return the answers to the edge server, which handles the working subtasks according to corresponding conditions and constraints. Next, the AI-assisted cost-evaluation algorithm calculates the cost of a producing group (i.e., cooperative manufacturing group) from the historical data. Then, the edge agents intelligently select suitable device agents to finish the product order after considering the whole cooperative group performances, such as producing time and product quality. Moreover, the edge agents send the selection result to the device agents, which are chosen to take part in the producing order. The main cooperative group is constructed based on the working steps.

After all the agents have been assigned with subtasks, they form two level-cooperative groups. The formation of these cooperative groups is beneficial to resource management. Then, according to the manufacturing task attributes, multiple agents complete the producing task. During this period, the corresponding device agents send their status data to edge servers timely, and the manufacturing process can be monitored by analyzing these data in the entire system. In contrast to the AI-driven cooperative operation between multiple agents, conventional methods often rely on human operators who participate in the whole process or computer-assisted operators also requiring human interventions. These methods greatly help to realize dynamic resource scheduling during the personalized production.

5.2.2 Industrial Internet of Things for Awareness

The IIoT provides a bridge for communication between the entities and the information in intelligent manufacturing. At present, the multiple agents are deficient in collecting and processing massive data. Recent advances in IIoT can meet this emerging need. A variety of decentralized manufacturing agents are connected to edge computing servers via high-speed industrial networks. It is industrial networks (e.g., wireless sensor networks) that can perceive physical resources in order to achieve reliable operation of the machine through real-time and effective monitoring in a smart factory. Therefore, IIoT realizes the coordinated allocation of resources and dynamic scheduling, improves manufacturing efficiency, and reduces production costs and resource consumption.

The primary issues of IIoT are the acquisition of real-time and efficient manufacturing data, as well as rapid heterogeneous information transmission in smart factories [10]. Controlling intelligent manufacturing equipment requires real-time control of complex dynamic behavior along with high-speed communication. Therefore, a standardized and unified transmission path is needed for the widespread distribution of resources in a smart factory. OLE for Process Control and Unified Architecture (OPC UA), which is an open platform with multiple functions for resources sharing services, manages to realize information interaction of the manufacturing resources in a smart factory. As a new technology of the OPC Foundation, OPC UA is applied to vendor-neutral transmission of preprocessing information or raw data from field manufacturing devices with safety and reliability. Generic services are defined by OPC UA, which follows the design paradigm of service-oriented architecture. With the support of OPC UA technology, we can develop a robust application architecture that withstands low latency and auto-detection of errors. Moreover, a single processor with sufficient redundancy between the client and the server application can achieve high-availability transmission. Consistent address space is provided by OPC UA, and a complete service model is built as well. That is the key factor to achieving information access in a uniform way between different machines, and it is something the previous system cannot do. The OPC UA also provides an open architecture of information transmission that is compatible with the existing technologies in manufacturing systems. In addition, OPC UA technology has many advantages, including reliability, security, compatibility, openness, and so on. These advantages satisfy the requirements of the integration of management while controlling continuous processes in modern industries. This is a convenient upgrade for manufacturing enterprises regardless of the geographical distance or domains of business.

However, due to limited traffic bandwidth, the network is more likely to have problems including congestion and latency. SDIN is introduced to bring new levels of network flexibility [11]. This makes the construction of the network more convenient. Furthermore, network resources are managed rapidly and configured properly according to SDIN, and network transmission becomes extensible and self-adapting. SDN is an innovative architecture that virtualizes the network. The core technology of SDN, OpenFlow, makes the network programmable by separating the control layer from the data transmission layer. Accordingly, the network facility becomes more intelligent with a management mechanism like this. In addition, the networking hub can access data resources and implement centralized control, traffic forwarding, and load balancing. The network control system is also able to plan the route between sensing and execution by considering factors such as network bandwidth, latency, redundancy, and security. Additionally, the network can monitor network anomalies to avoid malicious attacks by customizing security services. From the viewpoint of the IIoT field, SDIN provides the strategy of dynamic adjustment and routing according to the requirements of the network quality of service in CPPS. Therefore, the features of information transmission, such as occurring in real-time, reliability, and security, are ensured. One of the most important features of SDN is that it divides switch links as a set number of minimum independent paths in the whole network. In units of the free path, network resources are allocated according to real-time traffic situations. With the mechanism of SDN scheduling, the overall network load becomes much more balanced. Generally speaking, the demands of different applications (e.g., alert information, equipment log) for the real-time capability of the network vary in the intelligent manufacturing field. In order to avoid transmission delay or data loss caused by network congestion, SDIN sets up access rules based on the requirements in real-time. Moreover, the level of network openness is improved because SDIN virtualizes IIoT. In summary, SDIN provides a platform for network development and an innovative application.

What's more, the development of wireless network technologies such as Narrowband Internet of Things (NB-IoT), Third Generation Partnership Project (3GPP), and fifth generation (5G) make it possible for machine-type communication (MTC) devices to directly communicate with each other without core network relay. The emerging D2D communication technology makes it so that critical task forwarding is no longer dependent on service terminals. As a result, ubiquitous communication is also achieved among devices by D2D communication technology. D2D communication refers to a communication mode in which devices directly exchange information with neighbors under the control of a communication system. It provides higher data transmission rates and extends network capacity. General ground wireless nodes (e.g., enterprise eNodeB and workshop relay) are able to access the femtocell network with the support of LTE-Advanced, and the enterprise area or manufacturing workshop will be a small area of hotspots. Consequently, the wireless network covers the entire manufacturing environment. The MTC equipments communicate with each other through the air interface provided by 3GPP, which ensures high-speed data transmission. This technique specializes in distributed networks, where every node is able to send and receive signals. Therefore, MTC devices work as both servers and clients in communication networks. Once the D2D communication link is established, data transmission is free from relay and independent of the core network. In this way, stress on the core network is reduced, while at the same time throughput is promoted. Due to near-field gain and single-hop gain between intelligent devices, the D2D mode provides communication services with lower latency and greater reliability than the centralized mode in a cellular network environment. D2D technology enables robots with functional redundancy to negotiate with each other in an orderly and efficient manner during the manufacturing process. In addition, warehousing equipment and feeding equipment can communicate end to end, and they are able to achieve real-time adjustment of material reserves. As a foundational technology of D2D communication, LTE achieves communication authentication and admission control. In a wireless network's blind spots, an MTC device takes neighboring MTC devices as network relays and then transfers data to the manufacturing cloud platform or interacts with remote facilities via multihop network relays. To summarize, D2D communication technology offers high-quality network services for intelligent manufacturing. This mode is a convenient way to provide services (e.g., data transmission, instructions delivery, and resources sharing), particularly in a small-scale network.

5.2.3 Machine Learning for Decision and Cognition

ML refers to an intelligent behavior learned from the prior knowledge performed by a machine such as a computer or an intelligent controller. ML includes several processes which can be summarized as perception, understanding, learning, judgment, reasoning, planning, designing, and solving [12]. ML embedded in IIoT system helps the network have the cognitive abilities of representation, learning, and reasoning [13]. Although human beings have mastered more methods of data analysis (e.g., big data mining), it is difficult for labors to quickly respond to the change of analysis requirement when the amount of information increases constantly. ML embraces theories, methods, technologies, and applications to augment human intelligence. It includes not only ML techniques, such as perception, DL, reinforcement learning, and decision-making, but also ML-enabled applications, such as computer vision, natural language processing, intelligent robots, and recommendation systems. ML has outperformed traditional statistical methods in tasks, such as classification, regression, clustering, and rule extraction. Typical ML algorithms include decision tree, support vector machines, regression analysis, Bayesian networks, and deep neural networks.

As a subset of ML algorithms, DL algorithms have superior performance than other ML algorithms. The recent success of DL algorithms mainly owes to three factors: (1) the availability of massive data; (2) the advent of computer capability achieved by computer architectures and hardware, such as graphic processing units (GPUs); and (3) the advances in diverse DL algorithms, such as a convolutional neural network (CNN), long short-term memory (LSTM), and their variants. Different from ML methods, which require substantial efforts in feature engineering in processing raw industrial data, DL methods combine feature engineering and learning process, thereby achieving outstanding performance. However, DL algorithms also have their own disadvantages. First, DL algorithms often require a huge amount of data to train DL models to achieve better performance than other ML algorithms. Moreover, the training of DL models requires substantial computing resources (e.g., expensive GPUs and other computer hardware devices). Third, DL algorithms also suffer from poor interpretability, that is, a DL model is like an uncontrollable "black box," which may not obtain the result as predicted. The poor interpretability of DL models may prevent their wide adoption in industrial systems, especially in critical tasks, such as fault diagnosis, despite recent advances in improving the interpretability of DL models.

As ML technologies have demonstrated their potential in areas such as customized product design, customized product manufacturing, manufacturing management, manufacturing maintenance, customer management, logistics, after-sales service, and market analysis, industrial practitioners and researchers have begun their implementation. Edge devices (e.g., Nvidia Jetson TX) with ML algorithm can even mimic the brain's work manner. The cognitive technology, which is used to simulate human thought processes in a computerized model, can help the machine understand the environment abstractly. This is helpful for the empirical learning of an ML model. With the network data resources of the cognitive IIoT, ML enables the network services to respond to the external events adaptively, just like human beings. Therefore, the intelligent solutions that are superior to human beings can be carried out. Using the environmental perception, ubiquitous computing technology, and mobile communication technology, IIoT integrates many kinds of terminals into every phase of the industrial manufacturing. With the popularization of the low-cost sensors and intelligent, distributed terminals, intelligent IIoT has gathered these devices together with edge computing to integrate physical information. Edge intelligent IIoT makes it easier to transfer cloud services with low latency, high bandwidth, and low jitter [14] as shown in Figure 5.1. We can use the semantic association of information to perceive the changing industrial scenarios. The ML methods embedded into edge intelligent IIoT make the entire IIoT system have the ability of understanding, learning, and reasoning.

Therefore, the introduction of ML technologies can potentially realize personalized manufacturing. In summary, ML-driven personalized manufacturing has the following advantages: (1) Improved production efficiency and product quality: In factories, automated devices can potentially make decisions with reduced human interventions. Technologies such as ML and computer vision are enablers of cognitive capabilities, learning, and reasoning (e.g., analysis of order quantities, lead time, faults, errors, and downtime). Product defects and process anomalies can be identified using computer vision and foreign object detection. Human operators can be alerted to process deviations. (2) Facilitating predictive maintenance: Scheduled maintenance ensures that the equipment is in the best state.

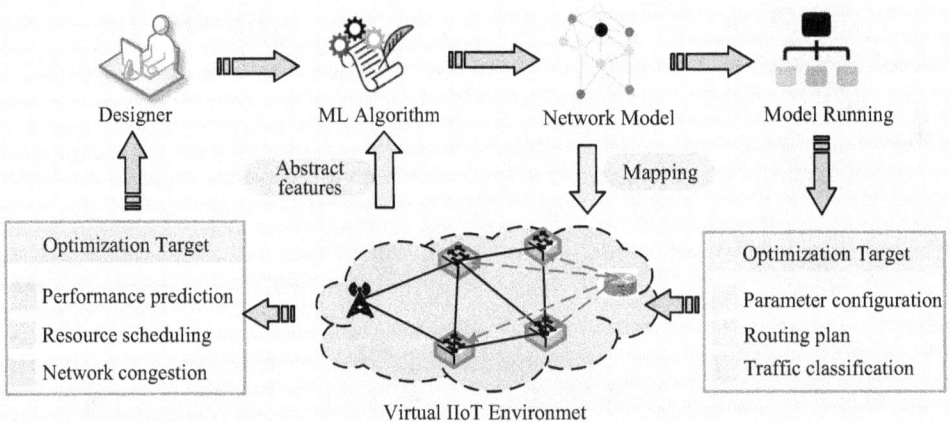

FIGURE 5.1 The ML-enabled network optimization method.

Sensors installed on a production line collect data for analysis with ML algorithms, including CNNs. For example, the wear and tear of a machine can be detected in real-time, and a notification can be issued. (3) Development of smart supply chains: The variability and uncertainty of supply chains for personalized production can be predicted with ML algorithms. Moreover, the insights obtained can be used to predict sudden changes in customer demands.

In short, the incorporation of ML and IIoT brings benefits to smart manufacturing. ML-assisted tools improve manufacturing efficiency. Meanwhile, higher value-added products can be introduced to the market. However, we cannot deny that ML technologies still have their limitations when they are formally adopted to real-world manufacturing scenarios. On the one hand, ML algorithms often have stringent requirements on computing facilities. For example, high-performance computing servers equipped with GPUs are often required to fasten the training process on massive data, while the existing manufacturing facilities may not fulfill the stringent requirement on computing capability. Therefore, the common practice is to outsource (or upload) the manufacturing data to cloud computing service providers who can conduct the computing-intensive tasks. Nevertheless, outsourcing the manufacturing data to the third party may lead to the risk of leaking confidential data (e.g., customized product design) or exposing private customer data to others. On the other hand, transferring the manufacturing data to remote clouds inevitably leads to high latency, thereby failing to fulfill the real-time requirement of time-sensitive tasks.

5.2.4 Digital Twin for Resource Management

As the core concept of Industry 4.0, the emergence of DT has provided a solution to connect the physical world with the information world while realizing data interaction and integration. Currently, the development of DT is still in its infancy. The definitions of DT concept are not unified yet. The popular explanation of DT is a dynamic virtual model from the design stage to the operation stage, which accepts real-time data for decision-making and optimization [15]. Modeling and simulation are the basis for DT implementation, but DT also includes data fusion, interaction, and collaboration, as well as a range of services. The DT-related technologies include identification, communication interfaces, DT models, data analysis, and human–machine interaction [16].

Although DT is now widely used in product life cycle management [17], including product design, manufacturing, virtual commissioning, robot assembly, and predictive maintenance, it has great potential in the resource dynamic scheduling. Through the digital model in virtual space and data interaction in physical space, the current system state can be monitored in real-time and the future state can be predicted, so as to effectively improve and optimize the resource allocation. More importantly, DT has great potential in reducing uncertainty, and several frameworks have been proposed. The DT applications involved in these studies are mainly concentrated in job shop and focus on the resource condition (such as equipment failure and worker loss), logistics, and order. Therefore, DT has become one of the key technologies for dynamic scheduling in the background of personalized flexible manufacturing.

Digital twin in design (DTD): Product design is based on the target function of the design object and user requirements. After discussion, analysis, and design processing, the requirements are transformed into a specific text or graphic expression to provide reasonable planning for the production stage. Driven by the product's DT data, the DTD provides a two-way interactive connection between physical and virtual products for data sharing. It also strengthens the synergy between these two and continuously explores innovative, unique, and valuable product optimization design schemes. Thus, it can achieve dynamic innovation that meets the customized requirements ahead of the time. The DT is not just a digital product display model; it can also improve the accuracy of the design and assign the behavior of the actual product to the corresponding virtual product. In addition, it can verify the performance of a product in a real environment through simulation. The constructed design originates from practice and serves practice. The DTD emphasizes the virtual–physical integration of the complete life cycle. It establishes links between products, production tools, environments, and processes and integrates information of different stages of the product life cycle into the design phase while providing a coherent overview of the development cycle. Therefore, the DTD can promote decision-making and improve design quality and efficiency in many aspects through its various functions, thus improving and optimizing products and their manufacturing systems.

Digital twin in manufacturing (DTMF): It is necessary to construct the DTMF before performing actual production tasks. It is also necessary to extract manufacturing-related process information from the DTD along with associated manufacturing resources information, such as information on processing equipment, process parameters, tooling, and molds. To realize the DTMF, it is important to simulate the production process of complex equipment under agreed conditions through virtual production in advance. This process can predict bottlenecks of resources, workpieces, and stages. Thus, in this way, bottleneck control can be achieved, and the speed and accuracy of a new product can be improved during the production process through the redistribution of bottleneck resources and optimization of component combination and operation sequence performed before production. The customized and sophisticated manufacturing process of complex equipment requires effective process planning and key indicator monitoring. In the manufacturing stage, the connection is established between the manufacturing entity and the virtual product by the DTMF. The measured data related to the product production, processing, and quality control processes, such as geometric measurement, vibration measurement, force measurement, and progress data, are mapped to the virtual product and displayed in real-time. This process ensures a comprehensive collection of production factors and manufacturing information and achieves online process control as well as monitoring based on the actual production data. The manufacturing process is characterized by dynamics and uncertainties, such as tool wear, aging of processing equipment, and changes in material properties, which are caused by environmental conditions. Despite careful planning, these situations can cause a gap between the actual production results and the expected results. The DTMF helps to integrate process planning data with the measured production data while monitoring key manufacturing parameters. Therefore, when an abnormal situation occurs, such as a violation of plans or discovering a better production plan through learning, the intelligent decision-making module provides the corresponding treatment and scheme adjustment timely. The new adjustment scheme is fed back to the manufacturing entity to improve the manufacturing quality and to realize dynamic control and self-optimization, thus achieving the purpose of the virtual control.

Digital twin in maintenance (DTMT): Complex equipment typically has the characteristic of high cost, which is related to costly maintenance. However, both users and enterprises expect to achieve the maximum value of the equipment at the lowest possible cost. In the use and service phase of a product, the DTMT is used to track and monitor the product status in real-time. The DTMT acquires real-time data that characterize the quality and functional status of the monitored equipment during operation, such as engine rotor speed, vibration value, lubricating oil temperature, and exhaust pressure, by using smart sensors. It also helps in constructing a remote monitoring system with quantitative indicators and establishes a hierarchical health management system for complex equipment parts, subsystems, systems based on equipment historical use and maintenance data. It analyzes and predicts equipment operating conditions, remaining life, and potential failures and assets failure time in advance. It also performs beforehand maintenance to improve the efficiency of spare parts management and to reduce or avoid customer losses and problems caused by unexpected equipment shutdown. A virtual product can adjust itself to adapt constantly to changes in the physical space conditions. It reflects and evaluates the health status of physical service products through virtual products. In this way, the risk of manual evaluation using a one-way physical information flow is reduced. The insights obtained by the application of the DTMT can also be used to improve product design and manufacturing processes. The application of complex equipment has a strong driving force for both enterprise production and social progress. The rationality of complex equipment design parameters and operating parameter settings, as well as the adaptability to different working conditions, determines the functional level, quality advantages, and customer satisfaction of equipment. Equipment providers collect real-time data during product service by the DTMT. Using these data, they construct an empirical model for parameter optimization for different application scenarios and provide customers with guidance on its configuration. This can improve user satisfaction, enhance user experience, and maximize the functionality of the equipment. At the same time, customers can connect with the product through the DTMT and thus provide positive feedback to manufacturers. By considering the customer preferences, manufacturers can obtain insight into potential, future, real-time demands of customers for products and thus can meet them in advance. Therefore, it can avoid making the research and development decisions from offsetting market demand, shorten the design time and introduction cycle of new products, and enhance market competitiveness.

5.3 DYNAMIC SCHEDULING BASED ON EDGE-CLOUD COLLABORATION

In the context of Industry 4.0, the smart factory has been effectively supporting the implementation of intelligent manufacturing. Namely, modern manufacturing requires high-efficient and flexible production to meet current market demands on small batch customization. Therefore, a self-organization dynamic scheduling is very important in smart factory [18]. Because the flexible manufacturing model of smart factory needs to be adaptive to diversification of the market, a dynamic scheduling of a job shop has become the key technology in advanced manufacturing. In order to effectively improve the adaptability and robustness of dynamic scheduling, the intelligent manufacturing systems integrate IIoT technology [19], high-performance computing, and other technologies. The dynamic scheduling based on edge-cloud collaboration allows connecting the machines, products, edge devices, and cloud for achieving the fast response, high stability, and self-organization [20].

In this section, we first implement the global resource awareness and distributed information processing of the factory through edge computing. Then, for a large number of automation equipment in the workshop, the concept of cloud robotic is introduced to provide adaptive services for different scenarios. Finally, in order to improve the robustness of dynamic scheduling, a cloud-assisted edge decision manufacturing architecture is proposed to meet the dynamic production needs of different stages.

5.3.1 Edge Awareness for Load Balance and Scheduling

With the development of industrial networking technology, the rapid increases of terminal devices, massive data analysis, and processing of things gradually raise up the burdens on manufacturing cloud platform [21]. The remote cloud platform can hardly perceive operating equipment status and realize the real-time multi-task and multi-object scheduling in smart factory. In that case, the cloud applications are required to migrate to the network edge nodes. The edge computing is located near the network edge, e.g., field-level devices, and it uses the distributed edge nodes such as converged network, computing, storage, application, and other core capabilities to provide edge intelligent service [22,23]. The industrial application of edge computing will make the full use of computing ability of an intelligent equipment to achieve the global perception and autonomous distributed information processing.

The platform of edge computing provides more complex computing, network, and storage functions, which achieve efficient information interaction to support resources scheduling and intelligent services. The edge nodes integrate functions of distribution, coordination, management, and protection of cross network resources, which provides intelligent tools for edge computing services (e.g., raspberry pie board, UDOO board, ESP8266). The introduction of OPC UA communication architecture enables intelligent equipment to exchange information to coordinate the operation. To ensure the minimum response time between detection and notification, the D2D communication mode is used to ensure the interaction between intelligent equipment, as shown in Figure 5.2.

5.3.1.1 Architecture of Edge Computing for ELBS

As an industrial example of edge computing, the energy-aware load balance and scheduling (ELBS) for the multi-task and multi-objects in smart factory will be introduced. The system architecture is shown in Figure 5.3. First, the energy-aware model related to the smart factory is established on edge nodes, and then it is used to quantify the relationship

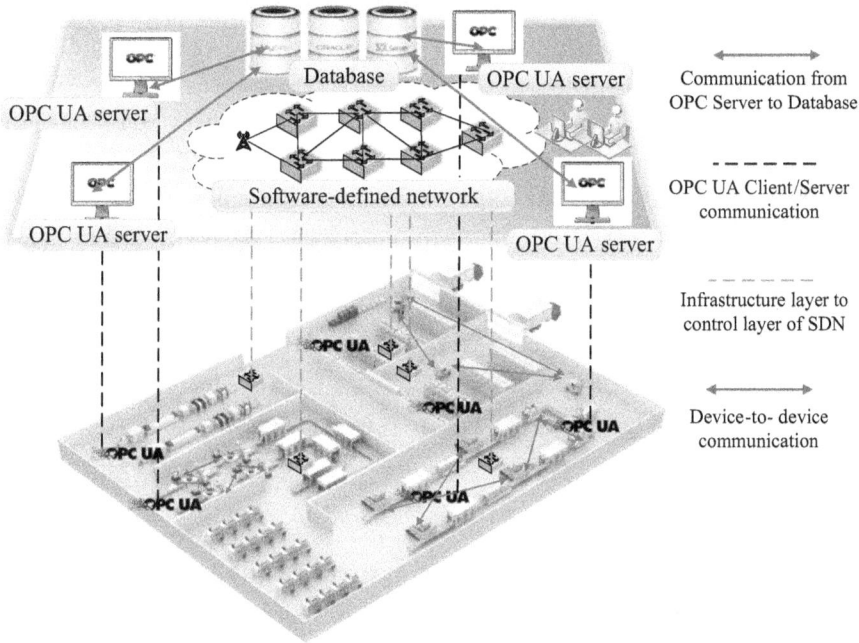

FIGURE 5.2   Information interaction in smart factory.

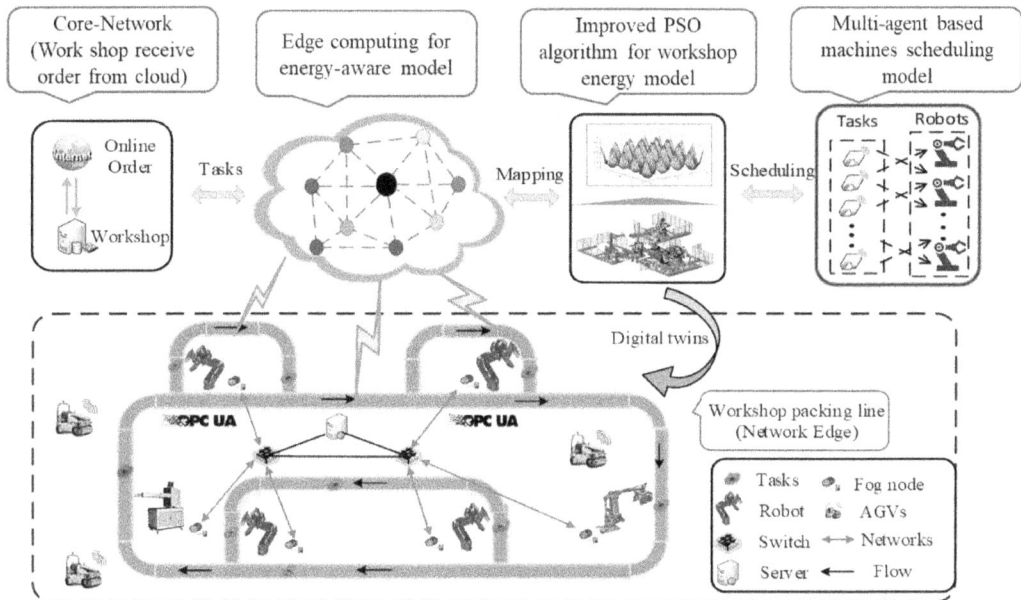

FIGURE 5.3   Architecture of edge computing for ELBS in smart factory.

between energy consumption and workload of the equipment. Second, a load balancing optimization function is established by taking into account the equipment energy consumption. The improved particle swarm optimization (PSO) algorithm is used to obtain an optimal solution. In this way, the task-related priorities of manufacturing clusters are determined. Finally, a MAS is introduced to guide the equipment workload balance with the task scheduling mechanism.

The edge computing has added a new dimension to the data analysis in the IIoT, which is a data processing platform on the network edge [24,25]. The fusion of many data types and sources is a crucial stage in the data collecting and intelligent control procedures to managing intricate industrial operations. Edge computing nodes can be used to deploy techniques like time series analysis, frequency analysis, and wavelet analysis to extract information from the sensor outputs. On the edge nodes, AI models and operation mechanism models can be run to create predictions using the characteristic data and then update the KB. The updated knowledge and data can be used to make dynamic and timely reasoning at the edge. This enables network edge decision-making to be intelligent.

5.3.1.2 Model Construction for ELBS

5.3.1.2.1 Equipment Energy Consumption Model   To realize the load balance based on edge computing, a quantitative equipment energy consumption model related to the smart factory order should be established. The platform of edge computing is built on the IIoT system using edge nodes (e.g., raspberry pie). Then, the energy consumption characteristics of intelligent equipment are abstracted by edge computing. Finally, the equipment energy consumption model is constructed on edge nodes.

The device considers only the certain type of tasks and can be reconstructed to achieve an immediate replacement of the equipment. The real-time power of device can be measured by voltage and current:

$$P_i(t) = U_i(t) I_i(t) \tag{5.1}$$

If M is the machine set defined as:

$$M = \{m_1, m_2, m_3, \ldots, m_j, \ldots, m_n\} (j \in [1, n]) \tag{5.2}$$

and S is the order set defined as:

$$S = \{s_1, s_2, s_3, \ldots, s_l, \ldots s_r\} (l \in [1, r]) \tag{5.3}$$

Then:

$$\begin{cases} \forall s_l, \exists! m_j \\ s.t. \ s_l \rightarrow m_j \end{cases} \tag{5.4}$$

The following binding rules are specified for $S \subseteq M$: task cannot be interrupted once it is started; task is not allowed to wait, and machine can remain idle when there is no task waiting.

The equipment energy consumption is calculated by edge nodes. The consumed energy includes energy needed for the following processes: equipment processing, workpiece delivery by a conveyor belt, and workpiece delivery to the next station in energy consumption structure. Therefore, the equipment energy consumption is defined by:

$$E_{\text{operate}, i}\left(\mu_{i1}, \mu_{i2}, t_{\tau 0}, t_{\tau 1}, t_{\tau 2}\right) = \omega_1 \int_{t_{\tau 0}}^{t_{\tau 2}} \mu_{i1}(t) P_{\text{conveyor}, i}(t) dt$$

$$+ \omega_2 \int_{t_{r0}}^{t_{\tau 1}} \mu_{i2} P_{\text{wait}, i}(t) dt + \omega_2 \int_{t_{\tau 1}}^{t_{\tau 2}} \mu_{i2} P_{\text{operator}, i}(t) dt$$

(5.5)

The energy consumption of the equipment in the idle state is defined by:

$$E_{\text{wait}, i}\left(\mu_{i1}, \mu_{i2}, t_{\tau 0}, t_{\tau 1}\right) = \omega_1 \int_{t_{\tau 0}}^{t_{\tau 1}} \mu_{i1} P_{\text{conveyor}, i}(t) dt + \omega_2 \int_{t_{\tau 0}}^{t_{\tau 1}} \mu_{i2} P_{\text{wait}, i}(t) dt \qquad (5.6)$$

where ω_1 and ω_2 are energy correction factors, which are constant and related to the equipment; P_{conveyor} is the conveyor belt operation power, P_{wait} is the equipment waiting power, and P_{operator} is the equipment operation power.

Further, μ_{i1} and μ_{i2} are the energy utilization rates that can be generally measured in a period of time.

$$\mu(t) = \left(E_v(t) + E_m(t)\right) / E_r(t) = \frac{\int P_v(t) dt + \int P_m(t) dt}{\int P_r(t) dt} \qquad (5.7)$$

In equation (5.7), $E_r(t)$ is the motor rotation power at moment t; $E_m(t)$ is the input power of the main motor drive system; and $E_v(t)$ is the motor non-rotating power.

The efficiency ratio of a single equipment is defined by:

$$\sigma_{mj} = \frac{\sqrt[T_n]{e^{T_0}}}{n} \sum_{i=1}^{n} E_{mj, i} \qquad (5.8)$$

where T_0 is the equipment operation time, T_n is the equipment running time, and n is the scheduling times of the involved equipment:

$$\theta\left(m_j\right) = e^{-\sigma_j} \in [0,1] \qquad (5.9)$$

Equation (5.9) defines the mathematical model of energy consumption ratio of m_j, where $\theta(m_j)$ is the machine energy intensity.

$$\text{Min}: \Phi_{\text{Loadmean}} = \sqrt{\frac{1}{n-1} \sum_{j=1}^{n} \left(\theta(m_j) - \tilde{\theta}(m_j) \right)^2} \tag{5.10}$$

$$\tilde{\theta}(m_j) = \frac{1}{n} \sum_{j=1}^{n} \theta(m_j) \tag{5.11}$$

Equation (5.10) represents the load balancing based on energy awareness, which is an objective function for scheduling optimization. The goal is to find the function minimum. In the following, an improved PSO algorithm is used to determine an optimal solution of the objective function. Afterwards, the current equipment energy intensity is arranged. The equipment dynamic scheduling is implemented using the MAS based on equipment energy intensity.

5.3.1.2.2 Optimal Solution of Energy Consumption Model   According to the order-oriented energy-aware model, the aim of scheduling algorithm is to obtain the minimum of Φ_{Loadmean}. The efficient optimization algorithm is the key factor to solve the scheduling problem of a flexible job shop. The proposed PSO algorithm is easy to implement, its search efficiency is high, and it uses a small number of parameters. Especially, the algorithm featured with the natural real code features is suitable for real-time optimization problems, which also has a profound industry background for intelligent manufacturing. The definition of notations is shown in Table 5.1.

With the basic PSO formula, we get:

$$V_i^{k+1} = \omega(k) V_i^k + c_1 r_1 \left(P_{\text{best}, i}^k - X_i^k \right) + c_2 r_2 \left(P_{\text{global}, i}^k - X_i^k \right) \tag{5.12}$$

$$X_i^{k+1} = X_i^k + V_i^{k+1} \tag{5.13}$$

To avoid the premature convergence of particles, the exponential decreasing weight formula is adopted such that global search ability and search accuracy of a discrete PSO algorithm can be improved:

$$\omega(k) = (\omega_{\max} - \omega_{\min}) Exp\left[-\frac{k^2}{\eta k_{\max}} \right] + \omega_{\min} \tag{5.14}$$

In equation (5.14), ω is the important parameter of PSO algorithm, which balances search ability of particles either in global or local way. The larger inertia weight can enhance the

TABLE 5.1 Notations

Notations	Definition
N	The number of robots
m	The number of particles
X_i	The position of particle i
V_i	The velocity of particle i
ω	The inertia weight
η	The random number and $\eta \in U(k, k_{\max})$
C_1, C_2	The acceleration coefficients (constant and equal to 2)
r_1, r_2	The pseudo random numbers, and $r_1, r_2 \in U[0,1]$
$P_{\text{best}, i}$	The best place for particle i in the history
$P_{\text{global}, i}$	The best position of all swarm particles

global search ability of particles, while the smaller inertia weight can enhance their local exploration ability. The inertia weight has a great influence on exploration ability in the early stage of algorithm execution.

In this study, m times $\theta(m_j)$ of equipment during different periods is the initial value. Further, Φ_{Loadmean} is the moderate function, which determines the fitness value used to judge the current particle position. When the maximal number of iterations or the minimum load balancing mean is achieved, an optimal particle position can be determined. The Energy-aware Load Balancing Priority Algorithm (ELBPA) is elaborated in Algorithm 5.1.

5.3.1.2.3 Load Balancing Scheduling Mechanism Based on the MAS   In regard to the equipment energy, achieving of dynamic scheduling is a key to realize an energy-aware load balancing in the smart factory. Edge nodes are deployed on the equipment side, which transforms equipment to the intelligent agents. The MAS consisted of intelligent equipment in smart factory. The MAS can share knowledge and information mutually through the negotiation and cooperation between agents, which can accomplish the task in the complex multi-ordered environment in a self-organized way. Aiming at the optimization of energy-aware model, we propose a dynamic scheduling system for the smart factory based on the multi-agent technology. According to the optimal solution calculated by the improved PSO algorithm, a task-related priority based on energy intensity of intelligent equipment is sorted. During the operation process, a load balancing of equipment based on the order-related energy consumption is achieved. In general, the management system receives the order from customer and then decomposes the order into several specific process tasks. Because the communication, negotiation, and collaboration in MAS require advanced communication and control protocols, the contract net protocol (CNP) is introduced. During the operation, the agent turns from the slave node to the master node, reads out the order information, and releases the bidding information. The standby agent, which sends task requirement to the master node based on the job characteristic, is the salve node. In order to enable agent to perform tasks efficiently and to avoid order waiting and conflicts between requests of multiple agents for the same task, the following definitions are given:

ALGORITHM 5.1   Energy-Aware Load Balancing Algorithm

Input: $X_i^0 = \{\theta_i(m_1), \theta_i(m_2), \theta_i(m_3), \ldots, \theta_i(m_n)\}$ //particle swarm

Input: ξ, λ // ξ is the minimal load balancing mean deviation

          // λ is the maximal number of iterations.

Output: X_{dest} //dynamic scheduling factors for the multi-agent;

Output: $\overline{Rank[m_j, i]}$ //sort by priority;

1   **begin**

2   **Initialization**   // initialized particle;

3   **for** $i \leftarrow 1$ **to** m

4   **for** $j \leftarrow 1$ **to** n

5         $X_{i,j}^0 = \theta_{\Gamma i}(m_j)$

6   **end for**

7   **end for**

8   **Do**

9   **for** $i \leftarrow 1$ **to** m //update particles' positions

10   **for** $j \leftarrow 0$ **to** k //achieve the previous best particles' positions

11   **if** $(\Phi_{\text{loadmean}}(X_i^j) \leq \Phi_{\text{loadmean}}(P_{\text{best},i}))$ **then**

12           $P_{\text{best},i} = X_i^j$ // update the best position for particle

13   //i in history.

14   **end if**

15   **end for**

16   **if** $(\Phi_{\text{loadmean}}(P_{\text{global}}) \leq \Phi_{\text{loadmean}}(P_{\text{best},i}))$ **then**

17         $P_{\text{global}} = P_{\text{best},i}$ //achieve the best neighbor particle

18   **end if**

19   V_i^{k+1} //calculate the velocity

20       $X_i^{k+1} = X_i^k + V_i^{k+1}$

21       $k{+}{+}$

22   **while** $(\Phi_{\text{loadmean}}(X_{\text{dest}}) \leq \xi$ or $k \geq \lambda)$ //the terminal condition

23   **end while**

24     $\overline{Rank[m_j, i]} = Sort(X_{\text{dest}})$

25   **end**

Definition 5.1

The attribute of agent is $S = \langle R, F, \delta \rangle$, where $R = [Fun_1, Fun_2, \ldots, Fun_n]$ represents the function set of agent, $F = [A_{neighnor,1}, A_{neighnor,2}, \ldots, A_{neighnor,n}]$ denotes the information mapping function, which is used to store communication sets of collaborative agent, and $\delta = [-1,0,1]$ has three kinds of operating state, namely bidding, standby, and tendering.w

Definition 5.2

$E = \langle R, B, \delta \rangle$ is the decision information model of agent, where B represents the communication information set for agent bidding, and when $\delta = 1$, the first agent of order entry is 1 and information on bidding agent label is returned. The improved PSO algorithm is applied to sort the bidding agent collection priority.

If we assume that:

$$M = \{m_1, m_2, m_3, \ldots, m_n\} \rightarrow \sum_n m_j = 1 \tag{5.15}$$

In equation (5.15), M is the task flow for order decomposition:

$$M_j = M - \bigcup_{i=1}^{j} m_i = \sum A_{\text{neighbor}, j} \tag{5.16}$$

Equation (5.16) determines the sequence of tendering agents. After tendering information is broadcasted, the winner agent is decided according to the bidding information and then task is forwarded through the logistics system.

Definition 5.3

B is the bidding information model of agent, and it is defined as $B = \langle R, U, \delta \rangle$, where U is the bidding communication information set of agents; and for $\delta = -1$, it returns the information on bidding. Since $R \otimes U = G$, if we set G as the agent bidding mapping set, we get:

$$G(X_i) = \sum_n U_p(\theta(m_j)) \tag{5.17}$$

The MAS is utilized to achieve dynamic scheduling of intelligent equipment, where the bidding evaluation strategy is a key factor affecting the task scheduling. After the equipment energy consumption model is established by edge nodes, the improved PSO algorithm is employed to solve the model and get the solution vector to obtain the priority of agent request task. After obtaining the task, the winner agent is transformed from the slave node to the master node. Then, the winner agent reads the information on the current job order, acquires the process information, and sends tender information to the relevant agent. When the order is completed, the tender information is sent to the logistics agent and the product will be exported. The load balancing algorithm based on the MAS (MA-LBA) is elaborated in Algorithm 5.2.

ALGORITHM 5.2   Load Balancing Algorithm Based on the
Multi-Agent System

1	**begin**
2	**Initialization**
3	**For** N_i // node i ←Intelligent agent
4	$N_i \leftarrow m_j$ // node i achieve the job and $m_j \in M_{rest}$
5	$\delta = 1$ // change bid status
6	$M_{rest} = M_{rest} \cap \{m_j\}$ // M_{rest} ← the rest of task set
7	**if** $(M_{rest} = \varnothing)$ **then** // change bid status
8	**break**
9	**for** $j \leftarrow 1$ **to** m // in set B and $Node_j \in B$
10	**if** $(\delta_{Node,\,j} = 1)$ **then**
11	$E + \{N_j\}$ //achieve the set of evaluation tender
12	**end for**
13	$U = \{X_{Node,\,1}, X_{Node,\,2}, \ldots, X_{Node,\,n}\}$
14	**call for** $ELBPA[U]$ //(Energy-aware Load Balancing- // Priority Algorithm, ELBPA)
15	$\overline{Rank[m_j, i]} = Sort(X_{dest})$ //get the task priority
16	$N_{winner} \leftarrow m_{j+1}$ //assign the task and $m_{j+1} \in M$
17	$\delta = -1$ // change bid status
18	call for MA-LBA[N_{winner}] //Load balancing algorithm based on multi-agent system (MA-LBA)
19	
20	**end for**
21	**end**

5.3.1.3 Case Study for ELBS

5.3.1.3.1 Prototype Platform   The candy packaging line is an intelligent manufacturing system integrating manufacturing and service, which has the typical characteristics of intelligent manufacturing, such as high connectivity, dynamic reconfiguration, and deep integration. The basic flow of used candy packaging line is as follows: first, customer chooses preferred candy and purchase it online; then, the order information is sent directly to the manufacturing system. Afterwards, the order-driven robots as agents complete the task in a self-organization way. Finally, the completed order is automatically transported by the logistical system of the smart factory.

As shown in Figure 5.4, in experimental setup, the raspberry pie board was attached to each robot. Namely, raspberry pie has the typical application characteristics of an edge node, which provides data storage, computing, and information interaction for field devices in plants.

In the experiment, every candy box that flowed through the packaging line was labeled with RFID tag containing order information written by the initial feeding equipment. Robots got detailed assignments through RFID reader, and the task model was set up by

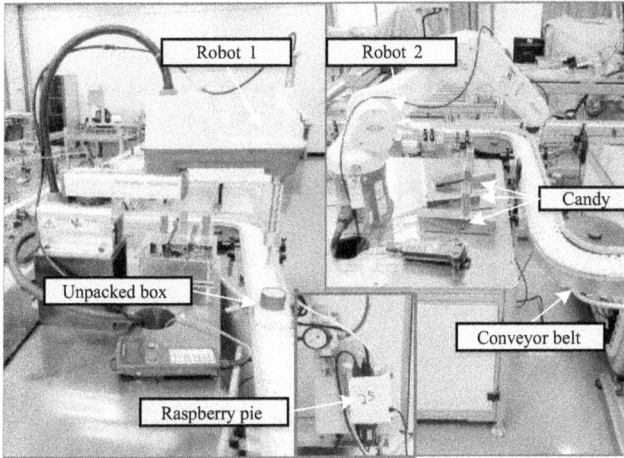

FIGURE 5.4   Prototype platform used for ELBS method in smart factory.

TABLE 5.2   Job Characteristics of Equipment in Workshop

	Job range	Reconfigurable	Performance	Intensity
Robot 1	●●●	No	Strong	0.67
Robot 2	●●●	Yes (slow)	Normal	0.54
Robot 3	●●	No	Weak	0.26
Robot 4	●●●	Yes(quick)	Strong	0.75
Robot 5	● ●	Yes(slow)	Normal	0.68

their attached raspberry pie boards to perform actions. As shown in Table 5.2, robots were set up to simulate different manufacturing processes according to the candy types. The relevant performance indexes of robots were defined based on the actual situation. If candy type had a cross, then the same process could be operated by different robots. In such a scenario, several agents were bidding the same task. Robots had the ability of reconfiguration, which can be regarded as a functional redundancy between machines. The discrete agent used a radio communication mode, which provided many-to-many communications. If we assume that there are enough orders, this platform can be used to test the proposed mechanism. Therefore, we simulated stochastic orders according to the throughput of the shop floor. Further, 5–10 candies were put into each candy box, and the types of candies were allowed to overlap. The number and the type of orders were randomly generated by the server. During the operation process, robots were not allowed to be suspended for a specific task. Besides, the task waiting was allowed when there was order saturation in the workshop.

5.3.1.3.2 Experimental Results   Using the production line for candy packaging, we conducted the 12-hour experiment. In the experiment, the situation of no task waiting was defined as a traffic moving state; the situation of 1–3 operation boxes waiting was specified as a traffic jam state; and the situation of more than 3 boxes waiting was defined as a traffic congestion state. To demonstrate the optimization characteristics, we compared

(a) CSCS-based method in smart factory

(b) ELBS-based method in smart factory

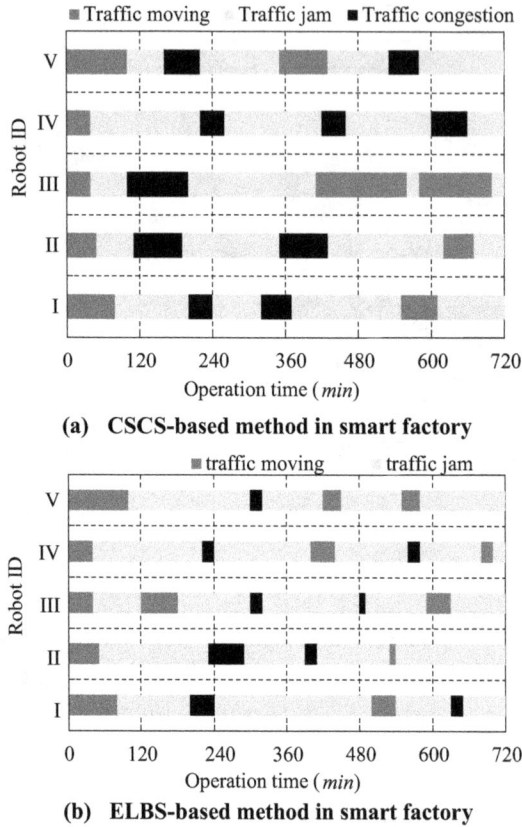

FIGURE 5.5   Gantt chart of the operating states of robots: (a) CSCS-based method in smart factory and (b) ELBS-based method in smart factory.

ELBS method with CSCS method that was applied in the platform previously. As shown in Figure 5.5, the load conditions of robots shown in Figure 5.5b are more balanced than those shown in Figure 5.5a. For instance, in Figure 5.5, robot 1 has better performance than other legacy robots, thus it is more likely to obtain jobs in CSCS method. Consequently, robot 1 is overloaded due to large number of tasks, the traffic congestion time is long, and the traffic moving time is obviously unbalanced. By applying the ELBS method, the operating status of robot 1 is greatly improved by changing priority. Furthermore, the time of traffic moving, traffic jam, and traffic congestion become more balanced.

As shown in Figure 5.6, in CSCS-based method, the workload of Robot 1 is equal to 0.602, which is much higher than the workload of other robots. In that case, only the operation mode that is predefined according to the equipment can lead to the excellent equipment overload, while the idle robots can cause waste of resources, but that mode is unfavorable to the optimal scheduling and planning of the job shop. On the other hand, in the ELBS method, the workload of Robot 1 is reduced to 0.309, while the workload of other robots is balanced in Figure 5.6.

Figure 5.7 reflects the mean difference in energy intensity between CSCS and ELBS methods. Compared to the CSCS method, the ELBS method can decrease the mean

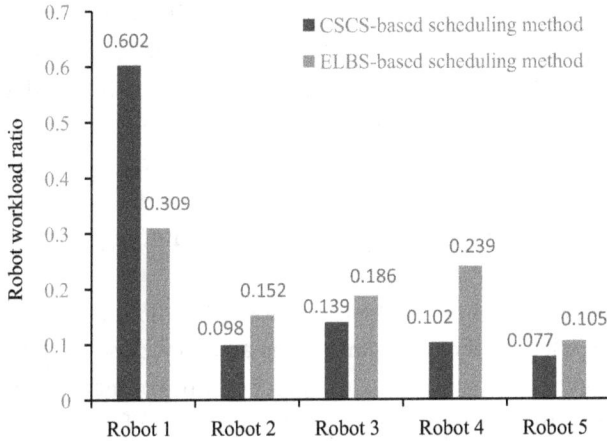

FIGURE 5.6   Load balancing performance of the operating robots.

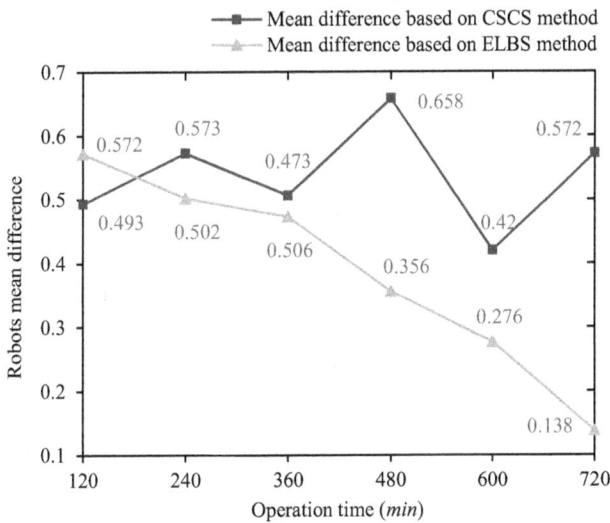

FIGURE 5.7   Comparison of equipment scheduling methods.

difference significantly, which means that the workload of robots is balanced. In general, the experimental results showed that the proposed ELBS method is superior over the CSCS method in load balancing of intelligent equipment in the smart factory.

The verification results showed that ELBS method significantly improves performance of robots in prototype platform. In this study, the raspberry pie was used to construct the edge computing platform, because it has typical characteristics of edge nodes. Compared to the cloud platform, the energy-aware models established on edge nodes are more precise. In contrast to the centralized energy consumption measurement, the energy-aware model maps the functional relationship between equipment energy consumption and workload, which set the basis for load-balancing scheduling of robots.

5.3.1.3.3 Discussion   According to the obtained experimental results, we can conclude the following. The edge computing enables robots to have terminal intelligence and respond quickly according to the product information. The tendering and bidding mode of robots has a demonstrable effect on complex scheduling problem in multi-task and multi-object system. Due to the discrete type of agent, the scheduling system has good portability, and the whole job cluster is equipped with a good scalability. In addition, the ELBS method has certain applicability in equipment fault diagnosis and task transfer.

As for the energy awareness of robots, the proposed ELBS method has good application effect. Although the used verification conditions are very close to the actual production, there still exists dynamic randomness in actual scheduling of manufacturing process, such as fuzzy delivery data, uncertain processing time, equipment fault maintenance, and emergency orders. The proposed ELBS method considered only two single factors, equipment energy consumption and equipment workload, which is more or less a shortcoming. In addition, intelligent agents in MAS interact with each other by many-to-many communication, and the efficiency of broadcast mode is low because of a large amount of communication information. The results presented in this paper provide an important reference for optimization scheduling strategy based on energy awareness.

5.3.2 Cloud Robotics for Service Provisioning

The introduction of automation equipment in industrial production in the past few decades has incurred many improvements in the industrial sector. With the development of industrial robots, programmed robots have reached high levels of performance in real-time applications, accuracy, robustness, and compatibility. Therefore, the robot network is formed to face the extreme environment such as earthquake, explore the unknown space, and operate fast and accurate grasp of the real demand.

A robotic network refers to a group of robots connected through a wired or wireless communication network [26]. An individual robot in networked robotics is regarded as a node. With sensing data and information shared among nodes, the operators can transmit command data remotely and receive measurement feedback, thus ensuring that a specific operation is carried out. The development of networked robotics has allowed them to be utilized in a variety of applications, such as long-distance medical surgery, disaster relief, and other specialized cases. However, robotics networks still face some inherent physical limitations: Due to the limitations of the robot's volume and other factors, there are obvious limitations to the computing and storage capacity of individual robots. This leads to a limited capacity of the traditional networked robotics when facing high complexity processing tasks. Also, the corresponding performance improvement of individual robots has obvious limitations as well.

With the development of cloud computing, big data, and other emerging technologies [27], the elastic resources provided by a ubiquitous cloud-based infrastructure can allow the transition from networked robotics to cloud robotics. At the 2010 Humanoids conference, James J. Kuffner proposed the concept of "cloud robotics" [28] and elaborated the potential advantages of robot clouds for the first time. The concept of cloud robotics soon caused extensive discussion and research, with researchers in Singapore presenting the construction of the DAvinCi framework [29], Japanese researchers building the business platform Rapyuta [30], and the development of the open source software package ROS (Robot Operating System) with efforts of Willow Garage's team [31] accelerating the development of cloud robotics. Kehoe et al. [32] presented a survey that introduced the most relevant work and organized it around the potential benefits of introducing cloud technologies for robotics and automation.

However, some technical challenges cannot be ignored. With the introduction of cloud technologies, the selection of the types of computation distributions and communication modes that should be applied in different scenarios is critical for overall performance. In addition, extracting relevant patterns from data in the cloud is a typical requirement in different applications, which poses the challenge of data format conversion when dealing with data uploading and downloading. Another important aspect is cloud security, especially the storage of important data in the cloud; this increases the requirements on various aspects of the systems. Finally, to ensure real-time performance, choosing the service quality guarantee methods and corresponding effect analyses is challenging.

5.3.2.1 Architecture and Application of Cloud Robotic

As previously mentioned, cloud robotics aim at transferring the high complexity of the computing process to the cloud platform through communication technology. This greatly reduces the computing load of a single robot and helps to provide multiple services in different scenarios. Considering these challengers in communication and security, the main architecture of the robotic cloud is proposed. As shown in Figure 5.8, the architecture of

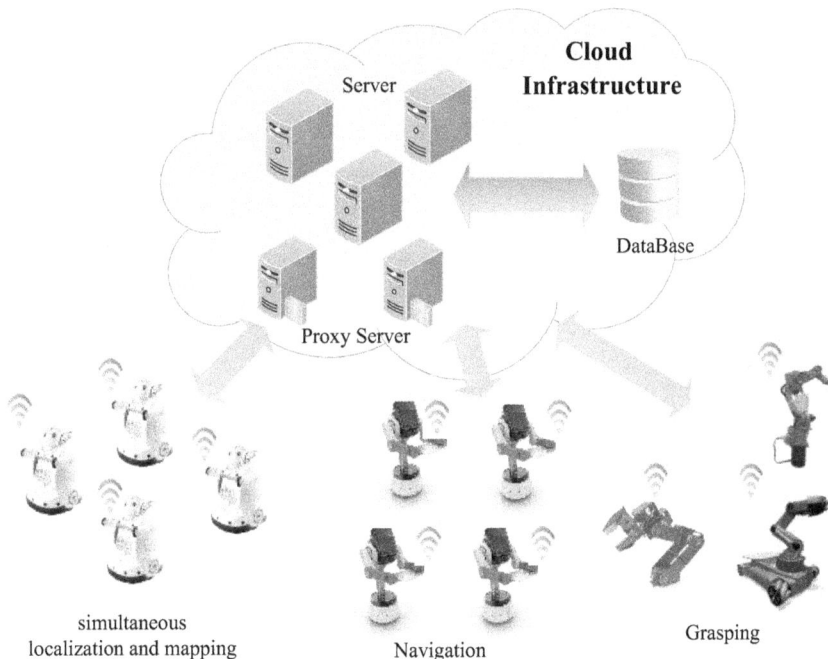

FIGURE 5.8 System architecture of cloud robotics.

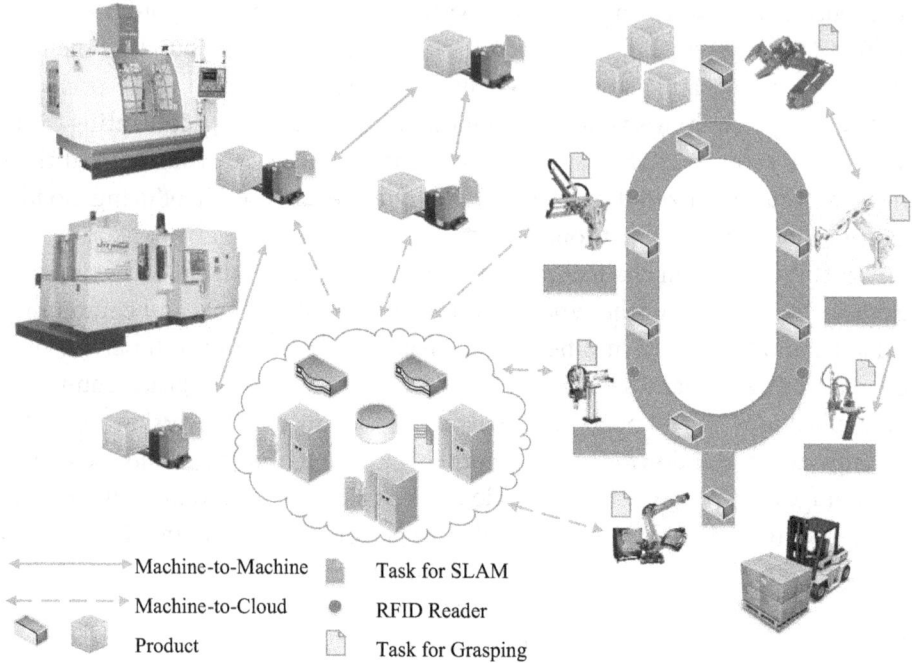

FIGURE 5.9   Implementation of cloud robotics in industrial environment.

cloud robotics is mainly composed of two parts: the cloud platform and its related equipment and the bottom facility. The bottom facilities usually include all types of mobile robots, unmanned aerial vehicles, machinery, and other equipment. Accordingly, the cloud platform is composed of a large number of high-performance servers, proxy servers, massive spatial databases, and other components.

Based on this architecture, cloud robots can provide services for a variety of complex tasks. In cloud robotic systems, the nodes can collaborate with spare nodes by transferring computing or storage tasks. Nodes which are not directly connected to the cloud resource can connect to the cloud through other nodes that have already established links to the cloud. This mechanism not only greatly expands the efficiency of multi-robot cooperative work but also improves the precision of task completion. Next, some classic applications of the cloud robotics are introduced, including Simultaneous Localization and Mapping (SLAM), grasping, and navigation. An example of the applications of cloud robotics including SLAM and Grasping is shown in Figure 5.9.

5.3.2.1.1 SLAM   SLAM is a fundamental topic since 1990s when Durrant Whyte first clearly defined SLAM [33] as exploring unknown environment problems. Extended Kalman filter (EKF) and particle filter (PF) are effective methods for SLAM. However, high-complexity tasks lead to more complicated state estimation and large-scale maps. The emergence of cloud technology allows SLAM to break the bottleneck caused by limited on-board computing and storage equipment. Besides RoboEarth, the relevant technical teams are also committed to continuously improving the SLAM-related code and platform construction. For example, the Kinect@Home project is working on collecting data to complete a RGB-D dataset [34]. In this project, Kinect users can use this platform to carry on the 3D mapping component through their home network browser, while at the same time, Kinect collects the data which the user uploaded. There are many similar platforms and projects around the world. Using a cloud platform, a significant portion of computing, map fusion, and filtered state estimates can be completed in the cloud, which provides strong support for the formation of the map.

5.3.2.1.2 Grasping   The grasping of unknown objects is of great practical significance of robotics in the industrial sector. Grasping problem was first proposed by Ferrari, Canny, and Mirtich et al. [35] to cope with geometric analysis of polyhedral objects, while the initial solution cannot deal with non-polyhedral objects. However, the fact that the unknown characteristics of objects imply a lot of data preprocessing and a large amount of computations lets along introducing learning ability. With the introduction of relative sensing devices, grasping has made progress, to some extent, but it still cannot meet the requirements for adequate response speed and precision. Under the support of big data, cloud computing capabilities, and storage capacity, a mechanical hand can send feature data obtained through a small number of sensors to the cloud using a specific data format. After feature analysis and model matching, data with operating characteristics will be downloaded to the manipulator so that the mechanical hand can grasp the object. It is worth mentioning that after the completion of the grasp, the relevant data can be stored in the cloud for sharing in cases of similar need.

5.3.2.1.3 Navigation   There are two types of navigation problems: the local navigation problem and the global navigation problem. The global navigation problem copes with the larger scale in which the robot does not know the state of its destination from initial position. In real-world robotic scenario, the environment is completely unknown so that the key is to find a collision-free path with high quality. To deal with Non-map-based scenario, several famous approaches are carried out such as fuzzy logic, sampling-based method. Moreover, the traditional sensor-based approaches and neural network models with optimized algorithm also play a role in the global navigation problem. However, due to the restricted onboard resources, these methods usually suffer from reliability problems; either local navigation problems or global navigation problems are vulnerable to large-scale computing and storage tasks which is challenging in large navigation areas especially the later. Cloud-related technologies provide a cloud-based navigation system for the future and are a very promising solution that avoids these two problems.

In summary, the main features of the cloud robotics architecture are as follows: (1) In the cloud infrastructure, where computing tasks are dynamic and resources are elastic and available on-demand. (2) The cloud robotics' "brain" is in the cloud. The results of processing can be obtained through networking technologies, while tasks are processed individually. (3) Computing work can be delegated to the cloud, which leads to a smaller robot load and greatly extended battery life.

5.3.2.2 Context-Aware Cloud Robotic for Material Handling

As a specific application example, cloud-enabled computation sharing for SLAM in advanced material handling is studied. The emerging autonomous navigation AGVs using laser-based localization in industrial environments can facilitate installation and enable guide-path modifications when new stations or flows are added [36]. The laser-navigated AGVs possess the characteristics of cognition and are important component of smart factories, can punctually satisfy the service requirements and facilitate decision-making, and can be effectively integrated into cloud-based management systems. Therefore, context-aware cloud robotics (CACR) can be achieved with the help of the RFID and cloud robotics.

For CACR, the critical first step is to perceive all the important information, such as device status, storage, demand information, and environmental parameters. The relevant data may be accurately gathered through IWNs and then forwarded to a dedicated industrial cloud. In this way, the specific information (e.g., demand information) may be analyzed, and context-aware services for cloud robotics may be realized. After gathering all the information, the critical second step of designing optimized decision-making algorithms under certain constraint conditions, such as energy efficiency or cost savings, needs to be carried out.

Figure 5.10 shows the cloud-based architecture for CACR in CIIoT. The device to cloud platform is cloud communication. In other cases, it is the M2M communication. All the robotics and other smart devices mutually collaborate and form a computing resource network. In this way, the robotics that cannot directly access the cloud can still access

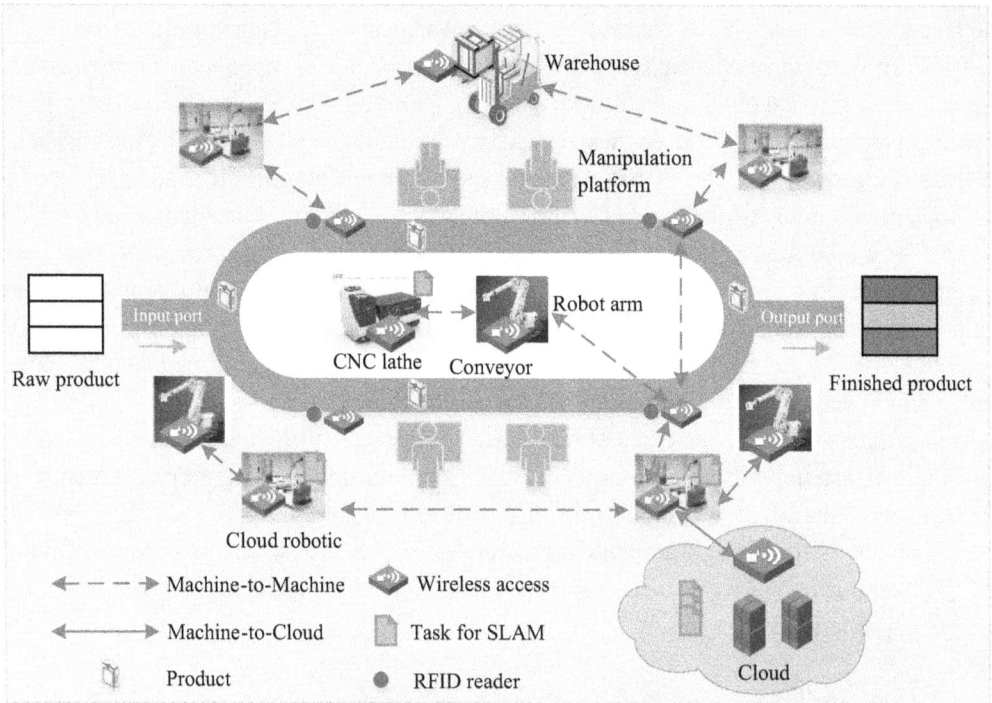

FIGURE 5.10 Cloud-based architecture for CACR in CIIoT.

TABLE 5.3   Advantages, Challenges, and Applications for CACR

Advantages	Challenges	Applications
Computationally intensive tasks' offloading to cloud	Communication bandwidth	SLAM
Access to vast amounts of data	Optimization architecture	Grasping
Knowledge and skills' sharing	Security	Machine learning
Context-aware service provision	Decision-making mechanism	Context-aware services

computation and storage resources through the robots able to access the cloud. The cloud provides a pool of shared computation and storage resources which may be allocated elastically according to real-time requirements. For example, the cloud robotics may offload complex computing tasks such as SLAM to the cloud and preserve their native computing resources. Through the cloud, information can be synthesized to form a new intelligent system. First, all the data from the devices and environments are analyzed in detail, which allows for the provision of some novel services (e.g., context-aware material handling). Second, the skills or behaviors of cloud robots may form a knowledge library containing information which is shared among the cloud robots for learning purposes. Third, big-data-based analytics can facilitate product design optimization and failure diagnosis.

Table 5.3 shows the advantages, challenges, and applications of CACR. Since computationally intensive tasks can be offloaded to the cloud, this will reduce hardware requirements on the robotics and save cost. Also, the CACR can acquire information and knowledge from the cloud, and easily share knowledge and new skills with each other. However, some challenges for CACR still exist, such as decision-making mechanisms, architecture optimization, communication issues, and security. The applications of CACR include SLAM, grasping, navigation, and context-aware services (e.g., material handling). In Sections 5.5 and 5.6, algorithms pertaining to these context-aware services are presented and validated using energy-efficient and cost-saving material handling.

5.3.2.3 Implementation Process of Cloud-Enabled SLAM

The implementation process of cloud-enabled SLAM is shown in Figure 5.11. This process involves two subsystems connected by IWN: (1) a robotic system and (2) a cloud platform. The robotic system consists of four modules, i.e., a data collection, a local processing, a wireless communication, and an actuator module. The data collection module acquires environmental data using a laser locator. The local processing module fulfils the preliminary data processing tasks. Information interaction between different subsystems is achieved through the wireless communication module, while the actuator module carries out the motion navigation.

The cloud platform incorporates two types of nodes, namely master nodes and slave nodes. In master nodes, the wireless communication module carries out data interaction between the robotic system and the cloud platform, while the job tracer module is used to decompose computing tasks and forward SLAM data to all the slave nodes. In each slave node, there is a task tracer which monitors its own task execution and sends the task status to the job tracer in real-time. Each subtask is completed using Map and Reduce functions, and the computed results are sent to the job tracer. Finally, the SLAM data are timely transmitted back to the robotic system.

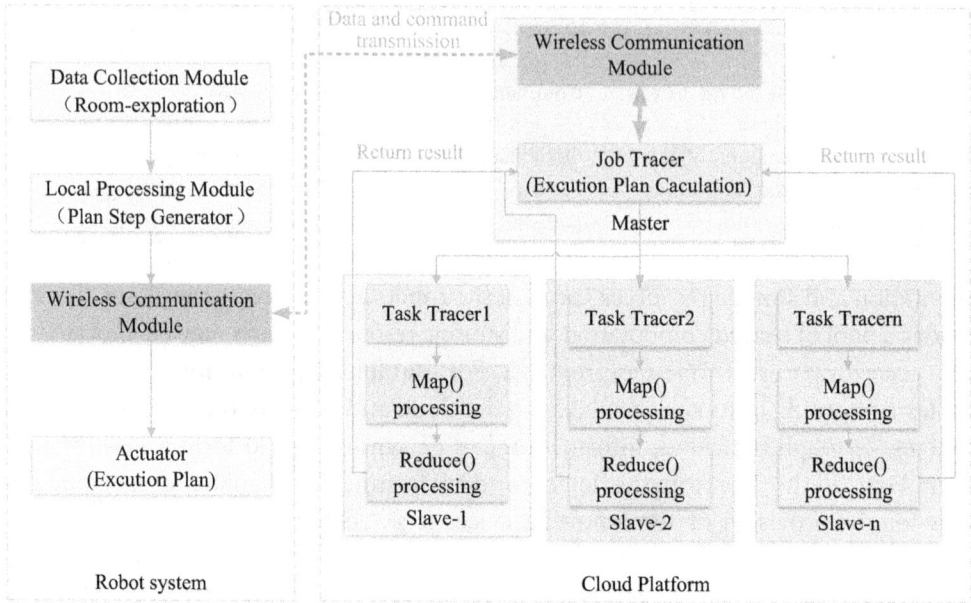

FIGURE 5.11   Implementation process of cloud-enabled SLAM.

In production environments, in order to ensure real-time performance, time-related parameters such as AGV data processing time, cloud infrastructure data processing time, and the network communication response time should be considered. The adoption of cloud robotics is meaningful only when the time parameters satisfy the following inequality:

$$T_p + \frac{C_1}{H} + \frac{C_2}{H} + T_c + \tau < T_R \tag{5.18}$$

where T_p is the preprocessing time for sending and receiving the data, C_1 and C_2 are the sent and received data sizes, H is the network bandwidth, T_c is the cloud infrastructure data processing time, τ is the network delay, and T_R is the robot data processing time when it is operating without access to cloud resources.

Another significant factor of real-time performance is communication protocol. Currently, ZigBee, WiFi, and Bluetooth are widely used and typically chosen for data transmission in practical applications. More concretely, although ZigBee and Bluetooth have better performance on power consumption, WiFi is more appropriate with significant superiority of data throughput and WiFi is able to directly approach the cloud. Considering poor connecting situation, an auxiliary device might be added to improve data acquisition. In addition, module of data transmission on AGV should also support adjustable sampling frequency and configurable application layer protocols.

5.3.3 Dynamic Scheduling Integrating Industrial Cloud and Edge Computing

In order to focus on deeper research on the realization of a flexible and scalable manufacturing system, an intelligent manufacturing architecture, called the Cloud-Assisted

Self-Organized Manufacturing Architecture, based on the Intelligent Production Edge (CASOMA-IPE) is presented.

The main principle of the CASOMA-IPE is the adoption of IIoT, that is, the connection of the IIoT-based elements including people, data, and things through transparent and interoperable communication, which aims to achieve the reconfiguration of a manufacturing system. Furthermore, this kind of information flow integration requires a multi-level information processing link and data transmission in different layers.

5.3.3.1 CASOMA-IPE Hierarchy

Figure 5.12 shows the CASOMA-IPE architecture. It includes three layers of the CASOMA-IPE, namely, physical resource layer, intelligent edge layer, and cloud layer. The physical resource layer includes physically controlled components such as machining devices and transporting devices. The machining devices include the computerized numerical control (CNC) devices, packaging devices, and storage devices, which can process materials. The transporting devices include the conveyor belts and Automatic Guided Vehicles (AGVs), which can transport materials. The intelligent edge layer contains many intelligent production edges (IPEs) as the IIoT adapters that aim to connect legacy systems to the cloud or to convert the legacy data from the third-party devices (e.g., KUKA robot) into the CASOMA-IPE common data model. Additionally, the IPEs can provide field devices with the ability of intelligent decision-making and communication and enable them to accomplish the production tasks through the negotiation process, which are the basis of the IIoT. The cloud layer can be divided into three parts, human–machine interface (HMI), manufacturing execution system (MES), and KB.

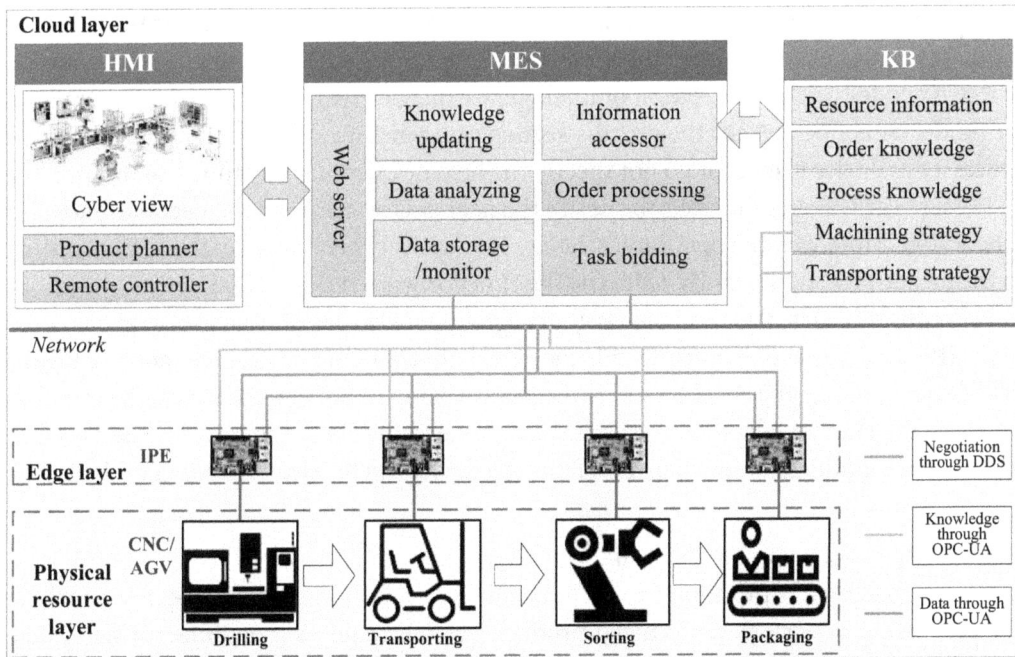

FIGURE 5.12 The Cloud-Assisted Self-Organized Manufacturing Architecture based on the Intelligent Production Edge.

On the one hand, the cloud is used to store the production data uploaded from the field devices and process the production data for updating the KB, including the resource information, order knowledge, planning knowledge, machining strategy, and transporting strategy. Meanwhile, the cloud forms the assisted strategies for the field devices to generate the execution parameters. On the other hand, it can provide users with the data interface from the MES to the HMI via the web. There are two scenarios for the HMI usage: (1) Customers place orders and provide their expectation on a product, or check the product processing progress and logistics status. (2) Managers monitor the manufacturing system in real-time or directly set parameters of field devices.

The main advantage of the CASOMA-IPE is the adoption of IIoT, that is, the connection of the IIoT-based elements including people, data, and things through transparent and interoperable communication, which aims to achieve the dynamic scheduling. Furthermore, this kind of information flow integration requires a multi-level information processing link and data transmission in different layers.

The interactions between the cloud and edge layers are represented by lines in Figure 5.12. The meaning of these lines is shown in the legend at the bottom right, including data interaction through OPC-UA, negotiating interaction through DDS and knowledge interaction through OPC-UA. Besides, the lines between the IPE and field device represent their interaction, indicating their one-to-one relationship.

During the production, the CASOMA-IPE adopts two kinds of intelligent strategies that are generated on two levels (e.g., locations): the process planning of production line level (e.g., the cloud) and the operation strategy of field device level (e.g., the IPE). The process planning on the cloud includes the order decomposition, process recognition, and process sequencing. While the IPE focuses on the operation strategies of field devices, including the allocation of machining parameters, trajectory design, path planning, and generation of PLC/G codes. Such a generation of intelligent strategies requires knowledge interaction for model computing. After the process sequence is generated, the cloud encapsulates the processes as device tasks that do not specify the field devices but only functionalities of the processes. The IPEs get the tasks to specify the field devices through the negotiation-based interaction. Moreover, the operation strategy is adaptively generated by the IPE according to the device task, significantly reducing the time for work redesign and device idling.

According to different path strategies, the programs for field devices manipulation are stored in the IPE database, and they are downloaded every time the field devices operate. During the field device running, the cloud monitors the job execution in devices and shares the information with the IPEs for adaptive decision-making. Due to the cloud supervision based on data interaction, the availability and current status of a device can be used for dynamic scheduling.

5.3.3.2 Reconfiguration Mechanism

In a small-batch customization production, the production line has to respond timely to the processing of various customized products, which requires a manufacturing system to be able to reconfigure the production logic quickly and dynamically. In order to meet this requirement, the CASOAM-IPE is modeled as a MAS to clarify the reconfiguration mechanism in the planning and scheduling phases.

5.3.3.2.1 Agent-Based Concept

The existing research results in the distributed artificial intelligence (DAI) field indicate that an intelligent manufacturing system equipped with the MAS technology is a promising approach since it can support flexibility, robustness, and reconfigurability [37]. In the MAS concept, manufacturing components in the CASOAM-IPE can be considered as autonomous agents. Thus, the inherent features of the MAS can naturally support the CASOAM-IPE requirements for reconfigurability and scalability.

According to the difference in functions and responsibilities, the agents in the MAS-based manufacturing system can be classified into three agent types. The first agent type is the Machining Agent (MA), representing the combination of an IPE and a connected machining device. The second agent type is the Transporting Agent (TA), representing the combination of an IPE and a connected transporting device. The third agent type is the Cloud Agent (CA), representing a cloud. In the following contents, the concept of Process Agent (PA) denotes the MA or TA for convenience.

The design of three agent types increases the system flexibility because agents can negotiate with each other to accomplish tasks instead of implementing the centralized controlling by a single-center, thus establishing a scheduling reconfiguration. The reconfiguration type depends on the corresponding relation between a logical layer and components in the CASOAM-IPE. So, this relation enables agents to achieve a better representation of system logic. In order to use offline and online data to perform the reconfiguration process, all agents need to interact with each other to share their information about status and knowledge via contract net protocol (CNP) and Foundation for Intelligent Physical Agents—Agent Communication Language (FIPA-ACL).

5.3.3.2.2 Reconfiguration Procedure   The reconfiguration procedure of the production logic is shown in Figure 5.13, wherein it can be seen that it is divided into two phases, namely planning phase and scheduling phase.

At the beginning of the planning phase, an order is built by customers and transferred to the CA. When the CA receives the order, it uses the feature-based tools to analyze the

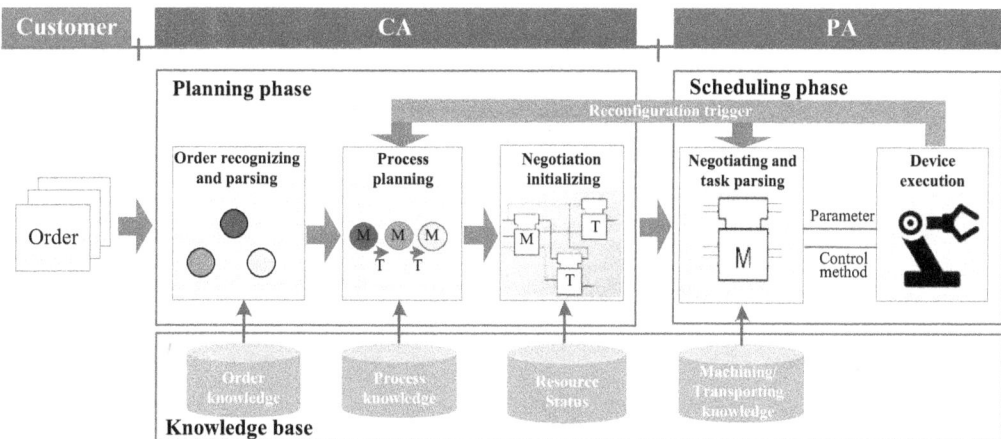

FIGURE 5.13   Reconfiguration flow of the CASOMA-IPE production logic.

product features and then extracts the processes from the product features in terms of the order KB. Then, the CA builds and optimizes the process sequence, including the machining and transporting processes through the process KB. After the process sequence is generated, the CA encapsulates the process as the device task, which does not specify the field devices but only functionalities of the processes. Finally, the CA publishes the task information to the PAs with a high matching degree of machining/transporting performance.

In the scheduling phase, the PAs compete with each other to get the task and specify the field device. After obtaining the task, the PA parses the task and converts it into the execution parameter and control method for the device operation according to the machining/transporting KB.

During the operation of the task-relevant devices, the availability and status of the devices are monitored in real-time by the CA. When an exception occurs, the CA receives the alarm and triggers the reconfiguration, which can be of two categories, planning reconfiguration and scheduling reconfiguration. The reconfiguration type depends on the exception type. The system makes the reconfiguration strategy according to the four reconfiguration principles: (1) When a new order is built, the planning reconfiguration is triggered. (2) When a new process sequence is generated, the scheduling reconfiguration is triggered. (3) When a task-relevant device is broken and another device regarding the process of the original process sequence is not available, the planning reconfiguration is triggered. (4) When the task-relevant device is broken and another device regarding the process of the original process sequence is still available, the scheduling reconfiguration is triggered.

5.3.3.2.3 Agents Interaction The sequence diagram of the interaction between agents is illustrated in Figure 5.14. In the planning phase, the CA subscribes to the status of MA and TA to coordinate the global system performance. Additionally, the CA obtains the orders and generates the process plan for responding to new product demand. In the scheduling

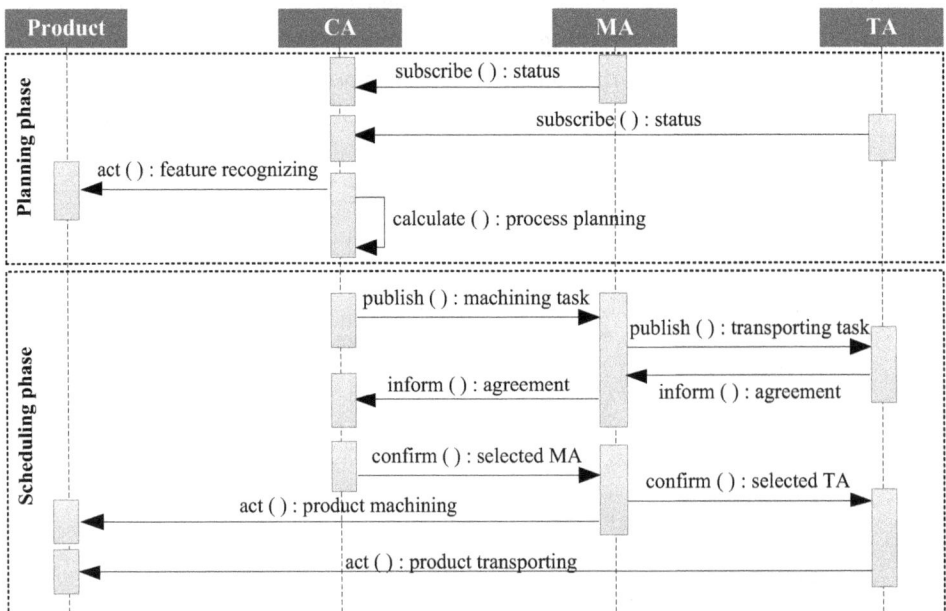

FIGURE 5.14 The sequence diagram of the interaction between agents.

phase, the CA publishes the machining task to the MAs to specify the target devices regarding the process sequence. Then, the selected MA publishes the transporting task to the TAs. After the agreements from both MAs and TAs are sent to the CA, the target devices are selected. The CA sends the confirmation to the selected MAs and TAs. Finally, the selected MAs and TAs begin to process and transport the product.

5.3.3.2.4 Agent Behavior   In the MAS, an agent can produce plans and make autonomous decisions to react to the external environment. Therefore, the first thing to consider is how to design the agent behavior according to its functionalities and objectives. Generally, the agent behavior is described as the transition relationship between different states. According to the current state and external input, the agent can move to the next state. Petri net [38] is one of the mathematical modeling languages convenient for the description of distributed systems, and it offers a graphical notation for stepwise processes that include choice, iteration, and concurrent execution. Thus, the Petri net is a good option to describe agent behavior.

The Petri net behavior model of the CA is presented in Figure 5.15. After the initialization, the CA enters a ready state. In this state, the CA is waiting for a message which can be of two types, namely, an order and a status. On the one hand, if a message is an order from a customer, the CA will enter the order-processing state. Then, the CA will

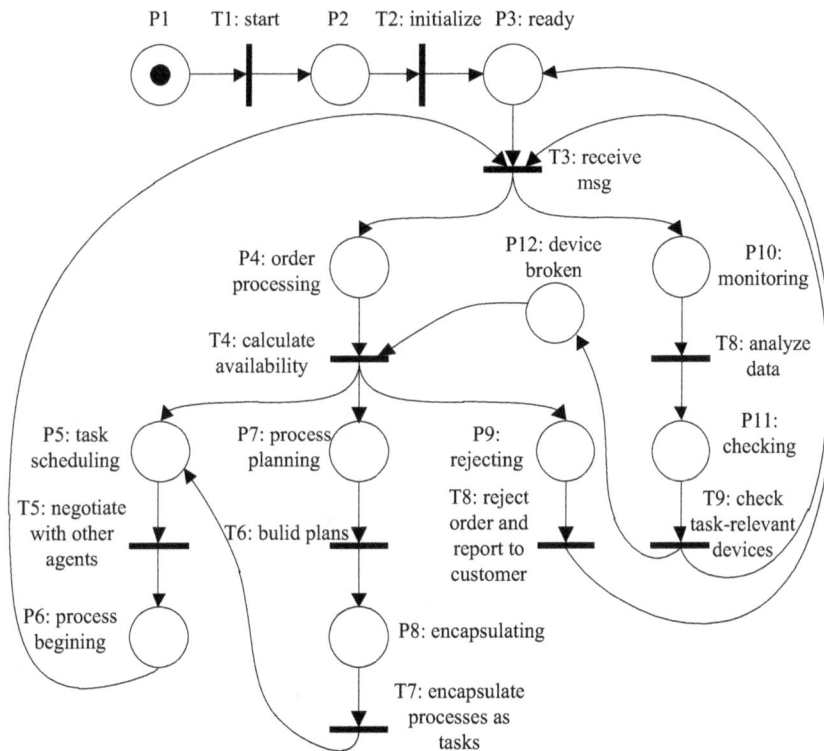

FIGURE 5.15   The CA behavior model.

calculate the availability set of the task-relevant devices to confirm the next state. If the availability set is not empty, the CA will build a process plan and negotiate with other agents to accomplish the production task; otherwise, the CA will reject this order and report that to the customer. On the other hand, if the message is a status message of another agent, the CA will enter the monitoring state. The CA analyzes data and checks the status of task-relevant devices, and the CA will go back to the ready state if there still are some devices can work for the task. Once the device is broken, the CA enters the device broken state and calculates the availability of task-relevant devices. According to the third and fourth reconfiguration principles, the CA can reconfigure the process planning and task scheduling.

The Petri net behavior model of the PA is presented in Figure 5.16. After the initialization, the PA enters a ready state. In this state, the PA is waiting for a message, which can be either task information or status. On the one hand, if the message is task information of the CA, the PA enters the negotiating state. The PA negotiates with other agents to decide whether to accept the task or not. If the PA accepts the order task, the process of the order task will begin. Otherwise, the PA rejects this order and reports that to the CA. On the other hand, if the message is a status from the field device, the PA will enter the monitoring state. The PA

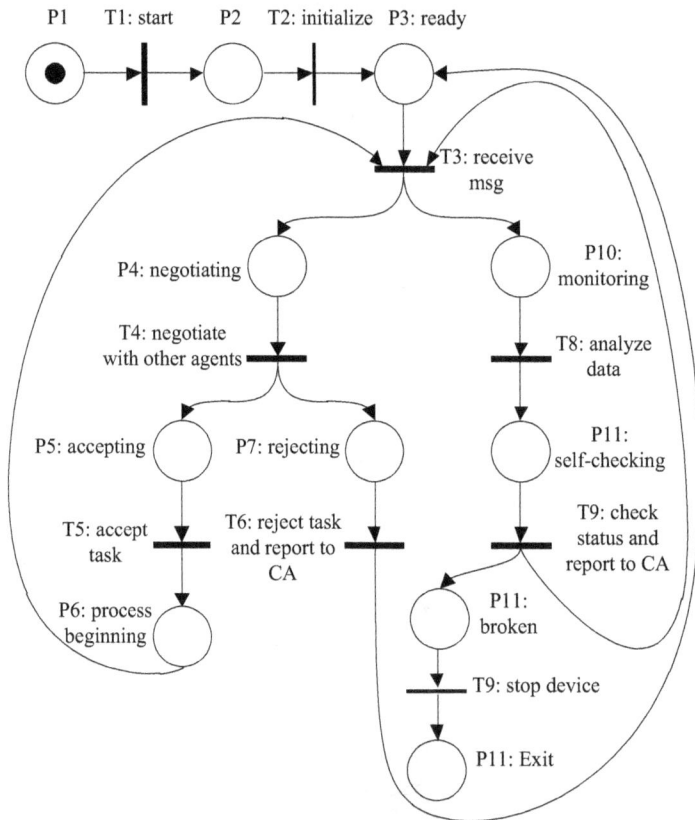

FIGURE 5.16   The PA behavior model.

analyzes data and checks the status of the device; the PA will go back to the ready state if the device can be working for the order task. Once the device fails, the PA enters the broken state and stops the device. Then, the PA sends the report to the CA and exits from the system.

5.3.3.2.5 Negotiation Mechanism   The time and resource constraints are not taken into account in the building of a process sequence, but these constraints denote an important basis for computation in the negotiation. Therefore, the idle time margin I_{Tmargin} is defined as a proportion of the devices' idle time in 10 minutes of a day, which excludes the task processing time. The larger the value of I_{Tmargin} is, the greater the time redundancy the device has; thus, the task is easier to complete, which can be expressed as follows:

$$
\left\{
\begin{array}{l}
I_{\text{Tmargin}} = 0, \left(T_{\text{procee}} > T_{\text{idle}} \right) \\[2mm]
I_{\text{Tmargin}} = 1, \left(T_{\text{idle}} - T_{\text{procee}} \geq 10 \right) \\[2mm]
I_{\text{Tmargin}} = \dfrac{T_{\text{idll}} - T_{\text{procee}}}{10}, \left(T_{\text{idle}} - T_{\text{procee}} < 10 \right)
\end{array}
\right.
\tag{5.19}
$$

where T_{process} denotes the task processing time and T_{idle} denotes the device idle time. With a limit of 10 minutes, the values of three intervals can be calculated using I_{Tmargin}.

The task matching degree I_{match} is defined as a matching degree of the device that is bidding for a task, and it is expressed as follows:

$$
I_{\text{match}} = I_{\text{Tmarigin}} \times I_{\text{ability}}
\tag{5.20}
$$

where I_{ability} is the serviceability, whose value is either 1 or 0; the value of 1 means that the bidding task can be processed and the value of 0 means that the bidding task cannot be processed.

The negotiation mechanism between the CA and the PA is shown in Algorithms 5.3 and 5.4. In the CNP, the CA acts as a contract contractor that publishes the task information and waits for the response of each PA. According to the idle time margin I_{Tmargin} and the serviceability I_{ability} of the task information, the PA calculates the task matching degree I_{match} and sends it to the CA. The CA receives the responses and sends the query information to the PA whose task matching degree $I_{\text{match}} > 0.1$. When receiving the query information, the PA calculates the comprehensive mental expectation proposed in and sends it to the CA. The comprehensive mental expectation is a criterion to evaluate the superiority–inferiority level of agents, which can be calculated by the mental parameters, namely, Perceptibility, Reliance, Activity, Friendship, and Capability. After the bids from the PAs are received, the

CA selects the most suitable PA based on the largest comprehensive mental expectation. Finally, the CA sends confirmation to the selected PA. The selected PA receives the task and processes it. If there are still unfinished tasks, these steps are repeated until all the tasks are done.

ALGORITHM 5.3   CA Model

1	Variables definition
	$INFO[n]$: Store the information of $task[n]$;
	$TSKc$: Store the number of the existing task;
	Rec: Store the number of response;
	$Tint$: Store the time interval between sending and receiving the message;
	$Tintthres$: Store the threshold of $Tint$;
	$Imatch[n]$: Store the matching degree of $PA[n]$;
	$Mp[n]$: Store the comprehensive mental expectation of $PA[n]$;
	$Tarid$: Store the target PA index;
2	**begin**
3	Initializes $i=0$;
4	**while** $TSKc>0$
5	Publish $INFO[i]$ as the task information to each PA;
6	$i=i+1$;
7	Wait for response of the PA;
8	**if** $Rec>0$ and $Tint<Tintthres$
9	Save match degree of $PA[n]$ as an array variable $Imatch[n]$;
10	**for** $j=Rec-1:0$
11	**if** $Imatch[j]>0.1$
12	Query $PA[j]$ the comprehensive mental expectation;
13	Receive $Mp[j]$;
14	**end if**
15	**end for**
16	**for** $k=Rec-2:0$
17	**if** $Mp[k+1]<Mp[k]$
18	
19	$Tarid=k$;
20	**end if**
21	**end for**
22	Send the confirmation to $PA[Tarid]$;
23	Send $task[i]$ to $PA[Tarid]$;
24	$TSKc=TSKc-1$;
25	**end if**
26	**end while**
	end

ALGORITHM 5.4   PA Model

1	Variables definition
	Imatch: Store the matching degree of PA;
	Reqc: Store the number of query response;
	Tint: Store the time interval between sending and receiving the message;
	Tintthres: Store the threshold of *Tint*;
	Recc: Store the number of competition response;
2	**begin**
3	Wait for the task information from the cloud;
4	Calculate the variable *Imatch* and send it to the cloud;
5	Wait for the cloud query;
6	**if** $Reqc > 0$ and $Tint < Tintthres$
7	Calculate the comprehensive mental expectation and send it to the cloud;
8	Wait for the competition result;
9	**if** $Recc > 0$ and $Tint < Tintthres$
10	Receive the task and process it;
11	**end if**
12	**end if**
13	**end**

5.3.3.3 Experimental Verification

In order to verify the adaptability of the IPE and the reconfiguring function of the CASOMA-IPE production logic, two groups of experiments were conducted. The hardware used in the experiments included a server that served as the cloud, five robots used for machining simulation, five PLC-controlled conveyor belts used for transporting the workpieces, ten microcomputers (Raspberry Pi) that were used as IIoT-based IPEs, and a router. The IPEs were connected via Ethernet through the router.

For a production line including five machining devices, the processing capacities of five machining devices are given in Table 5.4. In the experiment, there were orders available for processing four workpieces (A, B, C, and D), each of which accounted for 25% of the total number of the workpieces to be processed. The process set of the four types of workpieces is given in Table 5.5, which indicates the potential choices of the processes for each workpiece. As required, the type of workpieces entering the production line was random and the time interval between adjacent workpieces obeyed normal distribution.

The four groups of scheduling methods were as follows:

1. The pre-planned scheduling used the first-in-first-out (FIFO) rule to prepare the static execution plan before the production and did not change the execution plan during the production. The execution plan of the pre-planned scheduling is given in Table 5.6.

2. The centralized dynamic scheduling used the genetic algorithm-based scheduling rule to dynamically generate the execution plan of the random workpiece according to the device status.

TABLE 5.4   The Processing Capabilities of Five Machining Devices

Device Index	Process Index and Process Time (s)			
1	$P_{001}(10)$	$P_{002}(5)$	$P_{003}(8)$	$P_{004}(5)$
2	$P_{005}(6)$	$P_{006}(4)$	$P_{007}(7)$	
3	$P_{001}(13)$	$P_{004}(10)$	$P_{008}(4)$	$P_{009}(5)$
4	$P_{010}(3)$	$P_{011}(2)$	$P_{012}(5)$	
5	$P_{011}(5)$	$P_{012}(9)$	$P_{013}(4)$	$P_{014}(5)$

TABLE 5.5   The Process Set of Four Types of Workpieces

Workpiece Index	Process Index and Process Constraint (from Left to Right)			
A	P_{001}	P_{011}	P_{006}/P_{014}	
B	P_{003}/P_{004}	P_{008}	P_{012}	P_{007}
C	P_{008}	P_{010}/P_{012}	P_{005}	
D	P_{009}	P_{002}	P_{011}/P_{013}	P_{006}

TABLE 5.6   The Execution Plan of the Pre-Planned Scheduling

Workpiece Index	Device Index and Process Time (s)			
A	$M_1(10)$	$M_4(2)$	$M_2(4)$	
B	$M_1(8)$	$M_3(4)$	$M_5(9)$	$M_2(7)$
C	$M_3(4)$	$M_4(5)$	$M_2(6)$	
D	$M_3(5)$	$M_1(5)$	$M_4(2)$	$M_2(4)$

3. The distributed dynamic scheduling proposed in used the negotiation mechanism of the distributed resources to dynamically generate the execution plan. This scheduling method did not consider the reconfiguration in the planning phase.

4. The dynamic scheduling based on the CASOMA-IPE used the proposed reconfigurable method to generate the execution plan.

In order to verify the performance of the production line, two scheduling experiments were performed:

1. The four mentioned scheduling methods were used in the comparative experiment.

2. The dynamic scheduling experiment using the CASOMA-IPE was conducted for the scenario with an exception.

In the experiment, the delayed operation rate Z_j was used as the evaluation criteria of the system efficiency, and it was expressed as follows:

$$Z_j = J_t \div J_{\text{total}} \tag{5.21}$$

where J_t denoted the number of delayed jobs and J_{total} denoted the total number of jobs.

The relationship between Z_j and when the other parameters of the production line were kept unchanged is presented in Figure 5.17a. In Figure 5.17a, it could be seen that with the increase in the time interval of the workpiece, the job delay rate decreased because the excessively high workpiece density and slow processing speed resulted in high-rate job delay. On the contrary, when the workpiece density decreased and the processing speed of machining devices could match or even exceed the speed of workpieces feeding, the job delay rate decreased.

The longitudinal comparison showed that 38% of the workpieces to be scheduled as pre-planned were delayed at 8 seconds. 8% of the workpieces were delayed under centralized dynamic scheduling at 8 seconds. Almost no workpiece was delayed under distributed dynamic scheduling at 8 seconds, while the dynamic scheduling based on the CASOMA-IPE had no workpiece delays at 6 seconds. Thus, the job delay rate curves of the three mentioned dynamic scheduling moved more than three units of time leftward from the pre-planned scheduling. This means that the production performance of the dynamic scheduling was at least twice as much as that of the pre-planned scheduling. This was because pre-planned scheduling could not update the execution plan according to the devices' status, resulting in workpieces easily blocking in the key device. Meanwhile, the results also showed that distributed dynamic scheduling had more advantages in the job delay rate than centralized dynamic scheduling from = 1 second to = 10 seconds. The reason for the phenomenon was that centralized dynamic scheduling required a certain computing time to generate the execution plan. When <10 seconds, the delay of the workpieces occurred, which indicated that the computing time was longer than 10 seconds. However, the computing time of distributed dynamic scheduling was obviously less than that of centralized dynamic scheduling. In addition, the dynamic scheduling based on the CASOMA-IPE was more efficient than the distributed dynamic scheduling because the distributed dynamic scheduling could not reconfigure the production logic in its planning phase to guarantee global system performance. While the dynamic scheduling based on the CASOMA-IPE could reconfigure the production logic in both the planning and scheduling phases, it could make full use of the idle time and processing capacity of machining devices to improve the global performance.

In general, it is unavoidable that the device fails. The experimental scenario was as follows. We fed the production line with the workpieces at the time interval = 5 seconds. After the fifth workpiece had been fed, the machining device 1 failed and could not accept the processing task. Until the fiftieth workpiece had entered the production line, the machining device 1 was repaired and put into use again. The changing trend of the job delay before and after device failures caused by the changes in the number of workpieces to be processed is presented in Figure 5.17b. As can be seen in Figure 5.17b, when the machining device 1 failed at $n = 5$, the job delay rate was rising sharply; when $n > 50$, the job delay rate decreased accordingly and nearly coincided with that in the case of no failure, which indicated that the dynamic scheduling based on the CASOMA-IPE had strong robustness in the case of fault occurrence and recovery.

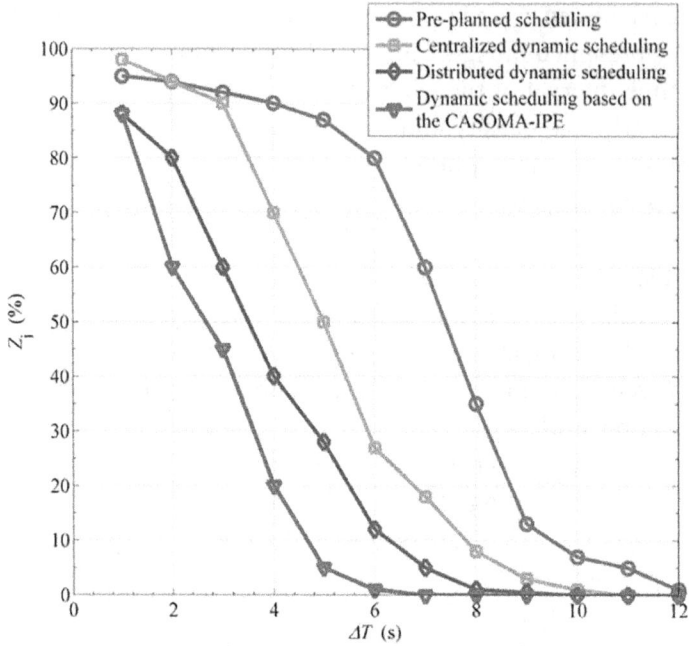

(a) The comparison of Zj of the pre-planned scheduling, distributed dynamic scheduling, and dynamic scheduling based on the CASOMA-IPE

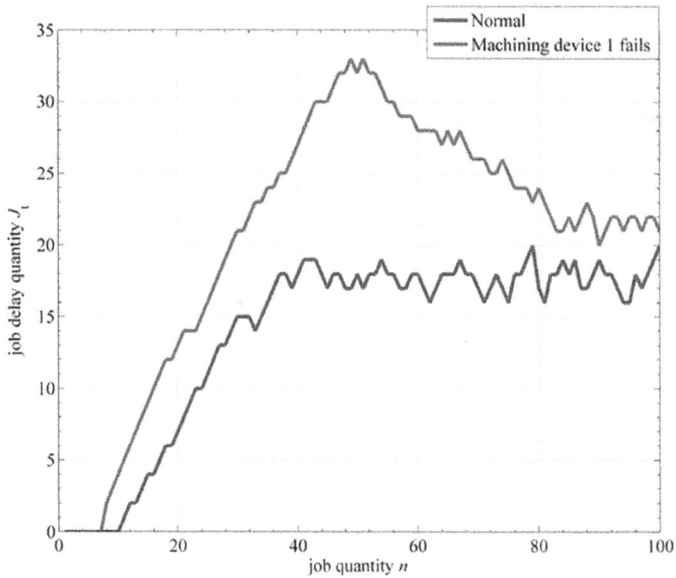

(b) The comparison of Zj before and after a fault occurrence

FIGURE 5.17　The scheduling results. (a) The comparison of Z_j of the pre-planned scheduling, distributed dynamic scheduling, and dynamic scheduling based on the CASOMA-IPE. (b) The comparison of Z_j before and after a fault occurrence.

5.4 RESOURCE RECONFIGURATION BASED ON ARTIFICIAL INTELLIGENT

At present, the increasingly significant personalization and diversification demand of customer results in the phenomenon that traditional large quantities and monotype production methods are becoming increasingly nonadaptive. At the level of the prior work, due to the numerous kinds of products, high-frequency change of production organization, large scale of production line, and complicated equipment resources in the production process, many problems appeared, such as perplexingly optimal allocation of manufacturing resources and complex production scheduling. In addition, there are some other shortcomings in the traditional manufacturing industry, such as long cycle of maintenance faulty, low utilization of equipment, and indigent energy saving. Therefore, there is an urgent demand transforming the dynamic scheduling into the reconfigurable resource allocation.

In the past decades, reconfigurable manufacturing system (RMS) [39] have emerged, which partially overcame the challenges caused by a great variety and uncertainty [40]. It can adjust its production capacity or functionality by reconfiguring hardware and/or software to respond to sudden market changes or internal system disturbances [41]. However, RMS needs to be stopped to change the configuration or strategy when an exception occurs. Thus, the responsiveness of a centralized control model is inadequate. The emerging of holonic manufacturing system (HMS) can also be built as a MAS, which allows agents to make decentralized decisions. Thus, the HMS can dynamically reconfigure its production logic to resist multiple disturbances by changing the interaction behaviors of the agents. Nevertheless, due to performance differences in manufacturing resources and the absence of global coordination, the existing manufacturing model cannot deal with the consequences of local convergence [42]. Also, the definition of dynamic behaviors of agents and the reconfiguration mechanisms of self-organization are still unanswered questions in the decentralized manufacturing field.

In this section, based on AI, resource reconfiguration method will be introduced in detail from the aspect of knowledge sharing and data-driven decision-making. It is then of immediate significance to construct the ontology for knowledge sharing, reuse, and reasoning. First, knowledge sharing mechanism for resource reconfiguration is studied. The ontology, which is a semantic representation of related concepts and their relationship in manufacturing resources, reveals the essence of these concepts and the links between them. At the same time, inference and query based on KB increase the cognition ability of resource agents, which ensures the context awareness and semantic understanding of edge intelligent. Then, data-driven ML methods overcome the weakness of ontology reasoning based predefined rules and have the ability of discovering new knowledge. The ML methods (such as pattern recognition, accurate modeling, knowledge processing, reasoning, and decision-making) provide important support for cognition enhancement. Further, the more and more data acquisition help to improve the decision-making of production plan for resource scheduling.

5.4.1 Knowledge Sharing for Resource Reconfiguration

In order to realize a user-centered intelligent manufacturing system, one important issue of the manufacturing cyber-physical systems (MCPS) is a reconfigurable technology of manufacturing resources [43]. In this context, the cloud-edge collaborative architecture has given stronger computing intelligence capabilities to edge nodes such as manufacturing resources [44], making it possible to share information and learn knowledge between manufacturing resources. The existing research on the manufacturing resources reconfigurable technology is focused on system scheduling strategy, network modeling, multi-agent, and so on. However, there are still some problems, such as low accuracy and slow response. Therefore, a resource KB is built to realize the sharing of processing experience between equipment, improving the efficiency of the resource reconfiguration.

The ontology, as the core technology of knowledge engineering, clearly specifies the conceptual model [45]. By modeling a domain-related knowledge, the ontology can effectively provide domain concept hierarchy, semantic structure, and semantic information sharing which strongly support system integration and interoperation [46]. Hence, an ontology-based reconfigurable method is proposed for knowledge sharing between manufacturing resources. Moreover, it effectively demonstrates the intelligent manufacturing resource reconstruction and related smart services. This knowledge sharing mechanism between manufacturing resources is based on cloud-edge collaboration. The ontology-based method for intelligent device semantic modeling and reasoning achieves device KB modeling and independent reasoning. The use of distributed data storage for industrial big data integrates the real-time data stored on the cloud with the ontology model to update the state of device. On this basis, the learning process between devices is realized by effectively screening, matching, and combining the existing knowledge primitives contained in the KB deployed on the cloud and the edge.

In order to achieve an interenterprise resource sharing and make companies more flexible to cope with the market change, the smart manufacturing exigently demands merging of decentralized core competencies including human, equipment, and technology. The equipment resources of CPS are integrated through the ontology-based intelligent manufacturing resource integration architecture, as shown in Figure 5.18. This architecture consists of four layers: data layer, rule layer, knowledge layer, and resource layer, which correspond to the concepts, instances, reasoning rules, and associated real-time data, respectively. Resource layer is the entity of CPS intelligent equipment, including robot, conveyor belt, and power line carrier. The knowledge layer is composed of the information model of an intelligent device, which is combined with the domain KB by the Web Ontology Language (OWL), wherein Sn ($n \geq 1$) represents the various submodules of the OWL [47]. The rule layer is used to gather the rules of intelligent features such as intelligent equipment independent decision-making and reasoning. The data layer covers a distributed database, Hbase, located on the cloud platform that stores the real-time data, and a relational database merged with the real-time data.

The mapping manager is the Uniform Resource Identifier of the intelligent device ontology, which offers mapping for a relevant information of the ontology, and resource positioning.

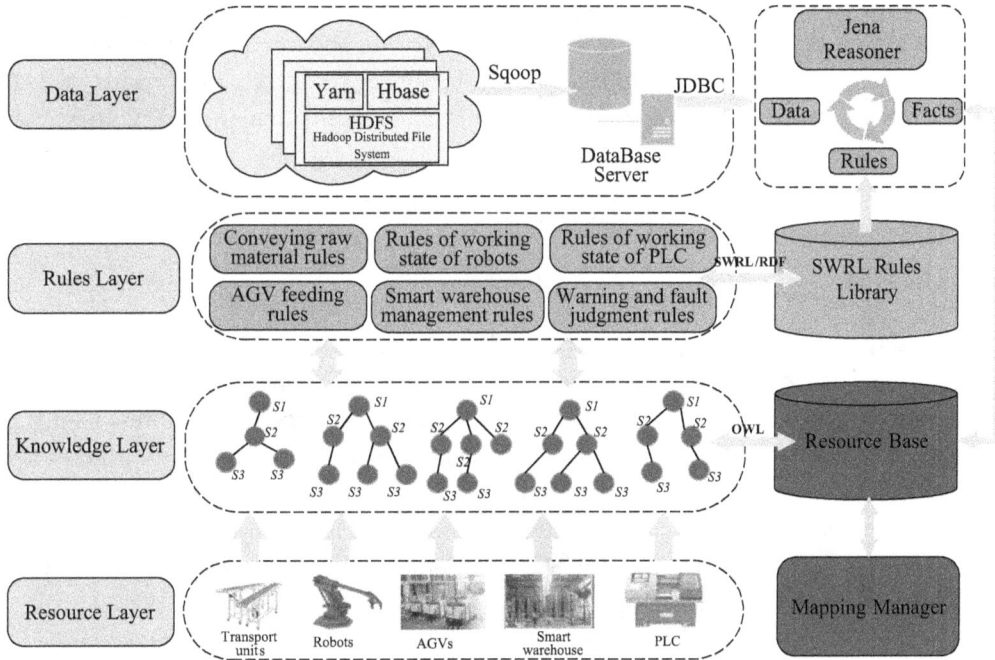

FIGURE 5.18 The intelligent manufacturing resource integration architecture based on ontology.

Compared with RAMI 4.0 [48], RAMI 4.0 ensures that all participants involved in industries 4.0 discussions understand each other. The idea of RAMI 4.0 highlights the product life cycle from production to usage, and this procedure embraces enterprise, work centers, station, control device, fields device, and product. However, our proposed architecture mainly emphasizes an ontology-based reconfiguration of manufacturing resources by device operation parameters, which means that the proposed framework is the subclass of RAMI 4.0.

In this framework, the manufacturing resources are expressed in the form of ontology, thus the real-time data of a device can be associated with the ontology model. The advantages of the proposed architecture are as follows:

1. The information of decentralized manufacturing resource in the CPS is stored in the KB to finish the centralized managing of resource.

2. The reversed and integrated ontology information actually achieves the separation of knowledge model and application. When the knowledge model is migrated to different applications, its integrity will not be destroyed, thus a good dynamic is guaranteed, which is conducive to the knowledge sharing and circulation.

5.4.1.1 Semantic Modeling of Manufacturing Resource

The ontology integration process is as follows. First, by discovering the relationships between entities, the mapping is generated. Then, the merging is achieved according to the application purpose. The intelligent manufacturing resources involve the processing of manipulator, AGV feeding, assembly stamping, labeling, product packaging, and other intermediate processes. These processes play a key role in the intelligent production line. In order to realize the reconfiguration of manufacturing resources, the various parts of manufacturing resources are organized into a unified and easy to manage KB. These constitute an effective knowledge-expression and organization form.

5.4.1.1.1 Manufacturing Resource Ontology Integration

Definition 5.4

Manufacturing resource ontology structure is a tuple (C, R, H_C, rel) [49], where C denotes the concept set and c signifies the concept $(c \in C)$; wherein the concept is a kind of equipment in the manufacturing field, and each concept can express its different aspects by attribute separately; R implies a set of relationships, r connotes a relation $(r \in R)$, and the relationship describes the connection between concepts or between attributes; relations can be divided into two categories: classification communications and connection links; H_C is a hierarchical relationship between two concepts, the concept of the parent class, subclasses, and so on, $H_C \subseteq C \times C$; rel is the connection relationship that does not include the other upper and lower hierarchy. $rel: R \rightarrow C \times C$.

Definition 5.5

(Device ontology category): The device ontology structure can be defined as an object, the device ontology mapping can be defined as a categorical morphism, and the ontology category *Ont* can be defined as a morphological function $(f, g): O \rightarrow O'$, where $O = (C, R, H_C, rel)$ and $O' = (C', R', H_C', rel')$ are ontology structures; $f: C \rightarrow C'$ and $g: R \rightarrow R'$ such that (1) if $(C_1, C_2) \in H_C$, then $(f(C_1), f(C_2)) \in H_C'$; (2) if $(C_1, C_2) \in rel(R)$, then $(g(C_1), g(C_2)) \in rel'(g(R))$. The morphism of the condition (1) defines the hierarchical structure of the concept, and the morphism of the condition (2) defines the relation between the concepts.

In Figure 5.19, ontology $O_1 = (\{x_0, x_1, x_2, x_3\}, \{r_1, r_2\}, \{(x_1, x_0), (x_2, x_0), (x_3, x_2)\}, \{(r_1, (x_0, x_3)), (r_2, (x_1, x_2))\})$, $O_2 = (\{z_0, z_1, z_2\}, \{t_1, t_2\}, \{(z_1, z_0), (z_2, z_1)\}, \{(t_1, (z_0, z_2)), (t_2,(z_1, z_1))\})$. According to the morphism $(f, g): \{x_0, x_1, x_2, x_3\} \times \{r_1, r_2\} \rightarrow \{z_0, z_1, z_2\} \times \{t_1, t_2\}$, we can get the ontology mapping: $x_0 \rightarrow z_0, x_1 \rightarrow z_1, x_2 \rightarrow z_1, x_3 \rightarrow z_2, r_1 \rightarrow t_1, r_2 \rightarrow t_2$. These mappings maintain the hierarchical relationship of the ontology; for instance, in ontology O_1, x_1 and x_2 are the sub-objects of x_0; correspondingly in $O_2, f(z_1), f(z_2) = t_1$ is a subclass of $f(z_0) = t_0$.

5.4.1.1.2 Reconfigurable Semantic Model Establishment

The ontology can provide a glossary of the entire vocabulary in the field of intelligent manufacturing and explain the relationship between the concepts. Thus, it can furnish efficient knowledge reuse and sharing between people and computers in the field of intelligent

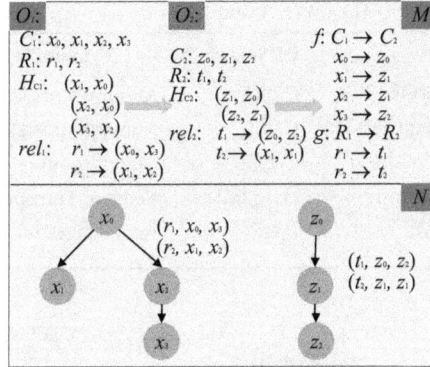

FIGURE 5.19 Ontology mapping.

TABLE 5.7 The OWL Class Constructors

Constructor	DL Syntax	Example
intersectionOf	$C_1 \cap ... C_n$	Robots$\cap$robots
allValuesFrom	$\forall P.C$	$\forall$speed.Manipulator
someValuesFrom	$\exists P.C$	$\exists$speed.Motor
maxCardinality	$\leq nP$	$\leq$1speed
minCardinality	$\geq nP$	$\geq$2states

manufacturing accordingly, which defines the structure of intelligent manufacturing equipment ontology and definition of formal description of device ontology. Considering that the graphical user interface is much easier to use and maintain, the famous open source software Protege is chosen to build the manufacturing device ontology. At the same time, the OWL describes the concept of intelligent manufacturing device resources, relationships, and constraints [50].

The Description Logic (DL) refers to the formal language of the object-based knowledge representation, which is a deterministic subset of the first-order predicate logic [51]. The DL terminology T-Box, corresponding to the OWL Ontology, contains class and attribute constructors as well as axiomatic relationships. Table 5.7 illustrates the constructors supported by the OWL, and Table 5.8 summarizes some of the axioms seconded by the OWL, which can be used to assert class, attribute inclusion or equality, class disagreement, individual equality, inequality, and attribute difference.

The intelligent manufacturing line is divided into device layer, data acquisition layer, transport layer, data processing layer, and application layer. The ontology library of the related intelligent production line according to the hierarchical logic is given in Figure 5.20. The model of the intelligent production line is widely used in the personalized customization environment, such as automobile and mobile phone.

5.4.1.1.3 Mapping between Manufacturing Resource Ontology and Data

The intelligent device ontology association technology maps the structured data to the Resource Description Framework (RDF) model [52], according to the ontology mapping

TABLE 5.8   The OWL Axioms

Axiom	DL Syntax	Example
subClassOf	$C_1 \subseteq C_n$	Manipulator$\subseteq$Device
equivalentClass	$C_1 \equiv C_n$	Manipulatior$\equiv$Robot
disjointWith	$C_1 \subseteq C_n$	Man$\subseteq$Robot
differentFrom	$\{x_1\} \subseteq \{x_2\}$	Sensor$\subseteq$Transport
inverseOf	$P_1 \equiv P_2^-$	hasVoltage$\equiv$hasPower

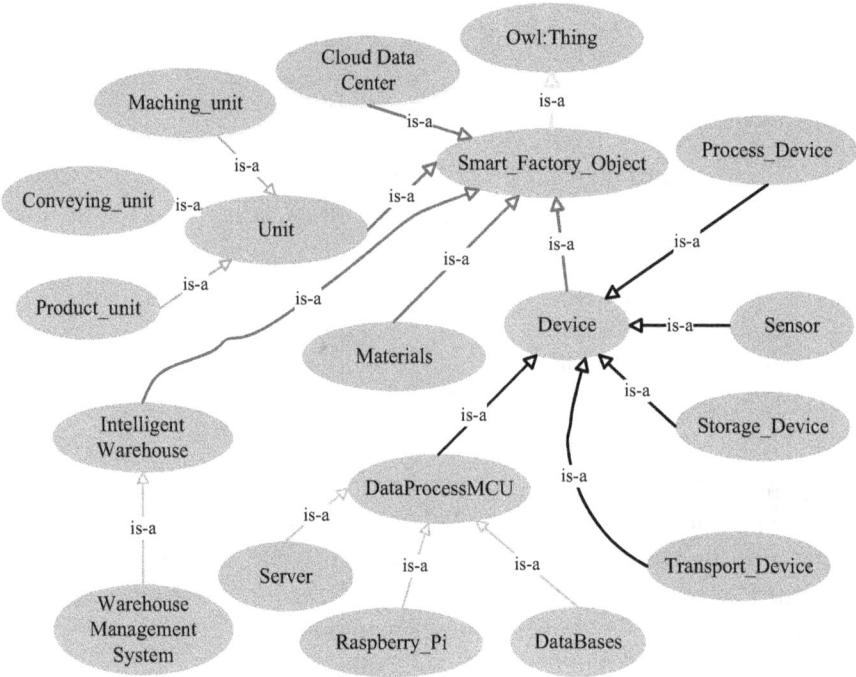

FIGURE 5.20   Device resource ontology library.

definition. The mapping from the relational database to the RDF data model includes semantic extraction, pattern mapping, and data mapping. Among them, semantic extraction is based on the characteristics of the relationship model that is used to extract the relationships type and the relations between these types. Pattern mapping reflects that the table inside the relational database maps to the RDF data model, whereas the class name must be the same as the relationship name. The column mapping completes mapping of column values in the relational database to the attributes in the RDF data model. Data mapping relates to the mapping of tuples in the relational database to instances of the RDF data model to form the KB. A data layer of the intelligent manufacturing resource integration architecture which uses Hadoop technology for distributed storage is established. Since the amount of data is larger than the processing range of the traditional database software, the data layer takes the large data processing measures. On the beginning, the real-time data of the acquired intelligent equipment are reserved on the distributed database and Hbase. Next, the data are migrated to the relational database by the Sqoop.

Definition 5.6

Ontology mapping refers to the process of discovering of a semantic correspondence between entities of two ontologies. The mapping function of the ontology [53] is: $map: O_1 \to O_2$, $map(e_1)=e_2$, $sim(e_1, e_2) \geq t$, where t is a threshold. When the similarity between e_1 and e_2 is greater than t, we can consider that they are semantically identical, indicating that e_1 is mapped to e_2.

Definition 5.7

Associative function: $f(a) \,^{\circ}\, A \to B \,^{\circ}\, f(b)$, $a \in A$, $b \in B$, $^{\circ}$ is a connection operation of the morphism, $a=(x_1, x_2,..., x_n)$, $b=(y_1, y_2,..., y_n)$; where A is a table inside the relational database, a is the line of data inside the table, set $(x_1, x_2,..., x_n)$ represents the data of a row with n columns; B shows a class in the RDF model, b is the class instance, and set $(y_1, y_2,..., y_n)$ indicates the instance attribute. In Figure 5.21, it can be seen that the device ontology instance M has power, speed, and state attribute, which correspond to the data of x_1, x_2, and x_3 inside the database, respectively. Using the association function, the manufacturing device ontology model can be well mapped with the data.

According to the above-mentioned Definitions 5.6 and 5.7, the main steps to achieve the manufacturing equipment model associated with the data are as follows: (1) application initializes the state of the intelligent equipment A through the Jena Application Program Interface (API) to read the ontology of the Dataproperty; (3) using the connection to the relational database, we can query the latest data in the specified period; (3) the latest data are compared with the initial value inside (1). Because some data may be wrong or missing in the process, it is necessary to determine whether the data type is correct. If the data type matches successfully, the subsequent operation will be finished; otherwise, the data of the next row in the database are queried until the matching data are found. We have associated the intelligent equipment ontology with the real-time data using the data semantics and reasoning based on rules, which plays an important role in ensuring the health of production line equipment and grasping the dynamic changes of resources to combine the real-time data with the intelligent equipment ontology.

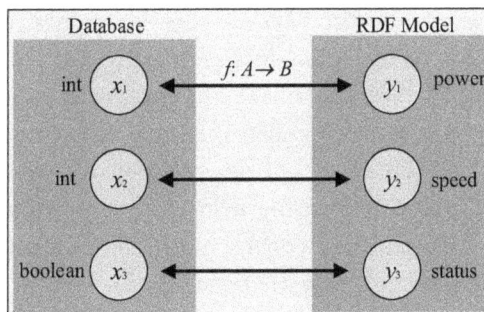

FIGURE 5.21   Data correlation process.

5.4.1.2 Manufacturing Resource Reconfigurable Knowledge Base

The ontology has a strong ability to show that knowledge can be a good representation of manufacturing equipment resources and individual concepts, but the ability for knowledge reasoning is relatively weak, which represents the bottleneck of OWL. As a result, with the help of the semantic web rule language (SWRL), we can determine an implicit relationship between manufacturing equipment ontologies, which improves reasoning ability of the equipment ontology and enhances intelligence of the equipment [54]. This section mainly comprises establishment of the semantic rule library and reasoning of the KB.

5.4.1.2.1 Reconfigurable Semantic Rule Base   In this section, we primarily explain the building of a rule layer in the manufacturing resource integration architecture based on the ontology using the SWRL language [55]. These rules are readable for humans and machines. Moreover, they will not have effect on normal operation when a new rule is randomly inserted into the manufacturing equipment reconfigurable KB. These inference rules can finish reconstruction between devices without interrupting the normal manufacturing process. Compared to the hard-coding rules (e.g., if-then statements), there are two advantages of the ontology-based expression. First, it can recoup the lack of OWL and manufacturing resources and services on the basis of intelligent manufacturing equipment resources ontology. Second, the SWRL can relate to the existing description of the rule reasoning engine.

The manufacturing environment device is mapped to the OWL ontology, while the SWRL rules that simulate from the expert system of manufacturing circumstances constitute the rule library. Manufacturing expert system contains numerous information for production environment, allowing us to stipulate more scientific rules. Both OWL and SWRL inevitably use the JESS tool, because the syntax of SWRL cannot be directly analyzed by an inference engine. Figure 5.22 illustrates the parsing process of SWRL. When the ontology model is combined with the SWRL, it may produce the erroneous results; hence, a judgment mechanism that tackles this error message is introduced. IEC 61499 standard defines a distributed model for splitting different parts of an industrial automation process and complex machinery control into functional modules called function blocks. These function blocks can be distributed and interconnected across multiple controllers. Therefore, IEC61499 function block is the control unit, and SWRL parsing rules are the power unit. The reconstruction of manufacturing resources is based on the ontology of knowledge, and the relationship between the reconfiguration of the execution unit depends on the IEC 61499 to complete the manufacturing.

The IEC 61499 standard has been established to make automation intelligent where the intelligence is genuinely decentralized and embedded into software components, which can be freely distributed across networked devices [56]. IEC 61499 has become a standard in industry, and researchers are using it in many aspects of manufacturing fields. The specification of IEC 61499 defines a generic model for distributed control systems and is based on the IEC 61131 standard. IEC 61131 has been widely accepted in the industrial automation domain. However, it is claimed that the standard does not address today's new requirements of complex industrial systems, which includes among others, portability,

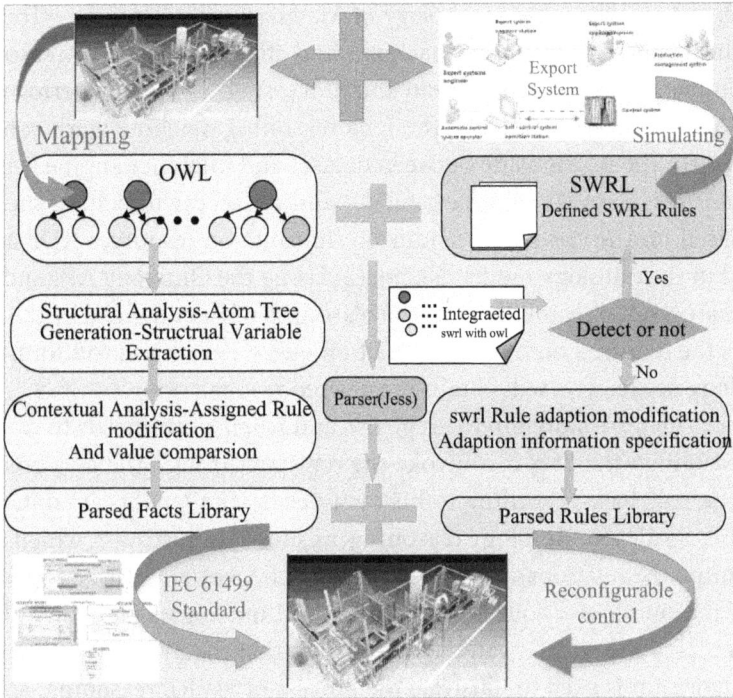

FIGURE 5.22 The parsing process of SWRL.

interoperability, increased reusability, and distribution. Therefore, the result of the references is combined with the IEC 61499 standard to accomplish reconfigurable control for the production of a circumstance equipment.

The SWRL rule base is created to implement the reconfigurable technology of manufacturing equipment resources in a better way. Due to the customers' personalized demands, a very frequent communication among devices needs to be established so that companies can confirm the completion of order tasks on time. In the meantime, enterprises should ensure quality of products and solve equipment health problems. Therefore, building of a reconfigurable semantic rule base should be considered from the following two aspects:

1. **Crossing definition of manufacturing equipment rules**: The manufacturing equipment ontology OWL accomplishes the initial reasoning by means of class and attributes characteristics, which should avoid a cross-definition with the rules. For instance: (1) for a subclass axiom $(C \subseteq D)$, the corresponding SWRL expression is $C(?x) \to D(?x)$; (2) for an equivalence axiom $(C \equiv D)$, the correlative SWRL is $C(?x) \to D(?x), D(?x) \to C(?x)$; and (3) for sub-attribute axiom $(Q \subseteq P)$, the correlative SWRL is $Q(?x,?y) \to P(?x,?y)$.

2. **Refining of manufacturing equipment rule base**: It is necessary for a rule base to wipe out redundant rules according to the deduction of predicate knowledge. The relevant refining rules are as follows: (1) if there is a rule $A \to B$, then there is $B \not\subset A$. Otherwise, it is an ordinary rule that should be removed from the rule base; (2) if there is a rule $A \wedge B \wedge C \to D, B \to C$, it can be reduced to $A \wedge B \to D$; and (3) if there is a rule $A \to B, B \to C, A \to C$, then $A \to C$ is redundant and should be removed from the rule base.

5.4.1.2.2 Fulfilling Reconfigurable Technology of Manufacturing Resource The KB update depends on the Jena [57] reasoning machine, and its generic logical reasoning engine enables the completion of RDF conversion and processing. This study performs reasoning through loading the rules format file. The Jena reasoning machine work can be summarized as follows: (1) The relationship between dataset and inference engine is implemented by the *ModelFactory* class. Then, a new model and an access reasoning mechanism are constructed based on the created or read triad information resources RDF and information contained in the ontology model [58,59]; (2) Using the Ontology API and Model API: The ontology model not only implements the reasoning of rules defined by the rules file but also completes the required ontology information query. As long as you want to query the updated ontology model, you will obtain the desired results.

The inference rules are stored in the registry, and when a user needs to call the rules to execute the reasoning, the user can invoke the registry through the Java object, while he can increase the rule base according to his own needs. The Jena itself contains a certain number of general rules for ontology reasoning, as shown in Listing 2, which are used primarily to examine the relationship of class relation, conceptual satisfaction, or attributes.

The SWRL reasoning is a collection of comprised ontologies and rules. The rules are edited by referencing the elements in the ontology so that the SWRL rules make decisions. The Tableau algorithm is used to illustrate the process of SWRL reasoning, and the SWRL reasoning algorithm is given in Algorithm 5.5. For the established device KB, the algorithm input includes T-Box, A-Box, and SWRL-Rules, which are used to describe declarations and rules of the device ontology. The algorithm not only gives the results of the reasoning but also returns a result set, which can be used to explain the relationship between atom and

---

LISTING 2   Generic Rules of Inference Engine

subClassOf(?x rdfs : subClassOf?y), (?y rdfs : subClassOf?z) → (?x subClassOf?z)
InverseWith(?x owl : InverseWith?y), (?a rdf : type?x), (?b rdf : type?y) → (?a owl : InverseWith?b)

ALGORITHM 5.5   The SWRL Reasoning Algorithm

Input: KB = (A-Box, T-Box, SWRL-Rules)
Output: resultSet(Atom, rule)
1: A-BoxChanged = True;
2: resultSet.init();
3: **while** A-BoxChanged == True **do**
4:   rules = Replace(SWRL-Rules, A-Box);
5:   **for** each instantiated Rule in rules **do**
6:     **if** !(consistent(Rule, T-Box, A-Box) && satisfied(Rule)) **then**
7:       **continue**;
8:     **end if**
9:     **if** (consistent(Rule, T-Box, A-Box || satisfied(Rules))
10:       **break**;
11:     A-Box.add(Rule.conAtoms);
12:     resultSet.add(Rule.conAtoms, Rule);
13:     A-BoxChanged = True;
14:   **end for**
15:   A-BoxChanged = False;
16: **end while**
17: **return** resultSet;

---

rule of the device instantiation, so that the reasoning process with any result can be further analyzed. In the informal case, the algorithm starts from the initialization of the result set and continues to reason until there is no new result in the A-Box. In each cycle, variables in the rule are replaced with instance individuals and data in the A-Box are used to generate the instantiated rules. Then, by checking each instantiated rule, we determine whether rule is consistent with the current KB and meets the corresponding constraints. Once the conditions are met, the result of the reasoning is added to the A-Box and result set.

In the production environment, companies need to monitor the working state of each device in a real-time. When the system faces with a warning and failure, various devices can autonomously deal with these anomalies. Therefore, enterprises need a real-time access to the working state data of the equipment in environment, so that staff can dynamically grasp operating parameters of the related equipment. However, when the system fails to perform the dynamic reconstruction, the related ones use these data to complete part of the reconstruction between devices. Moreover, the analyzed real-time data help enterprises to carry out work arrangements and assist staff to take appropriately preventive measures. Finally, with the SPARQL queries, we are capable of obtaining the current working state of a device. The SPARQL is a standard query language and data acquisition protocol developed for RDF, and it is similar to the select-from structure of the SQL query language, therefore a corresponding query processor is required to compile the SPARQL statement.

5.4.1.3 Case Study on Resource Reconfiguration

In order to verify the feasibility of the proposed ontology-based resource reconstruction method in the MCPS, in this section, we mainly discuss the reconfiguration process of the intelligent manipulator in the production environment. Based on the system architecture proposed in this paper, the reconfigurable process of robot equipment resource was explained in detail from four aspects: establishment of an intelligent manipulator ontology library, ontology model association database, creation of reconfigurable semantic rules, and fulfilment of manipulator reconfiguration. In Figure 5.23a, the production line manipulators were presented, and in Figure 5.23b, the ontology model was illustrated.

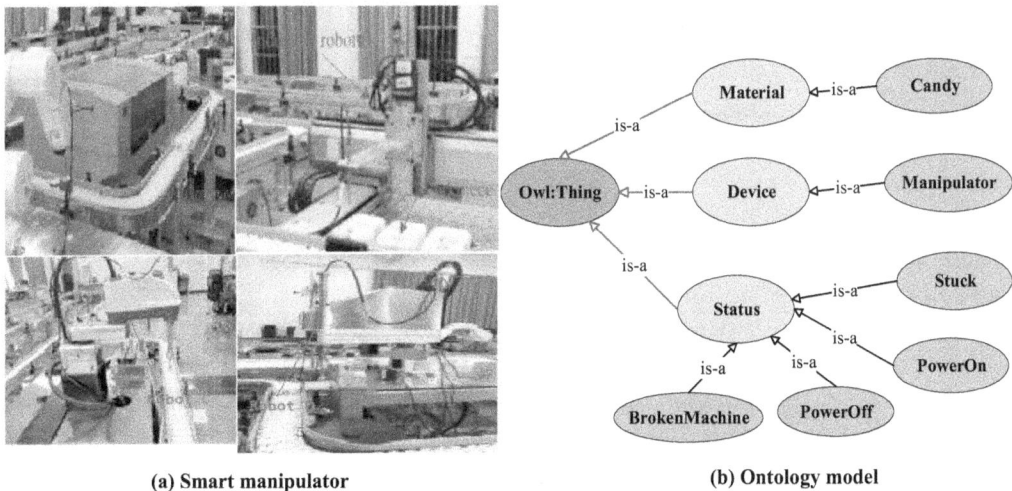

(a) Smart manipulator　　　　(b) Ontology model

FIGURE 5.23　Reconfiguration of device resource: (a) Smart manipulator and (b) ontology model.

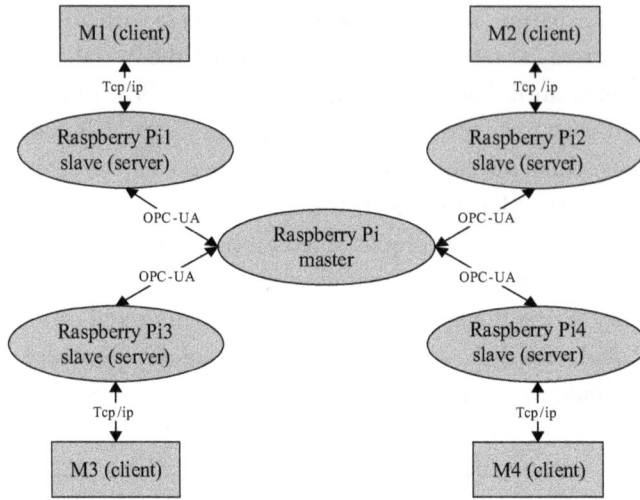

FIGURE 5.24   The communication of manipulator.

TABLE 5.9   Representation of the Instance Property

Attribute	Domain	Range	Description
power	Manipulator	xsd:int	Motor power
speed	Manipulator	xsd:int	Motor speed
condition	Manipulator	xsd:boolean	Current states
waiting	Manipulator	xsd:int	Waiting time
freeTime	Manipulator	xsd:int	Free time

In Figure 5.24, according to network function block of IEC 61499, we use the manipulator as a client and Raspberry Pi as a server. When tasks are allocated each server, the client will call on the server and the server return relevant task. In this way, the manipulator finishes the required task. Meanwhile, there is a master server, which assigns tasks to each slave Raspberry Pi. The structure of the communication of manipulators is distributed structure, which exactly meets the IEC 61499 network function block. The client and server use Transmission Control Protocol and Internet Protocol to establish communication, while Raspberry Pi uses OPC UA to establish communication each other. The OPC UA focuses on communicating with industrial systems for data collection and control.

We presented the definition of PowerOn, PowerOff, Stuck, BrokenMachine, and Manipulator types in Figure 5.23a, according to the robot working state in actual production environment. In Figure 5.23b, these five types were defined as separate classes, including the Manipulator class, PowerOn class, PowerOff class, BrokenMachine class, and Stuck class. Three instances M1, M2, and M3 were created in the Manipulator class. The attributes of manipulator instances are shown in Table 5.9. Figure 5.25 reflects the definition process of KB T-Box, wherein the class is represented by an ellipse. The instance of a class is revealed by a circle, and the rectangle blocks represent the individual attribute values (int and boolean) of the instance.

The program used the Java DataBase Connectivity to obtain the real-time state parameters from MySQL database, which associates the data with the corresponding attributes

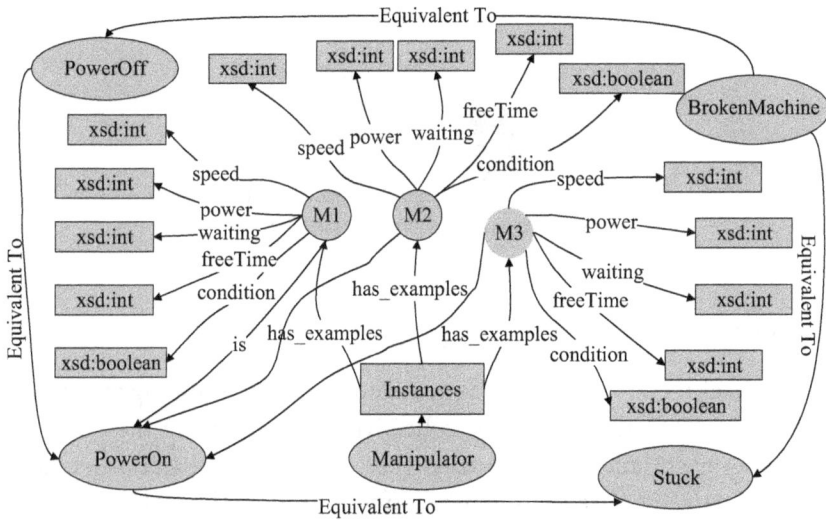

FIGURE 5.25   Knowledge base definition (T-Box) of manipulator.

LISTING 3   Reasoning rules
@prefix sf: http://www.semanticweb.org/smart#.@include <RDFS>.@include <OWL>. [rule1: (?M rdf:type sf:Stuck), (?M sf:waiting ?p)greaterThan(?p, 5) —> (?M rdf:type sf:BrokenMachine)] [rule2: (?M rdf:type sf:PowerOn), (?M sf:speed ?p)lessThan(?p, 2000) —> (?M rdf:type sf:Stuck)]

of a manipulator body through the built manipulator KB in the RDF format. When the program polled the database, we added the corresponding SWRL rules to the rules file. The corresponding rules shown in Listing 3, for instance, if the waiting on Dataproperty was greater than 5 minutes, it was defined as a Stuck machine state. If the state was greater than 10 minutes, it represented the BrokenMachine state; when robot did not have tasks and the state freeTime was greater than 30 minutes, the robot was in the PowerOff state. If robot started to work, the program called Jena's reasoning machine to load the rules file. According to the refreshed real-time data used to update the state parameter value, the ontology took the switch between the working states, which realized the reconfiguration between the manipulators. When tasks were issued to the robots, each robot performed its own tasks. If robot has gone to the abnormal state, the system would reassign unfinished tasks to the other machines with the defined rules. The system was able to automatically implement reconfiguration when faced with multiple order tasks. Using the SPARQL queried language, the real-time query of current robot situation could be obtained. The query results showed that the state of robot M1 changed from the original normal working state (true) to the fault state (false), and the states of robots M2 and M3 did not change, indicating that the ontology model updated the current robot state based on a real-time detection.

In order to evaluate the characteristics of ontology-based manufacturing resource reconfigurable method, we compared the resource reconfiguration method based on ontology with method based on multi-agent and method based on network. With the help of the method, the OWL described device attributes and their relationships improving

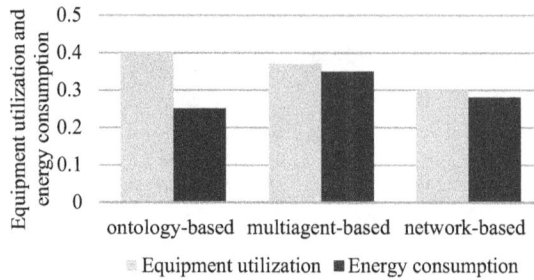

FIGURE 5.26   Comparison of methods.

physical and logical reconstruction of device resources. The multi-agent resource reconfiguration method made the task agent collective negotiate using the contract network method and other negotiation mechanisms. Thus, it could guide an optimal allocation of resources. In the network-based resource reconfiguration method, the application of network technology could quickly and flexibly organize manufacturing resources scattered in different regions together. However, the main limitation was the impact of network bandwidth on the system. The comparison of three methods in terms of equipment utilization and energy consumption is presented in Figure 5.26. In Figure 5.26, it can be noticed that using the proposed method, the equipment utilization rate is obviously improved and energy consumption is significantly reduced simultaneously. Consequently, the proposed ontology-based resource reconfiguration method facilitates the reconfigurable development of the MCPS manufacturing resources.

5.4.2 Data-Driven Cognition Enhancement and Decision-Making

In recent years, the fourth industrial revolution accelerated by IIoT has raised the global upsurge. Edge intelligence enables the IIoT to have the abilities of perception, computation, decision-making, and transmission. There are broad application prospects that integrate the edge intelligence into IIoT [60]. Cognitive technology-enabled IIoT realizes the semantic representation, sensor data correlation, and AI modeling in the network plane, which improves networking ability of context awareness and semantic understanding. However, cognitive technology lacks the theoretical basis on the upper-level decision-making and cannot recommend optimized operation scheme based on the IIoT data. With the rapid development of information science, computational intelligence will provide new solutions for the novel IIoT applications. Moreover, ML based on the intelligent computing methods already plays an important role in edge intelligent IIoT.

An ML-enabled framework of the cognitive IIoT is shown in Figure 5.27, which includes three layers, namely, the perceptual layer, transmission layer, and application layer. In the perceptual layer, IIoT collects the raw data easily from the whole life cycle of an industrial process with sensors to support the upper application in a smart factory. The transmission layer mainly includes the industrial Ethernet, short distance wireless communication technology, and a low-power wide-area network. Depending on the cognitive rules and the requirements of the network applications, the cognitive model is built on the network side to realize the semantic description for IIoT data. The developed

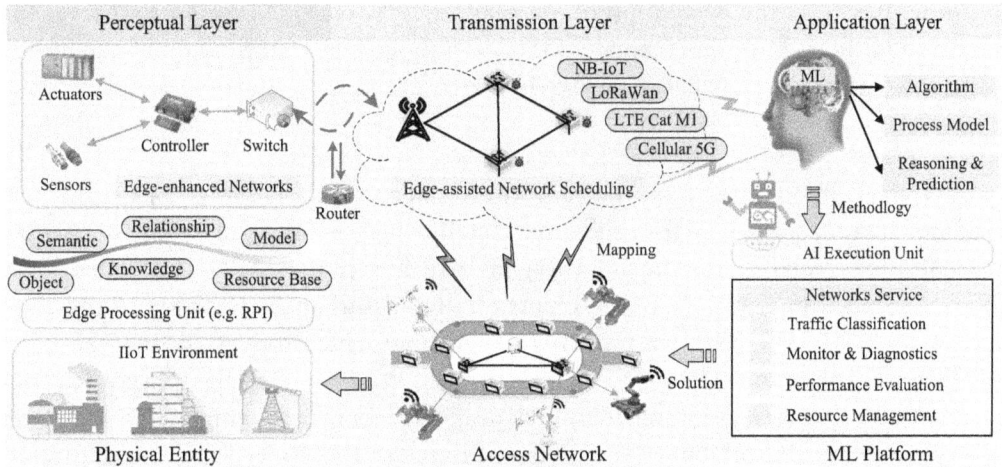

FIGURE 5.27 The ML-enabled framework of the cognitive IIoT.

model eliminates the data ambiguity mainly caused by different transmission protocols and forms a KB cross the accumulation of semantic data. The application layer mainly involves the analysis, computation, and knowledge mining of the IIoT data. It is also the main application field of the ML optimization algorithms which are employed in IIoT to help realize the intelligent applications. Based on easy-to-understand network resources, the ML theory provides more superior approaches for the construction and validation of the AI models. Moreover, it solves the shortage of the single cognitive technology in establishing the accurate models.

Due to the market demand for multi-product customization, smart factory often requires to rapidly respond to the order change. Dynamic adaptive planning (DAP) for the product processing path is necessary to achieve the product diversity in reconfigurable production line. In our prototype product line, there are four grabbing robots, a packing robot, and a warehousing robot. The experimental scenario can be described as follows: The candy orders are created by consumers on the Internet. The category of candy is different according to consumer preferences. Because each robot is responsible for grabbing different kinds of candy, the production line needs to adjust the packaging route of the candy packing line dynamically.

5.4.2.1 Learning Goals

As we know, the candy packaging line has the typical characteristics of multi-product customization. Facing with the production of multi-product customization, the scheduling of a candy packing line is difficult because of the frequent changes in task priorities; therefore, the original plan cannot always be carried out as scheduled. Nowadays the categories of materials stock have become more abundant, on the contrary, the critical materials are often absent. DRL, which combines the perception ability of DL with the decision-making ability of RL, is an intelligent method that is similar to the human reasoning. Based on the perception of the packing line states, DRL model achieves the goal through the maximum incentives. The learning goal is to optimize the scheduling strategy to realize the DAP based on the context awareness.

5.4.2.2 DRL Model for Dynamic Adaptive Planning

We built an RL model based on the context awareness which is shown in Figure 5.28. Based on the industrial scenario, the order details, material stock, and single machine workload were treated as monitoring variables, which were regarded as the states of the model. As shown in Figure 5.28, the inputs are the disturbances of action to the environment. For dynamic adaptive planning in reconfigurable production line, the input is the dynamic planning strategy to the DRL model. We evaluated the utility value considering these critical factors together, including the less order completion time, lower energy consumption, and higher equipment utilization. The utility value was an important parameter for the DRL model to select a proper optimization strategy. The intelligent agent changed the state of the candy packing line and accumulated the maximum utility value according to the action of the route planning decision. In the RL framework, the order quantity and planning decisions of the packing line were discrete, while the state of packing line and utility value of the intelligent agent were continuous. Considering these factors, DL was employed to approximate those continuous spaces. Specifically, the high-dimensional monitoring data were abstracted to approximate the real state of the packing line by prior learning.

Considering the high complexity of the training algorithm for the DRL model, we deployed the context awareness-based DRL model on a raspberry pi. The input parameters obtained from edge intelligent IIoT mainly include the order information, the

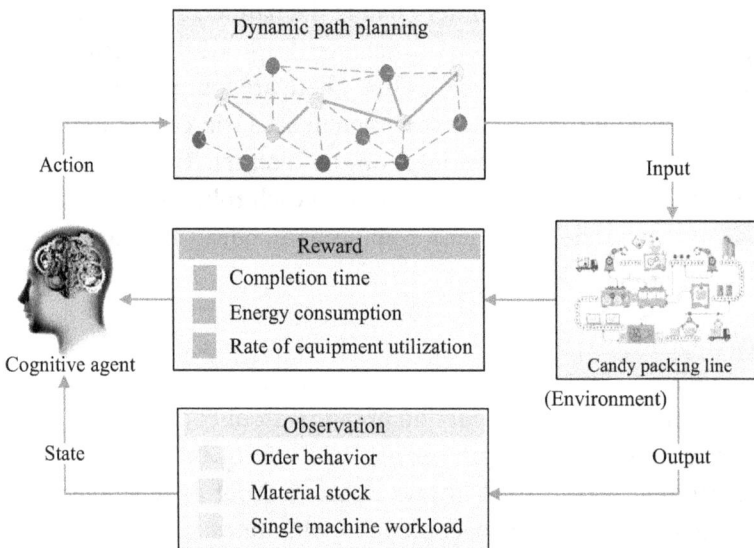

FIGURE 5.28 The RL model for dynamic adaptive planning based on the context awareness.

condition of material stock, and the single machine workload. And the reward parameters included the completion time, the energy consumption, and the rate of equipment utilization. According to these parameters, the cognitive agent took action to conduct the path planning for the product line. The cognitive agent can get the state of the packing line with sensors at each time moment [61]. Edge intelligent IIoT detected the information for the decision-making to obtain an effective observation of the task environment. The DRL model was trained offline with large amount of collected data previously stored on the manufacturing cloud. Afterwards, the model was optimized by online learning with the fresh collected data.

5.4.2.3 Experimental Results

We verified the DRL-based dynamic adaptive planning in the prototype platform. We imitated the consumers' behavior that the orders are randomly created toward the production line on the client side. We observed that the scheduling effect of the DRL-based DAP on the production line was increasing constantly with the increase of order quantity. As shown in Figure 5.29, we separately measured the completion time of the order, the energy consumption of the production line, and the utilization ratio of the equipment. The DRL-based DAP was compared with our previous work [62], namely, the central-station control scheduling (CSCS) mechanism and the self-organizing task scheduling (SOTS) mechanism. When the order quantity was small, the differences among the three mechanisms were not obvious. As shown in Figure 5.29a and b, when the order quantity increased to 2,000, the SOTS mechanism and the DRL-based DAP mechanism were superior to the centralized scheduling mechanism from both completion time and energy consumption. When the order quantity was 2,500, the completion time of the DRL-based DAP mechanism was 28.5 hours and the energy consumption was 57.2 kW. Thus, the DRL-based DAP mechanism had obvious advantages when the order quantity increased. The variance of the equipment utilization ratio in the production line is presented in Figure 5.29c, which shows that the SOTS mechanism was the lowest. The DRL-based DAP mechanism improved gradually as the order quantity increases; however, its effectiveness was not as good as that of the SOTS mechanism.

The CSCS mechanism was usually firstly considered the completion time of the order. The SOTS mechanism took into account the completion time and paid attention to the coordinate production lines and balanced the utilization ratio of the equipment. With the cognitive ability of edge intelligent IIoT, the DRL-based DAP considered more factors including the order behavior, material stock, and single machine workload. The experimental results showed that the DRL-based DAP mechanism had the lower utilization of the equipment than the SOTS mechanism. We observed that there was a weak correlation between the state and utility at the higher utilization ratio of the equipment, and the state of the material stock was also an important factor to determine whether the equipment participated in the operation.

In summary, the DRL-based DAP had a more comprehensive understanding of the production line. With the gradual maturity of the ML model, the proposed DRL-based DAP mechanism will become even more superior.

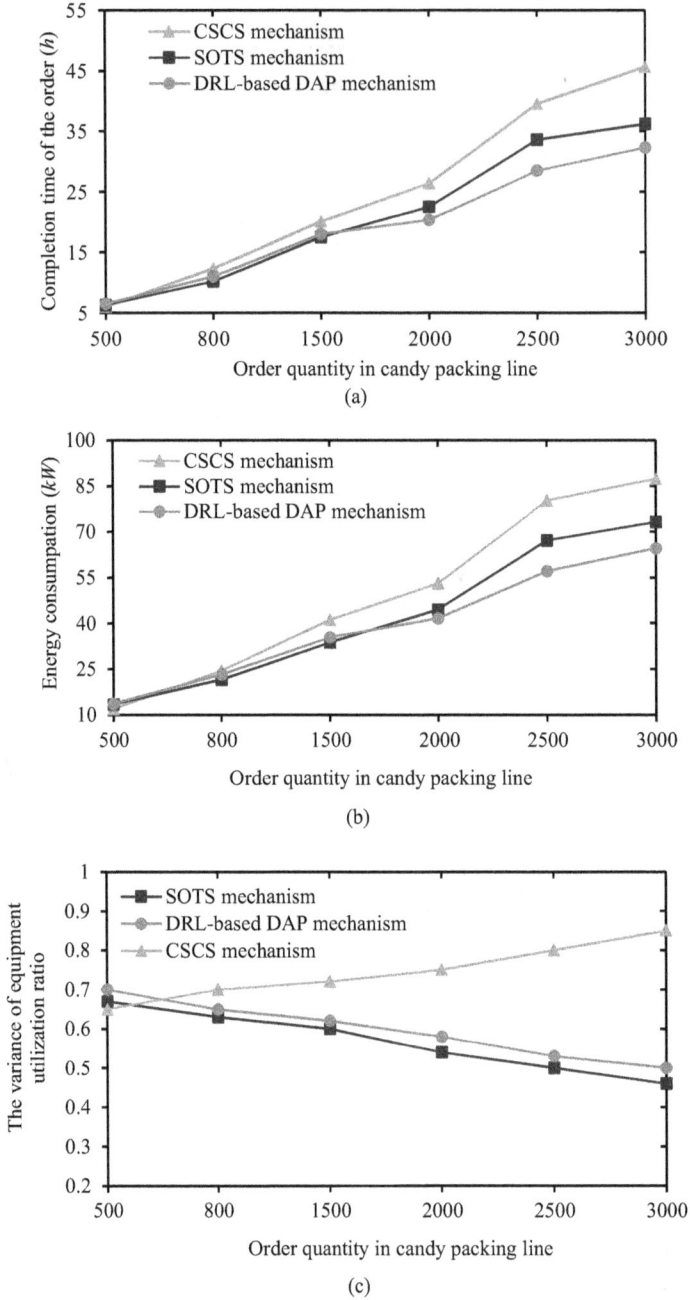

FIGURE 5.29   Comparison of (a) completion time, (b) energy consumption, and (c) variance of the equipment utilization ratio of the production line.

5.5 SUMMARY

In order to meet the current personalized demands, the manufacturing systems need to enhance their responsiveness and reconfigurability for manufacturing resources. This chapter fully explores the application potential of the IIoT from the aspects of edge computing and cloud computing. At the same time, the knowledge sharing and cognitive enhancement are regarded as the important promotion of AI to resource reconstruction. Several industrial application examples in different scenarios show that dynamic scheduling based on edge cloud collaboration and resource reconfiguration based on AI are beneficial to realize custom manufacturing.

However, limited to the unideal modeling environment, the current ML models are not mature enough to meet the harsh industrial requirements. More information collection is a crucial foundation to achieve ML applications in IIoT. In the future, we will consider ML models training under edge-cloud collaboration to support real-time scheduling. With the cloud server data, the past scheduling information is used to improve the future performance for the service of the edge intelligent. At the same time, DT technologies may be used to build more realistic simulation models, becoming the driving technologies to improve resource reconfiguration [63,64].

REFERENCES

[1] G.-J. Cheng, L.T. Liu, X.-J. Qiang and Y. Liu, "Industry 4.0 development and application of intelligent manufacturing," *Proceeding International Conference Information System Artificial Intelligence* (ISAI), pp. 407–410, 2016.

[2] W. Shi and S. Dustdar, "The promise of edge computing," *Computer*, vol. 49, no. 5, pp. 78–81, 2016.

[3] G. Hu, W. P. Tay and Y. Wen, "Cloud robotics: Architecture, challenges and applications," *IEEE Network*, vol. 26, no. 3, pp. 21–28, 2012.

[4] J. Wan, S. Tang, H. Yan, D. Li, S. Wang and A. V. Vasilakos, "Cloud robotics: Current status and open issues," *IEEE Access*, vol. 4, pp. 2797–2807, 2016.

[5] A. Rahman, J. Jin, A. Rahman, A. Cricenti, M. Afrin and Y.-N. Dong, "Energy-efficient optimal task offloading in cloud networked multi-robot systems," *Computer Networks*, vol. 160, pp. 11–32, 2019.

[6] Y. Lu, H. Wang and X. Xu, "ManuService ontology: A product data model for service-oriented business interactions in a cloud manufacturing environment," *Journal Intelligent Manufacturing*, vol. 30, no. 1, pp. 317–334, 2019.

[7] Y. Alsafi and V. Vyatkin, "Ontology-based reconfiguration agent for intelligent mechatronic systems in flexible manufacturing," *Robotic Computer Integrated Manufacturing*, vol. 26, no. 4, pp. 381–391, 2010.

[8] K. Li, T. Zhou, B. Liu and H. Li, "A multi-agent system for sharing distributed manufacturing resources," *Expert Systems with Applications*, vol. 99, pp. 32–43, 2018.

[9] Y. Zhang, C. Qian, J. Lv and Y. Liu, "Agent and cyber-physical system based self-organizing and self-adaptive intelligent shopfloor," *IEEE Transactions on Industrial Informatics*, vol. 13, no. 2, pp. 737–747, 2017.

[10] S. Jeschke, C. Brecher, H. Song and D. Rawat, *Industrial Internet of Things: Cyber Manufacturing Systems*. Springer, Cham, Switzerland, 2017, pp. 1–715.

[11] J. Wan, S. Tang, Z. Shu et al., "Software-defined industrial internet of things in the context of industry 4.0," *IEEE Sensors Journal*, vol. 16, no. 20, pp. 7373–7380, 2016.

[12] M. I. Jordan and T. M. Mitchell, "Machine learning: Trends, perspectives, and prospects," *Science*, vol. 349, no. 6245, pp. 255–260, 2015.

[13] C. Li and M. Qiu, *Reinforcement Learning for Cyber-Physical Systems: With Cybersecurity Case Studies*. CRC Press, Orlando, FL, 2019.

[14] Y. Wang, K. Wang, M. Toshiaki and S. Guo, "Traffic and computation co-offloading with reinforcement learning in fog computing for industrial applications," *IEEE Transactions on Industrial Informatics*, 2018, doi: 10.1109/TII.2018.2883991.

[15] F. Tao and Q. Qi, "Make more digital twins," *Nature*, vol. 573, no. 7775, pp. 490–491, 2019.

[16] G. N. Schroeder, C. Steinmetz and R. N. Rodrigues et al., "A methodology for digital twin modeling and deployment for industry 4.0," *Proceedings of the IEEE*, vol. 109, no. 4, pp. 556–567, 2021.

[17] Y. K. Liu, S. K. Ong and A. Y. C. Nee, "State-of-the-art survey on digital twin implementations," *Advances in Manufacturing*, vol. 10, no. 1, pp. 1–23, 2022.

[18] J. Wan, M. Yi, D. Li, C. Zhang, S. Wang and K. Zhou, "Mobile services for customization manufacturing systems: An example of industry 4.0," *IEEE Access*, vol. 4, pp. 8977–8986, 2016.

[19] X. Li, J. Wan, H. Dai, M. Imran, M. Xia and A. Celesti, "A hybrid computing solution and resource scheduling strategy for edge computing in smart manufacturing," *IEEE Transactions on Industrial Informatics*, vol. 15, no. 7, pp. 4225–4234, 2019.

[20] C. Xu, K. Wang, P. Li et al., "Making big data open in edges: A resource-efficient blockchain-based approach," *IEEE Transactions on Parallel and Distributed Systems*, vol. 30, no. 4, pp. 870–882, 2019.

[21] R. Bonomi, J. Milito, J. Zhu and S. Addepalli, "Fog computing and its role in the internet of things," *Proceedings of the First Edition of the MCC Workshop on Mobile Cloud Computing*, Helsinki, Finland, pp. 13–16, 2012.

[22] C. C. Byers, "Architectural imperatives for fog computing: Use cases, requirements, and architectural techniques for fog-enabled IoT networks," *IEEE Communications Magazine*, vol. 55, no. 8, pp. 14–20, 2017.

[23] L. Shu, Y. Chen, Z. Huo, N. Bergmann and L. Wang, "When mobile crowd sensing meets traditional industry," *IEEE Access*, vol. 5, pp. 15300–15307, 2017.

[24] F. Bonomi, R. Milito, P. Natarajan and J. Zhu, "Fog computing: A platform for internet of things and analytics," in *Big Data and Internet of Things: A Roadmap for Smart Environments*. Springer, Cham, 2014, pp. 169–186.

[25] M. Mukherjee, R. Matam, S. Lei, et al., "Security and privacy in fog computing: Challenges," *IEEE Access*, vol. 5, pp. 19293–19304, 2017.

[26] F. Li, J. Wan, P. Zhang, D. Li, D. Zhang and K. Zhou, "Usage-specific semantic integration for cyber-physical robot systems," *ACM Transactions on Embedded Computing Systems*, vol. 15, no. 5, p. 50, 2015.

[27] J. Wan, H. Yan, Q. Liu, K. Zhou, R. Lu and D. Li, "Enabling cyber-physical systems with machine-to-machine technologies," *International Journal of Ad Hoc and Ubiquitous Computing*, vol. 13, no. 3/4, pp. 187–196, 2013.

[28] J. J. Kuffner, "Cloud-enabled robots," *Proc. Of IEEE-RAS International Conference on Humanoid Robotics*, Nashville, TN, USA, 2010.

[29] R. Arumugam, V. R. Enti, L. Bingbing et al. "DAvinCi: A cloud computing framework for service robots," *IEEE International Conference on Robotics and Automation (ICRA)*, 2010, pp. 3084–3089.

[30] G. Mohanarajah, D. Hunziker, R. D'Andrea and M. Waibel, "Rapyuta: A cloud robotics platform," *IEEE Transactions on Automation Science and Engineering*, vol. 12, no. 2, pp. 481–493, 2015.

[31] M. Quigley, K. Conley, B. Gerkey et al. "ROS: An open-source robot operating system," *IEEE International Conference on Robotics and Automation (ICRA) Workshop on Open Source Software*, 2009, p. 2.

[32] B. Kehoe, S. Patil, P. Abbeel and K. Goldberg, "A survey of research on cloud robotics and automation," *IEEE Transactions on Automation Science and Engineering*, vol. 12, no. 2, pp. 398–409, 2015.

[33] H. Durrant-Whyte, D. Rye and E. Nebot, "Localization of autonomous guided vehicles," *Proceedings of International Symposium on Robotics Research*. Springer-Verlag, New York, 1995, pp. 613–625.

[34] R. Goransson, A. Aydemir and P. Jensfelt, "Kinect@Home: Crowdsourced RGB-D data," *Proc. IROSWorkshop on Cloud Robot*, 2013.

[35] B. Mirtich and J. Canny, "Easily computable optimum grasps in 2D and 3D," *Proceedings of the 1994 IEEE International Conference on Robotics and Automation*, 1994, pp. 739–747.

[36] S. Minaeian, J. Liu and Y. Son, "Vision-based target detection and localization via a team of cooperative UAV and UGVs," *IEEE Transactions on Systems, Man, and Cybernetics: Systems*, vol. 46, no. 7, pp. 1005–1016, 2016.

[37] H. Tang, D. Li, S. Wang and Z. Dong, "CASOA: An architecture for agent-based manufacturing system in the context of Industry 4.0," *IEEE Access*, vol. 6, pp. 12746–12754, 2018.

[38] M. Perše, K. Kristan, J. Perš, G. Mušič, G. Vučkovič and S. Kovačič, "Analysis of multi-agent activity using petri nets," *Pattern Recognition*, vol. 43, no. 4, pp. 1491–1501, 2010.

[39] Y. Koren and M. Shpitalni, "Design of reconfigurable manufacturing systems," *Journal of Manufacturing Systems*, vol. 29, no. 4, pp. 130–141, 2010.

[40] H. ElMaraghy, "Flexible and reconfigurable manufacturing systems paradigms," *International Journal of Flexible Manufacturing Systems*, vol. 17, no. 4, pp. 261–276, 2005.

[41] S. Huang, G. Wang, X. Shang and Y. Yan "Reconfiguration point decision method based on dynamic complexity for reconfigurable manufacturing system (RMS)," *Journal of Intelligent Manufacturing*, vol. 29, no. 5, pp. 1031–1043, 2018.

[42] S. Wang, J. Wan, M. Imran and C. Zhang, "Cloud-based smart manufacturing for personalized candy packing application," *The Journal of Supercomputing*, vol. 74, no. 9, pp. 4339–4357, 2018.

[43] G. H. Lee, "Reconfigurability consideration design of components and manufacturing systems," *The International Journal of Advanced Manufacturing Technology*, vol. 13, no. 5, pp. 376–386, 1997.

[44] Z. M. Bi, S. Y. T. Lang, W. Shen and L. Wang, "Reconfigurable manufacturing systems: The state of the art," *International Journal of Production Research*, vol.46, no.4, pp.967–992, 2008.

[45] T. Gruber, "Toward principles for the design of ontologies used for knowledge sharing," *International Journal of Human-Computer Studies*, vol. 43, no. 5, pp. 907–928, 1993.

[46] A. Maedche and S. Staab, "Ontology learning for the semantic web," *IEEE Intelligent Systems*, vol. 16, no. 2, pp. 72–79, 2001.

[47] D. L. McGuinness and F. Van Harmelen, "OWL web ontology language overview," *W3C recommendation*, 2004. https://www.w3.org/TR/2004/REC-owl-features-20040210/

[48] https://www.plattform-i40.de/I40/Redaktion/EN/Downloads/Publikation/rami40-an-introduction.pdf?__blob=publicationFile&v=4

[49] A. Steve, "Categories," in *Category Theory*, 2nd ed. Elsevier B.V., NewYork, 2006, pp. 1–23.

[50] S. Gardner, S. Beaumont, B. Davis, C. McMenamin, J. Chambers and J. Barnes, "System and method for creating, editing, and utilizing one or more rules for multi-relational ontology creation and maintenance," *U.S. Patent* 11/122, 069, May 5, 2005.

[51] T. Lukasiewicz, "A novel combination of answer set programming with description logics for the semantic web," *IEEE Transactions on Knowledge and Data Engineering*, vol. 22, no. 11, pp. 1577–1592, 2010.

[52] M. Wylot, P. Cudré-Mauroux, M. Hauswirth and P. Groth, "Storing, tracking, and querying provenance in linked data," *IEEE Transactions on Knowledge and Data Engineering*, vol. 29, no. 8, pp. 1751–1764, 2017.

[53] M. Ehrig and S. Staab, "QOM-Quick Ontology Mapping," *Proceedings of the 3th International Semantic Web Conference*, Hiroshima, Japan, Nov. 2004.

[54] H. Gennari, M. A. Musen, R. W. Fergerson et al., "The evolution of Protégé: An environment for knowledge-based systems development," *International Journal of Human-Computer Studies*, vol. 58, no. 1, pp. 89–123, 2003.

[55] A. B. Bener, V. Ozadali and E. S. Ilhan, "Semantic matchmaker with precondition and effect matching using SWRL," *Expert Systems with Applications*, vol. 36, no. 5, pp. 9371–9377, 2009.

[56] V. Vyatkin, "IEC 61499 as enabler of distributed and intelligent automation: State-of-the-art review," *IEEE Transactions on Industrial Informatics*, vol. 7, no. 4, pp. 768–781, 2011.

[57] Apache Jena. https://jena.apache.org/ documentation/index.html/

[58] M. Klein, "XML, RDF, and relatives," *IEEE Intelligent Systems*, vol. 12, no. 2, pp. 26–28, 2001.

[59] B. Quilitz and U. Leser, "Querying distributed RDF data sources with SPARQL," *European Semantic Web Conference*, Berlin, Heidelberg, 2008, pp. 524–538.

[60] K. Wang, Y. Shao, L. Xie, J. Wu and S. Guo, "Adaptive and fault-tolerant data processing in healthcare IoT based on fog computing," *IEEE Transactions on Network Science and Engineering*, 2018, doi: 10.1109/TNSE.2018.2859307.

[61] H. Park and N. Tran, "An autonomous manufacturing system based on swarm of cognitive agents," *Journal of Manufacturing Systems*, vol. 31, no. 3, pp. 337–348, 2012.

[62] J. Wan, B. Chen, M. Imran et al., "Toward dynamic resources management for IoT-based manufacturing," *IEEE Communications Magazine*, vol. 56, no. 2, pp. 52–59, 2018.

[63] F. Tao, J. Cheng, Q. Qi, M. Zhang, H. Zhang and F. Sui, "Digital twin-driven product design manufacturing and service with big data", *International Journal Advanced Manufacturing Technology*, vol. 94, no. 9, pp. 3563–3576, 2018.

[64] Q. Qi and F. Tao, "Digital twin and big data towards smart manufacturing and industry 4.0: 360 degree comparison", *IEEE Access*, vol. 6, pp. 3585–3593, 2018.

CHAPTER 6

Implementation of Customized Manufacturing Factory

T HE TRADITIONAL PRODUCTION PARADIGM of large batch production does not offer flexibility toward satisfying the requirements of individual customers. A new generation of smart factories is expected to support new multi-variety and small-batch customized production modes. For this, artificial intelligence (AI) is enabling higher value-added manufacturing by accelerating the integration of manufacturing and information communication technologies, including computing, communication, and control. The characteristics of a customized smart factory are self-perception, operations optimization, dynamic reconfiguration, and intelligent decision-making. The AI and cloud service technologies will allow manufacturing systems to perceive the environment, adapt to the external needs, optimize resource deployment, and expand service models. This chapter provides an overview of the key technologies of customized manufacturing through the multi-variety mixed-flow production line platform, followed by a discussion of the role of cloud computing in smart manufacturing, using candy packaging and manufacturing mobility services as examples. The reconfigurability and intelligent scheduling of smart factories are also explored, emphasizing the pivotal role of AI in manufacturing. Finally, this paper details the components of the customized production line platform, analyzes the application and deployment methods of cloud services and AI technology in customized manufacturing, and embodies the critical links and core technologies of customized manufacturing through the design of the intelligent platform's conveyor belt control system. This study aims to inspire enterprises to achieve the transformation toward intelligent manufacturing factories.

DOI: 10.1201/9781003460992-6

6.1 OVERVIEW OF CUSTOMIZED MANUFACTURING IN SMART MANUFACTURING FACTORY

Despite the progress made, the manufacturing industry faces a number of challenges, some of which are as follows: traditional mass production is not able to adapt to the rapid production of personalized products; resource limitations, environmental pollution, global warming, and an aging global population have become more prominent. Therefore, a new manufacturing paradigm to address these challenges is needed. The customer-to-manufacture concept reflects the characteristics of customized production where a manufacturing system directly interacts with a customer to meet his/her personalized needs. The goal is to realize the rapid customization of personalized products. The new generation of intelligent manufacturing technology offers improved flexibility, transparency, resource utilization, and efficiency of manufacturing processes. It has led to new programs, for example, the Factory of the Future in Europe, Industry 4.0 in Germany, and Made in China 2025. Moreover, the United States has also accelerated research and development programs.

Compared with mass production, the production organization of customized manufacturing is more complex, quality control is more difficult, and the energy consumption needs attention. In classical automation, the production boundaries were rigid to ensure quality, cost, and efficiency. Compared with traditional production, customized manufacturing has the following characteristics:

- **Smart interconnectivity**: Smart manufacturing embraces a cyber-physical environment, for example, processing/detection/assembly equipment and storage, all operating in a heterogeneous industrial network. The Industrial Internet of Things (IoT) has progressed from the original industrial sensor networks to the narrow-band IoT (NB-IoT), LoRa WAN, and LTE Cat M1 with increased coverage at reduced power consumption. Edge computing units are deployed to improve system intelligence. Cognitive technology ensures the context awareness and semantic understanding of the industrial IoT. Intelligent industrial IoT as the key technologies is widely used for intelligent manufacturing.

- **Dynamic reconfiguration**: The concept of a smart factory aims at the rapid manufacturing of a variety of products in small batches. Since the product types may change dynamically, system resources need to be dynamically reorganized. A multi-agent system (MAS) is introduced to negotiate a new system configuration.

- **Cloud service**: The smart factory contains a large amount of manufacturing data such as equipment operation data and order management data. The collection, processing, and analysis of these data, supported by edge computing, can be used to carry out a wide variety of manufacturing services. Cloud service is an important technology to tap the value of manufacturing plant data and improve the competitiveness of manufacturing plants.

- **Deep integration**: The underlying intelligent manufacturing entities, cloud platforms, edge servers, and upper monitoring terminals are closely connected. Data processing, control, and operations can be performed simultaneously in cyber-physical systems (CPSs), where the information barriers are broken down, thereby realizing the deep integration of physical and information environments.

6.2 CONSTRUCTION OF CUSTOMIZED FACTORY PROTOTYPES

The construction of a customized manufacturing plant requires the interconnection of production equipment and the development of strategic control schemes. This chapter will introduce the interconnection scheme of the equipment from the experimental platform of the candy packaging production line, describe the implementation method of the intelligent manufacturing architecture, and explain the realization technology and significance of customized manufacturing through the intelligent conveyor belt control system.

6.2.1 Intelligent Manufacturing Experimental Platform for Candy Packaging Production Line

6.2.1.1 Overview of the Platform Hardware

The overall layout of the experimental platform of the intelligent manufacturing candy packing line is shown in Figures 6.1 and 6.2. The production line includes multiple workstations, processing machines, and monitoring screens. Each camera position and the equipment used are shown in Figure 6.3. The layout of the device and the configuration parameters of the device are introduced in detail in Figure 6.3. The functions, uses, and methods of each module will be introduced in detail below.

1. **Product smart tray (in Figure 6.4)**: It is used to place products, integrated with embedded systems, store product information, and realize information interaction with private cloud; integrate ultra-wideband (ultra-wide band, UWB) position sensors, and on-site layout. The UWB positioning base station of the system forms a set of wireless indoor precise positioning system, as shown in Figure 6.5. The product intelligent tray is accurately positioned on the production line through the wireless indoor precise positioning system, and the positioning accuracy can reach within 10 cm.

2. **Processing, assembly, and conveying equipment**: It includes two sets of CNC engraving workstations with two sets of equipment, one laser marking workstation, one box loading workstation, one upper cover workstation, one loading workstation, one Unloading workstation, one assembly workstation consisting of a dual-arm robot, one packaging workstation including a four-axis robot, six-segment intelligent variable-speed conveyor belt unit, and the intelligent variable-speed conveyor belt unit is equipped with a set of path conversion devices at each branching node, each set of devices contains photoelectric sensors and Radio Frequency Identification Devices (RFID) readers for identifying product smart pallets.

   Figures 6.6, 6.7, and –6.8, respectively, show the workflow of CNC engraving workstation, laser marking workstation, and dual-arm robot. In Figure 6.6, the photoelectric

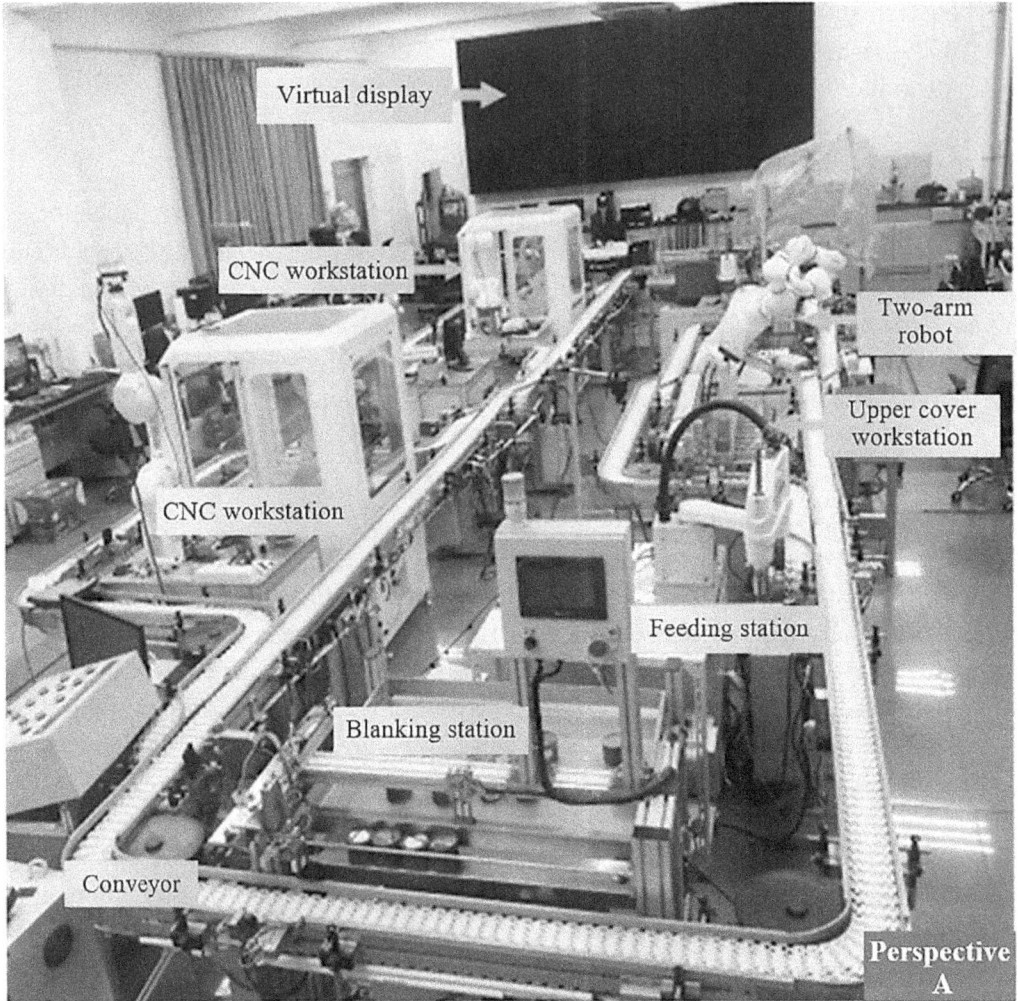

FIGURE 6.1 Intelligent manufacturing production line experiment platform of South China University of Technology (Perspective A).

switch of the CNC Jingdiao workstation first recognizes the arrival of the product, and the supporting robot of the workstation picks up the product to the CNC processing position and starts processing. There are two options for the process: engraving process or drilling process. After the CNC process is completed, the robot in the workstation will pick it up and put it back on the production line for the next step of the process. Figure 6.7 shows the working diagram of the laser marking workstation. After the photoelectric switch of the laser marking workstation recognizes the arrival of the product, laser marking is carried out at the present position. The content of laser marking is determined by the order. The optional patterns or text are shown in Figure 6.7. Figure 6.8 shows an assembly process workstation of a dual-arm robot, and the assembly process is assembled from two parts of the product. When a pallet comes carrying a product box and is recognized, the two arms of the dual-arm robot

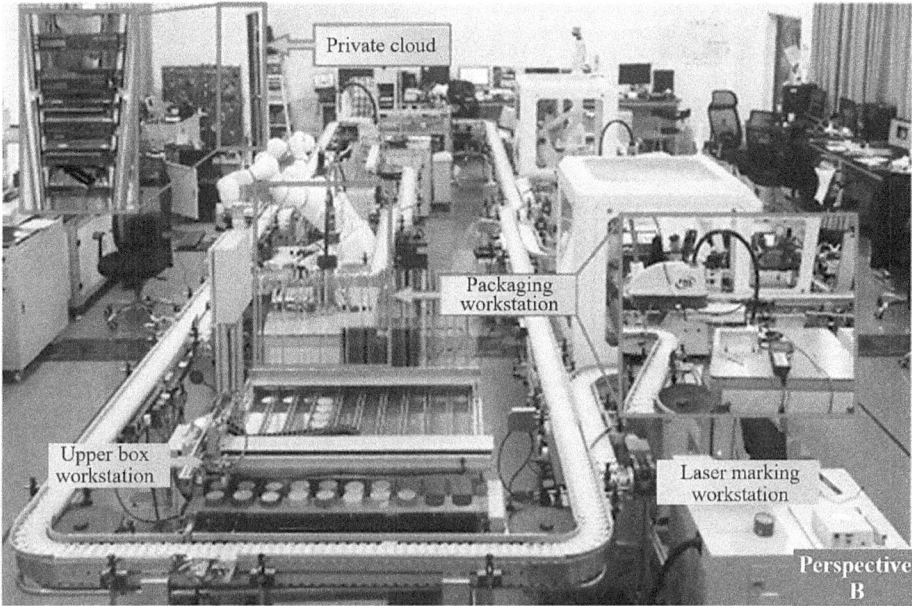

FIGURE 6.2 Intelligent manufacturing production line experiment platform of South China University of Technology (Perspective B).

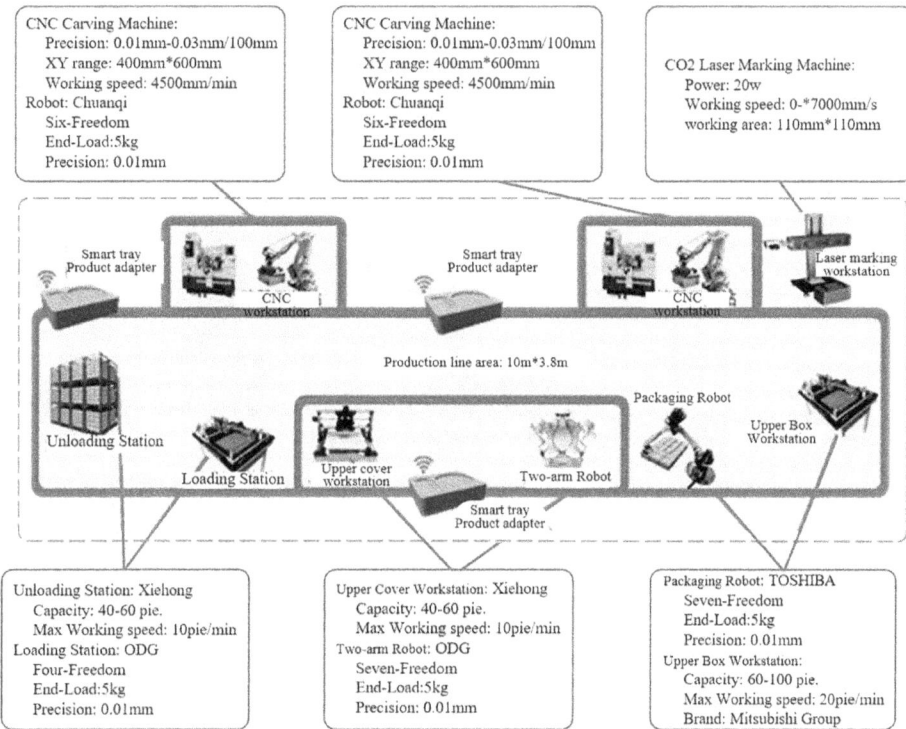

CNC Carving Machine:
  Precision: 0.01mm-0.03mm/100mm
  XY range: 400mm*600mm
  Working speed: 4500mm/min
Robot: Chuanqi
  Six-Freedom
  End-Load:5kg
  Precision: 0.01mm

CNC Carving Machine:
  Precision: 0.01mm-0.03mm/100mm
  XY range: 400mm*600mm
  Working speed: 4500mm/min
Robot: Chuanqi
  Six-Freedom
  End-Load:5kg
  Precision: 0.01mm

CO2 Laser Marking Machine:
  Power: 20w
  Working speed: 0-*7000mm/s
  working area: 110mm*110mm

Smart tray
Product adapter

Smart tray
Product adapter

Laser marking
workstation

CNC
workstation

CNC
workstation

Production line area: 10m*3.8m

Packaging Robot

Unloading Station

Upper Box
Workstation

Loading Station

Upper cover
workstation

Two-arm Robot

Smart tray
Product adapter

Unloading Station: Xiehong
  Capacity: 40-60 pie.
  Max Working speed: 10pie/min
Loading Station: ODG
  Four-Freedom
  End-Load:5kg
  Precision: 0.01mm

Upper Cover Workstation: Xiehong
  Capacity: 40-60 pie.
  Max Working speed: 10pie/min
Two-arm Robot: ODG
  Seven-Freedom
  End-Load:5kg
  Precision: 0.01mm

Packaging Robot: TOSHIBA
  Seven-Freedom
  End-Load:5kg
  Precision: 0.01mm
Upper Box Workstation:
  Capacity: 60-100 pie.
  Max Working speed: 20pie/min
  Brand: Mitsubishi Group

FIGURE 6.3 The hardware layout of intelligent manufacturing production line experiment platform for South China University of Technology.

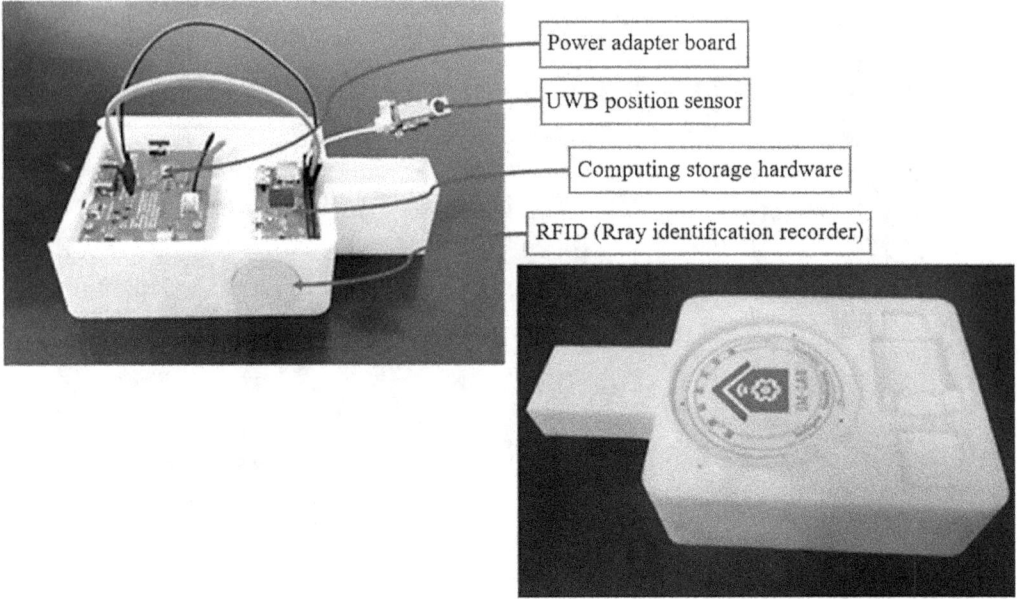

FIGURE 6.4   Product smart tray.

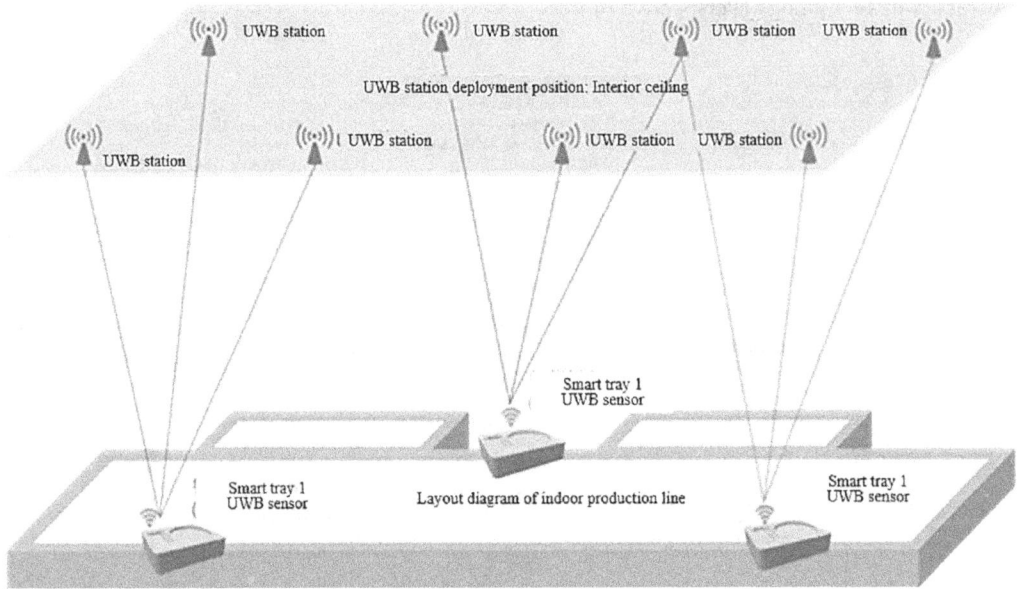

FIGURE 6.5   UWB positioning system.

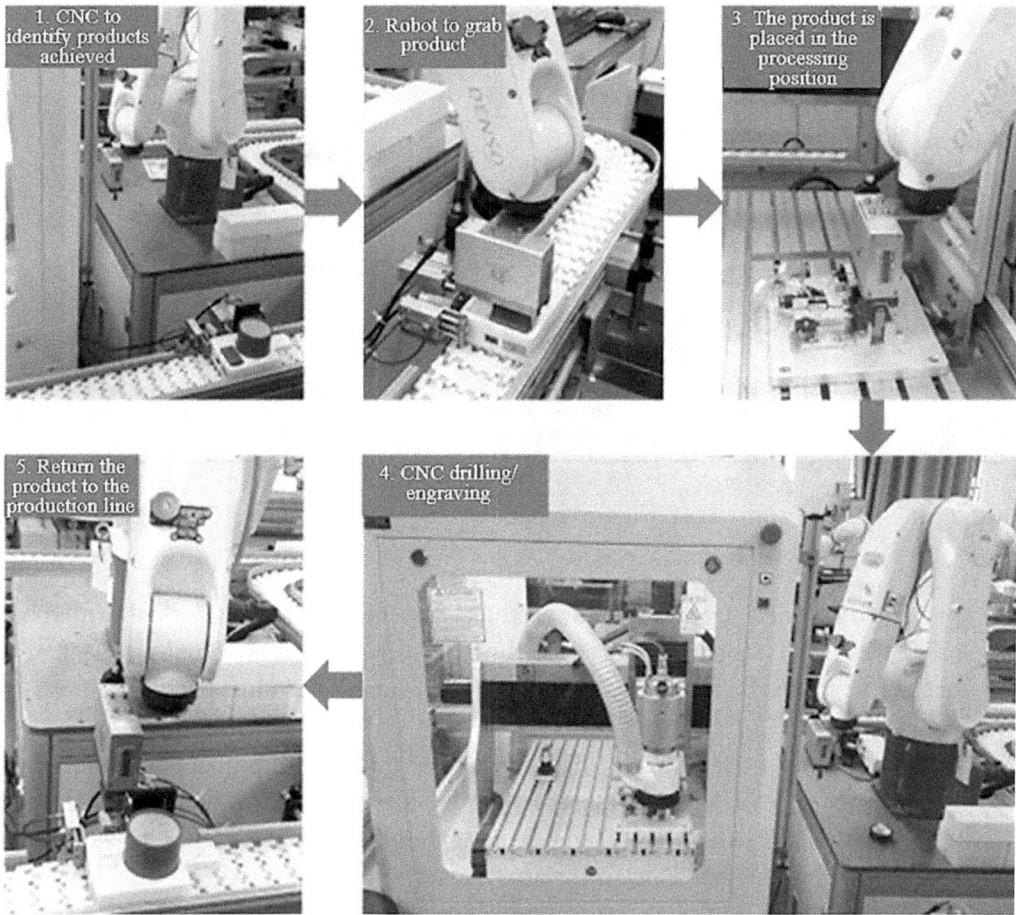

FIGURE 6.6   Processing process of CNC carving workstation.

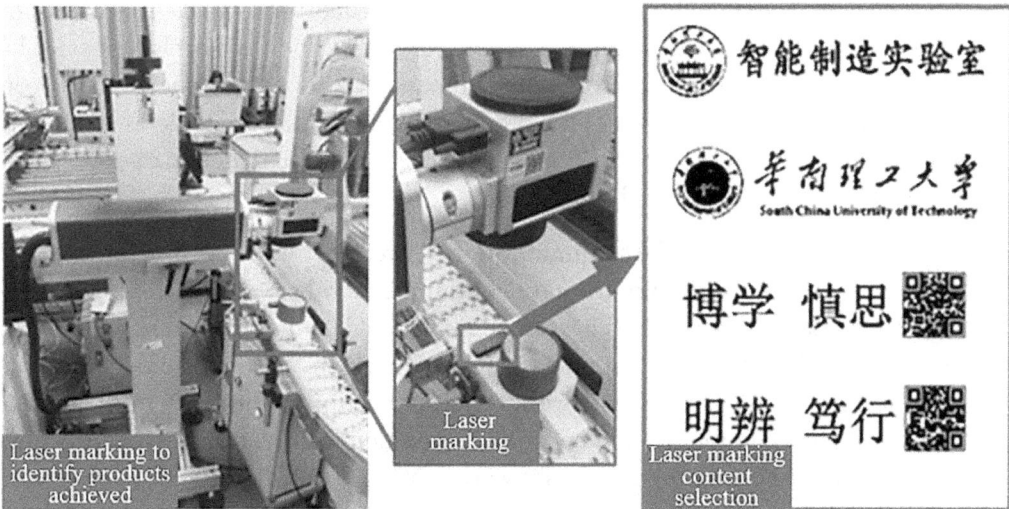

FIGURE 6.7   Laser marking workstation processing.

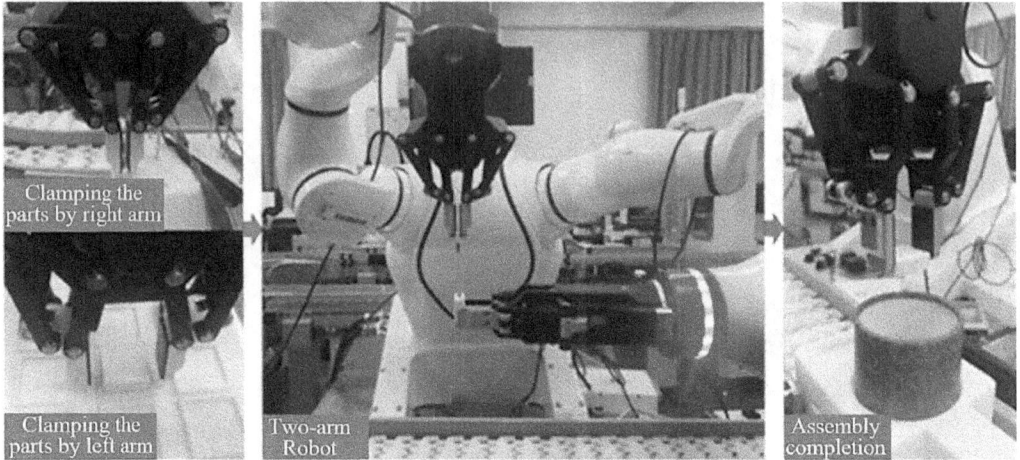

FIGURE 6.8   Assembly process of two-arm robot.

grip the two parts 1 and 2 of the product, respectively. Then, the dual-arm robot assembles the clamped products according to a certain process. Finally, the left arm of the robot puts the assembled product directly into the product packaging box.

3. **Resource adapter (in Figure 6.9)**: It is used to adapt the communication protocols of heterogeneous devices from different manufacturers, transform them into a unified communication method, and connect the underlying resources to the private cloud through wired or wireless networks.

4. **Communication facilities**: The entire intelligent production line system consists of three switches, one router and supporting software management and control tools to form a vertically integrated system that integrates data from software tools to hardware devices. The system is a multi-layer network structure that makes data identifiable between different layers. Due to different communication protocols supported by third-party hardware devices from different manufacturers, the connection between the middleware and machines of the production line in the private cloud adopts the OPC UA method to provide information models to ensure dynamic discovery and interaction between the cloud and devices.

5. As shown in Figure 6.10, the communication network of the production line is based on the requirements of point-to-point communication between machines in intelligent manufacturing and the characteristics of the data distribution service (DDS) without a central node and the interaction of publish and subscribe models. DDS is used to build a point-to-point network between machines' communication connections to build a communication network. In addition, considering the low real-time nature of the interaction between software tools and the characteristics of remote calls, middleware and software tools such as data visualization tools use Remote Procedure Call (RPC) communication. Between the controller with the strongest

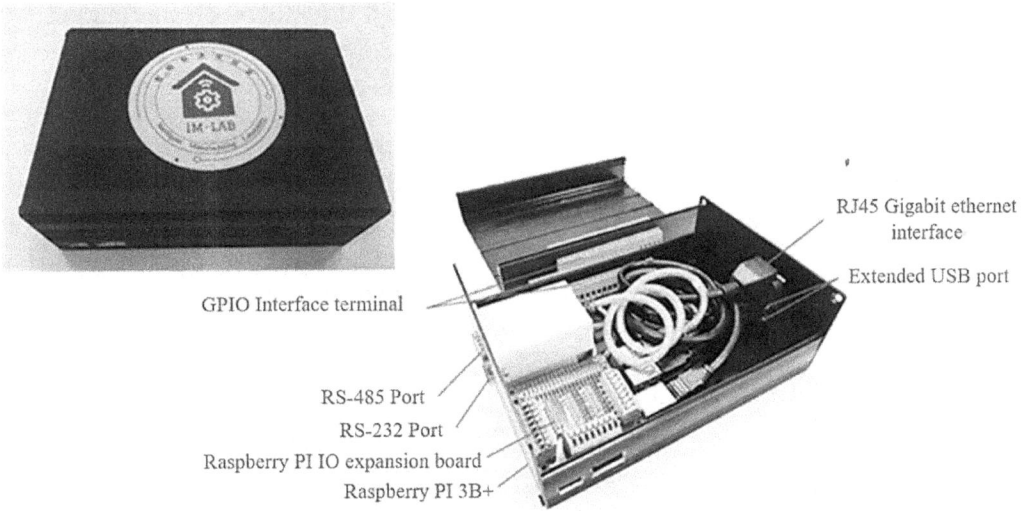

FIGURE 6.9 Resource adapter.

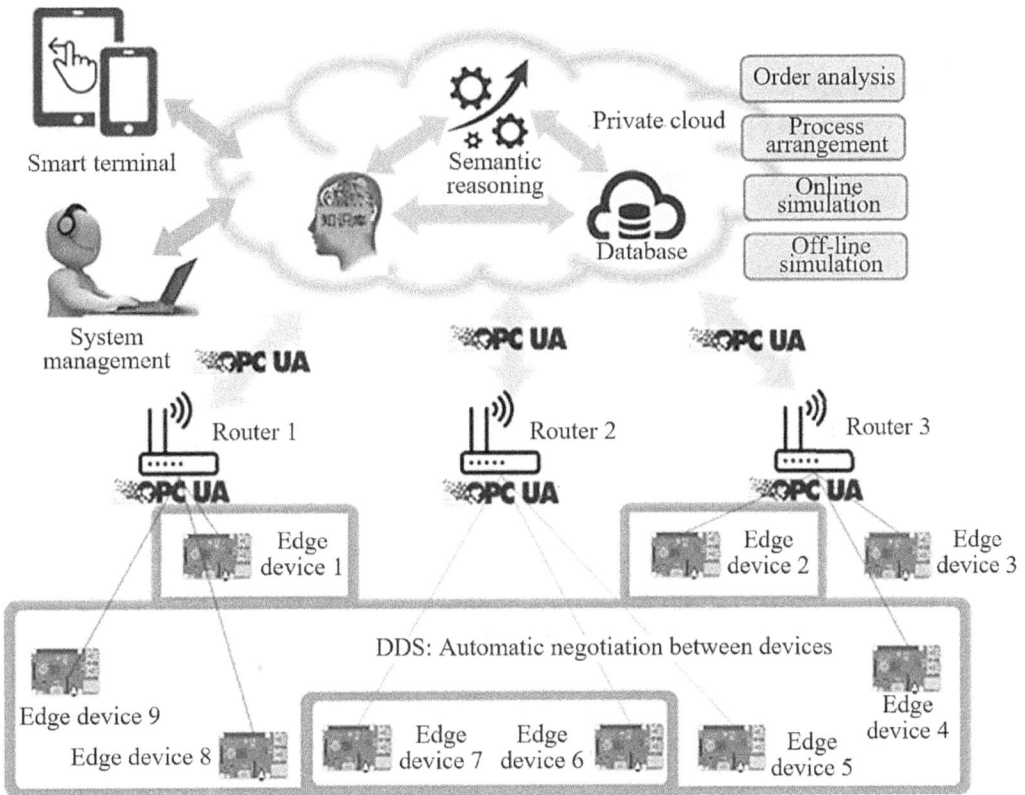

FIGURE 6.10 Communication facilities and network configuration.

real-time requirements and the motor or actuator, the EtherCAT communication protocol with strong real-time performance is adopted to ensure the logic reconstruction of the controller and the real-time control of the motor or actuator.

6. **Private cloud server**: It consists of five rack servers and serves as the central cloud platform of the entire production line. Through the cloud platform, the acquisition of real-time data and information of the system, system management, real-time scheduling, workflow control, and other related capabilities of intelligent manufacturing can be realized. For example, after the order system obtains the original information, the private cloud establishes machine processing procedures, processes, and resource utilization strategies and sends them to smart devices, robots, and conveying systems to complete product manufacturing. At the same time, manufacturing data and user interest information are also transmitted to the industrial private cloud. In addition, relevant data and graphics are also distributed to terminals for management and user verification and decision-making.

6.2.1.2 Intelligent Manufacturing Architecture and Its Implementation

The candy packing line mainly includes the production stations and the logistics transmission system. In the logistic transmission system, the packing box is continuously transferred by the conveyor belts or automatic guided vehicles (AGVs). The production stations are distributed discretely between the mainline and the branch line, and RFID tags are adopted to obtain the operation information. The equipment types of the production stations include materiel feeding, candy grasping, box delivery, and finished goods storage. The presented system meets the requirements of small-batch production. In particular, the packaged candy followed the taste, flavor, and color preferences given by the customers. The system includes four layers, all of which are connected by the industrial IoT with different link functions.

Device layer: The first layer is the device layer, including five robots, two AGVs, a conveyor system, and a warehouse. The device layer performs the basic functions of an intelligent production line, such as carrying, clipping, loading raw material, and unloading final products. Cognitive robots can be vertically integrated into a CPS in smart manufacturing.

Network layer: The industrial network layer (second layer) plays a key role in the information interaction and intelligent connection of different communication technologies, for example, industrial wireless local networks (WiFi and ZigBee), industrial Ethernet, industrial near-field communication (NFC), and mobile communications. There are three subnetworks for finishing different latency communication functions. Especially, wired industrial networks are employed as the inner equipment to achieve higher real-time performance. In this aspect, the wireless industrial networks were mainly adopted in the monitoring system, while the mobile wireless local networks also helped to achieve higher level flexibility; for instance, mobile wireless nodes were dynamically deployed to monitor the industrial environment status.

Computing layer: The third layer is the computing layer that is mainly involved with the analysis, computing, and knowledge mining of big data. A commercial solution has been adapted to build a cloud platform. XenServer developed by Citrix is used to realize the virtualization of the server cluster consisting of multiple virtual machines and the management of virtual machines. Meanwhile, we also establish a big data analytics framework, which is a software architecture based on a cloud platform for big data storage and distributed computing. Apache Hadoop, an open-source solution, is used to provide the nonrelational database HBase and the computing architecture of Yet Another Resource Negotiator (YARN). On top of the big data framework, the AI-assisted optimization algorithms (such as DL models) have been deployed to realize intelligent applications. To meet different latency requirements in the platform, a hybrid computing paradigm, orchestrating the cloud and edge computing paradigms, is adopted. Explicitly, edge computing is used to deal with real-time tasks, while cloud computing was focused on completing time-insensitive tasks, such as historical data processing. The deployment of edge computing enables cloud service characteristics, such as mobile computing, scalability, and privacy policy.

Service layer: The fourth layer is the service layer. At this level, a large number of manufacturing resources are stored at the cloud platform, which offers different AI services. Pattern recognition, accurate modeling, knowledge discovery, reasoning, and decision-making capabilities are provided.

The working process of the platform is as follows. First, customers select candy products according to their preferences, which included the color, taste, quantity, and variety of the candies in an AI recommender web service system. Then, these proposed schemes and candy order parameters are delivered to the manufacturing cloud through the web service, and the webserver was connected to the cloud via the Internet. The related product orders are created according to the submitting information. These orders were decomposed into different working steps by the ontology-based manufacturing system. Next, the multiple agents completed the production tasks in a self-organized way. After obtaining the working steps, the manufacturing devices are assembled into collaborative groups to finish all tasks. Thereafter, the platform finished the candy-wrapping task.

During the product manufacturing process, the manufacturing data are collected by sensors and then transmitted to the cloud or nearby edge servers. The analyzed results provide key information for product monitoring. More importantly, these results can be used to adjust the processes and procedures to ensure higher quality and increase the production efficiency of the whole system. The model-driven method with ontology proposed in was used to achieve interoperability and knowledge sharing in a manufacturing system across multiple platforms in the product life cycle. When multiple tasks were needed to be finished in the platform, the manufacturing resource reconstruction methods were employed for production scheduling. The cloud-based manufacturing semantic model proposed in was used to obtain general task construction and task matching. After implementation, three candy-wrapping tasks with ten different candies were processed in the AI-assisted platform at the same time, which represented a typical production line model for mass wrapping, and the first-in-first-out (FIFO) scheme was adopted accordingly.

6.2.2 Structural Design and Intelligent Control of Flexible Conveying System

In the intelligent manufacturing environment, based on cloud computing technology, server clusters are virtualized into logically unified huge resource pools, providing elastic computing and storage services for the system. On-site equipment, monitoring terminals, etc. are connected to the cloud platform through the industrial network and the Internet. Real-time process data at the manufacturing site, design simulation data such as CAD/CAE/CAM, and information system data such as ERP/MES are uploaded to the cloud platform in a unified manner to form complete, continuous, consistent, and real-time manufacturing big data, thereby providing transparency for the production process. It provides the basis for optimization and performance optimization. In addition, based on machine–machine communication and AI technology, processing equipment such as processing/testing/assembly, conveying devices, and intelligent workpieces can communicate and coordinate with each other to achieve self-organizing dynamic reconstruction, so intelligent manufacturing systems are highly interconnected, dynamic reconstruction, massive data, deep integration, and other notable features. A variety of processing equipment is distributed in the manufacturing workshop, and each equipment completes one or several processes. Usually, the relay of multiple devices is required to complete the processing of a workpiece, hence a material conveying system is required.

Material conveyor system transports workpieces between devices. Common conveying devices include conveyor belts, manipulators, and AGVs. The conveyor belt is a cost-effective conveying device with a wide range of applications. However, the existing conveyor belts are usually single-line or single-ring structures, and a single controller is used to carry out.

Centralized control, without decision-making and communication capabilities, is difficult to support the dynamic reconfiguration of resources in the intelligent manufacturing environment.

In order to meet the needs of intelligent manufacturing, a conveyor belt with a new structure and its intelligent control method are designed, so that the conveyor belt can not only change its structure flexibly but also communicate and coordinate with processing equipment and intelligent workpieces, so as to dynamically generate conveying paths. Support the self-organized dynamic reconstruction of the intelligent manufacturing process to meet the flexible manufacturing needs of small batches and multi-variety products.

6.2.2.1 System Design
The design of an intelligent conveying system needs to comprehensively consider the mechanical structure and control architecture, that is, it has the characteristics of modular structure, distributed control, and networked communication, so that it can flexibly change the topology of the conveying system and quickly realize multiple configurations

to adapt to highly dynamic manufacturing environment. The existing conveyor belt adopts a single-line or single-loop structure and single-controller control, which is tightly coupled in terms of mechanical structure and control method, so it cannot be reconfigured and lacks flexibility.

The conveying system in Figure 6.11 is composed of six sections of conveyor belts (respectively marked as L1, L2,…, L6), and each section is driven and controlled by an independent servo system (respectively marked as C1, C2,…, C6). There are three diversion stations in the clockwise direction, among which L1 and L3 intersect at diversion station S1, L2 and L4 intersect at diversion station S2, L2 and L5 intersect at diversion station S3, and each diversion station is equipped with an independent pneumatic system and controller. Various processing, testing, and assembly equipment, or storage systems for materials and workpieces can be arranged along the conveying system, as shown by M1, M2, M3, and M4 in Figure 6.11.

6.2.2.1.1 Modular Structure   Each section of the conveyor belt is spliced by several basic linear modules (divided into three types of 375, 750, and 1,500 mm) and right-angled arc modules (390 mm graduation circle), as shown in Table 6.1. The linear modules are used to adjust the length of the conveyor belt, while the arc module is used to change the direction of the conveyor belt. Based on this modular design, the conveying range and layout of the conveying system can be easily adjusted, thus strongly supporting the intelligent manufacturing system to support changes in workpiece varieties or production capacity.

6.2.2.1.2 Distributed Control   Each section of conveyor belt and each distribution station are controlled by an independent controller. These controllers are not only geographically dispersed but also equal in control logic. The conveyor belt controller is responsible for the start–stop control, speed regulation, and communication and decision-making. The diversion station controller is used to control the action of the diversion mechanism of the station and is responsible for related detection, communication, and decision-making.

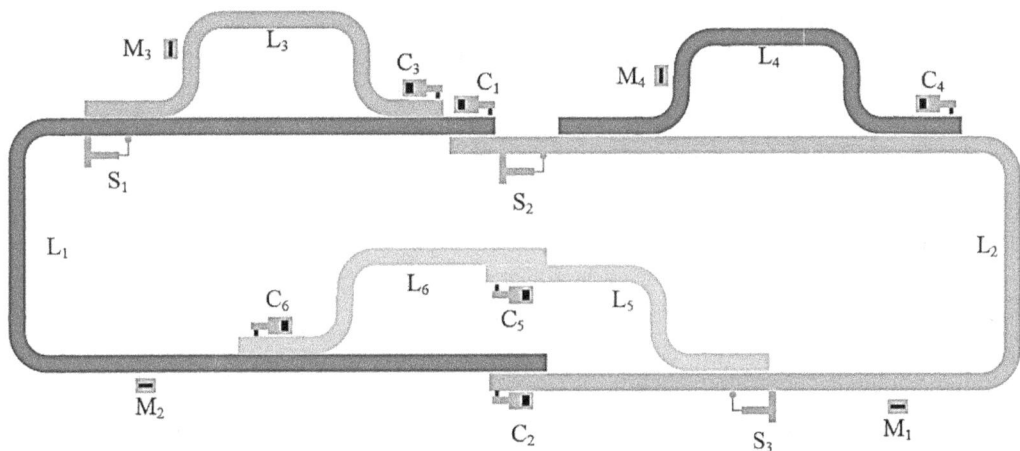

FIGURE 6.11   Schematic diagram of structural and functional modules of conveying system.

TABLE 6.1   Combination of Basic Modules of Each Conveyor Belt

Conveyor	Number of Linear Module (375 mm)	Number of Linear Module (750 mm)	Number of Linear Module (1,500 mm)	Number of Rectangular Arc Blocks
L_1	0	2	7	2
L_2	0	2	7	2
L_3	2	2	1	4
L_4	2	2	1	4
L_5	2	0	1	2
L_6	2	0	1	2
Total number	8	4	18	16

6.2.2.1.3 Networked Communication   In the intelligent manufacturing environment, conveyor belts, distribution stations, processing equipment, and intelligent workpieces are required to communicate with each other to support intelligent negotiation. In this design, controllers with Ethernet communication capabilities are selected, and each controller uses a switch for networking, so that it can quickly exchange information and lay a reliable foundation for the communication and coordination of various objects.

6.2.2.1.4 Cyclic Multipathing   In addition to the main circuit composed of L1 and L2 in this conveying system, L3, L4, L5, and L6 form three other small circuits, respectively, and each circuit intersects to form multiple paths. The diversion station is used for path selection. Since there are three diversion stations, there are eight different paths in this system. The conveyor belts included in each path are shown in Table 6.2. The value 0 of the diversion station S1, S2, and S3 corresponds to L1, L2 (corresponding to S2), and L2 (corresponding to S3) and L3, L4, and L5 are closed; the value 1 corresponds to L1, L2 (corresponding to S2), L3 (corresponding to S3) is closed, and L3, L4, and L5 are open.

Since the start and stop of each section of the conveyor belt can be independently controlled, in addition to the circular conveyor system shown in Table 6.2, it can also be configured as a single-line or multi-line open-loop conveyor system to support a variety of different application requirements. For example, only turning on L1 can form a single-wire open-loop path, while turning on both L2 and L5 can form a double-wire crossing open-loop path.

6.2.2.2 Negotiation-based Intelligent Control

The intelligent conveying system is composed of multiple independent conveyor belts, with multiple paths and dynamic structures and states. In addition, the intelligent manufacturing system is oriented to the mixed-flow manufacturing of multiple workpieces, and each workpiece requires a different path, so it is necessary to dynamically determine the path for each workpiece. However, the existing single-controller centralized control can only statically determine a fixed path, which cannot meet the needs of intelligent manufacturing. The intelligent control based on negotiation uses the interaction and negotiation of multiple conveyor belts, according to the workpiece. Because real-time and dynamic path determination is required, it has high flexibility.

TABLE 6.2 Path and Control Pattern of Conveying System

Path Number	Conveyor						Control Method		
	L_1	L_2	L_3	L_4	L_5	L_6	S_1	S_2	S_3
1	√	√					0	0	0
2	√	√	√				1	0	0
3	√	√		√			0	1	0
4	√	√			√	√	0	0	1
5	√	√	√	√			1	1	0
6	√	√	√		√	√	1	0	1
7	√	√		√	√	√	0	1	1
8	√	√	√	√	√	√	1	1	1

"√" indicates that the section of the conveyor belt is part of the corresponding path.

TABLE 6.3 Dynamic Varieties' Definition of Work Piece

Varieties	Value and Meaning	Description
CurOp	1,2,..., N is the operations that are going to happen. N is the total number of operations. $N+1$ indicates that all operations are complete	Represents the current operation to be performed, initialized to 1 and incremented by 1 for each operation completed
State	0: uncertain process equipment 1: uncertain transmission equipment 2: process equipment is not arrived 3: operation is not started 4: operation is starting	Working state

In the intelligent manufacturing environment, intelligent workpieces need to interact dynamically with processing equipment, conveying devices, etc. Embedded systems are a high-performance method for realizing intelligent workpieces, but in order to reduce costs, barcodes, two-dimensional codes, RFID tags, etc. are usually used which is a solution with a lower degree of intelligence [1]. The barcode can represent very little information and is usually only used as an identity identifier; although the two-dimensional code can carry more information, it is read-only like the barcode. Below, RFID tags are not only large in capacity but also readable and writable, so they can not only store static information but also dynamic (such as status) information. In summary, the use of RFID tags can achieve a balance between performance and cost. Therefore, this solution is increasingly becoming the most widely used solution in the intelligent manufacturing environment [2].

The processing technology requirements are preset in the RFID tag of the smart workpiece, and the dynamic information in the RFID tag will be updated according to the situation during the processing. The definition of the dynamic variables of the workpiece is shown in Table 6.3.

Negotiation-based intelligent control includes four stages. Based on the scenario shown in Figure 6.11, the basic process of each stage is described below.

6.2.2.2.1 Determine the Target Processing Equipment   Manufacturing, testing, assembly, storage, and other processing equipment use photoelectric sensors, RFID readers, network communications, and other means to interact with workpieces and conveyor belts. When workpieces move on the conveyor belt, they will trigger corresponding signals when they pass by sensors, and the processing equipment receives them. After receiving the signal, send a message to the conveyor belt. The conveyor belt stops moving after receiving the message and sends a confirmation message to the processing device. Then, the processing device controls the RFID reader to read the RFID tag information of the workpiece and performs corresponding actions according to the value of "State."

If the State is 0, the processing equipment that detects the workpiece acts as an intelligent workpiece agent to negotiate with other processing equipment and conveying devices to determine the target processing equipment. The negotiation process based on the contract network protocol [3] is shown in Figure 6.12, which mainly includes 10 steps. The intelligent workpiece agent publishes the operation requirements as the bidding party, and the processing equipment decides whether to bid according to its own ability and state and then determines the winning processing equipment through bid evaluation, and finally, the intelligent workpiece agent notifies the bid-winning equipment and assigns the value of State to 1.

Smart artifacts with computing capabilities (such as artifacts carrying embedded systems) can initiate negotiations autonomously without the need to deal with device agents.

6.2.2.2.2 Determine the Delivery Route   If State is 1, it means that the processing equipment has been determined, but the conveying path from the intelligent workpiece agent to the target processing equipment must also be determined. Each conveying device must know the positional relationship with other devices, including other conveying devices connected to the conveying device along the forward direction device and the processing equipment to which the conveying device can deliver. For example, the connection of each conveyor belt in Figure 6.11 can be expressed as: L1[M2, L3, L2], L2[L4, M1, L5, L1], L3[M3, L1],

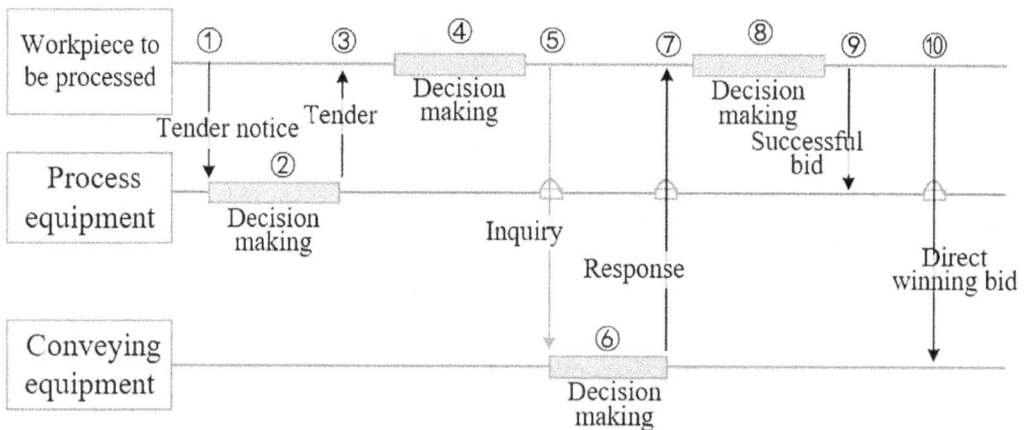

FIGURE 6.12   Interaction process of smart work pieces, processing machines, and conveyers.

L4[M4, L2], L5[L6], L6[L1], that is, each conveying device has a reachable set, representing other conveying devices and processing equipment that can be reached sequentially through this device. Processing equipment is also necessary to know its connection relationship with the conveying device. The equipment in Figure 6.11 can be expressed as: M1[L2], M2[L1], M3[L3], M4[[…]].

Without loss of generality, it is assumed that M1 is an intelligent workpiece agent and M4 is the bid-winning equipment. Now determine the processing path from M1 to M4. Although each processing equipment and conveying device only knows its own reachable set, appropriate delivery path can be finally determined after several rounds of negotiation. As shown in Figure 6.13, the required delivery path is determined to be M1→L2→L1→L2→L4→M4 after five rounds of negotiation.

Each round of negotiation includes three steps, consisting of an initiator and several responders. The following takes the first round of negotiation as an example to demonstrate the negotiation process:

1. **Initiation**: As the initiator, M1 checks its reachable set and knows that L2 is the only reachable delivery device, that is, the only responder, thus M1 asks whether L2 can reach M4.

2. **Response**: L2 checks its reachable set and then answers no.

3. **Confirmation**: M1 requires L2 to relay the inquiry process.

Next, L2 checks its reachable set as the initiator, determines that L5 and L1 are legal responders, and starts a new round of negotiation. Note: Since the conveyor belt is unidirectional, M1 is located behind L4, which means that the workpiece cannot be directly transferred from M1 via L2 reach L4, so L4 is not an optional responder. Since L2 did not fully use the delivery device in its reachable set in the second round of negotiation, it can appear as a legal initiator in the fourth round of negotiation. On the contrary, L1, and in the third round of negotiation, L5 have fully used its transport device in the reachable set, so they cannot appear as a legal initiator in the fifth round of negotiation to avoid circular

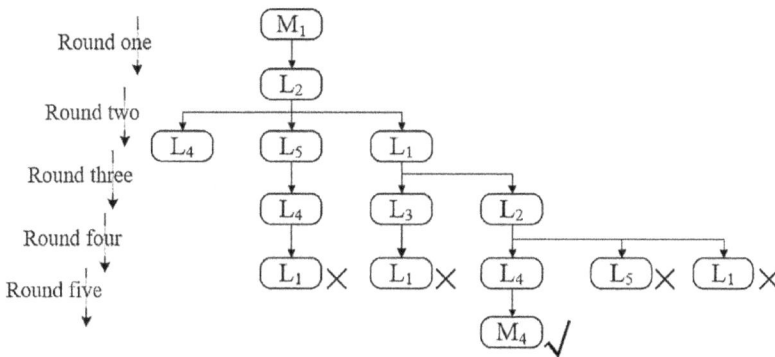

FIGURE 6.13   Relay negotiation process to determine conveying path.

negotiation. In this way, at least one new member is determined in each round of negotiation, until finally reaching M4.

6.2.2.2.3 Routing Control   The workpiece will pass through the diversion station during the movement and each diversion station uses photoelectric sensors, interception mechanisms, diversion mechanisms, RFID readers, etc. to interact with the workpiece. After the sensor detects the workpiece, the interception mechanism stops the workpiece, and the device reads and writes the RFID tag information of the workpiece and operates according to the "State" value. When the "State" value is 0, 1, 3, or 4, the diversion station does not change the flow direction and releases it directly.

If the value of "State" is 2, the distribution station queries the conveying path of the workpiece and performs corresponding operations. For example, for the path M1→L2→L1→L2→L4→M4, the distribution stations S3, S1, and S2 will turn on L2 in turn, L1, and L4 channels to realize the path 3 in Table 6.2, and then control the action of the diversion mechanism according to the calculation results to generate the required path, and release the workpiece after the operation is completed.

To sum up, the processing equipment needed to perform the operation is determined through negotiation between the intelligent workpiece agent and the processing equipment, and the transportation path from the intelligent workpiece agent to the target equipment is determined through negotiation between the intelligent workpiece agent and the conveying device and between each conveying device, while the splitter is responsible for constructing the required paths.

6.2.2.2.4 Operation Execution   When the processing equipment detects that the "State" of the workpiece has a value of 2, if the equipment is the target processing equipment, assign a value of 3 to "State" and perform operations such as positioning and clamping, and then assign a value of 4 to "State" and start the required operations such as processing, testing or assembly, assign "State" a value of 0 and add 1 to Cur Op. Then, the processing equipment acts as an intelligent workpiece agent to negotiate the next process. If the equipment is not the target processing equipment or "State" and the value is 3 or 4, it will be released directly.

The above four steps must be performed once for each operation in the operation sequence. After all the operations are completed, the automatic storage device arranged around the conveying system will remove the workpiece from the pallet and put it into the storage device.

6.2.2.3 Realization

The intelligent conveying system based on modular structure and intelligent negotiation needs to meet a variety of functional and performance requirements. A single conveyor belt must have the ability to start, stop, and speed regulation, and the confluence mechanism should operate normally. For a system composed of multiple conveyor belts, it must be able to communicate with each other. Communication and negotiation are used to dynamically determine the path according to the requirements of the workpiece; the shunt

FIGURE 6.14   Snapshot of flexible conveying system.

mechanism normally constructs the path according to the negotiation result; when multiple workpieces are processed at the same time, the system will not be deadlocked. The design rated speed is 20 mm/s, and the number of supported workpiece types is 20, which can run continuously for 24 hours. Based on the conveying system proposed in this paper, an intelligent manufacturing platform as shown in Figure 6.14 is constructed. The selection of main components and key design parameters is introduced as follows.

6.2.2.3.1 Main Body   The main frame of the conveyor belt is made of aluminum alloy profiles, the adjustable support column makes the height of the conveyor belt range from 500 to 800 mm, and the chain plate is made of white PVC material with a width of 105 mm. The system includes six sections of conveyor belt, each of which consists of several basic modules that are spliced together. Each conveyor belt includes a driving head (driving wheel) and a driving tail (driven wheel), and the rotating shaft of the driving head is connected to the output end of the reducer.

6.2.2.3.2 Diverting Mechanism and Converging Mechanism   The arrow in Figure 6.15 indicates the direction of workpiece movement, so the workpiece from conveyor belt B can continue to move along conveyor belt B under the control of the diversion mechanism and can also move to conveyor belt A. The diversion mechanism is mainly composed of a swing rod, an intercepting cylinder, and a photoelectric sensor. When the sensor detects the workpiece, the intercepting rod of the intercepting cylinder stretches out to stop the workpiece on the conveyor belt B. After a period of time (so that all the workpieces in the diversion area pass through), the swing rod mechanism moves under the drive of another cylinder, and the conveyor belt is selected according to the needs A or conveyor belt B. After the swing rod is in place, the intercepting rod of the intercepting cylinder retracts, and the workpiece continues to move forward. In the position shown in Figure 6.15, the workpiece will move from conveyor belt B to conveyor belt A.

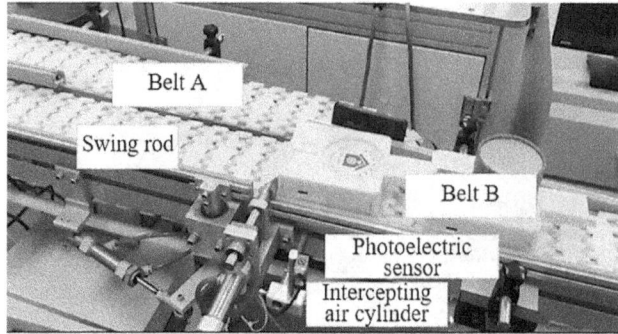

FIGURE 6.15   Snapshot of splitting mechanism.

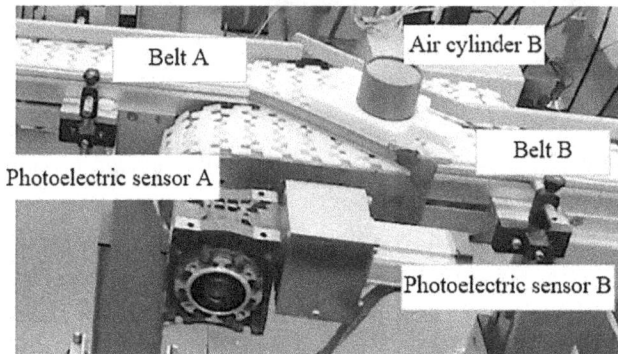

FIGURE 6.16   Snapshot of merging mechanism.

As shown in Figure 6.16, the workpieces from conveyor belt A and conveyor B can be safely merged to conveyor belt B under the control of the confluence mechanism. The confluence mechanism is composed of two pairs of oppositely arranged cylinders and photoelectric sensors. When sensor A detects the workpiece, the intercepting rod of cylinder A retracts, while the intercepting rod of cylinder B extends. In this way, the workpiece on conveyor belt B is stopped, and the workpiece on conveyor belt A can move smoothly to conveyor belt B without collision. Similarly, when sensor B detects the workpiece, the two cylinders cooperate to make the workpiece on the conveyor belt B pass smoothly.

6.2.2.3.3 Sensor and RFID System   The photoelectric sensor is a product of SICK company. The RFID system is a product of domestic Dongxin company, and the tag capacity is 2 kb.

6.2.2.3.4 Servo Drive and Control   As shown in Figure 6.17, the Panasonic servo motor and driver are selected, the motor torque used for conveyor belt 1 and conveyor belt 2 is 2.4 N·m, the driver power is 750 W, the motor torque used for the other four sections of conveyor belt is 1.3 N·m, and the driver power is 370 W. Compared with ordinary asynchronous motors, servo motors are more convenient in speed regulation and have higher

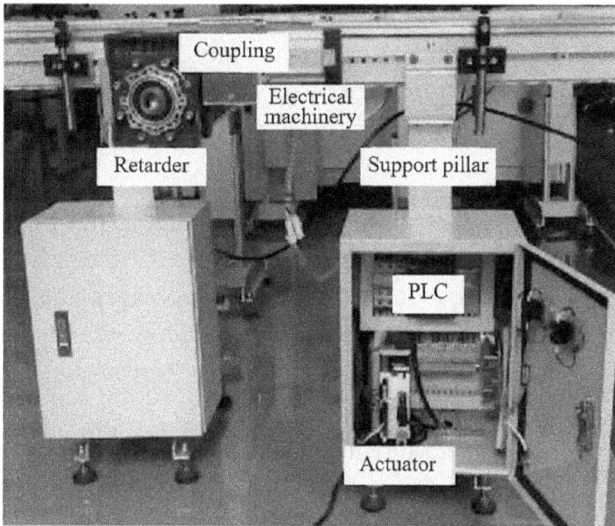

FIGURE 6.17   Snapshot of servo and control system.

position accuracy. The output shaft of the motor is connected to the worm gear reducer through a coupling. Siemens S7-200 SMART PLC controller, with Ethernet communication and pulse output function, is capable of inter-controller communication and servo motor motion control.

6.3   RESOURCE RECONFIGURATION AND OPTIMIZATION AND ITS APPLICATION

With the rapid development of AI and machine learning technology, the research on smart factories is gradually deepening. Smart factory refers to the intelligent manufacturing system realized through digital technology. It has the characteristics of reconfigurable and intelligent scheduling. It can adapt to market demand, improve production efficiency, and reduce costs. It has a very broad application prospect. The reconfigurability and intelligent scheduling of smart factories are the core elements of realizing smart manufacturing, and they have the following research significance:

Increase production flexibility: The production process of traditional manufacturing enterprises is often rigid and requires fixed equipment and processes. The reconfigurability and intelligent scheduling of smart factories make the production process more flexible, which can be adjusted according to orders and production tasks, and the production process is more efficient.

Realize production autonomy: The reconfigurability and intelligent scheduling of smart factories make the production process more autonomous. Through intelligent scheduling, the automation and intelligence of the production process can be realized, manual intervention can be reduced, and production efficiency and stability can be improved.

Reduce manufacturing costs: The reconfigurability and intelligent scheduling of smart factories make the production process more efficient, reduce the use of labor and equipment, and reduce production costs. By optimizing the production process, energy waste and loss can be reduced, resource utilization efficiency can be improved, and production costs can be reduced.

Increase productivity: The reconfigurability and intelligent scheduling of smart factories can improve production efficiency. Through intelligent scheduling and optimization, bottlenecks and delays in the production process can be reduced, production efficiency can be improved, error rates and product defect rates in the production process can be reduced, and product quality can be improved.

In short, the reconfigurability and intelligent scheduling of smart factories are the core elements for realizing smart manufacturing. With the continuous development of technology, smart factories will become an important trend in the transformation and upgrading of the manufacturing industry, which is of great significance for improving the competitiveness and innovation capabilities of the manufacturing industry and promoting the sustainable development of the manufacturing industry.

This chapter will describe the key technologies and solutions of intelligent manufacturing resource reconfiguration and scheduling from two cases, the reconfigurable method of resources in a pharmaceutical factory and the re-assignment method of workers in an engineering-to-order assembly island.

6.3.1 Reconfigurable Smart Factory for Drug Packing in Healthcare Industry 4.0

With the arrival of Industry 4.0, the market demands have been much more dynamic and challenging. Traditional pharmaceutical production, which is designed in rigid patterns as well as for limited products, can no longer adapt to new production trend of small batch and multi-variety. Specifically, the manufacturing control system of pharmaceutical production should be self-organized and flexible enough to external market's demands. Thus, functionality and structures reconfiguration of control systems is of great significance to make reconfigurable pharmaceutical production achievable. With the development of Information Communications Technology (ICT), cloud computing, CPS, reconfigurable automation, and modular machine tool, production reconfiguration has great potential to be much enhanced.

Over the past decade, the pharmaceutical manufacturing has paid much attention on the manufacturing of safe and quality products. While, in the aspect of production process, significant advancements have been achieved. Lee and Connor pointed out that continuous production is a promising alternative to batch processing in pharmaceutical manufacturing and concluded the control strategy for continuous manufacturing. Furthermore, Rehrl et al. [4] demonstrated the model-predictive control to a feeding blending unit used in continuous pharmaceutical manufacturing. Besides, recent research has been focused on integrated flowsheet modeling approach, which can predict the process dynamics affected and give insight on the characteristics and bottlenecks of the process. In data and communication aspects, Jiang et al. [5] focused on wireless medical CPS and proposed domain-specific language to specify vital real-time data. Tawalbeh et al. [6] concentrated on big data analysis for healthcare applications and proposed the mobile cloud computing model. In addition, Norman et al. [7] proposed to use three-dimensional printed technology for drug production. However, few works pay attention to production procedures as well as the dynamic market's demands of pharmaceutical manufacturing, which are of great urgency recently.

To cope with the emerging challenges, the proposal of model-integrated computation as well as its specific application based on domain-specific modeling language (DSML) makes it possible for pharmaceutical production to have more agility, flexibility, and less cost. IEC 61499 standard, which satisfies the features of DSML and extends from IEC 61131 standard, is introduced in this paper for industrial automation. The IEC 61499 standard provides high-level control paradigms of manufacturing resources and makes it applicable for accomplishing low-level reconfigurable machine control system. For example, Strasser et al. [8] discuss different design and execution issues of definition for execution models of the device, the resource, and the function blocks in IEC 61499 distributed automation and control systems. García et al. [9] present a low-cost embedded architecture by means of OPC UA services, integrated as IEC 61499 blocks. The IEC 61499 provides high-level capabilities by easily combining software components with independence of the hardware platform used. Moreover, Vyatkin et al. [10] address software design with the IEC 61499 standard enhanced with a time-stamping mechanism for cyber-physical automation systems. The proposed method enables invariant properties of the physical system in case of software reallocation to different hardware.

Another critical feature for reconfigurable pharmaceutical manufacturing is self-adaption for dynamic market's demands. However, traditional software control logic usually requires manual intervention which can hardly catch up rapidly changing demands and inefficient cost. Considering the emerging rapid development of knowledge-driven industrial control, the ontology-based knowledge base is introduced in this paper. Besides domain and relationship description, new facts behind the knowledge base can be inferred based on the reasoner [11,12] and queries (SPARQL [13], SWRQL [14], etc.) according to inner changes or outer requests. Furthermore, more complicated constraints between manufacturing resources can be achieved with semantic web rule language (SWRL) [15]. In fact, an ontology of manufacturing domain was proposed called Manufacturing's Semantics Ontology (MASON) [16] in 2006. In this paper, the reconfigurable control system is proposed based on the IEC 61499 standard and ontology. This study explores the data-driven reconfigurable manufacturing of Smart Factory for reconfigurable pharmaceutical production, from the perspective of the implementation of Healthcare Industry 4.0.

6.3.1.1 Overall Architecture

The data-driven reconfigurable Smart Factory for reconfigurable pharmaceutical production has a multilayered architecture consisting of three layers: the perception layer, the deployment layer, and the execution layer. The former layer depends on the latter, and the data flow between the three layers forms a closed loop.

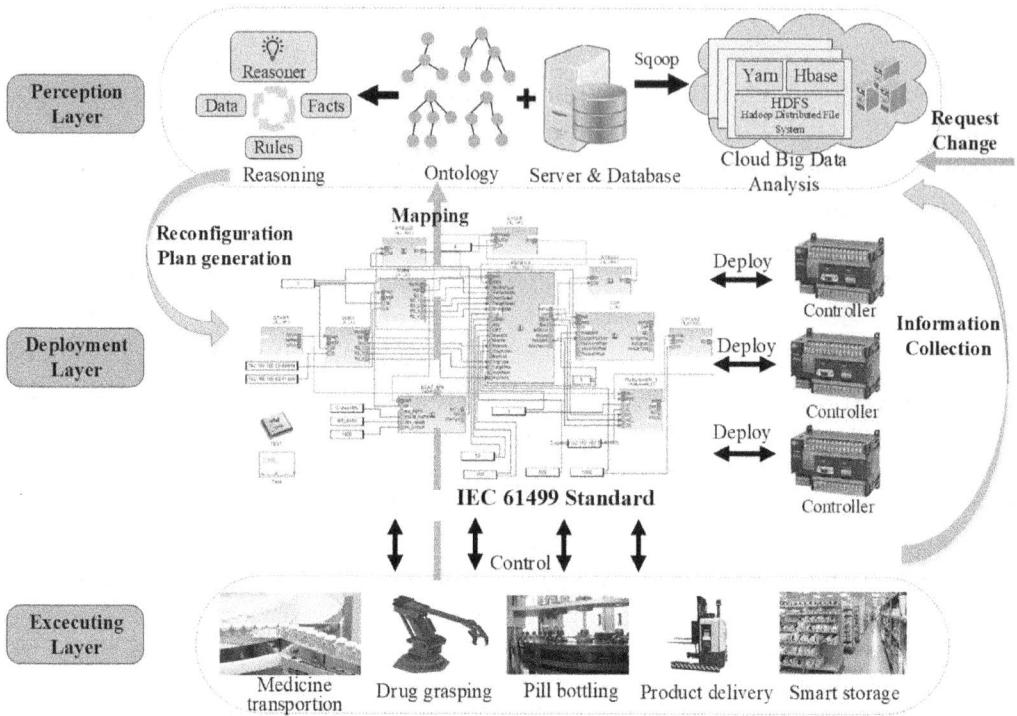

FIGURE 6.18 Overall architecture of the data-driven reconfiguration smart factory for pharmaceutical production.

Figure 6.18 shows all the three layers of the overall architecture as well as their communication for pharmaceutical production. The perception layer is responsible for knowledge representation and reconfiguration plan generation. Based on the specification of Smart Factory, such as attribution of devices, communication interfaces between devices, and operation sequence, an ontology is supposed to be established. The ontology model describes resources in classes and creates instances of the classes according to the specification of Smart Factory.

There are two kinds of data mainly affecting production plan generation, namely status data of low-level sensors, machines as well as external pharmaceutical production demands. The former reflects available functionality of pharmaceutical production devices and is directly connected to the cloud database, which updates the data properties of corresponding ontology instance. While, the latter is designed to be processed by the cloud server, which will be transformed into XML format. Together with the ontology model for pharmaceutical production and the standard expression of market's requests, the reasoner will reason out reconfiguration plan with a sequence of new fact queries which then passes the signal to the deployment layer.

The deployment layer is supposed to take the reconfiguration plan and schedule the function blocks accordingly. In this paper, we assume that all the experimental controllers are in compliance with the IEC 61499 protocol for its advantage on distributed controlling. In addition, the deployment layer consisted of two phases, namely design phase and implementation phase. The former undertakes a production plan from the cloud and then reorganizes function blocks of IEC 61499. Furthermore, in the implementation phase, the run-time environment of bottom machine is supposed to download the managing commands from the former with transmission control protocol/Internet protocol (TCP/IP). The run-time environment will generate and schedule the local function blocks into operation sequence, thus triggering the controller [e.g., power line carriers (PLCs)] to further pharmaceutical procedures.

To fully interpret the procedure of pharmaceutical production and the three layers above, the next three sections will focus on three key technologies of the architecture, namely Ontology establishment (knowledge representation), perception reasoning on the cloud, two-phase deployment layer, and IEC 61499 standard.

6.3.1.2 Knowledge Base Representation and Ontology Modeling

In this section, the ontology technology will be discussed which is responsible for knowledge representation of pharmaceutical manufacturing resources and acts as a basis for further applications. In this section, controllers of physical-level devices are assumed to be in compliance with the IEC 61499 protocol. To specify the manufacturing resources for further reasoning and reconfiguration, an ontology is established as the knowledge base to cover all significant devices for pharmaceutical production as well as their attribution.

As Figure 6.19 shows, manufacturing resources for medicine production (transport units, manipulators, controllers, etc.) are treated as resources in an ontology. Ontology is a

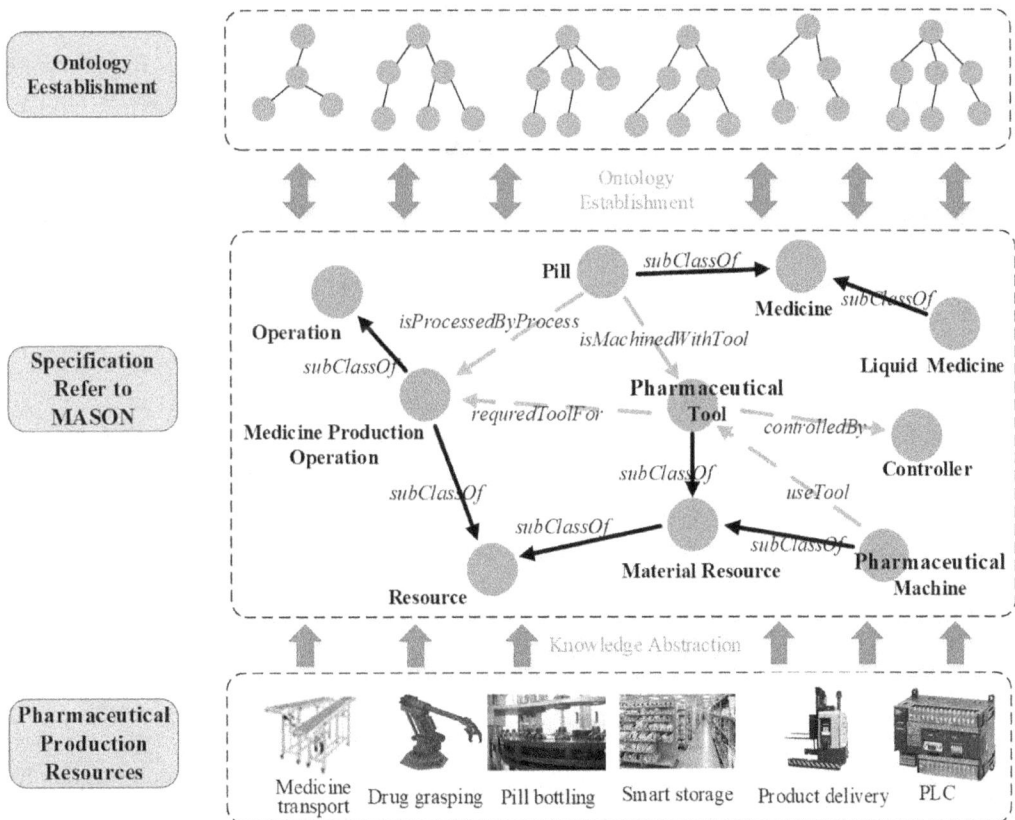

FIGURE 6.19 Ontology establishment for pharmaceutical production.

formal description of a domain, used for a particular purpose and expressed with controlled vocabulary. In manufacturing field, the ontological knowledge model can infer facts about the manufacturing environment and then decides whether the current environment can support the given manufacturing requirements. With low-level control component such as IEC 61499 function blocks, the ontology technology can achieve fast reconfiguration of modular manufacturing systems. In ontology, there are two critical components: terminological components (Tbox) and assertional components (Abox), where the former specifies concepts and the latter expresses the concepts with instances. In practice, concepts in Tbox are usually represented as classes, whereas elements associated with classes are expressed as instances correspondingly. In ontology, classes are in compliance with inheritance as the object-oriented language does, which objectively represent the general law in various domains. In addition, to objectively describe the association between classes as well as the intrinsic attribution of them, two kinds of properties called object property and data property are introduced in ontology. The object properties are designed to describe relationship between the instances of two classes. For example, in circumstances of medicine production, one class called *Pill_Gripper*, which is the tool to grasp the pills, may have no inheritance relationship with class *Drug_Loading_Machine*, but the two can be connected with the object property *use* (*Drug_Loading_Machine use Pill_Gripper*). Furthermore, the object properties support symmetric, transitivity, and reflexivity to further enriched description of complex relationships between resources. The data properties are designed to link instance with specific data, such as the abrasion condition of medicine gripper or left pill numbers of pill container. The data properties are critical when reasoning the available functionality of devices. Besides, to establish more complicated constraints, we introduce SWRL in this paper to enhance the ontology's ability for complex relationship expressing.

In pharmaceutical production environment, considering the varied market's demands and physical devices, in this paper, ontology establishment is based on MASON as the previous section mentioned. MASON, the ontology of manufacturing domain, was proposed a decade ago. The main high-level classes of MASON are *Entity*, *Resource*, and *Operation*. Based on these classes, more concrete subclasses such as *Machine Resource*, *Tool*, and *Manufacturing Operation* are generated, which are connected by properties such as *useTool* and *requireToolFor*. As Figure 6.19 shows, the established ontology inherited the three head concepts from MASON with more specific descriptions such as *medicine, pharmaceutical tool, Pharmaceutical Machine*, and *medicine manufacturing operation*. With additional object properties and data properties deployed, similar ontologies can be established for specific pharmaceutical production circumstances. In this paper, we establish the ontology referring to MASON and take advantage of its architecture design, such as classes of Machine resources, Tool as well as their relationships. Together with SWRL, more complicated relationships of the ontology can be achieved with more efficient expressions.

6.3.1.3 Reasoning Procedure and Cloud-Based Big Data Analysis Services

In this part, we will discuss the reasoning process with several critical technologies. As the reflection of reconfiguration, the reasoning procedures are designed in the cloud with additional resources. The cloud is responsible for collecting information from the executing layer and sends the reasoned logic to the deployment layer for devices' controlling.

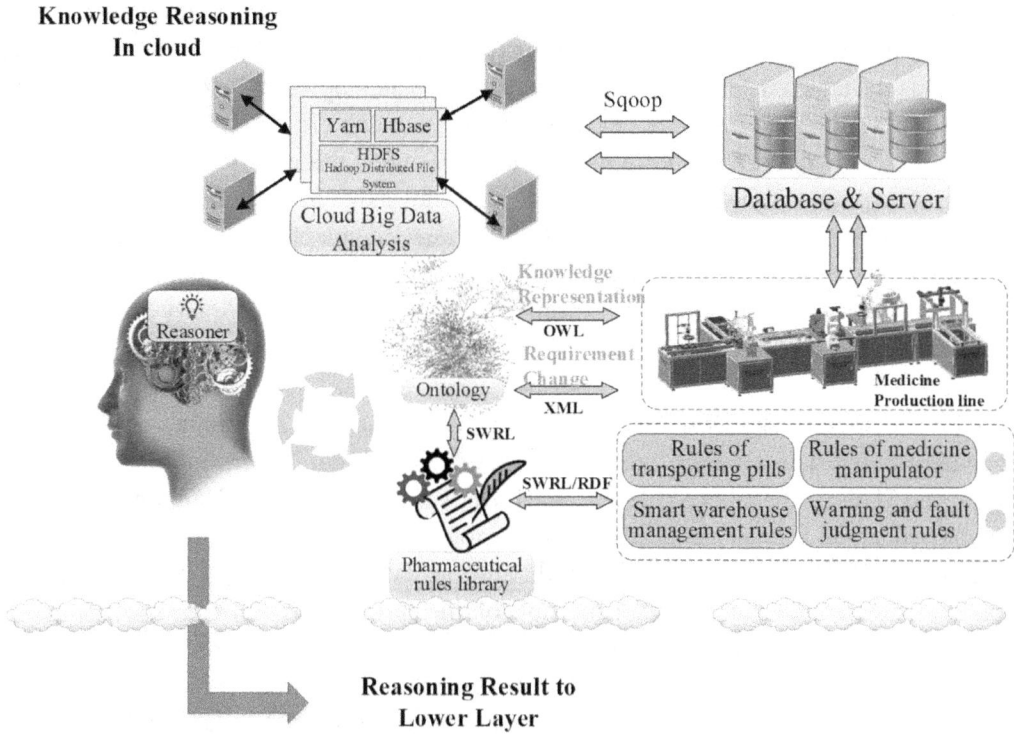

FIGURE 6.20   Knowledge reasoning and cloud resources.

6.3.1.3.1 Reasoning Procedure of Reconfiguration Pharmaceutical Production   As Figure 6.20 shows, all the significant components corresponding with knowledge reasoning are deployed in the cloud. As the previous section mentioned, concept abstraction and knowledge representation about the pharmaceutical production sensors, devices, and machines are expressed in ontology. In practice, we choose Web Ontology Language (OWL) developed by W3C Organization to express the MASON-based ontology as the previous section mentioned. However, in general, the restrictions of OWL are not enough to represent more complicated relationships. For example, transport rules for pills, working rules of medicine manipulator as well as warning and fault judgment rules are confined to specific manufacturing circumstances.

These restrictions are supposed to be represented by ontology but can hardly be implemented with object properties or data properties. With sufficient freedom describing attributions between pharmaceutical manufacturing resources, the SWRL can fully utilize properties to add constraint rules into the ontology.

When a new market's requirement for medicine production arrived, database and server in cloud first undertook it, conducted an analysis, and transformed the information to XML for further reasoning. Based on that, the following reasoning procedure requires several components work in coordination. First, a running process in the cloud will take the XML data then convert it into SPARQL, a set of specifications to query and manipulate resource description framework (RDF) graph content.

Then, the status data of pharmaceutical manufacturing resources are updated to the database. Specifically, in the database, several key statements of bottom machines are recorded in the form of the database table. For example, rpm of manipulator's motor, condition of photoelectric switch, and remaining number of the pills. This critical information is stored as the table's fields and inserted circularly to catch up with the newest statements. Based on that, the data property of these machines, which are represented as the ontology's instances, will correspondingly be updated. Finally, the reasoner, Jena reasoner in this paper, will query the appropriate expressed knowledge base (together with original ontology and SWRL rules) with the SPARQL. The query result contains information about which devices are available and chosen for the specific production demand as well as how these resources cooperate in an operation sequence. Eventually, the query result will be encapsulated as the high-level control logic to the deployment layer by the TCP/IP protocol.

6.3.1.3.2 Cloud-Based Big Data Analysis Services In the cloud, additional services such as industrial big data analysis, reliable persistent storage, data migration between databases, and Hadoop distributed file system (HDFS) are provided. The Hadoop framework is deployed as the fundamental infrastructure together with Hive, Sqoop, Pig, and other components. With Hadoop's powerful ability for big data distributed processing, it is possible to implement deep analysis for multiple offline applications. For the aspect of maintenance, utilization ratio and abrasion condition of manufacturing resources might be analyzed. Furthermore, medicine consumers' orders might be analyzed for the fine-grain analysis of pharmaceutical market's trend. In summary, the cloud big data analysis can act as additional services and provides computing support for further applications.

6.3.1.4 Two-Phase Deployment Layer and IEC 61499 Standard

In this part, we will deeply discuss the deployment layer with its two phases: design phase and implementation phase. The deployment layer is responsible for both high-level functionality reconfiguration and low-level devices' controlling for pharmaceutical production.

6.3.1.4.1 IEC 61499 Standard As the previous section mentioned, the current control systems are usually developed by IEC 61131-3 where the dynamic reconfiguration and real-time demand may not be satisfied. The appearance of IEC 61499 complements the dominant IEC 61131 standard with function block diagram and even-driven execution paradigm for application design, which much fits distributed control systems. Besides, the management model of IEC 61499 standard makes it possible for high-level reconfiguration. The controlling algorithms (proportion integration differentiation (PID) control, etc.) can be encapsulated in IEC 61499 function blocks. With add, delete, and various connection between function blocks, the dynamic reconfigurable control logics can be modeled. Furthermore, these reconfigurations even can be implemented during the operation by commands sending from the software tools. Thus, considering the ability for both low-level and high-level automation controlling, in this paper, the deployment layer consisted of two phases, namely the design phase and implementation phase.

6.3.1.4.2 Design Phase of Deployment Layer As Figure 6.21 shows, we apply two phases when deploying function blocks of IEC 61499 in deployment layer. The function blocks

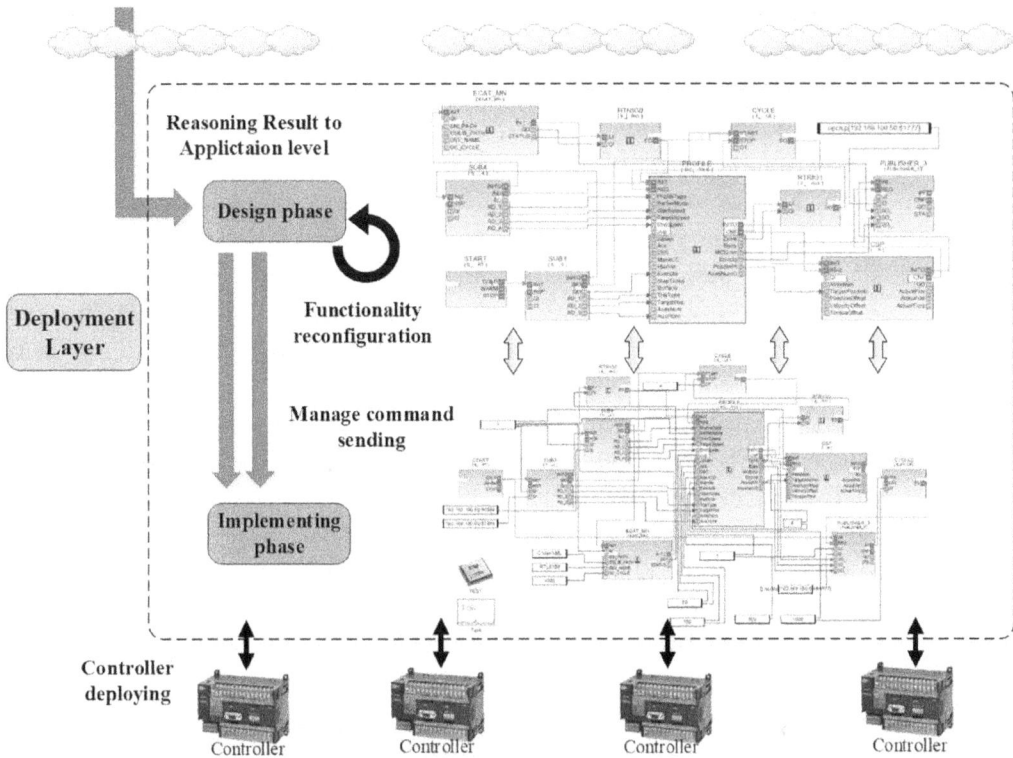

FIGURE 6.21   Two-phase IEC 61499 controlling in deployment layer.

modeling of the two phases can be modeled with the same function blocks edit environment. The upper phase called design phase is responsible for specific task scheduling according to pharmaceutical market's demand and implements user-defined functions with corresponding function block chains.

Generally, function blocks in design phase focus on abstract abilities, which contain several fundamental operations (Basic function blocks) and provide services of high reusability. In this paper, to automatically cope with uncertain pill-packing request, the operation sequences that are designed manually before are now accomplished by the knowledge base in the perception layer.

6.3.1.4.3 Implementation Phase of Deployment Layer   As for the implementation phase, one significant difference to the design phase is that the implementing layer is deployed on the run-time environment of corresponding controller and the manufacturing resource. The run-time environment is connected to pharmaceutical manufacturing resources, controllers (PLC, etc.), and it is typically with several fundamental components including operating system, several communication interfaces, and ability of computing as well as storage. Considering that the implementation phase within run-time environment is much closer to low-level devices, specific algorithms such as trajectory planning and devices' control law are supposed to be encapsulated in the function blocks within run-time environment.

In this paper, Raspberry Pi, the embedded tiny PC, is introduced as the run-time environment. In practice, the run-time environment within each pharmaceutical device has its own resource model for corresponding functionality, which maintains multiple function block models. Each function block model points to one function block chain, which contains a series of interconnected function blocks. Therefore, it is possible for different resource models to have identical execution semantics but vary in execution behaviors.

Under the implementation phase, the run-time environment (Raspberry Pi) undertakes the commands from the design phase as the input. With the execution semantics interpreted in the run-time environment, corresponding function block chain will be triggered. Based on that, controller connected to the run-time environment will accordingly drive the device for further pharmaceutical production.

In summary, the responsibility of the deployment layer is represented as follows:

When the deployment layer first undertakes coarse-grain medical production plan from the knowledge base, the design phase first analyzes it and schedules corresponding function blocks according to the operation plan. It should be noted that the scheduled functionality sequence in the design phase is not device dependent and is of more abstract. In this paper, one server is deployed to act as the container for the design phase. Furthermore, the design phase sends managing commands to the individual manufacturing resources, specifically each run-time environment by TCP/IP protocol. The run-time environment will interpret the command and trigger corresponding function block chain as the previous mentioned. Eventually, the controller linked to the run-time environment starts driving the devices to accomplish the medical production.

6.3.1.5 Reconfigurable Pharmaceutical Production

In this section, we will verify the proposed architecture with a scenario of reconfigurable pharmaceutical production. Specifically, in our experimental platform, there are seven kinds of medical pills existing. In our experiment, pharmaceutical market's trend is represented by medicine orders, which contain type of pills, type of medicine box, and numbers of the chosen type of pills, respectively.

6.3.1.5.1 Ontology Modeling of Reconfigurable Pharmaceutical Production   As the previous section mentioned, the ontology model used in the experiment refers to the MASON. Considering experimental circumstances and the architecture of MASON, the ontology model in the experiment is established in Figure 6.22. As shown in Figure 6.22, the right part is the class-level (Tbox) description of the ontology. Considering the architecture of MASON as well as the experimental reality, the concept of pharmaceutical machine (blue boxes) and pharmaceutical tool (green boxes) is introduced. Specifically, the classes of pharmaceutical machine are supposed to describe machines or electronic devices such as pill-packing agent, conveyor agent, and other general processing equipment, which are used to represent the manufacturing entities like MASON does. While the pharmaceutical tool is supposed to describe more fine-coarse status from components such as RFID reader and pneumatic cylinder. Referring to MASON, the connection between pharmaceutical machine and pharmaceutical tool is represented by several appropriate object properties

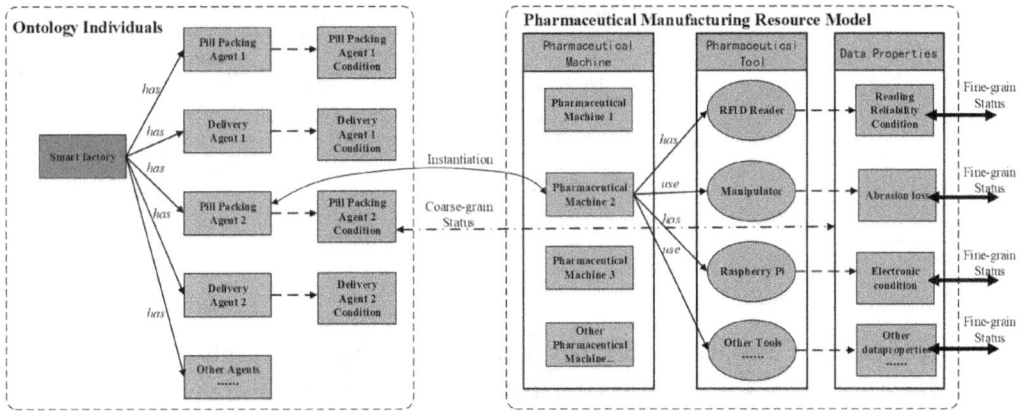

FIGURE 6.22 Ontology knowledge base of the experimental pharmaceutical manufacturing resources.

including *use* and *has*. To describe the attribution of the pharmaceutical tool, data properties are attached to corresponding pharmaceutical tool. For example, the abrasion loss of manipulator or electronic condition of embedded devices. The data properties can be Boolean, Integers, Double, and so on for appropriate description.

As for the left part of Figure 6.22, individuals (instances) of the right part are created as the member of Abox. Specially, the blue blocks of ontology individuals are the instantiation of the pharmaceutical machine in pharmaceutical manufacturing resource model, which are supposed to reflect the experimental resources. For the purpose of simplicity, we only show the instances of pharmaceutical machine, which embrace corresponding pharmaceutical tool as well as their data properties. When it comes to the reasoning procedure, the data properties attached to the pharmaceutical tool will be comprehensively valued for the pharmaceutical machine. The comprehensive valued result will be fully considered to determine whether the pharmaceutical machine is the candidate for a specific task. Therefore, a coarse-grain status will be encapsulated as the Boolean data property of each pharmaceutical machine's instance.

In summary, when a new pill-packing task arrives, the instance of pharmaceutical machine will reason out the coarse-grain status. Specifically, the coarse-grain status is decided by the several corresponding fine-grain statuses. For example, the coarse-grain data property of one pill-packing agent is updated to True (if Boolean type) only if all the low-level fine-grain data properties are ready (e.g., electronic condition of Raspberry Pi is good, abrasion loss of manipulator is lower than the threshold). Based on the update of coarse-grain and fine-grain data properties, the reasoner can decide whether the pharmaceutical machine can undertake the published task.

6.3.1.5.2 Capability of Pharmaceutical Manufacturing Resources and Experimental Platform   In our experiment, there are mainly four pill-packing agents (smart manipulators) and five delivery agents (smart conveyors) to cooperate with each other to accomplish the various

TABLE 6.4   Process Capacity of Four Pill-Packing Agents

Robot Index	Red	Orange	Yellow	Green	Blue	Pink	Purple	Run-Time Environment	RFID Reader
Robot 1	√			√			√	√	√
Robot 2	√	√			√			√	√
Robot 3			√	√			√	√	√
Robot 4		√			√	√		√	√

kinds of pill packing. The process capacity of the four pill-packing agents is shown in Table 6.4. As Table 6.4 shows, each smart manipulator can accomplish pill clipping for three different types. To make these machine resources smart, an embedded tiny computer called Raspberry Pi is connected to these manipulators together with sensors, controllers (PLC, etc.) as well as other devices, which acts as the run-time environment for the implement phases in the deployment layer. Equipped with Raspberry Pi, the pill-packing agents are capable of receiving managing commands from the design phase, communicating with connected data sources, and implementing controlling process to controllers (PLC, etc.). Eventually, to communicate with pill box attached with RFID tag, the pill-packing agents are armed with an RFID reader, which is responsible to obtain the procedures processed and will be processed from RFID tag attached to the pill box. Another significant agent is smart conveyors, called delivery agents in our experiment. The fundamental functionality of the delivery agent is to transport processing medicine within the pill box to the next pill-packing agent. Therefore, conveyor belt is required in the first place. Similar to the smart manipulator, Raspberry Pi as run-time environment, RFID reader, and PLC controllers are also equipped to delivery agent. When a new task arrives, RFID reader will verify the pill box. With functionality reconfiguration in the run-time environment, the PLC controller will eventually drive the conveyor belt for final execution. The experimental platform is shown in Figure 6.23.

As Figure 6.23 shows, the transport system is composed of five delivery agents. Each of these is made up of a conveyor, a PLC controller, a motor as the power source, an RFID reader to obtain the information from pill box, and finally a Raspberry Pi as the run-time environment as the brain. The RFID reader of the delivery agent is responsible for transporting path decision when a new pill box comes. With RFID tag read from pill box and process logic from the deployment layer, the controller of the delivery agent drives the conveyor to lead the pill box to appropriate pill-packing agent. In addition, four pill-packing agents are deployed in our experiment. Besides RFID reader, Raspberry Pi, which is equipped with a pill container for storing three kinds of pills, is also shown in Figure 6.23. One difference between pill-packing agent and delivery agent is the power source, where the power source of the former is pneumatic cylinder, while the latter is electric power.

In this experiment, the perception layer, which contains the knowledge base and responsible for production plan reasoning as well as big data analysis, is represented in four cloud servers, as shown in Figure 6.23. Another cloud server is deployed in the server rack for the design phase of the deployment layer.

FIGURE 6.23   Experimental platform of the proposed architecture.

Moreover, the statements of bottom machines and sensors as well as the medical orders from customers are stored in the cloud servers. Further statistics and machine learning application such as fault prediction of machines and customers' medial tendency can be analyzed.

6.3.1.5.3 Implementing Flow of Reconfigurable Pharmaceutical Production   The basic procedure of data-driven reconfiguration is illustrated in Figure 6.24. First, the monitoring module in the perception layer keeps waiting for new medical orders by circularly accessing the database with SQL queries. When satisfied, an instance of order will be generated and individuals of production machines will update their data properties synchronizing corresponding table of the database. The fine-grain status of machine resources is stored in database and the mapping relations are defined by *Ontop* platform [17]. Later on, the reasoner (Jena reasoner in the experiment) is responsible for reasoning the working condition of each pharmaceutical machine (the coarse-grain Boolean data property of machine resource's instance) which is updated again with SWRL rules. Specifically, the SWRL rules are defined to accomplish more complicated relations between classes in ontology. And the result after the reasoning processes will be deployed by SPARQL queries, and accordingly, the reconfiguration plans are inferred. Now that the perception layer on the cloud has accomplished its job, one server in the cloud servers' rack will undertake the operation plan from the cloud and schedule the function blocks. Corresponding agents with run-time

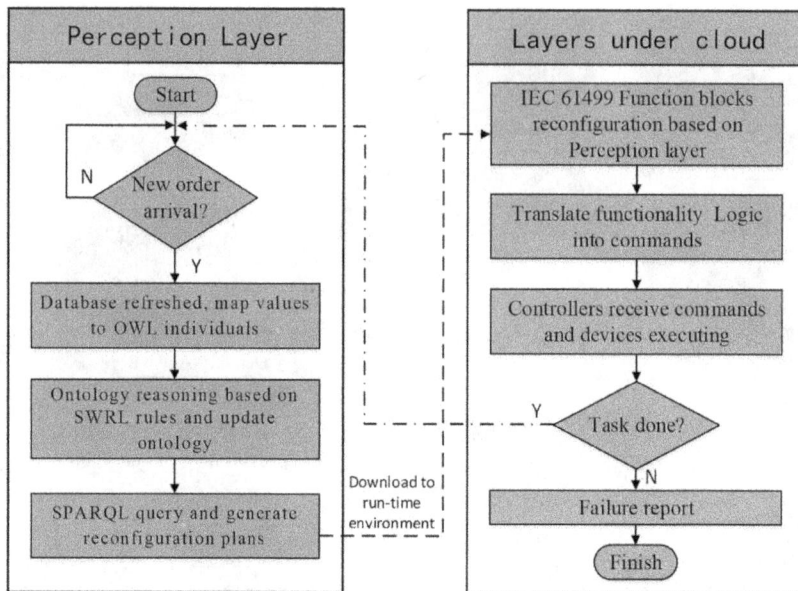

FIGURE 6.24   Execution procedure of reconfigurable pharmaceutical production.

environment (Raspberry Pi) will download the managing commands from the server before which are scheduled by generated reconfiguration plans. Based on that, the run-time environment translates the controlling logic to lower-level command which implements the controllers for device execution to further delivery or pill picking. Eventually, in case of operation failure, the agent would report the failure to the cloud.

6.3.1.5.4 Case Study   In this part, a case study of reconfigurable pharmaceutical production will be represented. With the workflow proposed in the previous parts, one medical production order of reconfigurable pharmaceutical production will be demonstrated.

In this part, we focus on executing sequence of relative agents to verify the effectiveness of this method. The medical production order is shown in Table 6.5 as follows. As Table 6.5 shows, the assumed order requires four types of pills: six red, four yellow, four blue, and five pink pills. For simplicity, in this part, we call pill-packing agent as PA and call delivering agent as DA.

When we come back to Figure 6.23, there are five delivering agents existing. Specifically, DA1 and DA5 are the main road and necessary for delivering to work station of any PA; DA2 is deployed through PA1 and PA2, which must be triggered if PA1 or PA2 is required; DA3 is deployed around PA3, while DA4 is responsible for delivering the pill box to PA4.

As the order received from the cloud, with reasoning procedure and two-phase deployment procedure accomplished, corresponding manufacturing resources are prepared for implementing. The specific workflow of the experimental platform for the assumed order is represented as shown in Table 6.6. The DA1 will first deliver the empty pill box from the warehouse. With the run-time environment (Raspberry Pi) deployed, sensor detecting, PLC of DA4 drives the conveyor to undertake the pill box to PA4. Having pink and

TABLE 6.5   Medical Request of Assumed Order

Color	Red	Yellow	Blue	Pink
Numbers	6	4	4	5

TABLE 6.6   Operation Orders of Reconfigurable Pharmaceutical Production Case Study

Operation Orders	PA1	PA2	PA3	PA4	DA1	DA2	DA3	DA4	DA5	Description
1					√					Pill box delivering from warehouse
2									√	Delivering to PA4
3				√						PA4 operation with Pink and Blue pills
4								√		Delivering for additional operation
5							√			Delivering to PA3
6			√							PA3 operation with Yellow pills
7								√		Delivering for additional operation
8						√				Delivering to PA2
9		√								PA2 operation with Red pills
...										Subsequent operations

blue candies added, DA5 and DA3 in sequence deliver the pill box to PA3. After operation (Yellow candies) of PA3, DA5 and DA2 deliver the pill box to PA2 for the last procedure, namely Red pill's operation. Table 6.6 shows the major operations for reconfigurable pharmaceutical production. Therefore, the remaining subsequent operations are omitted. The case study of a specific reconfigurable pharmaceutical production based on the experimental platform verifies the effectiveness of proposed architecture.

6.3.2 Data-Driven Reallocation of Workers in Engineering-to-Order Assembly Islands

Assembly islands with fixed-position layouts are typically found in factories assembling large, bulky, or heavy products, such as ships and planes, because they can provide the following advantages: reduced movement of work items, minimized damage or cost of movement, and more continuity of the assigned work force since the item does not go from one workstation to another [18]. A turbine valve is too large, bulky, and heavy to move. The total weight of a valve is close to 40 tons, and the structure is very complex and irregular. In order to move and assemble different components, both a bridge crane and an auxiliary scaffold are necessary. Considering safety, cost, and lead-time, the manager will decide whether the valve is to be placed on a fixed site and whether the workers and the required equipment will be moved to the assembly site, which is organized as an assembly island with a fixed-position layout.

With the development of industry 4.0, turbine valves for the power generation industry are typical engineer-to-order (ETO) products, which are manufactured and assembled in low volumes to satisfy the individual customer's specifications [19,20]. A turbine valve costing several million RMBs is a necessary component of a steam turbine, which is independently assembled and directly delivered to the customer's sites. Due to the large financial penalties for lateness, it is necessary to ensure that the assembly tasks of the valves are completed before the due dates, which are determined by higher level production plans, and are central to the contractual project agreement between companies and their customers [21,22]. However, the flow time of a valve assembly is long—more than 20 working days. During this time, there is an amount of uncertainty in executing the assembly schedule for valves, for example, the modification of the assembly process, equipment failure, quality problems, and delayed delivery of parts and sub-assemblies which prevent the execution of the existing schedule exactly as it was laid out. In order to meet the specified due dates and use human resources efficiently, a rescheduling procedure for the complex turbine valve assembly should be performed when unexpected events occur.

6.3.2.1 Overall Architecture

In this paper, we proposed the optimization model for the scheduling problem as shown in Figure 6.25, which aims to minimize the labor cost and reduce the disruptions of the heavy components and equipment supply plans. The industrial data and optimization objection are the preconditions which influence the model building. Turbine valve data is the input for the scheduling model.

The paper has focused on worker reallocation for the valve's assembly tasks to meet the specified due date, which determines the completion time of the valves for delivery. The objective was to optimize the reallocation of workers to tasks when an unexpected event occurs, to minimize the labor cost and the sum of the absolute differences between the start times of the remaining tasks in the modified schedule and the original start times of those tasks, which reduce the significant disruptions of upstream operations.

6.3.2.2 Solving the Problem

6.3.2.2.1 Problem Formulation   The turbine valve is typically capital goods and is highly customized to meet individual customer requirements. In order to reduce the throughput time, the early production planning is required to be defined before information on product customizations, and detailed production activities are completely disclosed. The influence of uncertainty or incomplete data on production systems performance can be severe, and the flexible scheduling is necessary to achieve the high agility [23]. In a turbine valve assembly plant, the sole bridge crane is definitely the bottleneck resource, which is required to move all kinds of heavy components and equipment. Because of the limited space of the turbine valve assembly island and the required short throughput time, the buffer is unavailable. In this environment, changes to the sequence of the assembly tasks can disrupt the materials and equipment logistic plan of the bridge crane, which can cause significant and frequent variation at other workstations. For a short time, the foreman cannot modify the sequence of assembly tasks and instead focuses on shortening the processing

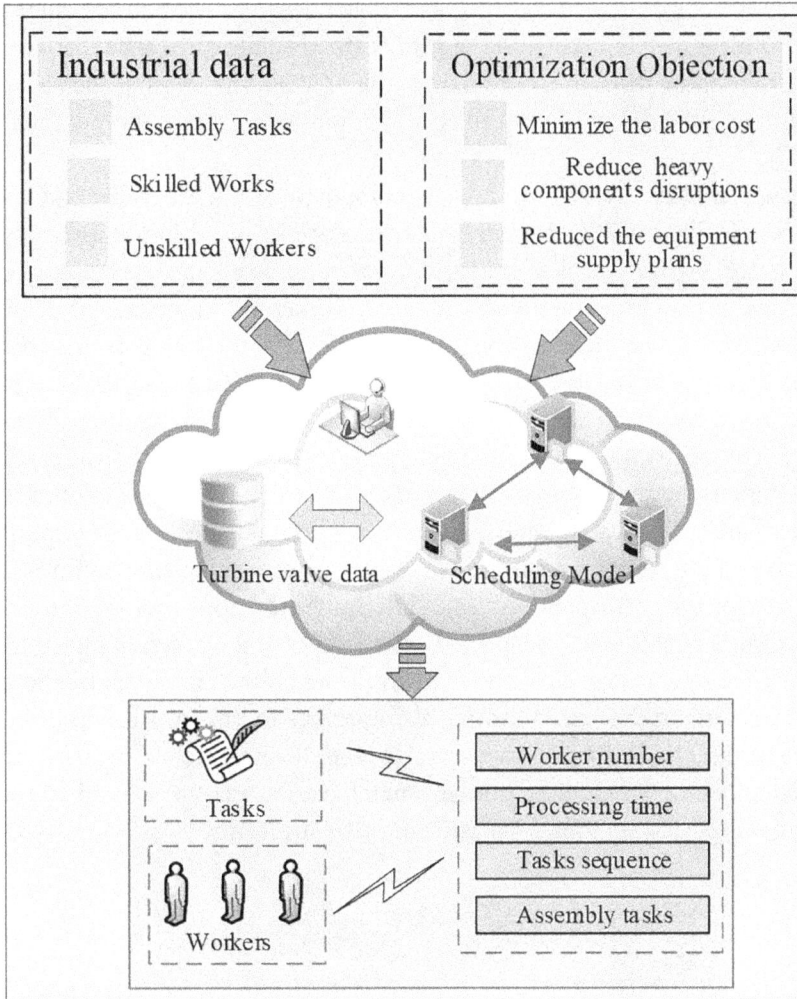

FIGURE 6.25 System architecture of the assembly tasks for turbine valve.

times of tasks by reallocating workers in the turbine valve assembly island. Consequently, the focus is on flexible workforce, which can be more easily adjusted than the heavy components and equipment [24].

The complicated tasks of the turbine valve assembly need the specific knowledge and the accumulated experience. Data mining is crucial to discovered knowledge [25]. Because of the high variation of the work content, the learning rate is very low. That is to say, training a new worker is difficult. In the turbine valve assembly island, there are two types of workers, unskilled and skilled workers.

In this work, we consider the case in which an original schedule has been generated for a given set of assembly tasks of the turbine valve. The schedule has been partially executed when an unexpected event occurs and delays the completion time of the turbine valve assembly under the tight constraint that tardiness is unacceptable. Then, the workers assigned to the succeeding tasks are reallocated to reduce the processing time of the tasks and ensure that the turbine valve can be assembled before the given due date. The objective is to minimize the labor cost and the sum of the absolute differences between the start times of the remaining tasks in the modified schedule and the original start times of those task, which reduce the disruptions of the heavy components and equipment supply plans.

6.3.2.2.2 Industrial Data   A turbine valve is composed of a main valve and a governing valve as shown in Figure 6.26; 24 indivisible tasks are involved in the assembly island of the turbine valve. Figure 6.27 is a real turbine valve. Table 6.7 has presented the assembly tasks, the initial processing times with the initial worker allocation, and the initial schedule. Since the sequence is unchangeable, for simplification, the task index has been used to describe the sequence of the assembly tasks, which means that the tasks are performed in order of the task index, from task 1 to task 24. Some large and heavy valve parts can be usually assembled by more than one worker, which determined the lower bound on the number of assigned workers, because it is difficult for one person alone to lift and orient components. On the other hand, considering safety, an auxiliary scaffold is necessary to perform some assembly tasks on the turbine valve due to its large size and complex structure. The limited available space for the corresponding tasks determines the upper bound of the number of the assigned workers. After an unexpected event occurs, the workers assigned to each task can be adjusted in the interval determined by the lower bound and upper bound mentioned above. When changing over to a new job, a worker has to put away any previous tooling, clean up the workplace, study the drawings and other documentation for the new job, fetch tools, and so on, all of which take time. A tangible variable cost is caused by the workers' reluctance to move between jobs, especially for just a short time [26].

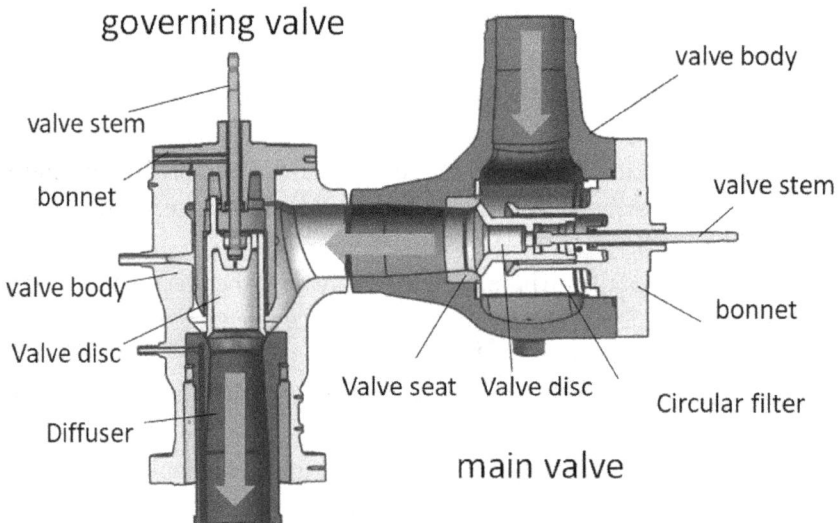

FIGURE 6.26   Components of turbine valve.

FIGURE 6.27   Assembly Island of a turbine valve.

TABLE 6.7   Data for the Problem

Task	Number of Workers		Changing Cost	Initial Assignment		Initial Processing Time	Initial Starting Time
	Lower Bound	Upper Bound		Type A	Type B		
1	2	4	10	2	1	12	0
2	1	3	15	1	1	23	12
3	1	3	15	0	1	31	35
4	1	3	10	0	1	31	66
5	2	4	10	3	0	10	97
6	2	2	0	2	0	10	107
7	2	3	30	1	1	33	117
8	2	3	15	1	1	25	150
9	2	2	0	2	2	10	175
10	2	4	20	1	1	30	185
11	2	3	20	1	1	27	215
12	2	4	20	1	1	24	242
13	2	3	225	1	1	36	266
14	2	4	40	2	1	45	302
15	3	4	15	2	1	15	347
16	2	2	0	2	0	10	362
17	2	3	20	1	1	30	372
18	2	3	20	1	1	23	402
19	2	3	35	1	1	47	425
20	2	4	20	1	1	24	472
21	2	3	25	1	1	31	496
22	2	4	40	2	1	42	527
23	1	4	15	2	1	49	569
24	1	4	10	3	0	55	618

In the turbine valve assembly island, there are two types of workers, type A (unskilled) and type B (skilled). The processing time of the assembly task is dependent on the skill level and the number of the assigned workers. Of course, some tasks are mainly carried out using mechanical equipment, and their processing times are independent of the skill and the number of workers. Table 6.8 has presented the processing time of each task with different worker assignments. It's obvious that the skilled workers take less work time to complete work. Thereby, the skill level is an important factor influencing the processing time of the assembly task. The optimized scheme for assembly tasks means to minimize the labor cost, reduce the disruptions of the heavy components and equipment supply plans.

6.3.2.2.3 Integer Programming Model   The integer programming formulation has been proposed for the worker reallocation problem in the assembly island of a turbine valve when unforeseen disruptions occur. The proposed model has been developed under the following assumptions:

TABLE 6.8   Processing Time of the Tasks with Different Worker Assignments

Task	Different Worker Assignments									
	1B	2A	1A1B	2B	3A	2A1B	1A2B	4A	3A1B	2A2B
1	-	15	15	14	12	12	12	10	9	9
2	45	24	23	23	20	19	18	-	-	-
3	31	19	19	18	15	15	14	-	-	-
4	31	20	20	19	16	16	15	-	-	-
5	-	12	12	11	10	10	9	8	7	7
6	-	10	-	-	-	-	-	-	-	-
7*	-	-	33	32	-	27	26	-	-	-
8*	-	-	25	24	-	21	20	-	-	-
9	-	10	-	-	-	-	-	-	-	-
10*	-	-	30	28	-	24	24	-	19	19
11*	-	-	27	26	-	22	21	-	-	-
12*	-	-	24	24	-	22	21	-	17	16
13*	-	-	36	34	-	30	29	-	-	-
14*	-	-	54	51	-	45	43	-	36	35
15	-	-	-	-	16	15	14	13	13	12
16	-	10	-	-	-	-	-	-	-	-
17*	-	-	30	28	-	25	24	-	-	-
18*	-	-	23	22	-	19	18	-	-	-
19*	-	-	47	45	-	39	37	-	-	-
20*	-	-	24	22	-	20	19	-	15	14
21*	-	-	31	30	-	26	25	-	-	-
22*	-	-	49	47	-	42	40	-	32	30
23	77	61	58	55	51	49	47	44	41	39
24	80	67	62	59	55	52	49	46	43	41

* The task is done by skilled workers.

- The workers allocated to the task cannot be changed before the completion of the task.

- Parallel working is not allowed because of the limited space around the valve.

- The sequence of tasks is unchangeable.

- The running task is executed until completion except for unforeseen interruption. In the case of an unexpected interruption, the part is assumed to be reassembled, and the new task processing time is redefined according to the modified work content and worker reallocation.

After defining the major symbols presented in Table 6.9, the objective function for the worker reallocation problem has been presented, followed by the corresponding problem constraints:

$$Minimizing : F_1 = \sum_{j=1}^{J}\sum_{w=1}^{W}C_w \cdot p_j^r \cdot X_{jw}^r + \sum_{j=1}^{J}R_j \cdot \sum_{w=1}^{W}\left|X_{jw}^0 - X_{jw}^r\right|. \tag{6.1}$$

TABLE 6.9   Principal Symbols

Symbol	Description
D	Due date of the turbine valve
J	Number of assembly tasks
i, j	Task index
W	Number of workers
WU_j	Upper bound of the workers assigned to task j, because of the limited space
LU_j	Lower bound of the workers assigned to task j, because of the limited space
w	Worker index
p_j^0	Initial processing time of task j
p_j^r	Processing time of task j after worker reallocation
S_j^0	Initial start time of task j
S_j^r	Start time of task j after worker reallocation
C_w	Labor cost per unit time of worker w
R_j	Cost of worker reallocation for task j
x_{jw}^0	0–1 variable of initial worker assignment. $x_{jw}^0 = 1$ means worker j is assigned to task w; otherwise, $x_{jw}^0 = 0$
x_{jw}^r	0–1 variable of initial worker assignment. $x_{jw}^r = 1$ means worker j is assigned to task w; otherwise, $x_{jw}^r = 0$

$$Minimizing : F_2 = \sum_{j=1}^{J} \left| S_j^r - S_j^0 \right|.$$

(6.2)

$$S_j^r + p_j^r \leq D, \quad \forall j$$

(6.3)

$$\begin{cases} S_j^r \geq S_i^r + p_i^r \\ S_i^r \geq S_j^r + p_j^r, \quad \forall i \neq j \end{cases}$$

(6.4)

$$S_j^r - S_i^r \geq P_i^r, \ \forall i, \ j \in \left\{ i, j \mid S_j^0 - S_i^0 \geq P_i^0 \right\}$$

(6.5)

$$WL_j \leq \sum_{w=1}^{W} X_{jw}^r \leq WU_j, \quad \forall j$$

(6.6)

$$p_j^r f = \left(X_{j1}^r, X_{j2}^r, \ldots, X_{jW}^r \right)$$

(6.7)

The objective function (6.1) serves to minimize the worker cost. In equation (6.1), the first term is the total labor cost and the second is the variable worker cost for the assembly tasks. Equation (6.2) serves to minimize the deviation between the start times of the remaining tasks in the modified schedule and the original start times of those tasks, which avoids any disruptions being amplified in the upstream operations. The constraints (6.3) ensure that the due date of the valve assembly can be respected. The constraints (6.4) ensure that only one task can be implemented at the same time, because the available space around the valve is limited. The constraints (6.5) mean that the sequence of assembly tasks cannot be changed during reallocation of the workers. The constraints (6.6) ensure that the number of the assigned workers for the assembly task cannot exceed the upper bound of the number of workers. Equation (6.7) indicates that the processing time of the assembly task is dependent on the worker allocation described in Table 6.8.

6.3.2.2.4 Experiment Results   The BP formulated was solved using LINGO 11.0 on a PC with a 3.6 GHz Core i7 CPU with 16GB RAM. The initial task schedule and the worker allocation, after preprocessing of the data, have been presented in Table 6.7, and the processing times of each task with different worker allocations have been presented in Table 6.8. There are a total of four type A workers and two type B workers in the turbine valve's assembly island. Assuming that every worker is always available during the schedule execution, the labor cost per unit time of workers A and B is 1 and 1.3, respectively. In the real scenario, the valve assembly process is organized and controlled by a foreman that schedules the assembly tasks and allocates workers to tasks. Furthermore, the foreman is responsible for problem solving when unexpected events occur.

Given the due date 674 and the fixed sequence of the valve's assembly tasks listed in Table 6.7, four real scenarios involving unexpected events have been tested to demonstrate the proposed approach.

In the first real scenario, it took more than 20 units of time to perform task 4, because a quality problem occurred. Since the due date was very tight and rigorous, the foreman immediately decided to reallocate the workers of tasks 7, 8, and 10, which ensured that the valve assembly could be completed before the given due date. The goal was to minimize the worker cost described in equation (6.1). The cost that needs to be addressed is exclusively the cost of the increased number of workers, since the cost of the assigned workers is fixed. Table 6.10 has presented the worker reassignments and the cost of the increased workers, including the increased labor cost and the variable worker cost. Only the tasks involving worker re-assignment have been presented in Table 6.10. The total reduction in the processing time of tasks 7, 8, and 10 is 21 units, which successfully compensated for the extra 20 units of processing time of task 4. The increased worker cost was a total of 102.7. Table 6.11 has presented the results computed by the proposed algorithm. The processing time of tasks 20 and 24 was reduced by 22 units, which ensured that the valve assembly could be completed before the fixed due date. The corresponding increased worker cost was 59.1. The optimal solution obtained by the proposed algorithm was a clear improvement on the solution implemented by the foreman: the worker cost was reduced by 43.6.

The comparison of the cost of the worker reallocation in the other real scenarios, including the assembly process modification, the logistical problem, and equipment failure, has been shown in Table 6.12. On average, there was a 36.8% reduction in the worker reallocation cost for the real valve assembly. The main reason for this was that the proposed algorithm could investigate a larger solution space than the foreman, who focused on the immediately following tasks in the fixed sequence when some unexpected events occurred.

TABLE 6.10   Worker Reallocation Cost for the Foreman

Task	Initial Assignment	Initial Processing Time	Re-assignment	Processing Time after Re-assignment	Increased Labor Cost	Changing Cost
7	1A1B	33	2A1B	27	13.2	30
8	1A1B	25	2A1B	21	11.8	15
10	1A1B	30	3A1B	19	12.7	20

TABLE 6.11   Worker Reallocation Cost for the Foreman

Task	Initial Assignment	Initial Processing Time	Re-assignment	Processing Time after Re-assignment	Increased Labor Cost	Changing Cost
20	1A1B	24	2A2B	14	9.2	20
24	3A	55	3A1B	43	19.9	10

TABLE 6.12   Comparison of the Proposed Algorithm with Foreman's Methods

| | | Worker Reallocation Cost | | | |
		Foreman	Proposed Algorithm	Cost Reduction	Percentage Change
Test	Unexpected Event				
1	Quality problem of task 4	102.7	59.1	43.6	42.4%
2	Process modification of task 7	127.4	75.3	52.1	40.9%
3	Logistical problem of task 9	153.2	95.5	57.7	37.7%
4	Equipment failure of task 11	176.7	130.4	46.3	26.2%
Average				49.9	36.8%

6.4 CLOUD SERVICE AND ITS APPLICATIONS

With the rapid development of AI technology, intelligent manufacturing is gradually moving to the cloud. As a new computing and service model, cloud computing technology plays a vital role in the development of intelligent manufacturing. In intelligent manufacturing, cloud computing is not only a supporting technology but also an important platform, which provides powerful data analysis and decision support capabilities for intelligent manufacturing and also brings many advantages to intelligent manufacturing. First, cloud computing can provide efficient, reliable, and secure computing and storage services. The cloud computing platform is highly scalable and elastic, and it can automatically allocate computing and storage resources according to user needs to ensure the high availability and efficiency of intelligent manufacturing systems. In addition, the cloud computing platform can also provide a high degree of data security and confidentiality, effectively protect various confidential information and key data involved in intelligent manufacturing, and reduce the risk of information leakage and data loss. Second, cloud computing can provide big data analysis and intelligent decision support for intelligent manufacturing. All kinds of data generated in intelligent manufacturing, including sensor data, equipment data, production data, and quality data, are massive and diverse. The cloud computing platform can effectively integrate and analyze these data through big data analysis technology, extract valuable information and knowledge, and provide intelligent decision support. In this way, the optimization and upgrading of the intelligent manufacturing system can be realized, and the production efficiency and product quality can be improved. Finally, cloud computing can realize the collaboration and sharing of intelligent manufacturing systems. Smart manufacturing involves collaboration among multiple departments and enterprises, and it requires the sharing and interaction of various data and information. The cloud computing platform can provide unified data storage and management, and realize data sharing and collaborative work among various systems. In addition, cloud computing can also support remote operation and monitoring, and realize remote collaborative work of intelligent manufacturing systems.

In summary, cloud computing plays an important role and significance in intelligent manufacturing. Through the application of cloud computing technology, intelligent manufacturing can realize efficient, intelligent, safe, collaborative, and shared production modes, promote the resource reconstruction and optimization of factories, and provide strong support for the sustainable development of enterprises. This chapter first introduces the application prospect of cloud in manufacturing. Then, it describes how to help customized manufacturing factories reconfigure and optimize resources based on the cloud. Finally, through the experimental platform of the personalized candy packaging production line and the intelligent control system of the conveyor belt, the important components and core technologies of the customized manufacturing factory are explained. Thus, the significance of the development of customized manufacturing factories is discussed.

6.4.1 Cloud-Based Smart Manufacturing Service in Personalized Candy Packing

Manufacturing industry keeps evolving to catch up with increasingly demanding consumption tendency. After three revolutionary promotions, today's automated machines and production lines can produce plenty of products with high quality and low cost at a superb rate. However, the production line is so rigid that it can produce only one type of products. And worse still, reconfiguring the old production line or building a new one is expensive and time-consuming. Therefore, it is suffering for manufacturers to cope with consumers' rapidly changing appetite. That is why we desire smart manufacturing systems to support personalized consumption demands by providing multi-type and small-lot (even one item) customized products with rapid response and without reducing quality or increasing expenditure.

The implementation of smart manufacturing depends on the integration of production process and organizations. Based on the philosophy of CPS, Industry 4.0 classifies three kinds of integration, i.e., horizontal integration through value networks, vertical integration and networked manufacturing systems, and end-to-end digital integration of engineering across the entire value chain. Smart production line and smart factory address the vertical integration. It leverages network, cloud computing, mobile cloud computing, big data, and so forth to implement deep integration of information technology and automation technology, along with the AI to enable self-organized reconfiguration. As a result, smart factory will exhibit high flexibility, high efficiency, and high transparency, making it appropriate to cope with personalized consumption demands.

Many existing technologies can be used to build smart manufacturing system. The previous research on flexible manufacturing system (FMS) [27,28], MAS [29,30], and other advanced manufacturing schemes provides us with the ability to construct a fully automated production line. Cloud computing services are already available from public market, such as Amazon Elastic Cloud Computing (EC2) [31]. Hadoop has almost grown up into a mature solution for building big data platform and has been used by some famous large business companies, e.g., Facebook, Twitter, and Yahoo, to manage mass data property. As the important infrastructure, the high-bandwidth network both in the wired and wireless forms becomes very usual.

Despite the aforementioned technical advancement, we can seldom see smart factories in the real manufacturing environment. It still needs framework, algorithms, and standards to wrap these technical components up. In this paper, we present a prototype system for packing assorted candy, rapidly responding to personalized demands.

6.4.1.1 Cloud-based System Framework

A unique packet of selected candies is usually a perfect gift for whom we care. At present, we should either take time to pack one ourselves or choose one that we are not quite satisfied with from market. Smart factory that can quickly pack customized candies is a promising solution to this kind of contradiction. In this section, we describe the imaginary process from submitting an order to getting the packet and explore the related supporting technologies, such as cloud platform, big data solution, physical configuration, and communication network.

6.4.1.1.1 Life Cycle Description   Figure 6.28 depicts the overall process on how the smart factory serves the personalized demands relying on a networked world. A consumer can access an online shopping Web site with a personal computer or smart mobile phone, and by the Web page, he can customize his packet in terms of:

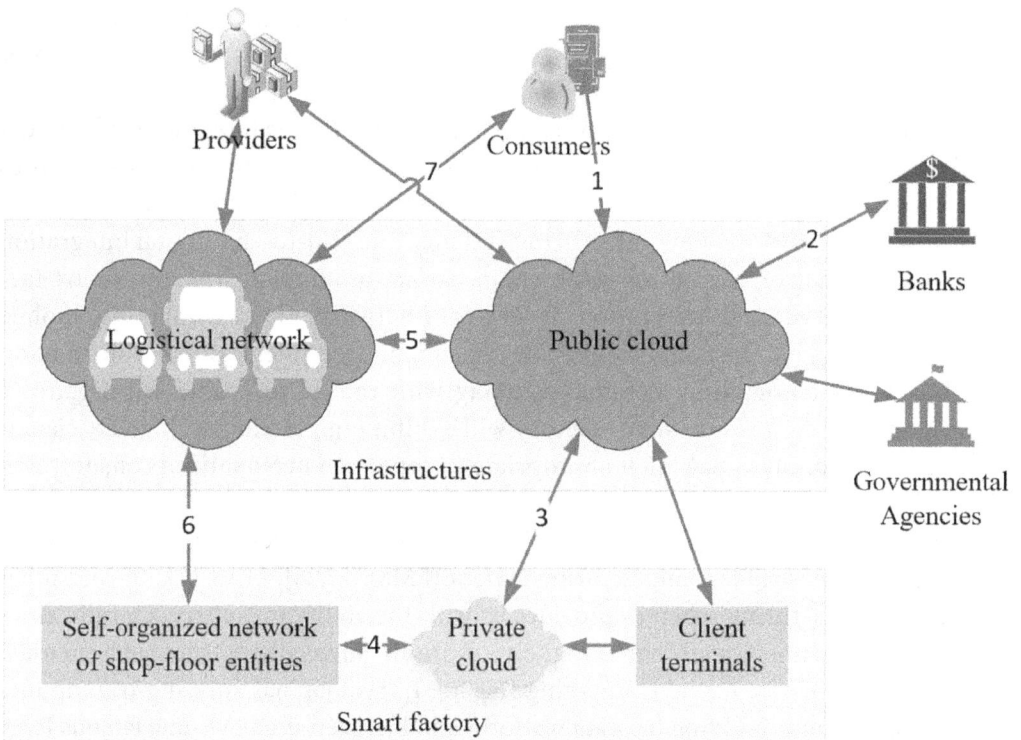

FIGURE 6.28   Life cycle for ordering the customized product.

- Box type,

- lid type,

- card type, e.g., birthday card, valentine card, and Christmas card,

- candy types, and

- the number of candies of each type.

After the consumer makes his final decision, he submits the order (line 1). However, only when he really pays the bill (line 2), his order is really converted and submitted to the shop-floor production line (lines 3 and 4). For our illustrative application, the order is encoded as the data structure shown in Figure 6.29. We use four bits each to encode box type, lid type, card type, and candy type, which means the selectable number of these types is 15. The maximum number of candies of each type is also 15. The sum of candy types chosen by the consumer is recoded in the four "Num candy types" bits followed by "Candy type x" and "Num candy type x" as a type-number pair to represent each selected candy type and its associated quantity. A group of state bits are appended in the end. Each bit corresponds to an operation for a type-number pair, and value "0" means the operation has not been finished, whereas value "1" means the operation is done.

In the shop floor, a box is selected and the data structure is written into the RFID tag that is attached to the box with all the state bits initialized to zero. Thereafter, the box is put onto a conveyer, and before that, the state bit corresponding to the "Box type" is set to 1. During the travel of the box, machines, conveyers, and the box itself negotiate with each other to organize needed resources dynamically. When the packet is completed, a deliver order will be submitted to a shipper (line 5) who is responsible for delivering the customized candy packet to the consumer through the logistical network (line 6 and line 7). Based on the uniform networked environment, information of both production process and logistical routing is transferred to the public cloud in real-time. Therefore, it is easy for the consumer to track the state change of his order.

6.4.1.1.2 Private Cloud and Industrial Big Data Platform   Within smart factory, we leverage cloud computing and Hadoop to build an elastic, reliable, and available platform to bear industrial big data and to facilitate interaction. We use the Citrix XenServer software to set up ten CentOS virtual machines (VMs) above five physical servers, i.e., two VMs operate on one physical server. As shown in Figure 6.30, five VMs are used to construct Hadoop cluster. VM acts as the master node to run NameNode, ResourceManager, and HMaster which are master part of HDFS, YARN, and HBase, respectively. Three VMs act

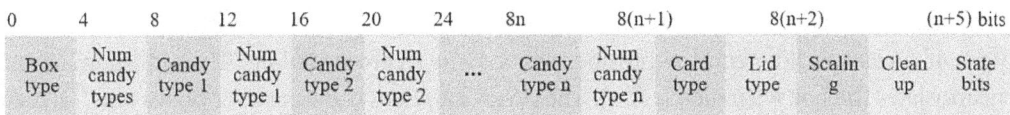

0	4	8	12	16	20	24	8n	8(n+1)	8(n+2)	(n+5) bits
Box type	Num candy types	Candy type 1	Num candy type 1	Candy type 2	Num candy type 2	...	Candy type n	Num candy type n	Card type	Lid type	Scaling	Clean up	State bits

FIGURE 6.29   Data structure for encoding user order.

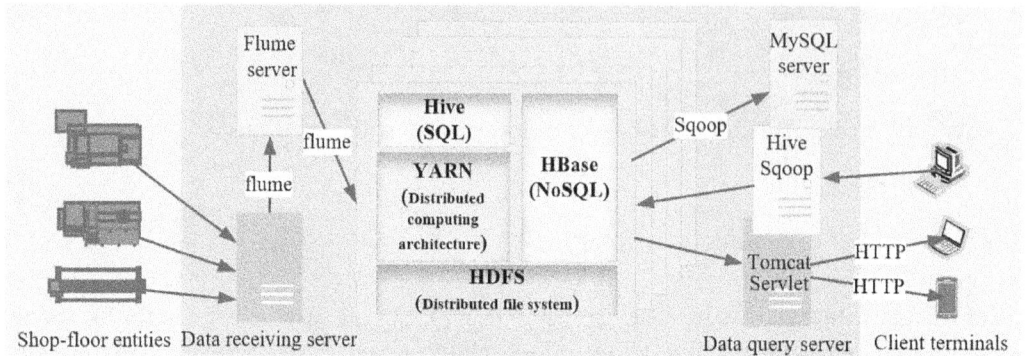

FIGURE 6.30 Hadoop-based big data solution on cloud platform.

as the slave nodes to run DataNode, NodeManager, and HRegionServer which are slave part of HDFS, YARN, and HBase, respectively. QuorumPeerMain, i.e., the ZooKeeper process, is also deployed in all the slave nodes to assist HBase. The last one VM hosts the SecondaryNameNode to work with the NameNode.

There are five VMs used as servers. To interact with shop-floor entities, data receiving server is used to accept data from physical entities, and the flume tool deployed in the flume server forwards the data to HDFS and HBase. To interact with clients, a Web server consisting of Tomcat and Java Servlet is built as data query server. Hive is a client tool that can be used with Hadoop to support SQL-like batch process, thus simplifying user programming. Hive is really deployed in a VM outside the Hadoop cluster to ensure safety by preventing non-qualified clients from directly accessing the cluster. One relational database system, i.e., MySQL, is deployed in the MySQL server to maintain the metadata of Hive. Because the Sqoop is also a client tool for exchanging data between Hadoop and MySQL, it is also deployed in the same VM used by Hive.

6.4.1.1.3 Physical Configuration Machine, conveyer, and product are three characters in shop floor for processing, transporting, and organizing production, respectively. In our prototype system (Figure 6.31), machines include four robotic arms, one sealing machine, and one automatic storage and retrieval system (ASRS). Five conveyor belts are conveyers that construct the cyclic conveying system with multiple branches. The box represents the smart product equipped with an RFID tag.

Robotic arms 1 and 2 are responsible for putting candies into the box. Three containers are assigned to each robotic arm, so this system can process six types of candies in maximum. While this is enough for an illustrative example, both containers and robotic arms can be added to the system to expand the number of candy types. Robotic arms 3 and 4 are used to load cards and lids, respectively. Our experimental setup supports three types of cards and lids, and the number can also be easily scaled up. The sealing machine encapsulates the box with plastic film. The ASRS manages the storage of boxes and finished packets. Initially, the ASRS selects a box, writes into order data, and loads the box onto the conveyor belt. Finally, it fetches the sealed packet and put it onto the shelf.

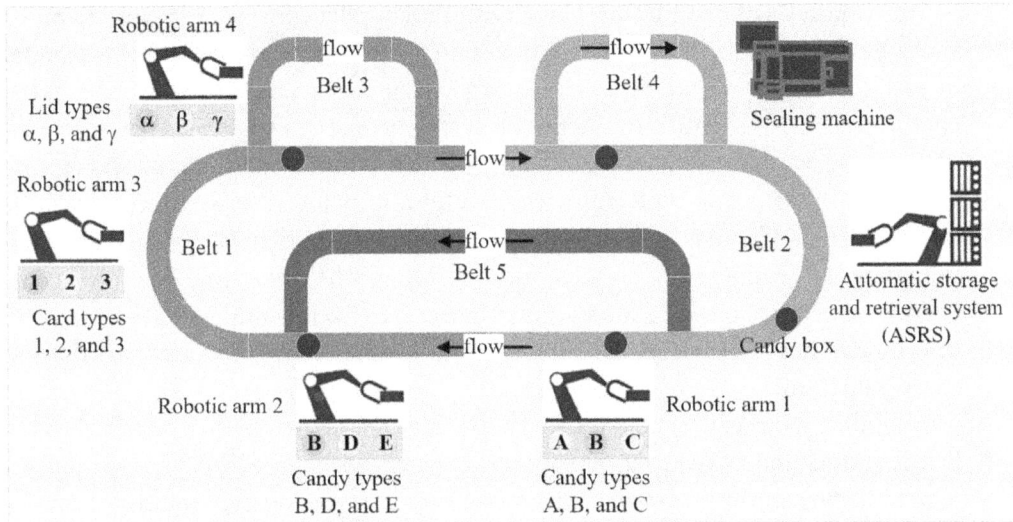

FIGURE 6.31   Physical configuration of shop-floor manufacturing environment.

Feature analysis of machines: We have four multi-functional machines, i.e., robotic arms 1–4. By multi-functional machine, we mean it can process multiple types. However, the sealing machine and the ASRS are considered as single-functional machines, even though they can process all the types of boxes. This is because, when doing sealing or storing operations, the sealing machine and the ASRS do not need to distinguish the types of boxes. Functional redundancy exists between robotic arms 1 and 2, as they both can process candies of type B.

Feature analysis of conveyors: We have a closed-loop conveying system other than an open line. This feature means much for dynamic reconfiguration, as the box that passes by a machine can be routed back for unlimited times. Another feature is that the conveying system has branches, probably resulting in multiple routes from a source to a destination.

Feature analysis of products: The RFID device makes the box a reduced-intelligence product, as it is not able to compute and negotiate with machines and conveyor belts by itself. By contrast, the embedded controller can bring a box with full intelligence, i.e., the box can conduct its production process, as it has computing, communication, and storing capabilities. The product can also be classified according to its operation sequence. For example, a packet needs three types of candies, ten of each, and the needed operation sequence is expressed as [1.A(5), 2.B(10), 3.A(5), and 4.C(10)]. The square brackets mean the operations should be performed orderly, so five candies of type A are put into box before ten candies of type B which is followed again by five candies of type A, and at last, ten candies of type C are processed. As both operation 1 and operation 3 process the type A candy, type A is defined as multi-occurrence (operation) type in this example, while the types B and C are defined as single-occurrence (operation) type.

Feature analysis of production paradigm: We support consumers to select among five types of candies, three types of boxes, lids, and cards, along with the number of each candy type. Therefore, a rather large variety of packets of assorted candies can be customized. The manufacturing system processes multiple types of products in two ways, alternative production versus hybrid production. Alternative production means one type is processed after another, and at any time, only one type of products exists in the system. By contrast, in hybrid production, multiple types of products can be seen in the system at the same time, and types may change from time to time. Considering the personalized candy packets with small or even one-item batch size, hybrid production will be more efficient than alternative production.

In summary, our experimental smart manufacturing system is characterized by the hybrid production of reduced-intelligence products with closed-loop conveying system. Moreover, multi-functional machines, functional redundancy, multi-occurrence operation types, and multiple conveying branches are in existence.

6.4.1.1.4 Network Solution   Three kinds of network technologies are adopted in the prototype system. The RFID, a kind of near-field communication technologies, is used to connect boxes with machines. Each box is equipped with a RFID tag, while each machine is equipped with an RFID reader–writer. When the box passes by a machine, the box and the machine can interact with each other through the RFID equipment. Machines use Ethernet to communicate with each other and interact with the cloud. The five conveyor belts use the wireless network products from Taiwan MOXA corp. for inter-communication and interact with the cloud.

6.4.1.2 Interaction around Cloud Platform within Smart Factory

In this section, we further describe the interaction process among shop-floor entities, cloud, and clients within the factory boundary that is previously shown in Figure 6.30. The interaction is around cloud. Therefore, we divide the interaction process into two parts: one is between cloud and clients, and the other between cloud and shop-floor entities. As a result, clients can fetch information from or apply some interventions to shop-floor entities through cloud.

6.4.1.2.1 Interaction between Cloud and Clients   As shown in Figure 6.28, people can link to our private cloud locally or remotely through the Internet, with some client terminals, such as personal computers and smart mobile phones. At present, we support clients to access cloud through Web page with a Web browser. As this is a general and popular communication method, it will not burden the clients with extra software. At the server end, we use Tomcat and Servlet to build the Web server. The main tasks are making Web pages using HTML, JSP, and JavaScript like programming languages. Specially developed APP is also applicable, which can implement complex functions while it causes some inconvenience concerning software downloading, installation, and use.

6.4.1.2.2 *Interaction between Cloud and Shop-Floor Entities* While the clients interact with the data query server through the standard HTTP protocol, the shop-floor entities interact with the data receiving server using the customized data formats. We have designed three types of interaction patterns with distinct datagram definition:

1. The cloud inquires shop-floor entities about some parameters. The shop-floor entities return the corresponding values to cloud after receiving the inquiry. This interaction occasion is due to the cloud. Table 6.13 depicts the datagram format.

2. The cloud subscribes to some events that will occur occasionally in shop floor. The related data are delivered to the cloud when these events occur. This kind of interaction is triggered by events, and the datagram format is illustrated in Table 6.14.

3. The cloud subscribes to some parameters, and shop-floor entities return the corresponding values periodically, as shown in Table 6.15. This interaction thus takes place in regular intervals.

6.4.1.3 Self-organized Mechanisms to Enable Dynamic Reconfiguration of Shop-Floor Entities

When a box is put onto the conveyor belt, it has a sequence of operations needed to be processed. We do not use a central controller to pre-allocate resources for the box. Instead, we desire robotic arms, conveyor belts, and boxes to negotiate with each other to reconfigure

TABLE 6.13   Inquiry and Response Datagram Format

Inquiry (0x01)	Object ID (16 bits)	Object type (8 bits)	Parameter ID (16 bits)	Timestamp (48 bits)	CRC16 (16 bits)	
Response (0x11)	Object ID (16 bits)	Object type (8 bits)	Parameter ID (16 bits)	Timestamp (48 bits)	Data (associated with Parameter ID)	CRC16 (16 bits)

TABLE 6.14   Event Subscription and Notification Datagram Format

Event subscription (0x02)	Object ID (16 bits)	Object type (8 bits)	Event ID (16 bits) Event ID (16 bits)	Timestamp (48 bits)	CRC16 (16 bits)	
Event notification (0x12)	Object ID (16 bits)	Object type (8 bits)	Event ID (16 bits)	Timestamp (48 bits)	Data (associated with Parameter ID)	CRC16 (16 bits)

TABLE 6.15   Parameter Subscription and Notification Datagram Format

Parameter subscription (0x03)	Object ID (16 bits)	Object type (8 bits)	Parameter ID (16 bits)	Timestamp (48 bits)	CRC16 (16 bits)	
Parameter notification (0x13)	Object ID (16 bits)	Object type (8 bits)	Parameter ID (16 bits)	Timestamp (48 bits)	Data (associated with Parameter ID)	CRC16 (16 bits)

in a dynamic, real-time, and online way. As the RFID tagged product can only store its information, it relies on machines or conveyers to negotiate. In this section, we describe the negotiation mechanisms to determine the machine and route (consisting of conveyers) for each operation. It is noted that the negotiation occurs in the operation level to deliver high flexibility, i.e., when an operation is completed, negotiation takes place again for the next operation. This design is very likely to determine a different set of machines and conveyers even for the products of the same type. Therefore, newly added devices can join system in time, and fault devices can be bypassed easily, exhibiting high robustness.

6.4.1.3.1 Negotiation Mechanism to Determine a Machine for an Operation   Suppose that a consumer customizes a packet expressed as [α, D(2), A(2), B(2), C(2), 1, β, Sealing, Cleanup], where α, 1, and β are box type, card type, and lid type, respectively; D, A, B, and C represent candy types along with the digits indicating the number of candies of each type; Sealing and Cleanup mean the sealing and the cleanup operation that are automatically appended to the end of the sequence. Certainly, this operation sequence will be encoded based on the data structure shown in Figure 6.29 as {0x1, 0x4, 0x42, 0x12, 0x22, 0x32, 0x1, 0x2, 0x1, 0x1, 0x0000}, where the prefix 0x is the hexadecimal notation, so that 0x1 means binary number "0001" and 0x42 means "01000010". The last number "0x00" are state bits, where the most significant bit (bit 7) indicates the state of the box operation, the bits 6–0 correspond to operations D, A, B, C, 1, β, and Sealing, respectively, and the bit 7 of the next byte corresponds to the cleanup operation.

The ASRS receives the data structure first. It selects a box corresponding to the code 0x1, changes the state bits to 0x8000, writes the data structure to the attached RFID tag, and, finally, puts the box onto the nearby conveyor belt. The box flows with the conveyor belt, and the electro-optic sensor associated with the robotic arm 1 will detect the box. The robotic arm 1 will request conveyor belt 2 to stop, and then, its RFID reader will read the box's data structure. Thereafter, the robotic arm 1 acts as the manager on behalf of the box and the other machines (including itself) serve as the contractors. The manager and the contractors negotiate with each other to determine a machine for loading two candies of type D. As shown in Figure 6.32, the negotiation process is designed based on contract net protocol, involving the following steps:

- **Announcing**: The manager announces the D(2) to all the machines, including robotic arms 1–4, sealing machine, and ASRS.

- **Contractors' decision-making**: Every contractor compares the requirements with its abilities. If it can load type D candy, if it has enough quantity, if it works well, and if it is idle, it will bid for the operation.

- **Bidding**: In this example, only the robotic arm 2 will bid for the operation.

- **Manager's decision-making**: After a predefined period of time expires, the manager analyzes the bidding results. If no one bids for the operation, the manager will restart the negotiation. If only one contractor bids, the manager will choose it as the

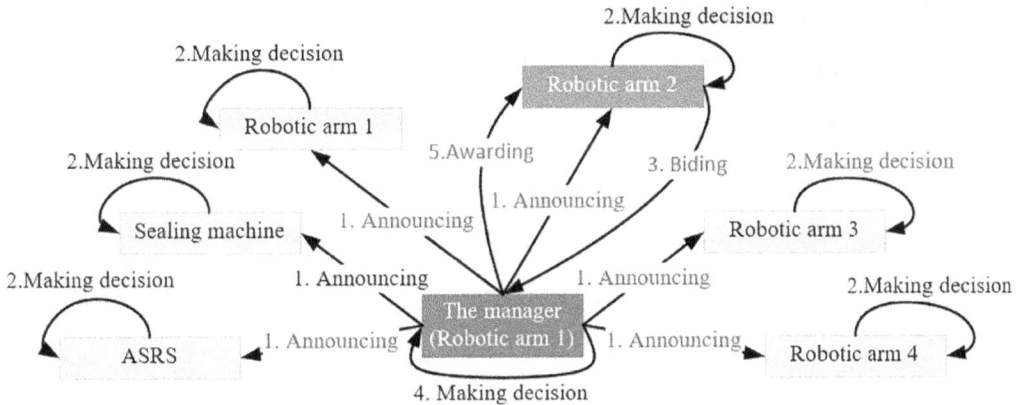

FIGURE 6.32   Negotiation process to determine a machine.

winner. If multiple contractors bid, the manager should select one as the winner. In our design, we use prioritized multiple rules, i.e., a group of rules are selected and each is assigned to different priorities. Rules with higher priorities are compared before those with lower priorities until one contractor wins out.

- **Awarding**: In this example, the manager should award the D(2) operation to the robotic arm 2, as it is the winning bidder for the operation.

6.4.1.3.2 Negotiation Mechanism to Determine a Route for an Operation   Now, a route should be determined to transport the box from robotic arm 1 to the robotic arm 2. The manager is still the robotic arm 1, but the contractors are the five conveyor belts instead. The prerequisites are that every conveyor belt knows which conveyor belts and machines it can reach successively following the flow direction, so does every machine. This is configured before system startup. The accessibility sequences of the five conveyor belts are:

- **Belt 1**: [Robotic arm 2, Robotic arm 3, Belt 3, Belt 2],
- **Belt 2**: [Belt 4, ASRS, Belt 5, Robotic arm 1, Belt 1],
- **Belt 3**: [Robotic arm 4, Belt 1],
- **Belt 4**: [Sealing machine, Belt 2], and
- **Belt 5**: [Belt 1].

The accessibility sequences of the six machines are rather simple, as each machine can interact with only one conveyor belt:

- **ASRS**: [Belt 2],
- **Robotic arm 1**: [Belt 2],

- **Robotic arm 2**: [Belt 1],

- **Robotic arm 3**: [Belt 1],

- **Robotic arm 4**: [Belt 3], and

- **Sealing machine**: [Belt 4].

The negotiation process to determine conveying route may involve several rounds. In each round, the manager interacts with one contractor. If the contractor can transport the box to the destination, a route is determined. Otherwise, the negotiation proceeds. Figure 6.33 indicates that the negotiation process to determine the route from the robotic arm 1 (both the source and manager) to the robotic arm 2 (destination) consists of two rounds, and a round consists of three steps.

Round 1:

- **Querying (line 1)**: The manager, i.e., the robotic arm 1, checks its accessibility sequence, and it is aware that the conveyor belt 2 is the only available contractor. Therefore, the manager asks the belt 2 if it is able to transport the box to the destination.

- **Reporting (line 2)**: On receiving the query message, the belt 2 checks its accessibility sequence. The result suggests that it is not able to transport the box to the destination. The belt 2 is aware that only the belt 1 can possibly transport the box to the destination. Note that conveyor belt is unidirectional, so the available contractors are those behind the source (robotic arm 1). Therefore, the belt 2 reports belt 1 to the manager.

- **Concluding**: On receiving the reporting message, the manager realizes that the negotiation should proceed.

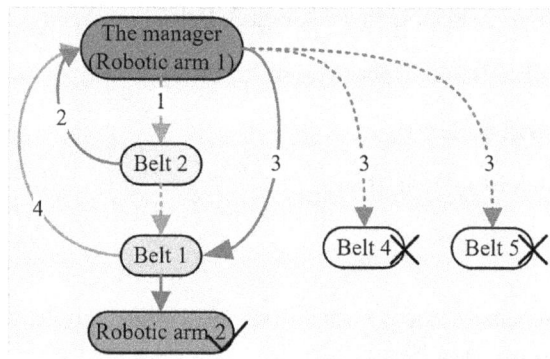

FIGURE 6.33   Negotiation process to determine a route.

Round 2:

- **Querying (line 3):** The manager asks the belt 1 if it is able to transport the box to the destination.

- **Reporting (line 4):** On receiving the query message, the belt 1 checks its accessibility sequence. The result suggests that it is able to do the job, so it reports the destination (the robotic arm 2) to the manager.

- **Concluding:** The manager constructs a route "robotic arm 1→ belt 2→bet 1→ robotic arm 2," and the negotiation ends.

6.4.1.3.3 Brief Summary of Negotiation By negotiation, machines agree that the robotic arm 2 is responsible for processing the operation D(2), and the chain of "belt 2→bet 1" is responsible for transporting the box from the source (the robotic arm 1) to the destination (the robotic arm 2). When the robotic arm 2 completes the operation, it will take over the role of manager to conduct negotiation for the next operation, i.e., A(2). Therefore, the machines and the routes for the packet may look like Figure 6.34. The robotic arm 1 processes D(2), the robotic arm 2 processes A(2), B(2), and C(2), the robotic arms 3 and 4 are responsible for the card and lid, the sealing machine devotes to sealing operation, and finally, the packet returns back to the ASRS.

It indicates that the robotic arm 1 wins the operation B against the robotic arm 2 for the first box. However, this is not always the case. As more boxes enter the system, the robotic arm 2 may acquire chance to perform the operation, especially when the robotic arm 1 is busy at that time. This is due to the negotiation mechanism, which says that a machine only bids for the operation when it is capable and idle, and the manager selects the winner

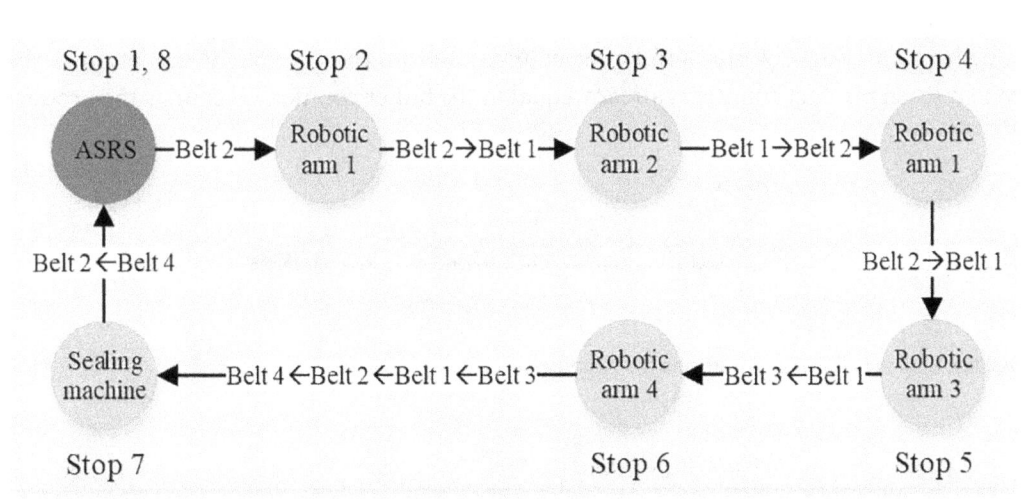

FIGURE 6.34 Resource sequence for the illustrative order.

from the bidders that is with higher precision and that is faster and nearer. As robotic arms 1 and 2 have equal speed and precision, the nearer robotic arm 1 will win when they both are idle.

6.4.1.3.4 Deadlock Free Conditions   As in the computer science, multiple types of packets concurrently existing in the system and competing for machines and conveyers will cause deadlocks. It is concluded that there are four necessary conditions that must hold simultaneously for deadlocks to occur, i.e., mutual exclusion condition, hold and wait condition, no-preemptive condition, and circular wait condition. Therefore, deadlocks can be prevented if we can break one or more necessary conditions. Considering the underlying properties of the physical shop-floor entities, it is the hold and wait condition that is easier to be broken. This is because machines are mutually exclusive and no-preemptive, circular wait can occur due to the loops caused by multi-functional machines, multi-occurrence operation types, or hybrid production. To break the hold and wait condition, buffers are necessary to enable a machine to release a box. In addition, flow control is also necessary to avoid saturating the buffers. This is summarized as Theorem 1 (Figure 6.35).

In other words, if we can make sure that the number of products does not exceed the number of buffers, the system is deadlock free. This is indeed intuitive because there are enough empty buffers for every machine to put back its finished product. The theorem also indicates that more products can enter the system by increasing buffers. Therefore, we have a simple deduction, as shown in Figure 6.36. With more products in the system, the vacancy rate of machine tools can be reduced. Moreover, machines can easily find buffers nearby, which helps to further increase the efficiency by reducing waiting time.

The buffers can either be trays or pallets running on conveyor belts, or carrying room provided by conveyers, such as AGV. Compared with machines for processing or testing, the buffers are much cheaper. Therefore, utilizing buffers to prevent deadlocks is a practical and economic solution. During system run-time, two counters should be maintained, i.e., product counter and buffer counter, and no more products are let enter the system when product counter is already equal to the buffer counter. Of course, the product

// Theorem 1:

The sufficient conditions to prevent deadlocks:
1) the least number of buffers equals to one; and
2) the sum of products in machines and buffers is less than or equal to the number of buffers.

FIGURE 6.35   Sufficient conditions for deadlock free.

// Theorem 2:

Increasing buffers contributes to higher efficiency.

FIGURE 6.36   Relationship between buffer quantity and production efficiency.

counter should be updated when finished products exit from the system, i.e., going back to the ASRS. The buffer counter should also be updated, but in practical, its value is rather invariable.

6.4.2 Mobile Services for Customization Manufacturing Systems

Given the new competition environment and technical background, modern-day global manufacturing industries and information technologies are deepening their degrees of integration with the support of IoT, industrial wireless networks, big data, cloud computing, mobile computing and embedded technology. This progress also brings increased hope for new applications, such as production customization and product life cycle management. Specifically, Industry 4.0 was proposed and launched in 2013. The integration of manufacturing technology, digital technology, and network technology can now be applied to the design-production-management-service. In addition, the manufacturing process now possesses the features (e.g., perception, analysis, decision, and control) needed to meet the product requirements of dynamic response, as well as the rapid development of new products.

In addition, with the continuous improvements in the level of products' lifespans, the personalized demands related to products are becoming more and more obvious. Customers are no longer the passive buyers of manufacturing process. Instead, consumers have become the possible designers, who wish to participate in the customization of their goods prior to purchase. As such, there is a need to meet the social element of consumer demands by developing flexible production methods which will meet the individual needs of multiple customers. However, at this present time, the traditional industry production methods (which follow the prescribed order and passive mode in the supply chain) can no longer meet the social aspect of manufacturing development requirements. In addition, an information barrier always exists between manufacturing enterprises and market supply chains. In this context, the proposed Industry 4.0 includes two major themes: (1) a smart factory and (2) intelligent production. Concretely, the machine groups will self-organize, and the supply chain will automatically coordinate.

To address the problems listed above and based on the theory and application of cloud computing and mobile computing, combined with the existing industrial control technology, this paper designs a Personalized Customization Manufacturing System (PCMS) with the ability of mobile services. Our PCMS adopts a cloud platform as an information processing means to form a flexible production mechanism. Today's smart mobile phones are adopted as a mobile terminal through which consumers can connect to the cloud platform. A PCMS exemplified by a customizable candy production system will be implemented. In our view, our proposed manufacturing production model conforms to the Industry 4.0 concepts, and this system is a representative example of Industry 4.0.

6.4.2.1 System Architecture

In this chapter, the system architecture design is shown in Figure 6.37. The structure of the system is divided into three layers: the manufacturing device layer, cloud service system layer, and mobile service layer. Each layer of the system depends on corresponding

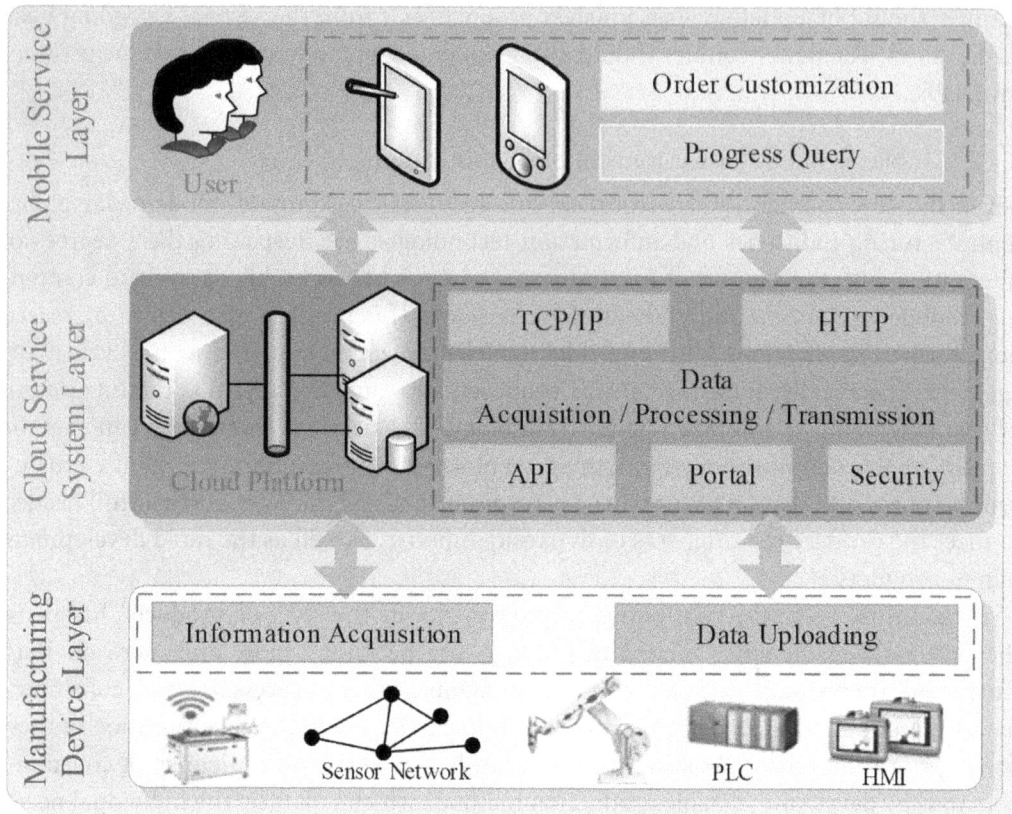

FIGURE 6.37   PCMS system architecture.

communication technologies and protocols to communicate with the other layers and coordinate production activities. According to the definition of Manufacturing Execution Systems (EMS) based on ISA95 standard, and the intelligent production pattern of Industry 4.0, in this chapter, the PCMS (depending on the information system) is capable of convenient order customization, as well as the optimization management of the whole production process. The PCMS can collect the relevant current data for the corresponding production instructions and processing, thus enabling the plant to achieve coordinated management of intelligent production.

6.4.2.1.1 Architecture Analysis   According to the PCMS system architecture, the function definition of the three system layers is as follows.

6.4.2.1.1.1 *Manufacturing Device Layer*   In this layer, a software-defined sensor network is used to collect resource information. The embedded module has assistant nodes for information conversion, and thus, the many devices of this layer can be associated with the cloud platform through this module. The intelligent robot can ensure the working process' characteristics of real-time and accuracy. Therefore, by using the above devices, an intelligent robot can automatically receive the cloud's decision to implement multi-aspect production activities, simultaneously improving the flexibility and efficiency of workshop production.

6.4.2.1.1.2 Cloud Service System Layer   This layer must qualify the CPS key characteristics. The cloud service system using cloud computing technology provides the APIs of data exchange, data storage, and data analysis to the heterogeneous resources of industrial IoT. In addition, the system provides web services as a means to receive personalized order information and send production management information. The cloud service system can fulfill information integration and resource sharing requirements. Therefore, the cloud service system can serve as the information hub center for the PCMS.

6.4.2.1.1.3 Mobile Service Layer   This layer is utilized to provide a personalized customization service and dynamic production process monitoring capabilities to consumers. To operate effectively, the mobile service layer needs a smart mobile terminal, which is used to access the cloud via the mobile Internet. The mobile services carry out remote operation so as to meet the intelligent matching of the factory's production resources. In order to achieve an effective personalized-oriented service for the client, this layer must provide portable user-friendly and reliable interaction methods.

6.4.2.1.2 Key Technologies   The key technologies involved in the PCMS can be broadly divided into the following: (1) collaborative technology of manufacturing unit, (2) network resources information processing technology, (3) cloud data processing technology, and (4) client terminal application development. By effectively combining the above technologies, a PCMS can break through the traditional manufacturing system's production mode limitations of information occlusion and single-line production style. Rather, a PCMS embodies the characteristics of intelligence, flexibility, and personalized-oriented service.

6.4.2.1.2.1 Collaborative Technology of Manufacturing Unit   This technology is mainly used to solve two aspects of the manufacturing device layer, namely, the object connection and the system reconstruction.

With regard to object connection, sensor network and RFID technology are used to realize the calibration and tracking of the processing object's position information. Industrial communication technology is used to provide real-time interaction of the processing object data information being supplied to the intelligent manufacturing units in the workshop. Together, the technology and information combine to achieve mutual coordination in the production process. By this means, the interoperability problem caused by resource heterogeneity in the workshop can be solved.

With regard to system reconstruction, a modular package is carried out, according to the functionality of the intelligent manufacturing cell. The cloud service system is used to make production planning decisions. By means of the above method, the intelligent manufacturing unit of the manufacturing system can automatically adjust the working mode and therefore quickly respond to customer's individual demands for products of various styles and types.

6.4.2.1.2.2 Network Resources Information Processing Technology   A PCMS consists of different working properties and various types of equipment (such as smart mobile phones and industrial robots) and networks (such as industrial networks and mobile networks) and so on. Thus, the network resource information processing technology must be able to facilitate the network resources' accessing and the format of data exchange.

From the aspect of accessing network resources, in the process of the information exchange between the customer terminal and the production workshop, the data exchange problem caused by the information heterogeneity of different equipment and the network is considered. Therefore, the web service technology is introduced to achieve the necessary cross network integrated application. A Uniform Resource Locator (URL) is used to locate different data resource processing centers in the cloud service systems. Therefore, different devices in the network can use the cloud service system's URL to access the corresponding data center and interact based on that data through the HTTP protocol.

With regard to the data exchange format, resource information interaction is mainly conducted through web services, in order to achieve the JSON data format description and output. We do this because this format is compressed, has low bandwidth occupancy, and is easily parsed. Therefore, every device in the PCMS depends on the JSON data parsing to obtain other devices' production information and/or to obtain the production planning decisions from the cloud service system.

6.4.2.1.2.3 Cloud Data Processing Technology   A PCMS needs to solve the problem of information occlusion inherent in traditional industrial manufacturing systems. Therefore, in this paper, the cloud service system adopts the Platform as a Service (PaaS) form as the PCMS's network information exchange hub. The database technology and the data mining technology are used for data storage and analysis. The API key is used to provide access permissions for data reading or writing operations between the cloud platform and the mobile terminal device. The cloud service system provides data channels for information communication between terminal objects. By this means, we may collect the data generated by equipment and send the relevant messages to the mobile terminal.

6.4.2.1.2.4 Client Terminal Application Development   Due to the fact that customer demand for information is becoming more and more diverse and detailed, a PCMS needs to be able to break through the traditional single-batch customer order interaction pattern. Namely, the client base now not only has wholesalers but also has ordinary end-user customers. Therefore, a convenient means of interaction and an appropriate communication platform must be considered. However, the PCMS makes a good choice in this regard by relying on the interactive mode of the mobile Internet and the portable mobile platform to provide the most direct personalized consumer demand information and interaction, whenever and wherever possible.

In order to achieve the requisite PCMS mobile services, we must carry out the development of mobile application software. Currently, the mobile application software development system can be used for mobile terminal APP development, such as Android and IOS. With relation to the demand of the enterprise's product supply and application environment, those same mobile application software development systems can be flexibly selected and adapted to develop a terminal APP.

6.4.2.2 Implementation Mechanism

A PCMS is a multi-closed-loop information system. As shown in Figure 6.37, the implementation mechanism of the system (according to message response objects) is divided into the cloud platform service mechanism, client access mechanism, and manufacturing system operation mechanism.

6.4.2.2.1 Cloud Platform Service Mechanism   Duo to the absolute need to ensure the safety and flexibility of data in any industry, every terminal in the system network uses information flow with built-in access rights to interact with the cloud service system. The cloud platform service mechanism mainly includes (1) client-oriented service mechanisms and (2) service mechanisms for the manufacturing system.

6.4.2.2.1.1 *Client-Oriented Service Mechanism*   The cloud service system will automatically allocate data channels and related API keys to registered customers to ensure the safety of those customers' data operations. The information loop is composed of the cloud service system layer and the client service layer, as shown in Figure 6.37. As such, and to deal with different information content of different customers' orders, the cloud service system will be based on a unified data storage format (as shown in Figure 6.38). The cloud service system will also process consumers' Write API Key related information, to-do aggregation, classification, modification, and other preprocessing tasks. With regard to customer information inquiries, the cloud service system will be based on consumers' Read API Key related information. Then, the related product processing information will be sent to the customer terminal in a JSON data format.

6.4.2.2.1.2 *Manufacturing System-Oriented Service Mechanism*   The cloud service system provides a production decision mechanism and production information access mechanism for the manufacturing system. The information loop is composed of the cloud service system layer and the manufacturing device layer, as shown in Figure 6.37. According to the consumers' order sequence and product style, the cloud service system will assign the

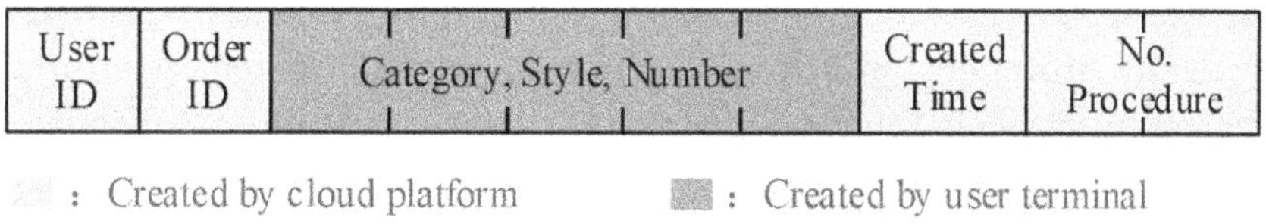

User ID	Order ID	Category, Style, Number	Created Time	No. Procedure

▨ : Created by cloud platform   ▮ : Created by user terminal

FIGURE 6.38   Data storage format.

appropriate production tasks and production modes to the manufacturing system. The manufacturing system uses information conversion assistant nodes to parse JSON data and obtain the production instructions. The manufacturing system then executes the production task. In the manufacturing system, distributed sensor nodes are used to perceive the location node information relating to the products in the manufacturing process. At the same time, the information conversion assistant node encapsulates the data and uploads that data to the cloud platform.

6.4.2.2.2 Client Access Mechanism   Enterprises (based on their own scope of business) provide customers with a dedicated APP. The customer can use this APP to register their related information, so as to obtain an API key to the cloud service system for their APP.

Customer terminals submit personalized demands via the selected mobile services. In this process, a terminal device must use WiFi or a 3G/4G signal to connect to the network and access the corresponding Internet cloud service system through the HTTP protocol. Accordingly, this type of application APP should have the corresponding order customization mechanisms and inquiry mechanisms to ensure operational safety.

The PCMS order customization function not only needs to provide users with an easily accessed and operated terminal interface, but also needs to provide the relevant information from the submitting mechanism to ensure the reliability of order delivery. The order customization information is shown in Figure 6.39. The APP excludes the consumer misoperation problem by judging whether or not the sent message is empty. In the process of message transmission, the APP needs to judge whether or not any abnormal network phenomenon exists or occurs. In addition, the APP also needs to provide users with intuitive information of user submission results, in order to ensure that successful orders have been submitted.

In order to make plans to sell products to customers easier, the mobile terminal APP needs to have a mechanism allowing the APP to exchange production information with the cloud platform, as shown in Figure 6.40. The APP also needs to provide a mechanism for detecting network anomalies, as well as being able to achieve data conversion and calculation mechanism performance, so that customers intuitively know the relevant information pertaining to the production progress.

6.4.2.2.3 Manufacturing System Operation Mechanism   In this paper, the workshop execution structure model of an intelligent manufacturing system is shown in Figure 6.41. This model is composed of a circular production line with several processing branches, a robot and a corresponding sensor. In order to create an intelligent and flexible manufacturing system, the design information of this model is as follows: (1) We assume that the production line has a total of $n(n \in N, n > 0)$ processing branches. This $B(i)\left(i \in [1,n]\right)$ stands for the ith processing branch. (2) Different colored blocks represent different processing degrees of semi-finished products. Specifically, the $P(k)$ color block indicates that a semi-finished product has been completed to the kth procedure. (3) When a semi-finished product is placed at BEGIN, this indicates that the semi-finished product is about to join this production process. Removing a semi-finished product from the END means that the

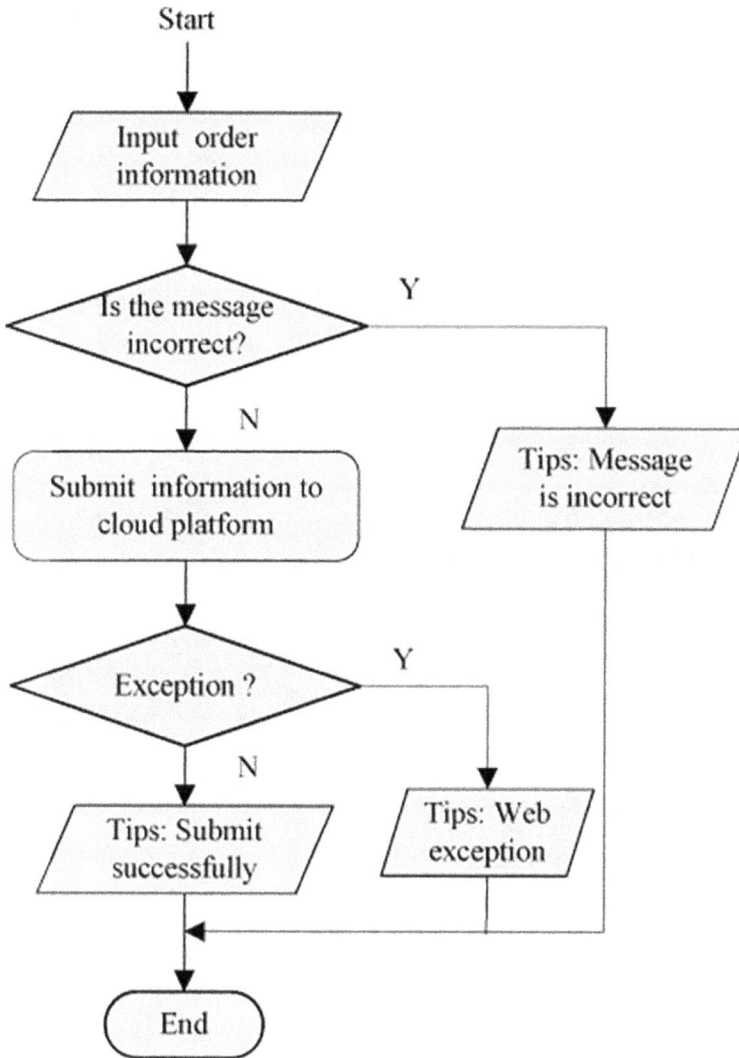

FIGURE 6.39 Order customization information.

process tasks of the semi-finished product have been completed. If a semi-finished product fails to finish a processing task, the product will continue on to the next cycle process in the production line.

In the production process, RFID tags are used to record the procedure information of semi-finished products on the processing branch. An RFID reader and a diverter are set at every intersection entrance of the circular production line and processing branch. The RFID reader will record the semi-finished products' procedure information and send this information to the manufacturing unit system of the processing branch. This information is then used to judge whether or not to adapt this processing branch to process. If the judgment result is yes, the semi-finished product will be pushed into this branch. Otherwise, the product will go to the next branch. Once the processing procedure of a semi-finished

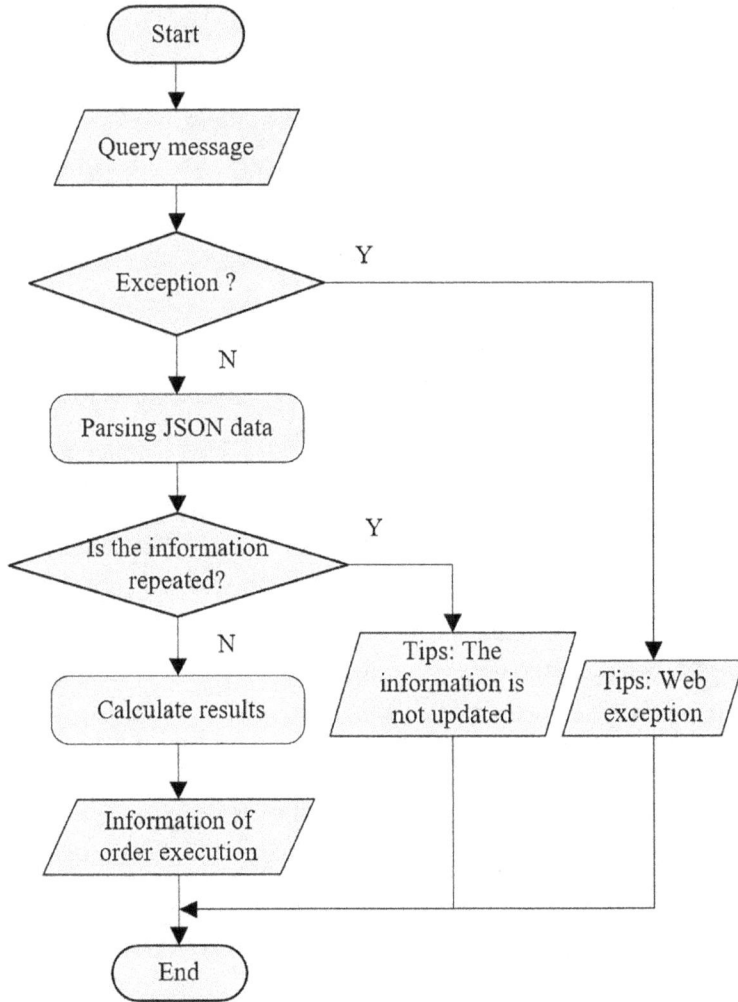

FIGURE 6.40 Procedure of status inquiry.

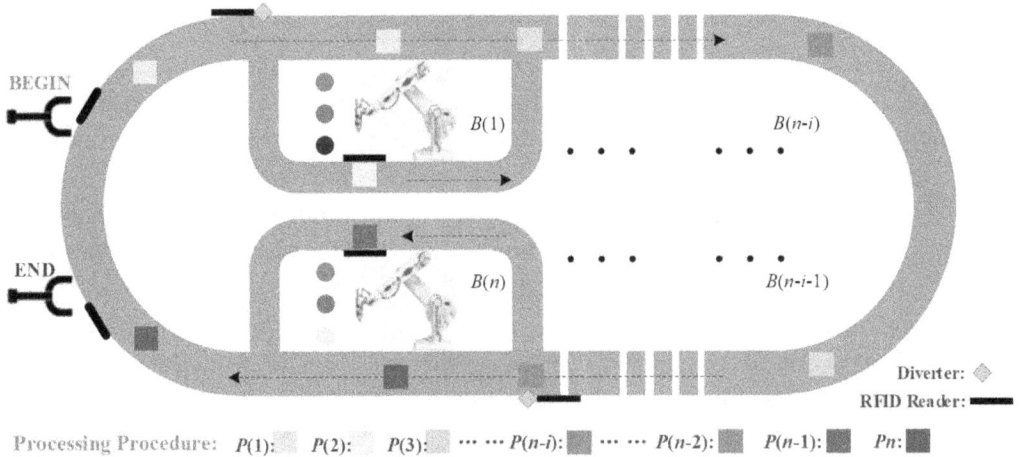

FIGURE 6.41 Production workshop sketch.

product is accomplished, the corresponding information will be sent to the cloud platform and written into each semi-finished product's RFID tag.

The processing branch adopts a processing style redundancy design method. This method means every processing branch can bear multiple types of parts. In addition, all of the processing styles in a processing branch can be replaced with a processing style, or combination of styles, used by other processing branches. This way, the processing branch can reduce the task amount of processing branches during other branches' idle time. The processing branch can also solve the substitution problem after the faults of other processing branches appear.

Assuming that every processing branch that can bear the quantity of the part style is three, the production line allows that the quantity of parts which can be used for processing is $m(m>0)$. Here, $Pa(z)$ stands for the style of the zth part, and the **Part** can be used to represent the set of parts on the production line, as follows:

$$\textbf{Part} = \{Pa(1),\ldots,Pa(m-i),\ldots,Pa(m)\}. \tag{6.8}$$

$Pb(i)$ stands for the set of a part's style where it is on the processing branch of $B(i)$,

$$Pb(i) = \{Pa(\alpha),Pa(\beta),Pa(\gamma)\},(z,\alpha,\beta,\gamma\in[1,m]); \\ Pb(i)\subseteq Part, i\in[1,n]. \tag{6.9}$$

According to the design of the processing style redundancy, the **Part** $=$ $Pb(1)$ $\bigcup Pb(2)\bigcup\cdots\bigcup Pb(n)$ is proved.

The production line can produce v kinds of products, and the part's style set of the jth kind of product is $C(j)$:

$$C(j) = \{Pa(\alpha),\ldots,Pa(\beta),\ldots\},\text{and}$$

$$C(j)\subseteq Part,\text{ and } C(j)\bigcap Part \neq \emptyset,\ j\in[1,v]. \tag{6.10}$$

In accordance with the knowledge that the manufacturing system can produce and meet many kinds of product requirements, $Part = C(1)\bigcup C(2)\bigcup\cdots\bigcup C(v)$ is proved.

Assume the production line is allowed to carry $N(N>n)$ semi-finished products. In the customer's order, the jth kind of product is booked, and the number is $M(M>0)$. The function $Num(X)\,(X\subseteq Part)$ is used to count the number of elements which are present in the $C(j)$ set, and the total number of the required processing steps is replaced with $Total(j)$ in this order:

$$Total(j) = Num(C(j)) * M. \tag{6.11}$$

The function $P(x)\,(x\in[1,v])$ is used to record the number where the xth semi-finished product has finished the product's processing steps. The function $SP(x)$ stands for the xth semi-finished product.

The remaining parts set of the xth semi-finished product stands for $Cx(j)'$, and therefore, $Cx(j)'\subseteq C(j)$ is proved.

Assume every processing branch may have three working states, namely:

$$B(i) = \begin{cases} -1, & fault; \\ 0, & idle; \\ 1, & busy. \end{cases}, \ i \in [1,n]$$

$$(6.12)$$

The function $Pr(SP(x))|B(i)$ stands for the ith processing branch dealing with the xth semi-finished product. When the xth semi-finished product accesses the intersection of the circular production line and the ith processing branch, the processing flow algorithm of the processing branch $B(i)$ is shown in Algorithm 6.1.

ALGORITHM 6.1 Processing Branch $B(i)$ Execution Algorithm

Input: Semi-finished product $SP(x)$
Output: The information of $P(x)$ and $Cx(j)'$

Begin

1	**switch** $(B(i))$ **do**
2	**case -1**
3	**Processing event is not allowed.**
4	$SP(i) \rightarrow B(i+1)$
5	**end**
6	**case 0**
7	$B(i)$ in standby state and $C_X(j)' \neq \varnothing,$
8	**if** $(Pb(i) \cap C_x(j)' \neq \varnothing)$ **then**
9	$\boldsymbol{Pr}(SP(x))|B(i)$
10	$C_x(j)' = C_x(j)' \setminus Pb(i) \cap C_x(j)'$
11	$P(x) = P(x) + Num\big(Pb(i) \cap C_x(j)'\big)$
12	**end**
13	**else**
14	Processing event is not happened
15	$C_x(j)' = C_x(j)'$
16	$P(x) = P(x)$
17	$SP(x) \rightarrow B(i+1)$
18	**end**
19	**end**
20	**case 1**
21	$B(i)$ is performing $\boldsymbol{Pr}(SP(y))|B(i), y \in [1,v]$
22	$SP(x) \rightarrow B(i+1)$
23	**end**
24	**endsw**

End

When a customer submits order information using mobile terminal, the cloud service system will receive and process the information. Meanwhile, the production command will be delivered to the manufacturing system. The manufacturing system will select the necessary production parts and adjust the production mode based on the production command. The production activities will also rely on the above processing algorithm to complete the corresponding task.

When a customer makes an inquiry regarding production information (using a mobile terminal), the cloud service system will sum up the uploaded data of the manufacturing system and push the data results to the consumer's mobile terminal. In the APP of the mobile terminal, that results information will be converted to the *Comp* and displayed to the consumer in an intuitive way:

$$Comp = \frac{\sum_{x=1}^{v} P(x)}{Total(j)} \times 100\%, \quad x \in [1, v]$$

(6.13)

6.4.2.3 Application Case

In order to verify the feasibility of the PCMS design, in this paper, a candy customization and package processing (as a simplified case model) was studied. This case was used to simulate candy production. In order to complete the simulation, this case experiment provided six different types of candies and three different types of packaging boxes, as individual parts of the finished product. According to the style of the candy box, we assumed that the application case could provide three types of products. In addition, every type of product can provide customers with a variety of candy style choices. In this paper, the experiment will be combined with the above key technologies and implementation mechanism, and the experiment will be carried out in the actual experimental platform.

6.4.2.3.1 Experimental Platform   In this paper, the experimental scene and cloud platforms of the application case are shown in Figure 6.42. The flexible production platform consists of several processing branches, as shown in Figure 6.42a. Every processing branch, as a

(a) Flexible production platform   (b) Cloud platform

FIGURE 6.42   Experimental platform.

production unit, can perform multi-processing tasks. The execution of the production unit needs to rely on a flexible transmission belt, industrial robots, sensors, and corresponding communication technology (such as WiFi and Ethernet) with each element of the unit coordinated with the others. The cloud platform consists of five servers, which constitutes a distributed cluster, as shown in Figure 6.42b. The Hadoop distributed system architecture is used to manage documents and data. In addition, this platform also provides web services for data reception and an inquiry service.

In order to meet the demand of customer orders with personalized customization and simulated information inquiries, this case provides a mobile terminal APP. This APP is suitable for the order processing operation of this manufacturing system. In order to simplify the description, the design of this APP omits the user registration interface and the operation of the number of products. The API and API key of the cloud platform also are directly set in this app. Moreover, this application case adopts single product customization to demonstrate the entire implementation process.

The mobile terminal APP's function interface is shown in Figure 6.43. The order setting mainly relies on the button options of the main interface, as shown in Figure 6.43a. The candy box type option is a radio button group, which in turn has three options. The option of candy type is shown by six candy buttons. When a candy button is pressed, the dialog corresponding with that candy will immediately pop up, as shown in Figure 6.43b. All selected options will then appear in the text view of the main interface, as shown in Figure 6.43c. Therefore, as mentioned in the experimental platform related settings, the application case can provide consumers with a flexible candy customization method and handle the implementation of production activities.

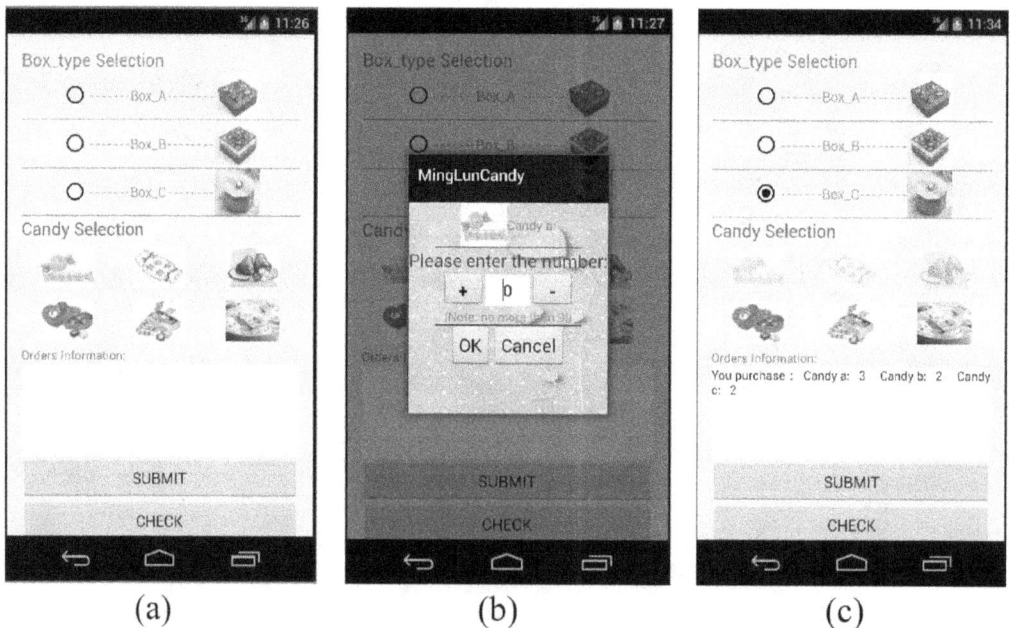

FIGURE 6.43 Mobile terminal APP: (a) main interface, (b) button dialog, and (c) selected button' schematic.

6.4.2.3.2 Experimental Analysis   Take the array of {2, 3, 2, 5, 4, 4} as an example. This array is used to represent the selected quantity from candy a to candy f. The customers' personalized needs are shown in Figure 6.44. The order information in the mobile phone APP is shown in Figure 6.44a. The results of the submitted information are presented in Figure 6.44b. After the consumer executes an inquiry operation, the mobile phone will obtain data from the cloud service system and translate that data into intuitive information, as shown in Figure 6.44c.

The experimental scene of the flexible production platform is shown in Figure 6.45. The candy is replaced with the marked yellow planchet in order to simulate a processing experiment. Every processing branch can carry on the processing task of multiple parts.

FIGURE 6.44   Mobile services for customized manufacturing: (a) customization settings, (b) order submission, and (c) information inquiry.

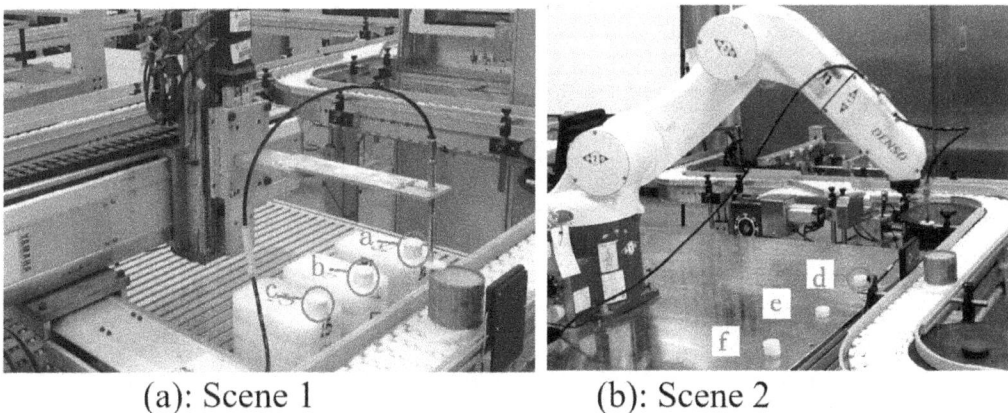

(a): Scene 1   (b): Scene 2

FIGURE 6.45   Processing scenes: (a) Scene 1 and (b) Scene 2.

The RFID reader is used to calibrate and track the candy and the information pertaining to the candy boxes.

In this paper, the PCMS can provide consumers with a mobile terminal APP, which in turn enables order customization and information inquiries. In the manufacturing system, the cloud platform is used for data processing, and the communication technology is used for purposes of information exchange and production coordination. The PCMS system can adopt big data technology as a means to process massive data for information extraction and application modeling. By analogy, the function of this PCMS system can be extended further. Big data technology can used for massive data analysis of IoT, and the results obtained during information extraction can be applied in industrial production and management. A mobile terminal APP which responds to consumer demands can be designed. A cloud platform service mode can read an industrial production model for reference, thus enabling the production model to be flexibly customized. The manufacturing device layer (according to the function feature of the manufacturing unit) can be modularized. The information flow can be uploaded to the cloud via network technology. By combining the cloud service model and the adjustment measures of enterprise managers, the processed information will be automatically delivered to the factory, so as to achieve the goal of intelligent production.

6.5 SUMMARY

Customized manufacturing factory is an important model to enhance personalized manufacturing capability and improve the competitiveness of enterprises. This section introduces their integration and convergence methods in terms of factory equipment, network, resources and services, and extremely key technologies. Integrated factory management is achieved through vertical integration of equipment and monitoring of production resources. Cloud services drive the factory to reconfigure and optimize resources and scheduling to achieve more efficient personalized manufacturing.

A candy personalization line platform for customized manufacturing trials is presented to validate the equipment integration and resource management approach. In addition, this section discusses intelligent control methods for conveyor belts for customized manufacturing. The application methods of equipment integration, resource reconfiguration, and scheduling optimization in cloud services are illustrated through practical cases of candy packaging, personalized recommendation, pharmaceutical packaging, and worker island assembly, and their importance for achieving customized manufacturing is verified, illustrating the key technological components for developing customized manufacturing factories. Ultimately, it promotes the development of national and industry customized manufacturing models.

REFERENCES

[1] G. Herzog and A. Kröner, "Towards an Integrated Framework for Semantic Product Memories," in *SemProM: Foundations of Semantic Product Memories for the Internet of Things*, ed: W. Wahlster. Springer, Berlin, Heidelberg, 2013, pp. 39–55.

[2] C. He, Y. Zheng and L. Wang, "A review of RFID applications for discrete manufacturing processes," *Computer Integrated Manufacturing System*, vol. 20, no. 5, pp. 1160–1170, 2014.

[3] R. G. Smith, "The contract net protocol: High-level communication and control in a distributed problem solver," *IEEE Transactions on Computers*, vol. C-29, no. 12, pp. 1104–1113, 1980.

[4] J. Rehrl, J. Kruisz, S. Sacher, J. Khinast and M. Horn, "Optimized continuous pharmaceutical manufacturing via model-predictive control," *International Journal of Pharmaceutics*, vol. 510, no. 1, pp. 100–115, 2016.

[5] Y. Jiang, H. Song, R. Wang, M. Gu, J. Sun and L. Sha, "Data-centered runtime verification of wireless medical cyber-physical system," *IEEE Transactions on Industrial Informatics*, vol. 13, no. 4, pp. 1900–1909, 2017.

[6] L. A. Tawalbeh, R. Mehmood, E. Benkhlifa and H. Song, "Mobile cloud computing model and big data analysis for healthcare applications," *IEEE Access*, vol. 4, pp. 6171–6180, 2016.

[7] J. Norman, R. D. Madurawe, C. M. V. Moore, M. A. Khan and A. Khairuzzaman, "A new chapter in pharmaceutical manufacturing: 3D-printed drug products," *Advanced Drug Delivery Reviews*, vol. 108, pp. 39–50, 2017.

[8] T. Strasser, A. Zoitl, J. H. Christensen and C. Sünder, "Design and execution issues in IEC 61499 distributed automation and control systems," *IEEE Transactions on Systems, Man, and Cybernetics, Part C (Applications and Reviews)*, vol. 41, no. 1, pp. 41–51, 2011.

[9] M. V. García, F. Pérez, I. Calvo and G. Moran, "Developing CPPS within IEC-61499 based on low cost devices," *2015 IEEE World Conference on Factory Communication Systems (WFCS)*, 2015, pp. 1–4.

[10] V. Vyatkin, C. Pang and S. Tripakis, "Towards cyber-physical agnosticism by enhancing IEC 61499 with PTIDES model of computations," *IECON 2015-41st Annual Conference of the IEEE Industrial Electronics Society*, 2015, pp. 001970–001975.

[11] S. Abburu, "A survey on ontology reasoners and comparison," *International Journal of Computer Applications*, vol. 57, pp. 33–39, 2012.

[12] J. Pérez, M. Arenas and C. Gutierrez, "Semantics and complexity of SPARQL," *ACM Transactions on Database Systems (TODS)*, vol. 34, no. 3, p. 16, 2009.

[13] J. Pérez, M. Arenas and C. Gutierrez, "Semantics and complexity of SPARQL," *The Semantic Web - ISWC 2006*. Springer, Berlin, Heidelberg, 2006, pp. 30–43.

[14] M. J. O'Connor and A. K. Das, "SQWRL: A query language for OWL," *OWL: Experiences and Directions*, 2009.

[15] A. B. Bener, V. Ozadali and E. S. Ilhan, "Semantic matchmaker with precondition and effect matching using SWRL," *Expert Systems with Applications*, vol. 36, no. 5, pp. 9371–9377, 2009.

[16] S. Lemaignan, A. Siadat, J. Y. Dantan and A. Semenenko, "MASON: A proposal for an ontology of manufacturing domain," *IEEE Workshop on Distributed Intelligent Systems: Collective Intelligence and Its Applications (DIS'06)*, 2006, pp. 195–200.

[17] D. Calvanese *et al.*, "Ontop: Answering SPARQL queries over relational databases," *Semantic Web*, vol. 8, pp. 471–487, 2017.

[18] G. Q. Huang, Y. F. Zhang and P. Y. Jiang, "RFID-based wireless manufacturing for walking-worker assembly islands with fixed-position layouts," *Robotics and Computer-Integrated Manufacturing*, vol. 23, no. 4, pp. 469–477, 2007.

[19] A. N. Carvalho, F. Oliveira and L. F. Scavarda, "Tactical capacity planning in a real-world ETO industry case: An action research," *International Journal of Production Economics*, vol. 167, pp. 187–203, 2015.

[20] B. Chen, J. Wan, L. Shu, P. Li, M. Mukherjee and B. Yin, "Smart factory of industry 4.0: Key technologies, application case, and challenges," *IEEE Access*, vol. 6, pp. 6505–6519, 2018.

[21] D. H. Grabenstetter and J. M. Usher, "Developing due dates in an engineer-to-order engineering environment," *International Journal of Production Research*, vol. 52, no. 21, pp. 6349–6361, 2014.

[22] H. Vaagen, M. Kaut and S. W. Wallace, "The impact of design uncertainty in engineer-to-order project planning," *European Journal of Operational Research*, vol. 261, no. 3, pp. 1098–1109, 2017.

[23] A. Alfieri, T. Tolio and M. Urgo, "A two-stage stochastic programming project scheduling approach to production planning," *The International Journal of Advanced Manufacturing Technology*, vol. 62, no. 1, pp. 279–290, 2012.

[24] J. Hytonen, E. Niemi and V. Toivonen, "Optimal workforce allocation for assembly lines for highly customised low-volume products," *International Journal of Services Operations and Informatics*, vol. 3, pp. 28–39, 2008.

[25] C. W. Tsai, C. F. Lai, M. C. Chiang and L. T. Yang, "Data mining for internet of things: A survey," *IEEE Communications Surveys & Tutorials*, vol. 16, no. 1, pp. 77–97, 2014.

[26] E. Niemi, "Worker allocation in make-to-order assembly cells," *Robotics and Computer-Integrated Manufacturing*, vol. 25, no. 6, pp. 932–936, 2009.

[27] O. O. Balogun and K. Popplewell, "Towards the integration of flexible manufacturing system scheduling," *International Journal of Production Research*, vol. 37, no. 15, pp. 3399–3428, 1999.

[28] P. Priore, D. de la Fuente, J. Puente and J. Parreño, "A comparison of machine-learning algorithms for dynamic scheduling of flexible manufacturing systems," *Engineering Applications of Artificial Intelligence*, vol. 19, no. 3, pp. 247–255, 2006.

[29] P. Leitão, "Agent-based distributed manufacturing control: A state-of-the-art survey," *Engineering Applications of Artificial Intelligence*, vol. 22, no. 7, pp. 979–991, 2009.

[30] W. Shen, Q. Hao, H. J. Yoon and D. H. Norrie, "Applications of agent-based systems in intelligent manufacturing: An updated review," *Advanced Engineering Informatics*, vol. 20, no. 4, pp. 415–431, 2006.

[31] "Amazon elastic compute cloud," https://aws.amazon.com/ec2/. Accessed 12 Dec 2015.

For Product Safety Concerns and Information please contact our EU
representative GPSR@taylorandfrancis.com
Taylor & Francis Verlag GmbH, Kaufingerstraße 24, 80331 München, Germany

www.ingramcontent.com/pod-product-compliance
Lightning Source LLC
Chambersburg PA
CBHW080129220326
41598CB00032B/5005

*9 7 8 1 0 3 2 6 0 9 0 1 0 *